The senses

Edited by

H. B. BARLOW

ROYAL SOCIETY RESEARCH PROFESSOR IN THE PHYSIOLOGICAL LABORATORY, UNIVERSITY OF CAMBRIDGE

and

J. D. MOLLON

LECTURER IN EXPERIMENTAL PSYCHOLOGY, UNIVERSITY OF CAMBRIDGE

The right of the University of Cambridge to print and sell all manner of books was granted by Henry VIII in 1534. The University has printed and published continuously since 1584.

CAMBRIDGE UNIVERSITY PRESS

CAMBRIDGE

NEW YORK PORT CHESTER

MELBOURNE SYDNEY

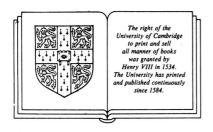

Published by the Press Syndicate of the University of Cambridge
The Pitt Building, Trumpington Street, Cambridge CB2 1RP
40 West 20th Street, New York, NY 10011, USA
10 Stamford Road, Oakleigh, Melbourne 3166, Australia

First published 1982
Reprinted 1984, 1987, 1988, 1989 (with amendments)

Printed in Great Britain at the University Press, Cambridge

Library of Congress catalogue card number: 81-17007

British Library cataloguing in publication data

The Senses. (Cambridge texts in the physiological sciences; 3)

1. Senses and sensation
I. Barlow, H. B. II. Mollon, J. D.
612′.8 QP431

ISBN 0 521 24474 9 hard covers
ISBN 0 521 28714 6 paperback

Contents

List of contributors

Dr M. Alpern, Department of Ophthalmology, University of Michigan, Ann Arbor, Michigan 48109, USA

Dr J. Atkinson, The Psychological Laboratory, Downing Street, Cambridge CB2 3EB, UK

Professor H. B. Barlow, The Physiological Laboratory, Downing Street, Cambridge, CB2 3EG, UK

Dr A. J. Benson, RAF Institute of Aviation Medicine, Farnborough, Hants, UK

Dr O. J. Braddick, The Psychological Laboratory, Downing Street, Cambridge CB2 3EB, UK

Professor E. F. Evans, Department of Communication and Neuroscience, University of Keele, Keele, Staffordshire ST5 5BG, UK

Professor A. Iggo, Department of Veterinary Physiology, Royal (Dick) School of Veterinary Studies, Summerhall, Edinburgh EH9 1QH, UK

Dr E. B. Keverne, Department of Anatomy, Downing Street, Cambridge CB2 3DY, UK

Dr A. Knowles, Department of Biochemistry, University of Bristol, University Walk, Bristol BS8 1TD, UK

Professor M. Millodot, Department of Ophthalmic Optics, UWIST, Arlbee House, Greyfriars Road, Cardiff CF1 3AE, UK

Dr J. D. Mollon, The Psychological Laboratory, Downing Street, Cambridge CB2 3EB, UK

Professor E. N. Willmer, Clare College, Cambridge CB2 1TL, UK

Dr M. Woodhouse, Department of Ophthalmic Optics, UWIST, Arlbee House, Greyfriars Road, Cardiff CF1 3AE, UK

Preface

This book is the third in the series of Cambridge Texts in the Physiological Sciences and the reader foremost in our minds is a medical student in the second or third year of a preclinical course. We hope, however, that this text may equally recommend itself to those in the several other disciplines that concern themselves with the senses – to students of psychology, optometry, education and art, as well as postgraduates in ophthalmology and otology. The medical student will find here a more extensive treatment of the cognitive and behavioural aspects of perception than is usually found in medical textbooks, whereas the psychologist will find reference material on topics, such as physiological optics and the vestibular system, that are not normally treated in detail in psychological textbooks. Our aims are: to be interdisciplinary and comprehensive without entering into exhaustive detail on every topic; to be up-to-date while yet giving the student a grasp of the historical sequence of discovery; and to be accurate while not excluding all issues that are controversial and uncertain. In the tradition of Cambridge preclinical teaching, the approach is that of the natural scientist, but clinical examples are used throughout to illustrate basic mechanisms.

In order to secure expert treatments of each of the senses, we have asked a number of colleagues to cover specific topics, but the drafts of individual chapters have been widely and recurrently circulated among the contributors and we hope the reader may find the present text more coherent than most multi-authored books. Common to all our contributors is the working assumption that the human senses can be modelled by physical instruments and systems, and that perceptual performance will ultimately be explicable in terms of physiological mechanisms; phenomenological observations are discussed not as an end in themselves, but are introduced to throw light on underlying mechanisms and, of course, to help define what is to be explained. The book is organised classically, in that we treat each sensory modality in turn, but we have endeavoured to draw attention to many common features of sensory systems, such as topographic maps, feature detectors, efferent control, and the presence of parallel subsystems that extract different types of information from the same receptor array. A number of techniques are common to the study of different modalities and in particular in many chapters the reader will come across 'contrast–sensitivity–functions', obtained by stimulating a given sensory system with sinusoidal stimuli of varying frequency and modulation; a short, general primer on this approach has been incorporated in Chapter 1.

In the interest of readability, we have outlawed references from the main text; but suggestions for further reading will be found at the end of each chapter and we have added annotated bibliographies on specific topics, so that the book may serve as a reference source for those who wish to delve further into the seductive world of sensory science.

Afferents from muscles and joints will be dealt with in a forthcoming volume in this series on motor control and are therefore not considered here.

We should like to thank the following colleagues who have given guidance on specific sections of the book: L. Bartoshuk, F. W. Campbell, R. H. S. Carpenter, H. J. A. Dartnall, P. J. Lachmann, B. C. J. Moore, M. J. Neal and L. T. Sharpe. We are indebted to A. Hills for much kind help throughout the preparation of the text, to F. Hake and R. Overhill for drawing, and to B. Clifton and K. Knights for typing sections of the manuscript.

H. B. Barlow
J. D. Mollon

Cambridge,
July 1981

CHAPTER ONE

General principles: the senses considered as physical instruments

H. B. BARLOW

When a person falls ill, he is likely to become aware of it through his own senses, for he will experience something like a toothache, a pain in the abdomen, or a sudden chill. If the condition progresses to the point where he seeks help, his advisor, even while listening to the patient's story, will use his own senses to pick up what he can of the cause of the illness, and in very many cases what he sees, hears, feels, and smells will be enough for him to diagnose the condition. X-rays, laboratory tests, and monitoring devices have decreased what was previously a total reliance on direct sensory information, but the patient's sensations, and the messages from the doctor's sense organs, still provide the most important, potentially most reliable, and most up-to-date, sources of information about the illness.

Of course it is not just in sickness that the senses tell one about one's own body. The *interoceptors* constantly monitor its chemical and physical state, and these are key elements of the homeostatic mechanisms that stabilise the internal environment. Complex forms of life are completely dependent on accurate interoception. Similarly it is the *exteroceptors* that enable one to assess the surrounding environment, and so to comprehend the tasks one confronts. It is worth noting, however, that these so-called exteroceptors are really specialised interoceptors; they sense the outer world only by means of its physical and chemical influence on the special sense cells of the nose, the ears, or the eyes.

Thus the senses are the bodily mechanisms for gathering up-to-date information, and as such it is hard to exaggerate their importance. Wars are won and lost by the intelligence system, and if one considers the survival of individuals and species in more natural environments one can see at once that superiority of sight, hearing, or smell must confer an immense advantage in the competitive struggle for food resources, habitat, and mates. The interoceptive system, which monitors bodily functions, is less spectacular for its task is one of maintenance. But an army marches on its stomach, and faulty internal information, for instance to the supply department, can lose a war as decisively as erroneous intelligence.

This book describes what is known of the physiology of the senses: that is, the physical, chemical, and biological mechanisms of this information-gathering system. The principal methods for investigating these processes are very similar to those employed in other branches of physiology. One can examine the anatomical structure of the sense organs, and in some cases this gives immediate insight into their mode of operation, but this method is not as powerful as one might suppose when it is used by itself. The analogy between the structure of the eye and that of an image-forming device such as a camera seems obvious to us, but before image-forming devices were familiar, and before the image cast by the lens on to the retina had been directly observed, the anatomical cues were misinterpreted. The optic nerve was dissected down to the eye, where it was found to spread out over the retina. Following the retina round the vitreous humour one comes to the ciliary body suspending the lens, and it was thought that the 'quivering' of the lens as the light passed through it excited the sensitive terminals of the optic nerve. The apparent anatomical continuity of optic nerve and ciliary body was misleading, but the acceptance of this mistaken teaching emphasises the importance of functional models when the anatomy is interpreted, and the desirability of detecting and recording the physiological process at every possible point. Nowadays one can do this, for one can record the electrical responses of cells at almost every level in the visual, auditory, and other sensory pathways, and information from this source is prominent in this book. Much information has also been obtained from biochemical investigations, especially of the photosensitive pigments underlying vision. But sensory systems possess special interest as the objects of physiological study because one can receive guidance from two types of knowledge not available for other systems. In the first place, a branch of engineering has grown up over the past few decades that deals with the problems encountered in man-made instruments and information-gathering systems. Such knowledge tells one the physical limits and natural difficulties likely to be encountered in sensory systems, and this provides insight rather similar to that given by the law of conservation of energy and other thermodynamic principles in understanding metabolism. The next section of this chapter discusses some general aspects of information-gathering by the senses from this viewpoint.

Sensory systems are also of unique interest because they provide the information on which our conscious sensations are based, and this makes it possible to obtain guidance from a second source of knowledge. A human can be asked to report his sensations in experiments where the physical properties of the stimuli are accurately controlled. The study of how the subject's report varies with the physical parameters of the stimulus is called *psychophysics*. By this means an objective and rather complete account can be given of what information the human senses can and cannot convey to the brain. This overview of their performance is enormously useful when analysing their anatomical and physiological mechanisms, for one knows what these mechanisms must achieve with a precision that is often lacking in other systems. As a result of this body of psychophysical knowledge, and through applying some principles of communication engineering, we are beginning to understand the mechanism of the sensory parts of the brain as well or better than any other part.

1.1. PRINCIPLES OF OPERATION OF SENSORY PATHWAYS

Plan of sensory pathways

Fig. 1.1 shows a diagram of a peripheral sensory pathway. The sensory nerve fibre at the left has terminals in the skin at the top, and when these are appropriately stimulated they are depolarised, thus initiating action potentials that are propagated down the axon. When a stimulus is applied and held constant at a steady value these impulses are initiated after a short latency, and then reach a peak frequency which varies with intensity. In most sensory nerves the frequency then declines, even when the stimulus remains constant, a property which is known as *adaptation*. Many pathways also have a *maintained discharge* at a low rate which results from spontaneous depolarisation of the terminals occurring in the absence of a deliberately applied stimulus. Centrally the fibre terminates on second order neurons, and this part of the system will be considered later.

For such an information-gathering system to be effective there are certain definite requirements. It must have elements *selective* for different types of physical or chemical stimulus, these elements must respond *speedily*, they should be *sensitive* to small changes in the input, and they must respond *reliably* and not give spurious information. What we gain from the analogy with man-made instrumental systems is the knowledge that selectivity, speed, sensitivity and reliability cannot be improved indefinitely but are limited by physical factors. Furthermore they are usually interrelated, so that one of

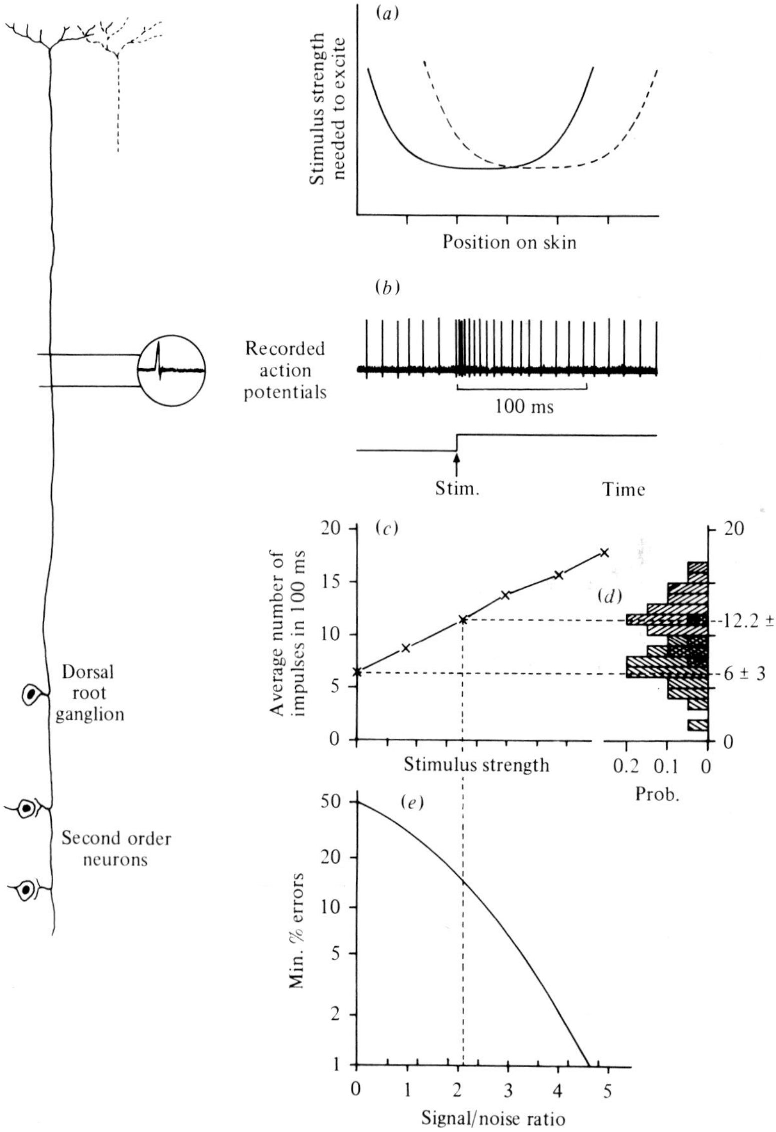

Fig. 1.1. Diagram of a peripheral sensory pathway. A nerve fibre running from sensory terminals in the skin to the synaptic endings on second order neurons in the spinal cord is shown to the left. On the right (*a*) illustrates the selectivity of the fibre for the position of the stimulus on the skin, (the dotted curve shows the corresponding curve for a neighbouring fibre); (*b*)

them cannot be improved without detriment to another. The physical factors limiting performance in living tissues are not necessarily the same as those encountered by the communications engineer, but it is a challenge to find out what the biological factors are, and how they affect performance. These aspects of sensory pathways will now be discussed, with special emphasis on how they can be measured and how they are interrelated.

Selectivity

Almost all sensory fibres show selectivity for the anatomical position of a stimulus. The schematic nerve fibre in Fig. 1.1 is distributed to only a portion of the skin and thus responds to physical stimuli applied in that region and no other. If all nerve fibres had the same receptive area, or *receptive field* as it is called, this system would convey no information at all on the whereabouts of an excitatory stimulus, so the transmission of such information entirely depends upon the fact that different nerve fibres exhibit different positional selectivities.

Sense organs are also selective for other characteristics of the stimulus. For instance in the skin, some respond to temperature change instead of mechanical deformation, others respond to products released from damaged cells and thereby signal injury. Qualitative selectivities of this sort lie behind the major modalities of skin sensation. In some cases the selectivity can be plotted as a graph, not of position as in Fig. 1.1, but of some other variable. Examples of these are the tuning curves of auditory nerve fibres (Fig. 14.16) or the spectral absorbance curves of the different categories of cones subserving colour vision (Fig. 9.4).

---

shows a train of action potentials resulting from application of a stimulus; (*c*) plots the average number of impulses in 100 ms following stimuli of varying strengths; (*d*) shows histograms of the frequency of obtaining 2 to 17 impulses in 100 ms when no stimulus was applied (lower peak), and following a moderate stimulus causing an average of 12.2 impulses (upper peak); both these distributions have a standard deviation of ± 3 impulses. The reliability of detecting a stimulus depends on the signal/noise ratio, which is $(12.2-6)/3 = 2.1$ for a stimulus of the strength shown here. Because these distributions overlap, errors are unavoidable when judging whether a stimulus was present or not from the number of impulses in a pulse train, and (*e*) shows that for a signal/noise ratio of 2.1 the error rate would be at least 15% on average. A stronger stimulus would separate the distributions shown in (*d*), and the error rate would fall as shown in (*e*).

There are some important terms and concepts associated with selectivity curves like those shown in Figs. 9.3*c* and 14.16, because the degree of selectivity is the most important factor determining how a sensory pathway combines and separates the physical stimuli acting on the endings. To take the combining aspect first, it is a matter of practical importance to know how effectively mixtures of different spectral lights stimulate the eye. Suppose, for example, that a lamp for illuminating streets is made by mixing a narrow band of yellow from sodium, which has a wavelength near 589 nm, with a violet at 436 nm from mercury; will these lights act entirely independently, or will the eye be able to combine them? Under dark-adapted conditions, the visibility of a light depends solely on the total number of quanta absorbed in the photosensitive pigment rhodopsin (see Chapters 5 and 7). As a result the effectiveness of the combination can be calculated by multiplying the energy of each wavelength, E_λ, by a factor conventionally labelled V'_λ representing the relative effectiveness of light at that wavelength, and adding the products (i.e. calculating $E_{436} \times V'_{436} + E_{589} \times V'_{589}$). Similarly the effectiveness of a broad spectral band, or of white light containing all wavelengths, can be calculated by integrating the energy in each narrow band weighted by the relative sensitivity in that band, that is calculating $\int E_\lambda V'_\lambda \, d\lambda$. Now one can see that a very broad curve of spectral selectivity will cause light over a broad band of wavelengths to be combined, with obvious advantage to the sensitivity of the system because the total amount of light energy utilised is increased.*

Now consider the spectral selectivities of the cones, which take over from the rod system at higher light levels. There are three different types, and this enables them to transmit information about the spectral composition of the light as well as the total quantity, so that objects such as ripe fruit can be distinguished by colour as well as by other characteristics. The spectral sensitivity curves of the three types differ from each other, and the effectiveness of the system in analysing and transmitting information about the spectral compo-

1 * When the photopic visibility function V_λ is substituted for the scotopic function V'_λ (see Figs. 5.8, 7.3) the resulting integral expresses the effectiveness of the radiant energy in stimulating the eye at the light levels where most visual tasks are performed. It is therefore the quantity that needs to be measured in many practical situations, such as determining if the light is good enough for reading easily, and for this reason it is used to define *photometric* units of measurement. These are based on the *lumen*, the unit of luminous flux which is given by $L = 680 \int_0^\infty E_\lambda V_\lambda \, d\lambda$, where L is lumens, E_λ is in watts/nanometre, and V_λ is the photopic visibility function.

sition of the light depends on two properties; first the number of different classes of cone, and second the degree to which the spectral sensitivity curves differ from each other. In the case of the colour system there are only three different classes, and their spectral sensitivity curves overlap a great deal. A system with improved capacity to analyse the spectral composition of light would require more than three different spectral sensitivity curves; their peaks would be closer together, and it would help if they were narrower so that the overlap of the different spectral bands was not increased. In such a system the same amount of light would have to be divided among more receptors, and the narrower absorption curves would absorb less energy from white light, so this is an example where selectivity and sensitivity are opposed; one cannot improve the capacity of the system to separate, or resolve, without sacrificing sensitivity.

Many of the same considerations apply to the spatial resolution of the eye – its capacity to separate out the different parts of an image. High resolution requires a large number of nerve fibres each with a narrow receptive field, whereas if the resolution is poor the receptive fields can be large to achieve sensitivity and fewer of them are needed. The retina of a cat provides a beautiful example of this: in the periphery the ganglion cells are sparse and their receptive fields large, while towards the centre of the field of vision the ganglion cells are crowded together and also have smaller receptive fields, so that the amount of overlap between neighbouring receptive fields stays roughly constant.

Before leaving the topic of selectivity, a source of confusion that arises in discussions of resolving power must be discussed. If one were asked to measure the resolving power of the colour system, it would not be unreasonable to find out how small a difference of wavelength a person could discriminate. Such tests yield an astonishingly low figure, under 1 nm in the yellow region of the spectrum, which is sufficient to distinguish between the double lines of sodium near 589 nm! One might conclude from this fact that, in the visible spectrum stretching from 400 to 700 nm, there are about 300 distinguishable hues; the actual number is rather less than this because a 1-nm difference cannot be detected across the whole spectrum, but about 200 hues are genuinely discriminable. One can be confused by the question 'Does the eye resolve colours into 3 classes, or 200?' There is no doubt that 3 is the more fundamental answer, for it indicates the number of separately variable aspects of the spectral distribution of light that the eye transmits to the brain; the excitation of a red, a green, or a blue cone can each vary while the excitation of the other two is kept constant. The figure of 200 discriminable hues results from different combinations of activity of these three types of cone in the same sort of way that 1000 different numbers can be represented by three decimal digits. One has to distinguish carefully between the number of classes of cone with separately variable outputs and the number of possible ways of combining those outputs in various proportions.

The same distinction is important in considering the spatial resolution of the eye; the position of a dot or line can be judged with a precision of a few seconds of arc, which is ten times better than the distance between the parts of the image that can be truly resolved and signalled separately. The improved positional accuracy can be achieved by the mathematical techniques of interpolating and averaging, and in long-evolved sensory systems there are many features that an engineer knowledgeable about instrument design can help one to understand. It is often remarkable to see how natural selection and evolution have led to designs that approach limits set by the physical nature of the tasks being performed.

There are some exceptions to the obvious good design of many sensory systems; in these the physiological facts point to arrangements that seem far from ideal. For instance the chemical senses use a peculiar method of signalling in which each nerve fibre has a mixed sensitivity to several distinct chemical modalities (see Chapter 19). Consequently it is a complex and difficult logical problem to make discriminations between different chemical stimuli, whereas it would seem much easier if each nerve fibre signalled only a single type of chemical. Perhaps we underestimate the ease with which the nervous system can perform the tasks of pattern recognition called for in this instance, and hence overestimate the advantages of keeping the modalities separate. Examples like this are intriguing because they point to areas of ignorance where it may be possible to make important discoveries.

Speed

The second desirable characteristic of a sensory system is speed of response. The great survival advantage of rapid action needs no emphasis, and a speedy sensory message is the first requirement to achieve this. Nerve fibres from the ear respond within a millisecond or less, and the same is true for fibres from the vibration-detecting Pacinian corpuscles, but many other systems respond at a surprisingly leisurely pace, and to understand this one must realise the advantage of a long summation time. Just as a large receptive field may have the advantage in sensitivity over a small one, so a mechanism that integrates in time and reacts to the total stimulus energy over a long period can be more sensitive than one that integrates over a short time only. The point can be illustrated by considering a pair of galvanometers of identical construction except for the strength of the return springs: the one with the weaker spring will be more sensitive, but the pointer will take longer to reach a steady reading. In Fig. 1.1 the latency and duration of the impulse discharge shown in the second line will set limits to the sensitivity of the system as well as determining how quickly the animal can react.

Sensitivity

It has already been pointed out that high spatial resolution requires small receptive fields and such small fields can collect only a small amount of energy from the applied stimulus. A speedy response requires a short integration time, which also reduces the available energy. High sensitivity of the basic receptor mechanism – the capacity to respond to a small amount of energy – is thus a key requirement for attaining good resolution and quick responses. We shall see later that high sensitivity is also needed for *reliability*. In addition, high sensitivity may itself have protective value because it might, for instance, enable an animal to hear or smell a predator at a greater distance. In the case of vision, it may also enable an animal to hunt or graze for extra hours at dawn and dusk.

It is worth noting, however, that high sensitivity is not always advantageous. A hyper-sensitive skin would be distracting and might lead to constant grooming behaviour or scratching, and if one's ears were too sensitive one might be deafened by the throb of the pulse and roar of blood flowing in the artery of the basilar membrane, or by molecules bombarding the tympanic membrane under Brownian motion. These thoughts make one reconsider what is meant by sensitivity: what is it that limits the process of gathering *important* information?

The third figure from the top in fig. 1.1 shows the result to be expected when the average number of impulses occurring in a sensory nerve in, say, 100 ms, is plotted against the intensity of an applied stimulus. Note first that some impulses occur *before* the stimulus is applied. This is the *maintained discharge*, which occurs in the absence of stimulation, and its rate varies greatly in different systems, and in different types of nerve fibre. For individual optic nerve fibres it varies from over 50 s^{-1} to under 5 s^{-1}, but is rarely completely absent. Auditory nerve fibres also usually have a maintained discharge, as do those from vestibular organs. Others, such as those from Pacinian corpuscles and many touch receptors, are quiet until excited.

When the sense organ is stimulated more strongly the number of impulses increases, linearly at first but usually at a lower rate as the stimulus intensity is raised further. Sensitivity can be defined by the slope of this increase, that is by the average number of extra impulses elicited per unit of applied energy. Notice, however, that there is something missing from such a definition, useful though it is. If the nerve is quiet before stimulation, a single extra impulse is easily distinguished and is obviously a significant event. But if there is an irregular maintained discharge of, say 60 impulses s^{-1}, then in a counting period of 100 ms there will be about 6 impulses even without any stimulation, and a single extra impulse may easily pass unnoticed.

Reliability and noise

Communication engineers are of course familiar with the situation in which unwanted disturbances obscure the detection of the signal of interest, and it has come about that these unwanted disturbances are always called 'noise', even with a video television signal or when considering erroneous operations in a computer, because the problem was first encountered with acoustic signals. For sensitivity, what is important is not the amplitude of output signal achieved by a given input signal, but the relationship between the output signal and the unwanted signals, or noise level. This concept of signal/noise ratio can now be incorporated into the discussion of a sensory pathway.

If one goes back to the case where the maintained discharge was obscuring the detection of an extra impulse, a moment's consideration will show that it is not the average *rate* of discharge that matters, but its *variability*. If exactly 6 impulses occurred every 100 ms, then when 7 or more impulses occurred we should know that a stimulus had been applied. But if 6 impulses is only an average, and the number is sometimes 3, sometimes 9, then one will need many extra impulses before one can reliably detect a change, for the number must be raised beyond the upper limit of the maintained discharge.

To the right of the plot of the average number of impulses, two histograms are shown (Fig. 1.1*d*). These represent the actual numbers of impulses counted in the 20 trials used to determine two of the average numbers. The lower one is for the maintained discharge, and it shows that the number ranged from 2 to 12 in the hypothetical example. The average was 6, but what is more important is the variability. The spread of the numbers is best expressed by their standard deviation, $(\Sigma(\bar{x}-x)^2/(n-1))^{\frac{1}{2}}$, which is 3.0 impulses in this case. This is the 'noise' of the neural message over the 100-ms period, and it means that 3 extra impulses are required to achieve a signal/noise ratio of 3/3, or unity.

The upper histogram shows the distribution of the numbers of impulses obtained on the 20 trials used to determine the response, which averaged 12.2 impulses in 100 ms. This is an increase of 6.2 over the maintained discharge and corresponds to a signal/noise ratio of 6.2/3, or 2.1. Note that, although the means of the two histograms are quite well separated, the tails of the distributions overlap. If a response of 9 impulses occurred, this could be an unusually high value of the maintained discharge, or it could be an unusually low value resulting from a stimulus whose mean response is 12 impulses. It follows that a response of this value cannot be assigned with certainty to either histogram. When it occurs it is impossible to know whether a stimulus was applied or not; the best that could be done would to be choose some criterion, say of 10 impulses, and decide that when this value or more occurred, the stimulus was present. But even then there would be a proportion of 'false alarms', when non-existent stimuli appeared to occur, and a proportion of 'misses', when genuine stimuli remained undetected.

The lowest part of Fig. 1.1 shows how this proportion of errors (false alarms and misses) changes with stimulus strength, assuming that in each case the criterion has been adjusted to minimise the total number of errors. Even with zero signals, half the answers are right by guesswork, and the proportion of errors decreases as the stimulus is increased. The scale of stimulus strength is the same as in the figure showing response amplitude, directly above, but since the variability of the maintained discharge (noise), as well as the number of extra impulses (signal), is known for each stimulus strength we can add another scale, the signal/noise ratio. The important point is that the number of errors bears the same relation to signal/noise ratio in all communication systems. This statement would require careful qualification to be exact and universal, but as an approximation one can see that signal/noise ratios of 2 or 3 are needed to reduce the errors to about 10%, and ratios of 4 or 5 to reduce them to 1%. It should also be noted that an analysis of the errors made by human subjects when performing a sensory task enables one to derive a figure, referred to as d' by signal detection theorists, that is a lower limit to the signal/noise ratio in the sensory pathway involved. This method provides the firmest means of connecting sensations and physiological mechanisms.

One can now see both why sensitivity is important for reliability, and why the simple definition of sensitivity is inadequate. If more impulses were obtained per unit of stimulus energy, this would separate the two distributions shown in Fig. 1.1*d* and at first sight would appear to improve reliability. But there would be no gain if, as a result of increasing responsiveness to the stimulus, the responsiveness to the unwanted disturbances causing the fluctuating maintained discharge had also increased, so that the breadth of the distributions increased in proportion to their separation. Whether an increase of sensitivity would necessarily cause such changes depends upon the source of the maintained discharge. It is clear that the origin of noise in sensory systems is an important question.

1.2. TRANSFORMATIONS OF SCALE AND INTENSITY

The senses have been considered so far as a set of physiological mechanisms for gathering up-to-date information, but as we follow the sensory messages centrally we would like to be able to explain the ways in which this information is utilised. How does a tennis player predict the future position of the ball so that he can return it? How does my secretary decipher my handwriting? What goes on between the ears and the mouth of a simultaneous translator? It must be realised at the outset that we can only go a very short distance towards answers to questions like these, but several clues have already come from communication engineers who handle electronic information, and it is likely that we shall get more from those who try to program computers to perform tasks of a complexity comparable to the ones mentioned above.

The ideas from electronic engineers are mainly concerned with the advantages obtained from mathematical transformations of the quantities conveying the information. To take a simple example, it is well known that the pitch of a note is determined by the fundamental frequency, F, of the source of the sound. The natural musical intervals of octave, fifth, fourth, major third and minor third,* correspond to simple multiples of F, namely $2F$, $3F/2$, $4F/3$, $5F/4$, $6F/5$. Now define P as the logarithm of F ($P = \log F$). In terms of P, the natural intervals correspond to constant additions, namely 0.301 for the octave, 0.176 for the fifth, and so on. Addition is a simpler operation than multiplication, so if one is interested in musical intervals there is something to be said for dealing with the transformed quantity P instead of the actual frequency F. It is therefore interesting to find that the keys of a piano are placed according to P; that is the position of a note on the piano, measured from the lowest note, is nearly proportional to the logarithm of its frequency; likewise in musical notation the vertical position of a note follows P, not F. Also, if you look at a graphical representation of the frequency response of a Hi-Fi set, you will almost always find that frequency has been plotted on a logarithmic scale. A physiological counterpart can be found in the position of maximum response of the basilar membrane.

There are perhaps even greater advantages in expressing the intensity of a sensory stimulus on a logarithmic scale, because the physical values of stimuli tend to vary over enormous ranges; furthermore relative values are usually more important than absolute intensities. Thus an acoustical frequency–response curve usually has response given in decibels, which is a logarithmic scale. It used to be thought that subjective sensory intensities were, in general, proportional to the log of the physical intensity, and this is still a useful idea to have in mind. It is certainly generally true that the added stimulus (ΔI below) required to elicit the smallest perceptible change in sensation corresponds, very roughly, to a constant fractional increase (k) in the physical stimulus (I). This can be expressed in a

* These are called the Pythagorean intervals and it was known in Pythagoras' time that they corresponded to lengths of a vibrating string bearing to each other the ratio of small integers. In the equal-tempered scale used on modern fixed-pitch instruments like the piano, only the octave is strictly accurate. On this scale a semitone interval is the ratio of frequencies corresponding to the twelfth root of 2: repeated 12 times this gives the octave, 2, but for the fifth (7 semitones) the result is 1.4983 instead of 1.50, for the major third it is 1.26 instead of 1.25, and for the minor third 1.189 instead of 1.2.

form known as Weber's law: $\Delta I = kI$. If it is then assumed that such a threshold change in sensation corresponds to the addition of a constant quantity of sensation ΔS, one obtains $\Delta S/\Delta I = 1/kI$ and when this is integrated one gets the so-called Fechner law: $S \propto \log I$. But this treatment can be criticised on many grounds: for instance it neglects important temporal factors in adapting to new stimulus levels, and it is far from clear that increments of sensation can be integrated to predict the value of a steady level of sensation. Furthermore Weber's law has not been found to hold at all accurately, so these relations have not proved to be useful in physiology except as approximate general guides.

A rather different type of transformation is the topographical distortion in the mapping of sensory surfaces on to the cerebral cortex: in man, the cortical area for the thumb is many times larger than that for the knee, while in the pig the snout area is larger than that for the whole front leg. Again, the cortical representation of the central part of the visual field is enormously enlarged relative to the periphery, and interestingly enough an approximately logarithmic relation is also encountered here: in the monkey the distance on the cortical surface to the representation of a point in the visual field at eccentricity $e°$ is nearly proportional to $\log e$. This may be the direct consequence of the high resolution at the fovea, for this would require proportionately more central machinery which would necessarily occupy more space. However the particular mapping that is found has an interesting property, for it is approximately 'conformal'. Lines at a fixed orientation in the visual field such as vertical have an orientation in the map that is not constant but varies with position in the visual field. However, in spite of this distortion the angle between two short lines lying close to each other is unchanged; as a consequence the shape of an *L*, for instance, is preserved, and this may have functional importance.

The transforms so far considered are monotonic distortions of the physical scales encountered in specifying sensory stimuli. Much more radical transformations are known which would change the whole scheme of representation of the sensory stimuli. Though we do not know for certain what reorganisations of sensory information *actually* occur in the brain, it is important to realise that they can occur, and the next section describes the Fourier transformation. This has long been recognised as important in hearing, and the application of the same ideas to vision has attracted much attention.

1.3. FOURIER TRANSFORMS AND CONTRAST SENSITIVITY FUNCTIONS

The basic idea is that a quantity such as sound pressure, which varies in time, can be expressed as a sum of unending sinusoidal waves of constant amplitudes which persist indefinitely. Leading mathematicians disbelieved Fourier's theorem when he proposed it in 1807 and it is not easy to accept the fact that one can represent a transient event, such as a single sharp click, by a set of superimposed sine waves that have persisted indefinitely in the past and will persist indefinitely into the future. These aspects lie at the root of the utility and power of the Fourier transform, and perhaps the radical change of time scale also explains its biological relevance, but before returning to this theme some illustrations will be given.

Fig. 1.2 shows a sine wave at top, and a square wave of the same frequency at bottom. In between are the waveforms created by adding up the first three and the first eight terms of the series:

$$\sin \Omega t + (\sin 3\,\Omega t)/3 + (\sin 5\,\Omega t)/5 + \ldots + (\sin n\,\Omega t)/n$$

where Ω is the frequency in radians s^{-1} and the frequency in cycles s^{-1} is $F = \Omega/2\pi$. Notice that only odd harmonics are involved. Even with only three terms, up to $5\,\Omega$, one can see that an approach to the square wave is being made. The third figure is a good approximation, and here the highest term has a frequency 15 times the fundamental Ω, and an amplitude 1/15 as great. The square wave at the bottom has all the higher terms; notice that the amplitude of the square wave is actually less than that of the fundamental at the top; the ratio of their peaks is $\pi/4$.

In this example only odd harmonics were involved because only these happen to be required to synthesise a square wave. Changing the coefficients (1/3, 1/5, 1/7, etc.) would obviously change the shape of waveform that was formed, and the variety of synthesisable waveforms can be further increased in three important ways:

(1) The even harmonics (2Ω, 4Ω, etc) can be included: this will have the slightly unexpected effect of making the top and bottom halves of the waveforms unlike each other. For instance in order to synthesise a rectangular wave with the flat tops shorter (or longer) than the flat bottoms one would require even harmonics.

(2) Cosine as well as sine terms can be included: both terms are necessary when the waveforms to be synthesised do not have mirror symmetry about a vertical axis.

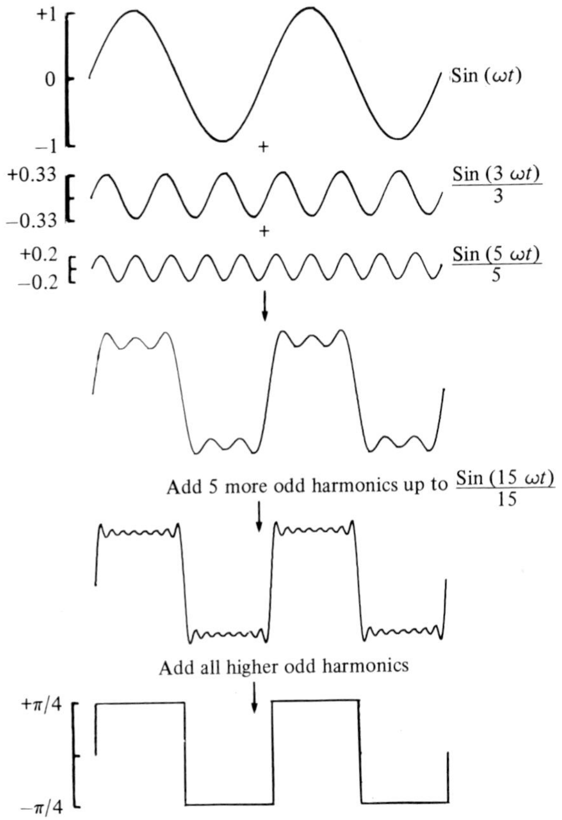

Fig. 1.2. Synthesis of a square wave (bottom) from sine waves. Any repeating waveform can be represented as the sum of the sine and cosine waves of appropriate amplitudes at the repetition frequency of the waveform and integral multiples of it. For a square wave, the coefficients defining the amplitudes of the sine waves are 1, 1/3, 1/5, 1/7, ... for frequencies 1 ×, 3 ×, 5 ×, 7 ×, ... the repetition frequency; all others (i.e. cosine waves and all even multiples of the fundamental) are zero. The figure shows the result of adding the first three and the first eight terms of the series. Notice that the square wave synthesised by the full series has an amplitude 0.785 (i.e. $\pi/4$) times that of the fundamental sine wave.

(3) The value of the fundamental F ($F = \Omega/2\pi$) can be lowered, and the number of harmonics increased. In Fig. 1.2 the frequency chosen for F was that of the original square wave that was to be synthesised, but exactly the same approximations could have been achieved by choosing another frequency, say 1/10 of F, and then picking out tenth, thirtieth, fiftieth, etc., harmonics of appropriate amplitude. By lowering the fundamental a large number of harmonics are introduced that are not required for that particular square wave and have zero coefficients; they are available, however, and they could be used in an interesting way, for they allow the synthesis of a waveform that repeats every $10/F$ sec, instead of every $1/F$ sec as with the square wave in Fig. 1.2. Thus a single cycle could be synthesised, with nothing happening for the next nine cycles, and then a repeat of the single cycle. If this process were carried to the limit, one could represent an event which *never* repeated, but this would require that F be zero and the harmonics would have to be spaced at infinitesimal intervals. It is only in this sense that one can represent a single click by a set of unending sine waves, but that is remarkable enough.

There is an obvious parallel between the mathematical representation of a complex waveform as a set of superimposed sine waves, and the musician's ability to decompose a chord into its component notes. This parallel should not be pushed too far, but the important lesson of the Fourier transform is that one can represent a rapidly changing sound pressure as the superposition of functions, in this case sine waves, that have a fixed, unchanging, description. Sine waves of constant amplitude are not, however, the only possible set of functions, called 'basis functions', that one can use for such a decomposition, and for the ear the most realistic basis functions would be sine waves that die away after half a dozen or so vibrations. Since these do not persist indefinitely, transformation using them does not give a static, unchanging, representation of a waveform, but one that changes in time. However, the important point is that it changes more slowly than the sound pressure itself; the time scale of the representation has been slowed down to a point where nerve fibres can transmit it and the nervous system react to it. That is presumably the main biological advantage derived from analysing sounds in terms of frequency rather than as instantaneous sound pressure.

Fourier-type transformations are not confined to one-dimensional variables like sound pressure, but can be employed on two-dimen-

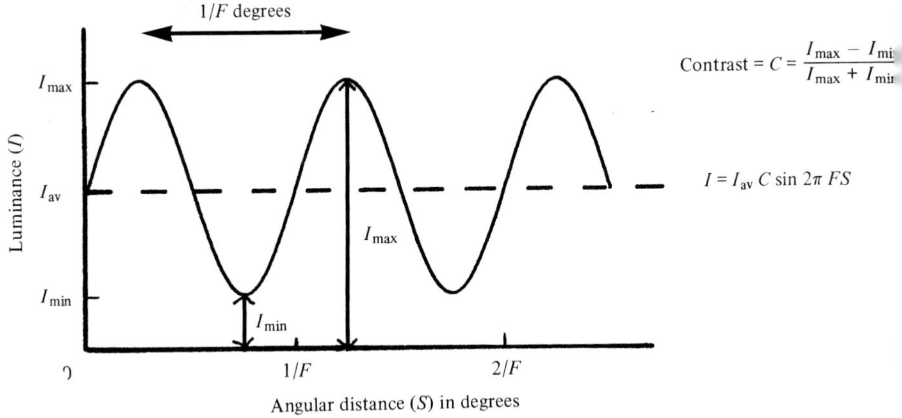

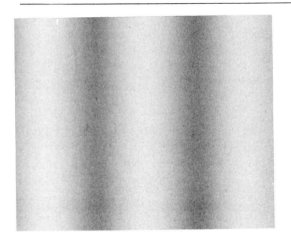

Fig. 1.3. A spatial sinusoid (top) and a cross-section of the luminance along it. The amplitude of modulation is expressed as *contrast*, given by $(I_{max} - I_{min})/(I_{max} + I_{min})$. The frequency, F, is usually expressed in cycles per deg, so one period of the grating is $1/F$ deg. At 57.3 cm from the eye 1 cm subtends 1 deg. At 6 m this grating would have spatial frequency 3.6 cycles deg^{-1}, and you could detect a contrast of about 0.003 (contrast sensitivity $1/0.003 = 333$, see Fig. 8.1).

sional variables like the pattern of luminances that make a picture. A photograph can in theory be represented by the sum of a harmonic set of sinusoidal spatial waves that are uniform over the whole picture area and run across it in a variety of directions. (These wave patterns must not of course be confused with the wave nature of light itself, where frequency determines colour.) A spatial sinusoid is shown at the top of Fig. 1.3, with the light luminances along a cross-section below. Note that luminance must always be positive, so these spatial sinusoids are modulations of an average luminance. It is convenient to refer to the *contrast* C of such a grating, which is the maximum deviation from the average luminance divided by the average luminance; this is given by

$$C = (I_{\text{max}} - I_{\text{min}})/(I_{\text{max}} + I_{\text{min}}).$$

The frequency of a spatial sinusoid in a visual experiment is usually given in cycles deg^{-1}. This is simply the number of complete cycles of the grating that subtends 1 degree at the eye; the grating shown in Fig. 1.3, would have a frequency of 3.6 cycles deg^{-1} when viewed from a distance of 6 m.

Since any picture can be broken down into sinusoidal components like this it is instructive to see how sinusoidal gratings are handled by an optical system. Fig. 1.4 shows this for the human eye in a form known as the *Modulation Transfer Function* (MTF). Suppose one looks at a sinusoidal grating, such as that of Fig. 1.3 but with a contrast of 1.0. If the optics are good one naturally expects the image on the retina to preserve the full contrast, but this is not the case; the ordinates show the contrast in the image for different spatial frequencies, and it will be seen that the high spatial frequencies suffer severe loss of contrast, as shown by the decline of the curve to the right. Indeed there is a total loss of contrast for spatial frequencies above a certain limiting value, known as the *limiting resolution*: spatial sinusoids of frequency above this value are completely flattened out.

One might think that an MTF like that of Fig. 1.4 represents a poor performance, but this is not so at all. The dotted line shows the best possible performance for an optical system with 2 mm entrance pupil, the diameter for which the eye's optical performance is best. The limiting performance is set by diffraction of light at the pupil; the period (in radians) of this *cut-off frequency* is actually given by λ/D, where λ is the wavelength of light and D the diameter of the pupil. The eye does not perform as well as this perfect instrument, but it is not much worse and the major part of the loss at high frequencies is unavoidable.

Fig. 1.4 was obtained by an objective technique described in Chapter 3 in which the quality of the image falling on the retina is determined from the small fraction of light that is reflected back out

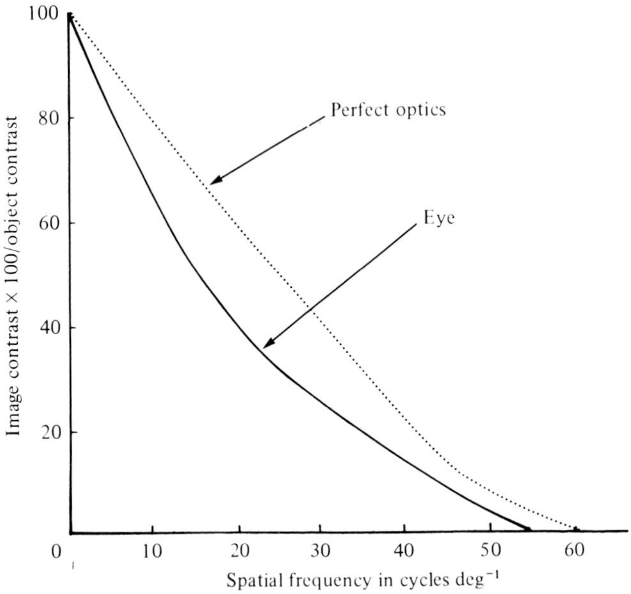

Fig. 1.4. The modulation transfer function (MTF) of the human eye with 2 mm pupil compared with a diffraction-limited optical system of the same aperture. The curve shows the contrast in the image formed of a spatial sinusoid of 100% contrast. Spatial frequencies of both object and image are expressed in cycles deg^{-1}. It will be seen that there is a progressive loss of modulation in the image as the spatial frequency is increased. At 62 cycles deg^{-1} for perfect optics, all contrast is lost; this is the cut-off frequency resulting from diffraction for this aperture. The eye loses all modulation just below this limit at 55 cycles deg^{-1}. (From Campbell & Gubisch, 1966.)

of the eye again. One would expect the optical loss of contrast at high spatial frequencies to make them less easily visible, and this will be discussed in Chapter 8, where Fig. 8.1 shows a *contrast sensitivity function*. To determine this a subject looks at a spatial sinusoid of variable contrast which is adjusted until it is barely distinguishable from a uniform, unmodulated grey. The contrast sensitivity is the reciprocal of this setting, so if at threshold $(I_{max}-I_{min})/(I_{max}+I_{min}) = 0.01$, the contrast sensitivity is 100. Note that in this figure the scales for both frequency and contrast are logarithmic, whereas linear scales were used in Fig. 1.4.

Modulation transfer functions are a very good way of characterising the performance of physical instruments such as camera lenses and television systems, because they give such a complete description of their performance. To a large extent the same is true for contrast sensitivity functions, and for this reason sensory thresholds have been measured for many types of sinusoidal stimulation. The reader will come across curves characterising the eye's ability to detect flickering lights (temporal contrast sensitivity curves), lights of varying wavelength (spectral sensitivity curves), how the ear detects sounds of different frequency (audiograms), and how the vibration of skin is sensed. The technique has also been used on isolated components of sensory pathways such as the retinal ganglion cells, or cortical neurons in visual, somatic or auditory sensory areas. However, a word of warning may be needed when interpreting these curves because it is only in linear systems* that knowledge of the frequency response gives complete information about the response to an arbitrary stimulus: the brain is certainly not a linear system, so the Fourier transform does not necessarily give important insight into how the brain interprets sensory messages. Nevertheless the theoretical possibility of decomposing pictures into spatial sinusoids of different frequency and orientation introduces the idea of representing sensory messages in unfamiliar, non-obvious ways. If a picture, or a sound, can be represented as the sum of a set of sinusoids, we need no longer think of the representations in our brains as *copies* of the picture or sound; we must be prepared for surprises and subtlety in the transformations of sensory messages in the nervous system.

1.4. SELECTIVITY FOR PATTERN; TRIGGER FEATURES

Some of these surprises were encountered when sensory pathways were investigated by recording the activity of single neurons more central than those connected to the receptors themselves, for such neurons were found to have unexpected properties resulting from the complicated interactions that occur at the synapses leading to second and higher order neurons. The important new property achieved as a result of these synaptic interactions is selectivity for *pattern*, rather than for simple physical or chemical characteristics of the stimulus. Moreover, much of the complexity of the mechanism is concerned

* The linearity referred to is between stimulus strength S, and response magnitude R, i.e. $R = KS + \text{Constant}$ for the condition to hold. When this is true, 'superposition' holds, so that the added response ΔR resulting from an added stimulus ΔS is the same regardless of any steady level of stimulation.

with achieving *invariant* responsiveness to a specific pattern, in the face of variations in the physical attributes of the stimulus and its position.

Retinal ganglion cells

These properties of pattern selectivity and invariance of response are best illustrated by examples from the visual system, because it was there that the importance of pattern excitation was first discovered, and where the analysis has progressed furthest. The technique is to place a recording microelectrode in the ganglion cell layer (see Fig. 2.2*b*) and move it about until the action potentials from a single ganglion cell can be picked up separately from those of its neighbours. A search is then made for a region in the visual field where excitation by light causes a response. Most of the ganglion cells have a maintained discharge rate of 5 to 60 impulses s^{-1}, and the effect of stimulation is to increase or decrease the firing rate. The first step is usually to turn a small spot of light on and off and mark the positions in the visual field from which responses are elicited, thus mapping the *receptive field* of the particular ganglion cell one has isolated. Fig. 1.5 shows examples of such maps; a plus sign means that a spot turned on at that position increases the firing rate, and a minus sign means that it decreases the firing rate; when a decrease occurs when the spot is turned 'on', there is usually a rebound above the maintained discharge rate when it is turned 'off'. An '0' means that no change occurred with a spot at that position, and this would also be so outside the 0s. It is usually found that plus (or 'on') and minus (or 'off') regions inhibit each other so that a light turned either on or off simultaneously on a pair of such regions causes a smaller response than the same stimulus applied to reach region separately.

Now consider the maps at the top of Fig. 1.5. To the left a roughly circular region of the visual field subtending about 1° at the eye has plus signs in it, and this is surrounded by minus signs. The field shown to the right has the opposite arrangement, minus signs surrounded by plus signs. Receptive fields with this concentric arrangement of regions that respond in the opposite sense are very common in the retinae of all species, and the same arrangement is found in second and higher order neurons of other modalities such as touch and hearing. It was first discovered by Hartline in his investigation of the compound eye of the primitive crustacean *Limulus*, and is called *lateral inhibition*. It has the effect of making the cell compare the amount of light at one position with the amount at neighbouring

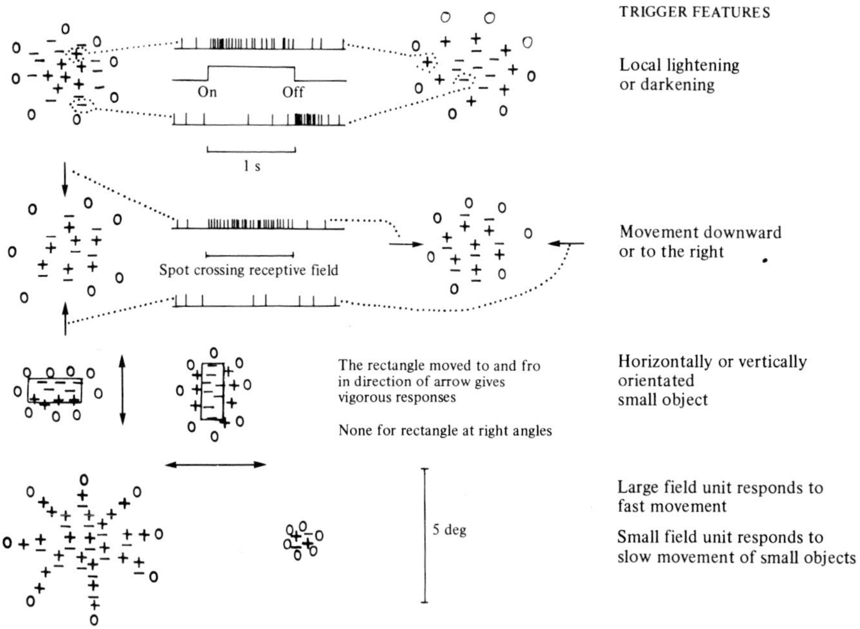

Fig. 1.5. Receptive fields and trigger features of five types of ganglion cell found in the retina of rabbits. + and − signs mean that a stationary spot of light turned on at that position causes an increase (+) or decrease (−) of the maintained discharge; when it is turned off the opposite change occurs, namely a decrease (at +) or an increase (at −). Examples of these responses with indications of the receptive field regions that would produce them in an on-centre and off-centre ganglion cell are shown in the centre. No responses result from stimulating at positions marked 0, or outside the ring of 0s. Ganglion cells with receptive fields in which 'on' (+) and 'off' (−) zones are concentrically arranged, as in the top pair, are commonly found in many species, and they respond best to small areas that are lighter or darker than the surrounding region in the visual field. However, maps of the receptive field made with stationary spots often fail to reveal important pattern-selective properties of the ganglion cell. When mapped in this way the second pair of ganglion cells responded at both 'on' and 'off' throughout their receptive fields, but testing with a moving spot showed that the left-hand one responded best to downward movement and not at all to upward movement, while the right hand one responded to rightward movement and not to leftward movement; this is indicated in the second pair of responses in the centre. The ganglion cells with the other four receptive fields mapped here also responded selectively to specific trigger features (see text).

regions. The signal from a cell with an 'on' centre thus indicates that the illumination of a particular region in the visual field is greater than the illumination of its neighbours; the cell with minus signs at its centre correspondingly signals that the illumination is less than in neighbouring regions, i.e. it is darker. It is important to realise that the retina is densely covered with ganglion cells of these two types, and that their receptive fields overlap and collectively cover the whole visual field.

The map of the receptive field is a useful preliminary stage in analysing high-order neurons in sensory pathways and one can proceed in three ways. The first is to analyse the cellular mechanisms whereby selectivity is brought about, and some success has been achieved along this path by recording intracellularly from the cells of the retina (see Chapter 6).

The second way to proceed is to make a map in the Fourier or frequency domain: instead of determining the sensitivity to spots of light in different spatial positions the sensitivity to spatial sinusoids of different frequencies and orientations is measured. This approach is described in the chapter on spatial resolution and analysis.

The third approach is to make explorations of the response properties of the neurons when faced with the kind of visual stimuli encountered in real life, in the hope of discovering what part these neurons ordinarily play in perception. These concentric cells respond very vigorously to white (for the on-centre type) or black (for off-centre type) spots of appropriate size moved into their receptive field centres. They also respond well to white or black bars of about the same width as their field centres, and to black–white edges crossing their fields, but they do not respond well to uniform illumination. It is probably correct to regard them as performing the first step in emphasising borders and edges in the visual image. However, one could have reached this conclusion equally well by inspection of the receptive field map, or even from the contrast sensitivity function; the importance of further qualitative exploration in revealing the pattern selectivity of cells only becomes fully evident with the other types of retinal ganglion cell. The retina of the rabbit provides some good examples.

Consider the receptive field map for the unit shown in the second row in Fig. 1.5. When explored with a spot turned on and off it gives a brief discharge at all the points marked with a plus and minus sign, and no responses elsewhere. It was only by exploring with a *moving* spot that the striking result shown by the records was found. A small spot of light moved downwards through the receptive field gives a vigorous burst of impulses sustained for the whole time the spot moves in the field; however, when the spot retraces its path in the opposite direction, only a few impulses occur. Other examples of this class of cell respond preferentially to motion to the right, as shown in a second receptive field map. Yet others prefer movement in other directions. As with all classes of ganglion cell there are large numbers of them, and together these cover the whole visual field for all four directions of movement with their overlapping receptive fields.

Invariance of pattern selectivity

The directional selectivity of such retinal ganglion cells persists in the face of large changes in the physical properties of the stimulus (see Fig. 1.6). First one can change the overall luminance level by a factor of 10000 or more, and the retinal ganglion cell will persist in responding to an object moving in one direction but not in the reverse. Second, one can keep the background illumination constant and vary the contrast of the spot; the ganglion cell responds preferentially to the same direction of movement, regardless of the brightness of the spot, and it even persists in that preference when a dark spot is used instead of a light one. Finally a small stimulus spot can be moved to and fro almost anywhere within the receptive field and the cell will respond to the same direction of movement at all positions. Thus it can be shown that a ganglion cell's directional selectivity is invariant over a wide range of luminances and contrasts, and also over the limited range of positions in the visual field covered by its individual receptive field.

One naturally starts off by thinking that the physical properties of the light stimulus must be the important variables determining the response of a retinal ganglion cell. Fig. 1.6 shows that this is too simple a view, for a response occurs only when the spatio-temporal *pattern* of stimulation characteristic of movement in a particular direction occurs; the light intensities and the position are relatively unimportant.

One needs a term to describe the pattern of stimulation that would normally excite such a cell, and this is sometimes called its *trigger feature*. In the rabbit's retina further exploration reveals ganglion cells with a wide variety of trigger features. For instance velocity of movement is an important variable for some cells which respond vigorously to objects moving so rapidly that they fail to excite other cells. These are called 'fast movement detectors' (see Fig. 1.5). In the

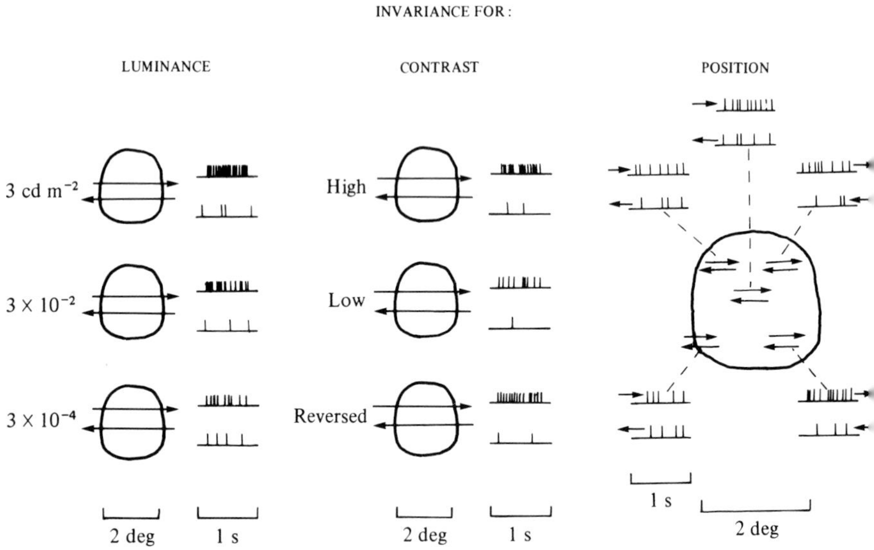

Fig. 1.6. Invariance of pattern selectivity over luminance, contrast, and position. This retinal ganglion cell responded well to rightward movement and poorly or not at all to leftward movement over a wide range of luminances, contrasts, and at most positions within its receptive field. Only the outline of the receptive field is shown here. The left column shows responses for two directions of movement at three different luminances. The centre column shows them for spots of high, low and reversed contrast. The right hand part shows responses for to-and-fro motion confined to different subdivisions of the receptive field. In all cases, rightward movement causes a vigorous response, leftward movement little or none.

region of the rabbit's retina that is normally aligned with the horizon one can find cells that respond best to elongated dark or light objects, and also cells which refuse to respond to anything but very slowly moving objects.

This list of trigger features of ganglion cells found in the rabbit is not complete. Some known types have been omitted and there are types of cell, for instance those subserving colour vision, which we are confident the rabbit possesses, but which have only recently been recorded from neurophysiologically. Incomplete though it is, the above description illustrates the extraordinary diversity of pattern-selective elements that may be found at a low level in the visual pathway of a single species. Knowledge of the *anatomical* diversity of ganglion cell types was actually available almost 100 years ago from Ramon y Cajal's histological work on the retina, but only now are we beginning to correlate the physiology with the anatomy. A start has also been made in uncovering the mechanisms whereby pattern selectivity is achieved, but here too there is a long way to go.

Knowledge of the diverse types of ganglion cells and trigger features is less complete in other species. The most prominent ganglion cells in the cat retina are the concentric type, which were in fact first discovered there by Kuffler before the work on rabbit was done. Much later it was found by testing with sinusoidal gratings that these concentric cells are not all alike but belong to two subtypes called X or sustained, and Y or transient. The properties of these subtypes are discussed further in Chapter 8. Later still a host of other types, collectively called W-cells, were discovered, and some of these appear to be similar to those in the rabbit; their properties, however, are less well known because they are harder to record from. Colour opponent cells are prominent in monkey (see Chapter 9) and the division into sustained and transient types probably also holds; it is not known whether the retinas of monkey or man possess the more strongly pattern-selective elements of the rabbit.

Neurons at higher levels in the visual pathway

Hubel and Wiesel pioneered the study of neurons at more central points in visual pathways, starting with cats and then going on to Rhesus monkeys. At the level of the lateral geniculate nucleus (see anatomical plan of visual pathway, Fig. 2.5) the predominant type of neuron has a concentrically arranged receptive field as at the retina, and we know from subsequent work that the division into sustained (X) and transient (Y) subtypes is maintained. Fig. 1.7 shows these receptive fields diagrammatically. At the next level, in the primary area of visual cortex (area 17), new properties appear; the single cells here require an elongated stimulus at a particular orientation for optimum response. Different cells require different orientations, and in addition the stimulus often has to be moved in the right direction and has to be about the right size. Furthermore this is the first point in the pathway where cells are found that can be stimulated through either eye.

As well as describing these new properties of orientational selectivity, directional selectivity, size selectivity and binocularity, Hubel and Wiesel divided the cells into two classes they called 'simple' and 'complex'. The main basis for the distinction was whether a map of the receptive field, with its 'on' zones and 'off' zones, gave evidence

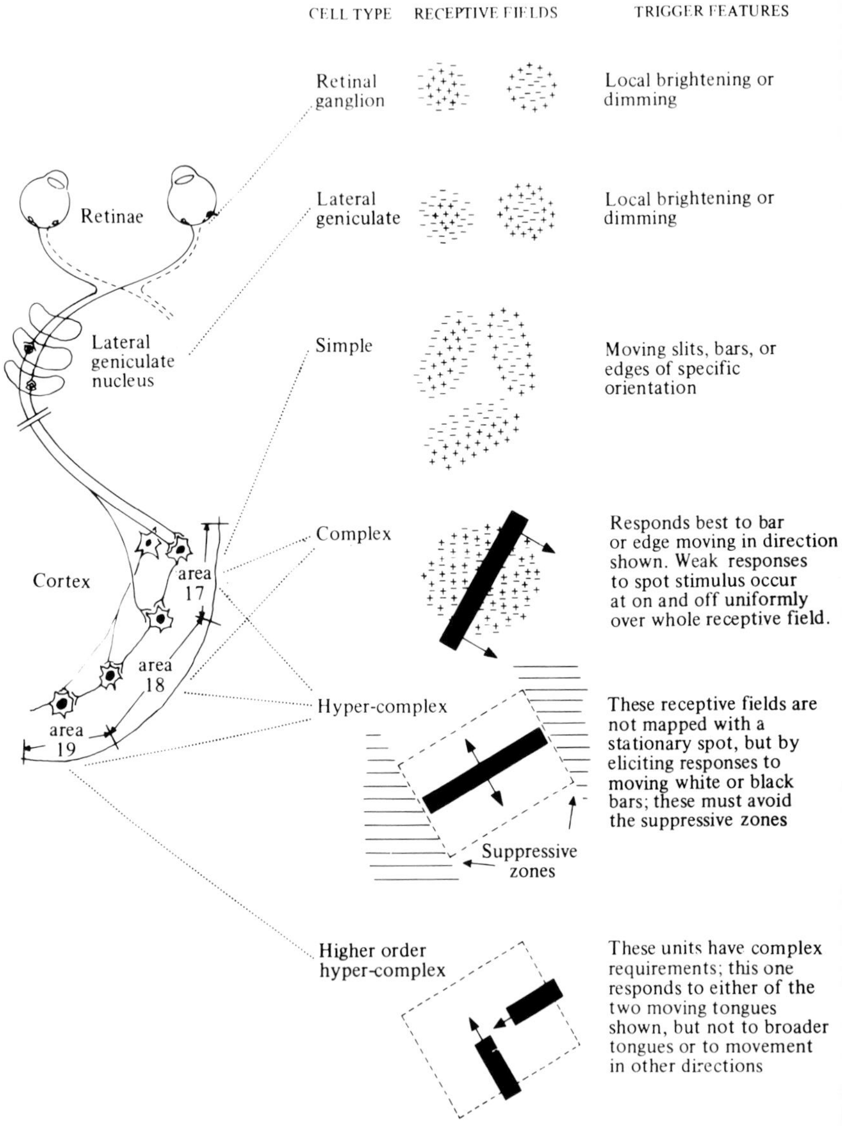

Fig. 1.7. Cell types and trigger features at different positions in the visual pathway of the cat, shown diagrammatically. Monkey cortex is mainly similar (see text). The elongated receptive fields in area 17 are sensitive to bars and edges of the appropriate orientation and size. This pattern selectivity results from the excitatory and inhibitory connections to cortical

of their pattern-selective properties. This was the case for simple cells; for instance an elongated 'on' zone showed the orientation and approximate size of the bright bar that would optimally stimulate it. Complex cells, however, often gave both 'on' and 'off' responses throughout their receptive fields, and even when this was not so their orientational or directional selectivity could not be predicted from the map alone (compare the directionally selective cells of the rabbit's retina).

At first Hubel and Wiesel thought that the complex cells received their input solely from the simple cells, but it is now thought that there is a direct input from the lateral geniculate nucleus (LGN). Thus it seems likely that there are parallel X and Y pathways from the retina through to the primary cortex. This discussion will be continued in Chapter 8; the significance of this dual system may be that the X system is concerned with analysis of form, whereas the Y or transient system is concerned with movement and hence with the precise timing of visual stimuli. One must also remember the W system in cats; these neurons are thought to project to the cortex but their influence has not yet been worked out.

Next to area 17 lie areas 18 and 19. These are often called secondary visual areas, though in the cat area 18 receives a direct projection of Y type fibres from the LGN. These areas have neurons with more complex type of selectivity than those of area 17, and two of these are shown in Fig. 1.7.

The situation in the monkey is broadly similar to that in the cat; two differences worth noticing are the reduced direct projection from the LGN to area 18, and the presence of an extra stage of densely packed cells with concentric receptive fields in a subdivision of layer IV of area 17. These cells are all monocular, and it is only in other layers that binocular neurons are found.

It has long been known that there is an orderly mapping of the retina on to area 17, so that the cells in one part of the cortex have their receptive fields lying close to each other in one part of the visual field. Hubel and Wiesel discovered an orderly microstructure within this mapping. Cells in the monocular lamina of layer IV are arranged in alternating strips about 0.5 mm wide, one strip connecting to one

---

neurons made by lateral geniculate neurons (LGN) with concentric receptive fields. The more complex properties of cells in areas 18 and 19 result from connections made to these areas from cells in area 17, though there is also a direct input to area 18 from LGN.

eye, the other to the other. Furthermore, cells of different orientation selectivities are not arranged at random: they lie in orderly sequences of gradually changing orientation, a shift through 180° occurring in about 0.5 mm. Some of this microstructure is pictured in Chapter 2, Fig. 2.6.

This work on the cortex has led to interesting work on development, for the properties of the cortex at birth in monkeys, or at eye-opening in kittens, are not altogether like those of adults. Furthermore deprivation of vision during the first three months by rearing in darkness or by suturing the eyelids interferes with the development of the cortex, and this sensitivity to deprivation of normal visual input during early life is thought to be the causative factor in human *amblyopia*, and also the much commoner condition of *stereoblindness*. These conditions are discussed in Chapters 10 and 20, and the role of area 17 in the preliminary sorting of visual stimuli according to their disparity is also discussed in Chapter 10.

It is natural to ask 'What is the role of the primary visual cortex, area 17, in the processing of sensory information?'. The primary visual cortex certainly acts as a redistribution centre, and this is the clue to its six-layered lamination, for the cells are largely segregated according to the destination of their axons. Thus many lamina VI cells send their axons to the thalamus, including the LGN, whereas many lamina V cells send axons to the superior colliculus, the mid-brain visual centre. Lamina IV contains the granule cells on which incoming fibres from the LGN mainly terminate, and these cells and those of laminae II and III send axons to other regions of the cortex. These other regions include secondary visual areas 18, 19 and 20 (also called V II, V III and V IV), and also the contralateral visual cortices by way of the corpus callosum.

Unfortunately there is no agreement about what is achieved by the selectivity of cortical neurons and the redistribution described above. In contrast, the pattern-selective properties of the rabbit retina make sense almost intuitively, for a small herbivore that is popular fodder for many carnivorous predators *needs* special ganglion cells to detect the slowly stalking beast approaching over the horizon, or the suddenly swooping hawk descending from above. But the specificity of many of these retinal cells is as great or greater than that of the cortical neurons, and it is achieved after two or three synapses instead of five or six for most neurons in area 17.

The more orderly arrangement of cells according to eye preference and orientation preference suggests a more methodical analysis of the information that reaches the cortex, and perhaps the most plausible hypothesis is that its role is to detect *linking features* such as motion, colour, disparity, and texture which are local properties of each part of the image. The multiple secondary visual areas (V II, V III, V IV) surrounding the primary cortex, and to which V I relays information, would then have the task of using this information to identify the parts of the visual scene that possess a common linking feature and thus probably belong to an object that is to be recognised. Such segregation of *figure* from *ground* was a familiar theme of the Gestalt school of perceptual psychologists who attacked the problem of how we see objects as a whole rather than as a multiplicity of separate parts. It is a tempting hypothesis that the primary cortex detects the characteristics (linking features) that allow this synthesis to be achieved, but it will require more experimental knowledge of the secondary areas to put the hypothesis to the test.

Finally it is worth remarking that the cortex, which is thought to be responsible for man's intellectual pre-eminence, is a remarkably homogeneous structure throughout its extent. In the visual system we have progressed further than anywhere else towards discovering the mechanisms of some simple examples of pattern selectivity, such as the detection of motion, orientation, and disparity. However this is only a beginning and there is much more to be discovered before we gain the insight into its overall operation that the uniformity of structure of the cortex tempts one to believe may be attainable.

1.5 CONSCIOUS PERCEPTION

Some account must be given of the relation between the information reaching the brain and our conscious perceptions, if only to dispel the natural belief that the relationship is a simple one. Of course stimulating sense organs does usually result in a conscious perception, but the relation should not be regarded as a direct, causal one. The occurrence of an earthquake may result in the appearance of headlines in the newspaper, but the causal chain is complicated and involves so many conditional factors that it would be a bit absurd to say that the earthquake caused the headline. The same is true for sensory excitation causing perception. Conscious perceptions are best regarded as *interpretations*, made in the light of previous experience, of the evidence provided by our sense organs. This interpretation occurs unconsciously and the existence of this step is apt to be denied, for one instinctively places great reliance on the validity and directness of perceptions. But one cannot persist in this denial after experiencing illusions (see Chapters 7 and 12), and those such as the tilt after-effect (Fig. 7.1) and others requiring pre-adaptation (Fig. 8.7) give a very direct demonstration of the importance of immediately previous experience; it becomes a little easier to accept that a life-time of previous experience must influence what one perceives.

Our perceptions, then, are not always valid and they are not the direct appreciation of the environment; they are interpretations of sensory messages, and this has important consequences. For instance two people will often given different reports when they witness the same scene, not because one is a liar, unobservant, or perverse, but simply because the past experiences of the two people are different and hence their interpretations in the light of it lead to different results; in other words, they genuinely have different perceptions of the scene. One need not cease to accept that 'seeing is believing', but one comes to realise that seeing is *only* believing, and beliefs are based on prejudice as well as fact.

It was said at the beginning of this chapter that the patient's and the doctor's senses provide the most important, potentially the most reliable, and the most up-to-date information about an illness. That is true not only for the diagnosis of an illness, but for all situations where you use the evidence of your own senses and those of others. However, one needs to be aware of the complexity of the sensory pathways and the fallibility of perception in order to make optimal use of this evidence.

1.6. SUGGESTIONS FOR FURTHER READING

General references

Wiener, N. (1948) *Cybernetics, or control and communication in the animal and the machine.* The Technology Press. New York: John Wiley. (A look at biological control mechanisms by the mathematician who invented cybernetics; inspiring but not very accurate.)

Cherry, C. (1957) *On human communication.* New York: Science Editions Inc., MIT Press and John Wiley. (An engineer's viewpoint on human communication; discursive but instructive.)

Pierce, J. R. (1962) *Symbols, signals and noise.* London: Hutchinson. (Another engineering viewpoint biased towards psychology.)

Jones, D. S. (1979) *Elementary information theory.* Oxford: Clarendon Press. (A useful and clear account of the basics of information theory.)

Swets, J. A. (ed.) (1964) *Signal detection and recognition by human observers*. New York: John Wiley. (A collection of early papers on signal detection theory.)

Laming, D. (1973) *Mathematical Psychology*. London and New York: Academic Press. (Mathematical rigour applied to problems of decision and choice.)

Barlow, H. B. (1972) *Perception*, **1**, 371–94. (A speculative attempt to account for the performance of complex perceptual tasks in terms of known properties of nerve cells.)

Mollon, J. D. (1977) Neurons and neural codes; Neural analysis. Chapters 3 and 4, in *The Perceptual World*, ed. K. Von Fieandt & I. K. Moustgaard. London and New York: Academic Press. (Discussions of the relation between nervous activity and perception.)

Specific topics

Photometry. Le Grand, Y. (1968) *Light, colour and vision*. London: Chapman and Hall.

Human signal/noise discriminations. McNicol, D. (1972) *A primer of signal detection theory*. London: Allen and Unwin.

Reliability and noise of nerve responses. Barlow, H. B. & Levick, W. R. (1969) *Journal of Physiology*, **200**, 1.

Physical limits of vision. Barlow, H. B. (1986) Ch. 1, pp. 3–18 in *Visual Neuroscience*, ed. J. D. Pettigrew, K. J. Sanderson & W. R. Levick). Cambridge University Press.

Cortical maps. Adrian, E. D. (1947) *The physical background of perception*. Oxford: Clarendon Press.

Maps of visual field on cortex. Cowey, A. (1979) *Quarterly Journal of Experimental Psychology*, **31**, 1.

Fourier transforms. Bracewell, R. (1965) *The Fourier transform and its applications*. New York: McGraw-Hill.

Modulation transfer function. Hopkins, H. H. (1962) *Proceedings of the Physical Society (London)* **79**, 889.

Optics of the human eye. Campbell, F. W. & Gubisch, R. W. (1966) *Journal of Physiology*, **186**, 558.

Receptive fields and trigger features in rabbit retina. Barlow, H. B., Hill, R. M. & Levick, W. R. (1964) *Journal of Physiology* **173**, 377.

Invariant properties of trigger features. Maturana, H. R., Lettvin, J. Y., McCulloch, W. S. & Pitts, W. H. (1960) *Journal of General Physiology* **43**, Suppl. 2, Mechanisms of Vision, 129.

Histology of retina. Dowling, J. E. (1970) *Investigative Ophthalmology*, **9**, 655.

Sustained or X and transient or Y ganglion cells Lennie, P. (1980) *Vision Research*, **20**, 561.

Responses of cells and organisation of visual cortex. Hubel, D. H. (1988) *Eye, brain, and vision*. Scientific American Library, New York.

Mechanisms of pattern selectivity in retina. Barlow, H. B. & Levick, W. R. (1965) *Journal of Physiology* **178**, 477.

Perception as interpretation. Gregory, R. L. (1973) (with Gombrich, E. H.) in *Illusion in Nature and Art*. London: Duckworth.

Linking features and Gestalt perception. Barlow, H. B. (1981) *Proceedings of the Royal Society* B, **212**, 1–34.

CHAPTER TWO

Anatomy of the visual system

E. N. WILLMER

The diagrams in this chapter are designed to illustrate the essential structural features of the visual system of Primates. Since most of the experimental work has been done on macaque monkeys, many of the diagrams are based on the monkey rather than on man. The systems are, however, believed to be basically similar.

In all the diagrams of the eye the light is deemed to be entering from the top of the diagram (arrow). It is focussed by the refracting cornea and lens so that after passing through the vitreous body and inner layers of the retina it is absorbed by the visual pigments in the outer segments of the rods and cones. Light unabsorbed by the retina is absorbed by the melanin of the pigment epithelium and choroid. The apparent inversion of the cell layers in the retina results from its embryological development (see Fig. 2.5*c*) and probably depends on the evolutionary history of the eye in early vertebrates.

(a)

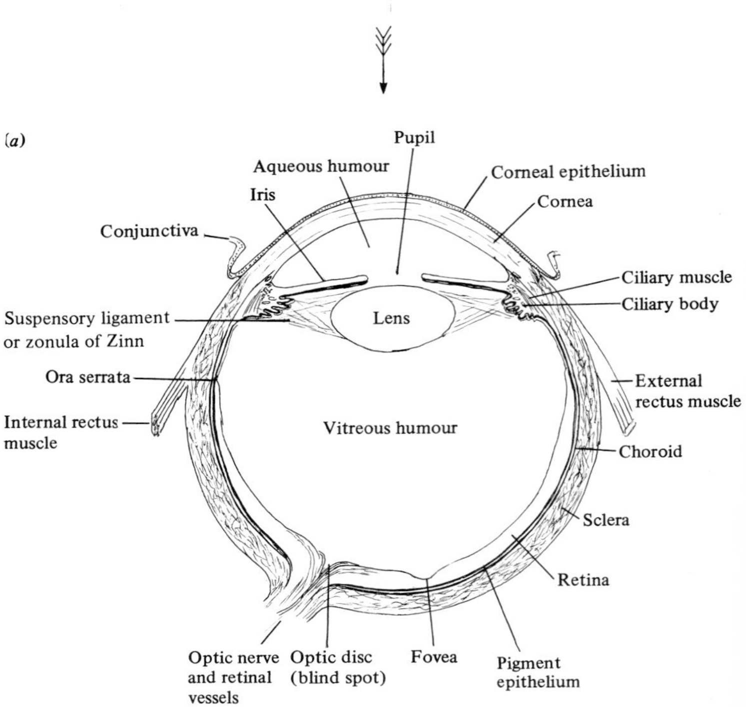

Fig. 2.1. (*a*) Equatorial section through the eye, in diagrammatic form to illustrate main features. The thickness of all the layers has been greatly exaggerated.

(*b*, *c*) Extrinsic muscles of the right eye: (*b*) from above, (*c*) from below. The superior rectus, inferior rectus, medial rectus and inferior oblique are supplied by branches of the third (oculomotor) nerve. The superior oblique is supplied by the fourth (trochlear) nerve. Note that the tendon passes through the trochlea. The lateral rectus is supplied by the sixth (abducens) nerve. The levator palpebrae receives fibres from the 3rd nerve and the sympathetic system.

(*d*) Enlarged diagram of a section through the anterior part of the eye. The cellular lens is slung in a hammock of fine collagen fibres (zonula of Zinn) between which the aqueous humour, produced from the epithelium over the ciliary processes, percolates before it passes through the pupil into the anterior chamber (arrows), from which it drains away into the canal of Schlemm through the trabecular meshwork, and thence into the veins. The optical properties of the lens are caused by special intracellular proteins (crystallins) and the presence of hyaluronic acid, probably between the cells.

The ciliary muscles reduce the focal length of the lens by causing the ciliary processes to approach the pectinate ligament and so reduce the tension on the fibres of the zonula. This allows the elasticity of the lens to cause it to become more spherical. Some of the muscles run radially to the pupil, others circumferentially. When both sets contract the perimeter of the ciliary processes is reduced. (See Chapter 4.)

The collagen fibres of the cornea are arranged in alternating layers. In each layer the fibres run parallel to each other but at right angles to the fibres in the layers above and below. The fibres are embedded in a medium rich in special mucopolysaccharides (keratosulphate, chondroitin sulphate and hyaluronic acid). This medium normally has the same refractive index as the fibres, but is sensitive to its water and ionic content. All these factors contribute to its transparency or otherwise.

The vitreous humour owes its transparent and gelatinous character to hyaluronic acid.

(*b*)

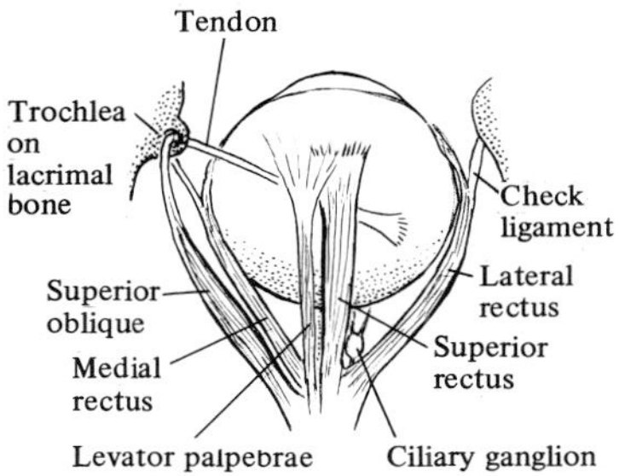
Tendon
Trochlea on lacrimal bone
Check ligament
Superior oblique
Lateral rectus
Medial rectus
Superior rectus
Levator palpebrae
Ciliary ganglion

(*c*)

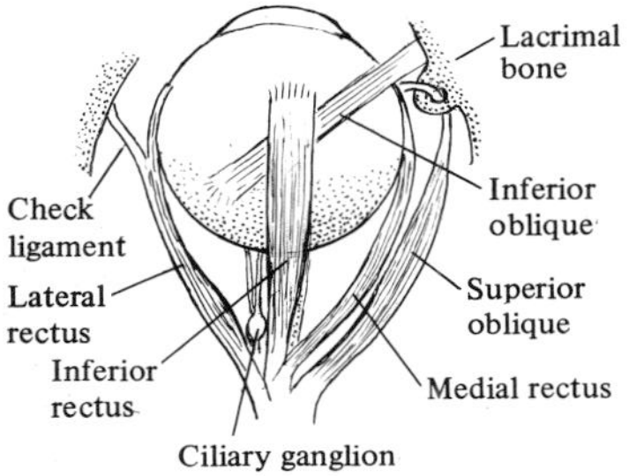
Lacrimal bone
Check ligament
Inferior oblique
Lateral rectus
Superior oblique
Inferior rectus
Medial rectus
Ciliary ganglion

(*d*)

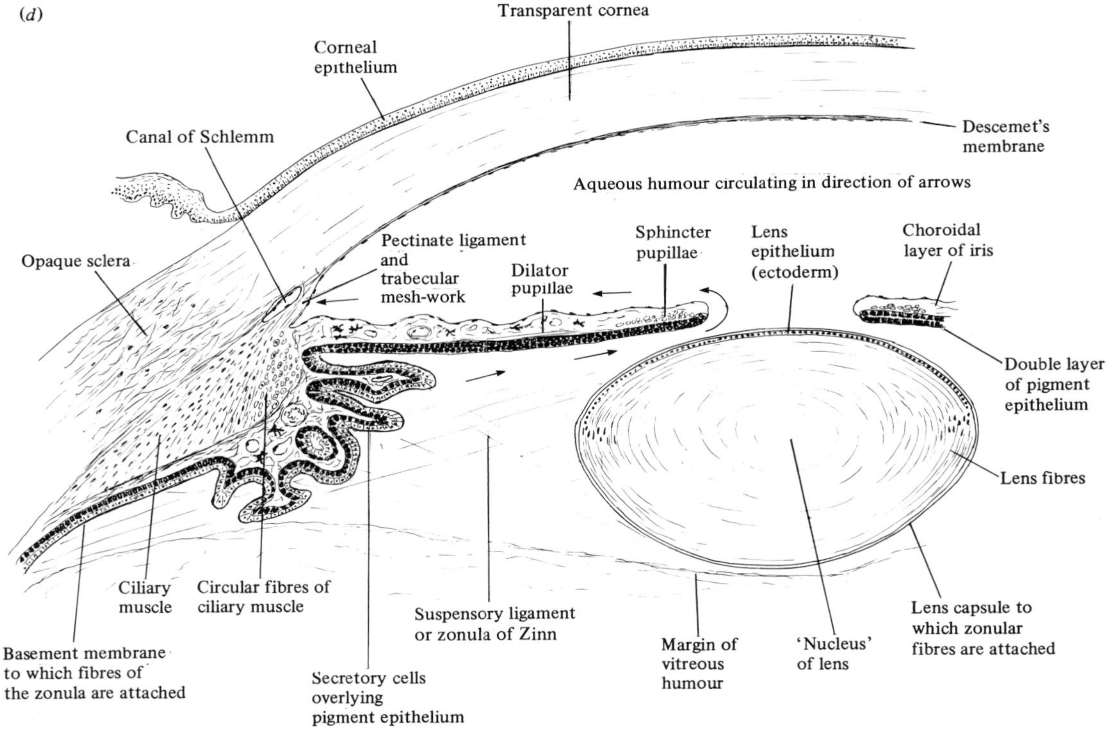
Transparent cornea
Corneal epithelium
Canal of Schlemm
Descemet's membrane
Aqueous humour circulating in direction of arrows
Pectinate ligament and trabecular mesh-work
Sphincter pupillae
Lens epithelium (ectoderm)
Choroidal layer of iris
Opaque sclera
Dilator pupillae
Double layer of pigment epithelium
Lens fibres
Ciliary muscle
Circular fibres of ciliary muscle
Suspensory ligament or zonula of Zinn
Lens capsule to which zonular fibres are attached
Basement membrane to which fibres of the zonula are attached
Margin of vitreous humour
'Nucleus' of lens
Secretory cells overlying pigment epithelium

Fig. 2.2. Histological features of the retina.
(*a, b, c*) Radial sections of the retina in three regions as seen in fixed and stained preparations, schematised. The following points should be noted. (i) Lengthening and narrowing of the cones from periphery to fovea, and their increasing concentration. (ii) Corresponding decrease in the number of rods and their complete absence from the centre of the fovea. (iii) Progressive changes in the character and numbers of ganglion cells and bipolar cells from the periphery to the foveal slope. The fovea ('pit') is actually the central depression from which these cells are pushed aside. (iv) The cone pedicles near the fovea form an almost continuous line. (v) The outer processes of the rods and cones interdigitate with the processes of the pigment cells of the retina. (vi) The pigment layer of the retina rests on the choroid coat which is very vascular and contains pigment cells of a different type (cf. *d, e*). (vii) Blood vessels that enter the eye with the optic nerve at the 'blind spot' ramify over the inner surface of the retina and they also extend between the ganglion cells and bipolar cells. (viii) The elongated nuclei of the fibrous supporting cells (Müller's fibres) can be seen among the nuclei of the bipolars.
(*d*) Two mesenchymal pigment cells (melanophores) from the choroid (or the anterior layers of the iris). These cells occur as isolated mobile units.
(*e*) Surface view of the pigment epithelium of the retina. These cells are fully adherent to each other in a single layer. Both types of pigmented cells contain melanin, a black pigment that is not photosensitive.
(*f*) The tip of a ciliary process is shown. Note that the pigment epithelium rests on vascular connective tissue closely akin to the choroid and containing melanophores. Over the pigmented epithelium there is another layer of unpigmented epithelium. This is believed to produce the aqueous humour by controlled filtration from the tissue fluid and the blood. The superposition of these two epithelia has an embryological explanation (see Fig. 2.5). The collagen fibres of the zonula are attached to a basement membrane on the surface facing the zonula.
(*g*) The tip of the iris is shown. The anterior portion is again composed of tissue akin to that of the choroid. The combination of its fibrous character and differing numbers and distribution of melanophores gives the iris its characteristic colour, for it is backed by the same double epithelium as the ciliary processes but behind the iris both layers are pigmented. The sphincter pupillae muscle, activated by the short ciliary nerve (parasympathetic) is derived from the epithelial layer; so too are the fibres of the dilator pupillae, whose cell bodies may remain in the epithelium and which are activated by sympathetic fibres running in the long ciliary nerve from the superior cervical ganglion.

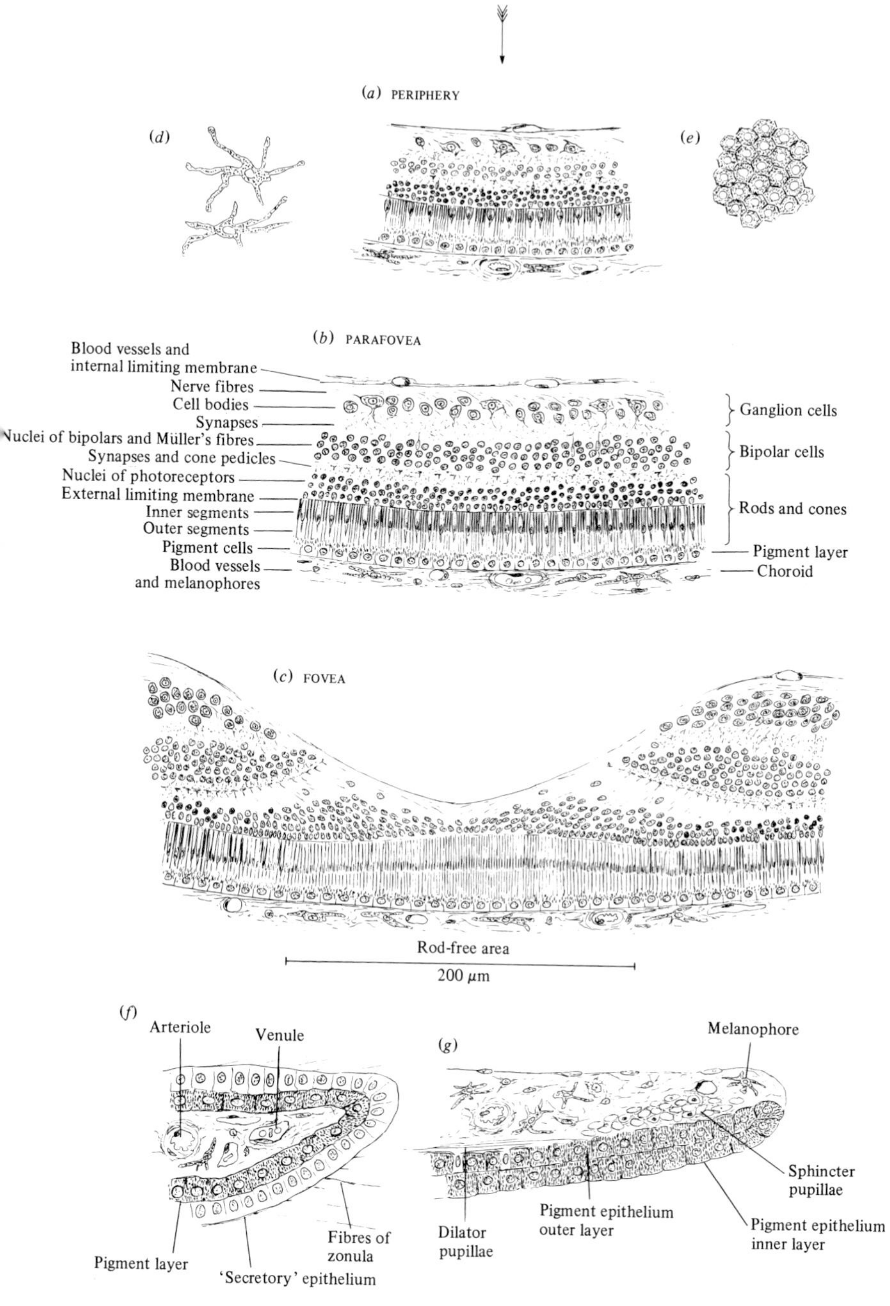
(a) PERIPHERY
(d)
(e)
(b) PARAFOVEA
Blood vessels and internal limiting membrane
Nerve fibres
Cell bodies
Synapses
Nuclei of bipolars and Müller's fibres
Synapses and cone pedicles
Nuclei of photoreceptors
External limiting membrane
Inner segments
Outer segments
Pigment cells
Blood vessels and melanophores
Ganglion cells
Bipolar cells
Rods and cones
Pigment layer
Choroid
(c) FOVEA
Rod-free area
200 μm
(f)
Arteriole
Venule
Pigment layer
'Secretory' epithelium
Fibres of zonula
(g)
Melanophore
Sphincter pupillae
Pigment epithelium outer layer
Pigment epithelium inner layer
Dilator pupillae

Fig. 2.3. Cytological features of the retina.
(*a*, *b*, *c*) Schematic diagrams of the supposed arrangements of cells based on observations on sections of the fixed retina treated by the Golgi technique of silver impregnation. This technique, of which there are several modifications, usually impregnates completely those cells that take the silver at all, though occasionally the axons may remain unimpregnated.
(*a*) Foveal centre; (*b*) parafovea; (*c*) periphery.
(*d*, *e*, *f*) Sections of the same parts of the retina as they would appear if fixed and stained by standard histological methods, (e.g. Susa fixation and iron haematoxylin). For the sake of clarity the cells that correspond to those depicted in the Golgi preparations have been drawn more heavily. Shrinkage during the various processes makes precise measurements on histological preparations of dubious value.
(*g*) The arrangement of the foveal cones as seen in transverse sections at three different levels: (i) through the inner segments which probably contain the macular pigment; (ii) through the ellipsoids, which are concentrations of mitochondria; and (iii) through the outer segments, which contain the photo-sensitive pigments (see Chapters 5 and 9). The long central processes of the foveal cones probably also contain the macular pigment.
(*h*) A ganglion cell, impregnated with silver, as it would appear in a retina spread out flat on a slide and subjected to a Golgi technique.
(*i*) Transverse sections of the peripheral rods and cones at different levels. Such a gradation is frequently seen in sections that are not absolutely transverse to the rods and cones.

Note that rods terminate centrally in end-knobs or 'spherules', while cones end at a slightly deeper level in end-feet or 'pedicles'.

Horizontal cells and amacrine cells do not show typical axons though the former may have one much longer process than the rest. S. L. Polyak, who made extensive studies of retinae impregnated by the Golgi technique suggested that ganglion cells could be classified into several different groups according to the pattern and extent of their dendritic trees. In these diagrams, midget ganglion cells and two other types of ganglion cell can be distinguished by the single or multiple origin of their dendrites, the 'receptive field' of the dendrites and also by the distribution and size of the Nissl's granules (ribosomal particles) in their cell bodies.

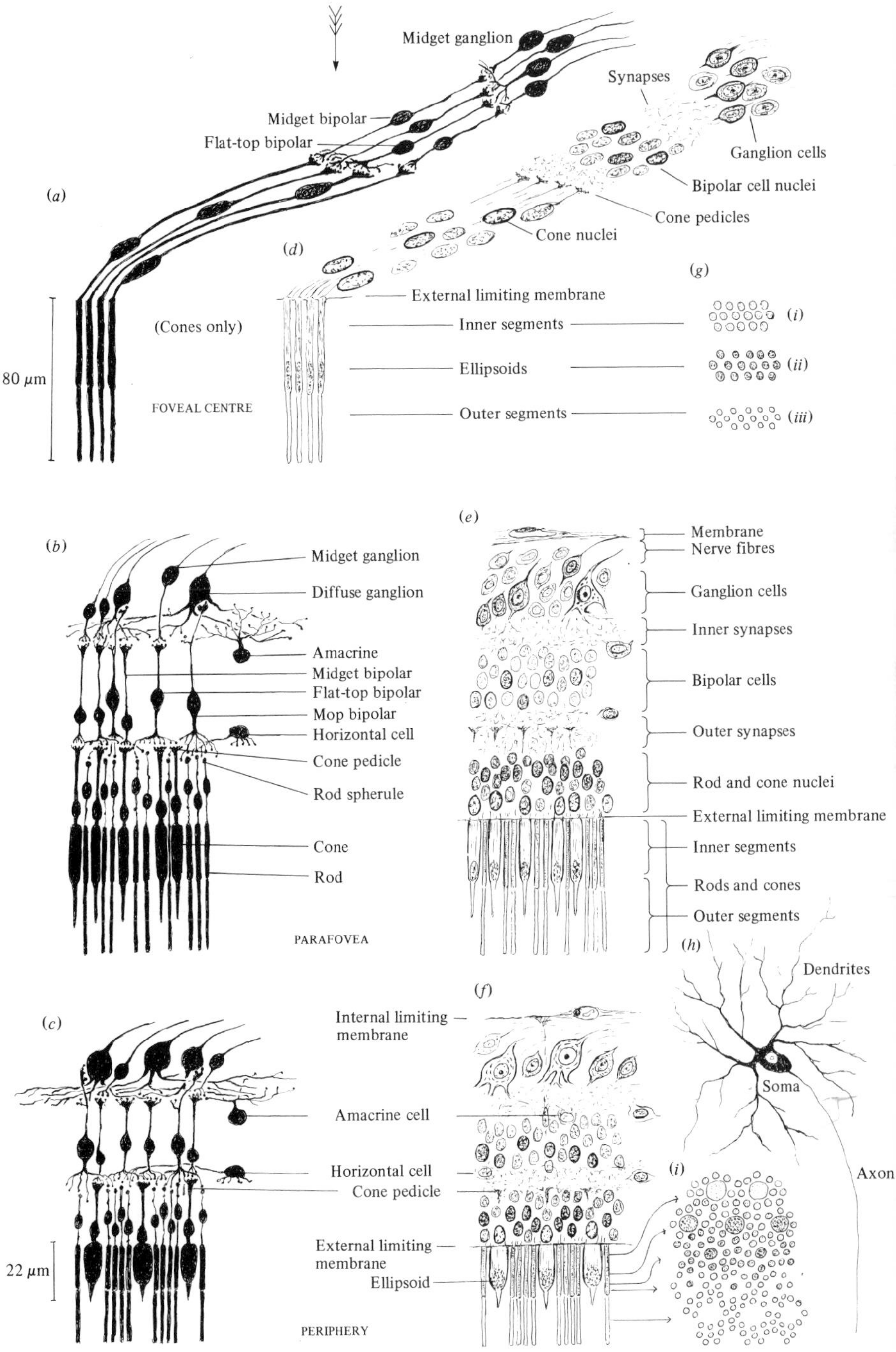
Midget ganglion
Synapses
Midget bipolar
Flat-top bipolar
Ganglion cells
Bipolar cell nuclei
(a)
Cone pedicles
Cone nuclei
(d)
(g)
External limiting membrane
(Cones only)
Inner segments
(i)
80 μm
Ellipsoids
(ii)
FOVEAL CENTRE
Outer segments
(iii)
(e)
(b)
Membrane
Nerve fibres
Midget ganglion
Diffuse ganglion
Ganglion cells
Inner synapses
Amacrine
Midget bipolar
Flat-top bipolar
Bipolar cells
Mop bipolar
Horizontal cell
Outer synapses
Cone pedicle
Rod spherule
Rod and cone nuclei
External limiting membrane
Cone
Inner segments
Rod
Rods and cones
Outer segments
PARAFOVEA
(h)
Dendrites
(f)
(c)
Internal limiting membrane
Soma
Amacrine cell
Horizontal cell
(i)
Axon
Cone pedicle
External limiting membrane
22 μm
Ellipsoid
PERIPHERY

Fig. 2.4. Retinal structure reconstructed from studies with the electron microscope (largely derived from the observations of J. E. Dowling & B. B. Boycott, (1966). *Proceedings of the Royal Society Series B*, **166**, 80–111). Colour coding is used throughout this figure to differentiate cells as follows:

Blue: Rods, rod bipolar cells, diffuse ganglion cells.

Red: Cones, midget bipolar cells, midget ganglion cells.

Green: Flat-top bipolar cells.

Black: Processes of pigment epithelium cells, horizontal cells, amacrine cells.

In the main diagram, which shows a general plan of the retina, the colour coding serves to draw attention to the apparently direct pathways suggested by the anatomy: (i) numerous rods → rod bipolar → diffuse ganglion (all blue in the diagram); (ii) single cone → midget bipolar → midget ganglion (all red); (iii) several cones → flat-top bipolar → diffuse ganglion.

On the left are enlarged diagrams of the main types of synapse so far recognised.

(1) Synapses in the outer (first) synaptic layer.

(*a*) Cone pedicle with process from midget bipolar and horizontal cell and also a process from a flat-top bipolar. (Note that there is no synaptic ribbon nor any aggregation of vesicles in this type of synapse.)

(*b*) Cone pedicle with processes from midget bipolar and horizontal cells.

(*c*) Rod spherule with process from rod bipolar and processes from horizontal cells.

(2) Synapses in the inner (second) synaptic layer.

(*a*) Midget bipolar with process from midget ganglion and process from diffuse ganglion (both with synaptic ribbons) and with processes from amacrine cells. Note the position of vesicles in the last.

(*b*) Flat-top bipolar with diffuse ganglion cell and with amacrine cell processes. Note positions of synaptic ribbons and vesicles.

(*c*) Rod-bipolar knob with processes from diffuse ganglion cell and amacrine cell.

(*d*) Terminal process of rod bipolar and process of amacrine cell on cell body of diffuse ganglion cell.

In the main diagram note also the following points.

(*e*) Processes of the pigment epithelium cell (with pigment granules) interdigitating with the outer processes of rods and cones, for the phagocytosis of the tips of which they are believed to be responsible.

(*f*) The junction between the inner and outer segments of a rod. Note that the outer segment is composed of a stack of flattened vesicles.

(*g*) The junction between the inner and outer segments of a cone. Note that the more basal part of the outer segment is composed of flat processes resembling flattened and laterally extended microvilli. In both rods and cones note the centrioles and the basal bodies from which ciliary fibres project into the outer segment, suggesting that the outer segments are modified flagella, or cilia, comparable with those on ependymal cells which are embryologically related to rods and cones (see Fig. 2.5*c*).

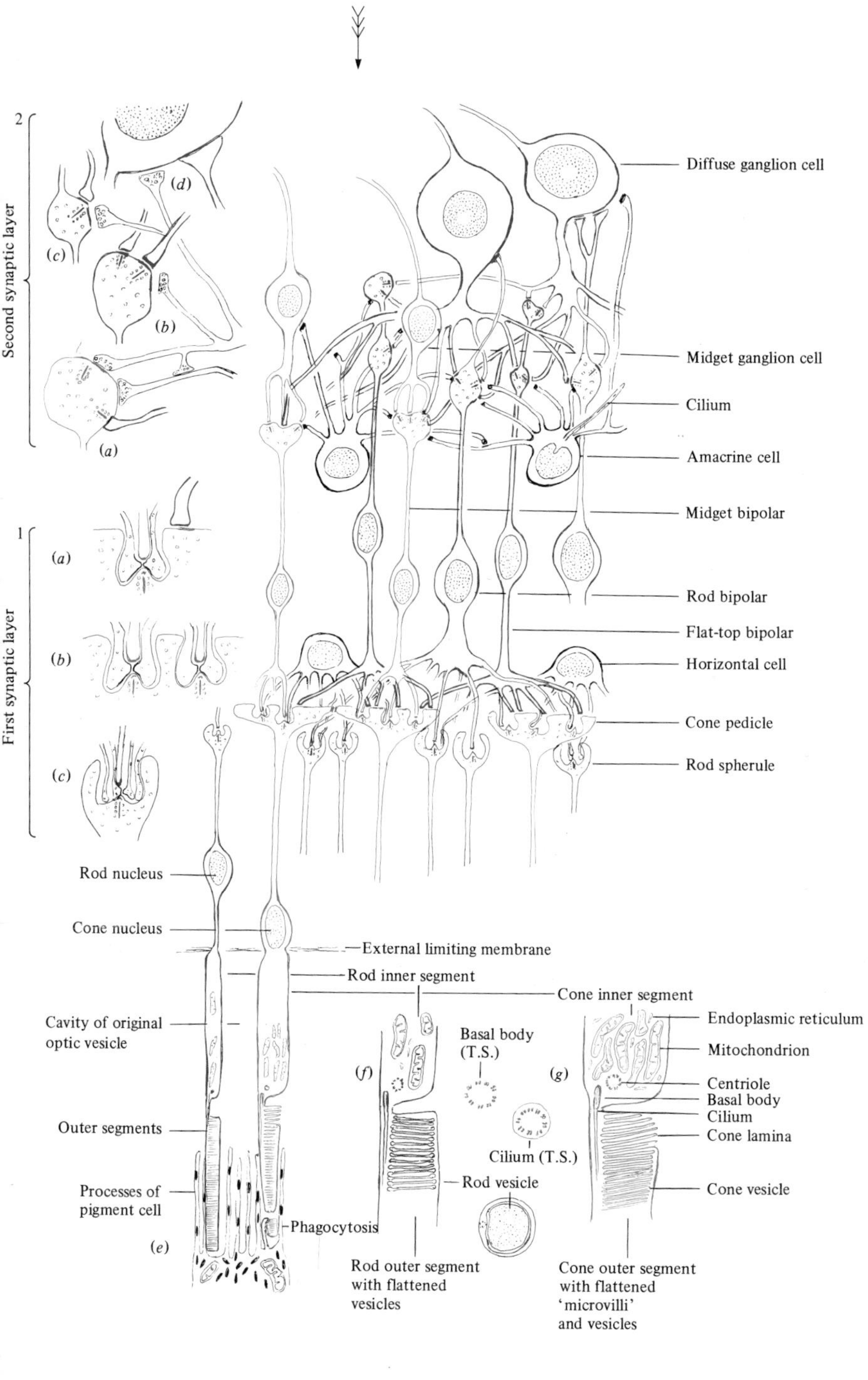
2
Second synaptic layer
(a)
(b)
(c)
(d)
1
First synaptic layer
(a)
(b)
(c)
Diffuse ganglion cell
Midget ganglion cell
Cilium
Amacrine cell
Midget bipolar
Rod bipolar
Flat-top bipolar
Horizontal cell
Cone pedicle
Rod spherule
Rod nucleus
Cone nucleus
External limiting membrane
Rod inner segment
Cone inner segment
Cavity of original optic vesicle
Outer segments
Processes of pigment cell
Phagocytosis
(e)
(f)
Basal body (T.S.)
Cilium (T.S.)
Rod vesicle
Rod outer segment with flattened vesicles
(g)
Endoplasmic reticulum
Mitochondrion
Centriole
Basal body
Cilium
Cone lamina
Cone vesicle
Cone outer segment with flattened 'microvilli' and vesicles

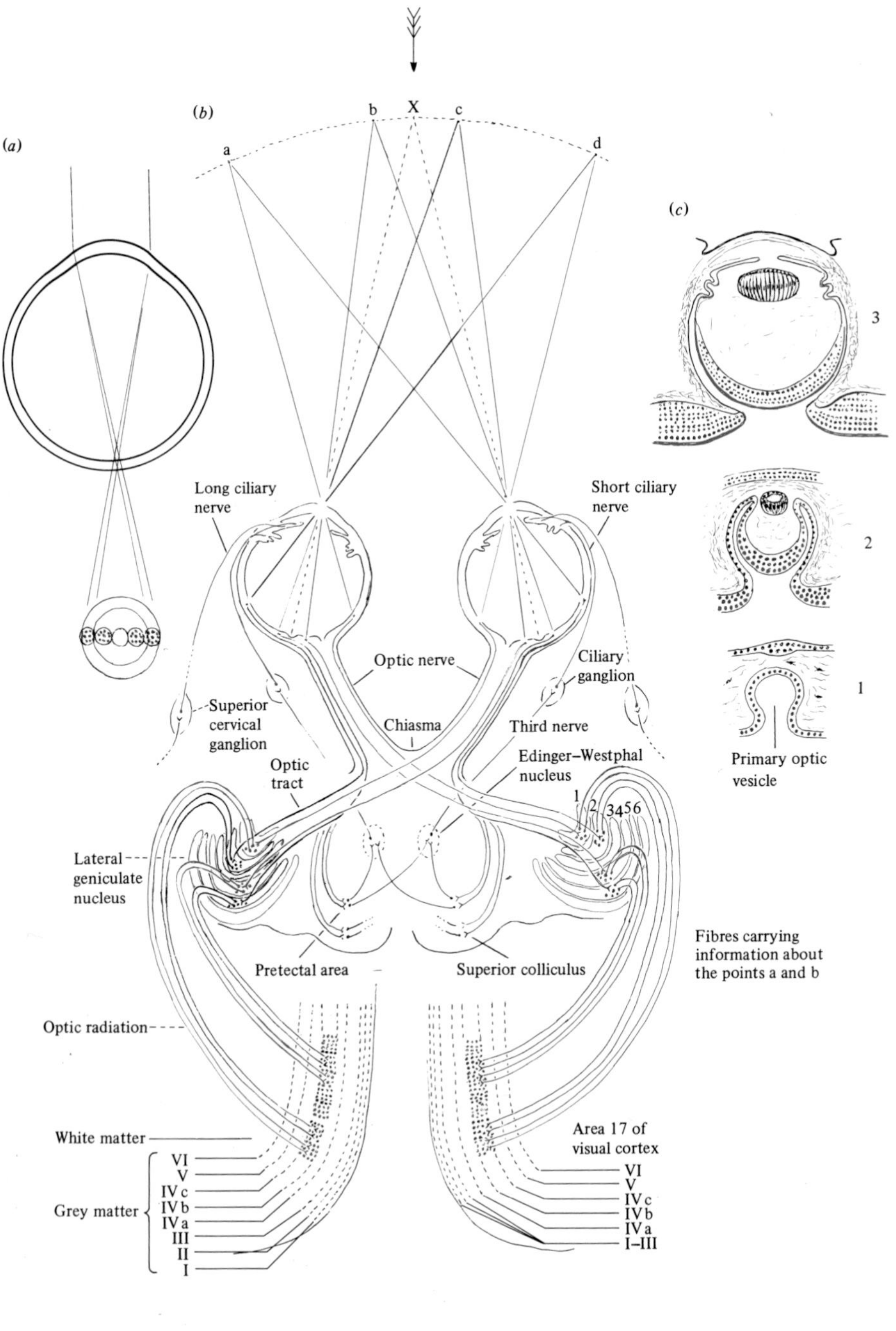

(a)
(b)
a
b
X
c
d
(c)
3
2
1
Primary optic vesicle
Long ciliary nerve
Short ciliary nerve
Optic nerve
Ciliary ganglion
Superior cervical ganglion
Chiasma
Third nerve
Optic tract
Edinger–Westphal nucleus
1
2
3456
Lateral geniculate nucleus
Fibres carrying information about the points a and b
Pretectal area
Superior colliculus
Optic radiation
White matter
Area 17 of visual cortex
Grey matter
VI
V
IV c
IV b
IV a
III
II
I
VI
V
IV c
IV b
IV a
I–III

Fig. 2.5. Neural pathways; chromatic aberration; embryology.
(*a*) Diagram to illustrate the effects of chromatic aberration. When yellow-green rays are focussed on the retina, blue rays come to a focus in front of it and red rays behind it. In consequence, a point of white light, if nominally focussed on a cone, would spread out at the level of the receptor processes as a 'blur circle' with a diameter of about five cone diameters, the outside receiving only blue and violet rays.
(*b*) A schematic diagram of the main neural pathways involved in vision. The eyes are deemed to be directed towards the point X of the object abcd. The images of the points a, b, c and d fall as indicated on different parts of the retinae. If the neural paths (appropriately coloured) are followed from the stimulated points along the optic nerves, optic tracts, lateral geniculate nuclei and optic radiations it will be seen that information from the left visual field ends up in the right visual cortex and from the right field goes to the left cortex. Synapses are made in the lateral geniculate nucleus with cells in specific layers (ipsilateral in layers 2, 3 and 5, contralateral in layers 1, 4 and 6). The fibres from the cells in the geniculate nuclei terminate in layer IV of the cortex.

Other fibres from the retina travel to the superior colliculus and the pretectal area. From the latter, neurons lead to the Edinger–Westphal nucleus, from which the third nerve takes origin to supply some of the extrinsic eye muscles (see Fig. 2.1) and to relay in the ciliary ganglion to activate the ciliary muscles and the dilator pupillae, by way of the short ciliary nerve (parasympathetic). The dilator pupillae is supplied by the long ciliary nerve (sympathetic) from the superior cervical ganglion.
(*c*) Schematic and simplified diagrams to illustrate the main features of three stages in the embryogenesis of the eye.
Black: ectoderm and its derivatives, the lens and conjunctiva. The pigment epithelium, derived from the neural ectoderm.
Blue: neural ectoderm and neural tissue, including that of the retina.
Green: ependyma and the rod and cone layer of the retina.
Red: 'secretory' epithelium over the ciliary processes, derived from neural ectoderm. The mesenchyme which gives rise to the cornea, the sclera, the ciliary muscles, the choroid, the anterior part of the iris and the vitreous body.
1. Outgrowing optic vesicle inducing thickening of the ectoderm.
2. Invagination of the optic vesicle (simplified); separation of the lens 'vesicle'; beginning of the differentiation of the neural ectoderm.
3. Main features of the maturing eye and neural tissue. The conjunctival epithelium overlies the developing cornea. The lens fibres (epithelial cells) obliterate the cavity of the lens vesicle. The iris and ciliary processes are formed and the layers of the retina are beginning to differentiate.

Note that the pigment layer lies on one side of the occluding optic vesicle and the rod and cone layer on the other. It is the processes of the cells on the apposing walls of the vesicle that interdigitate, so that the rods and cones and the processes of the pigment cells are bathed in fluid that was originally cerebrospinal fluid.

Fig. 2.6. The lateral geniculate nucleus and the striate cortex.
(*a*) Model of the lateral geniculate nucleus, showing the relationship between the fibres of the optic tract and those of the optic radiation.
(*b*) The laminated structure of the lateral geniculate nucleus as seen when a section is stained to show the Nissl's granules in the cells. The cells in layers 1 and 2 are noticeably larger than the rest. Fibres from the fovea end predominantly in the central portions of layers 6, 5, 4 and 3 and very rarely if at all in 1 and 2. From the parafovea they end in all layers, and from the periphery the fibres end on the lower margin of the body where the layers tend to be much less distinct.
(*c*) A superficial view of the left visual cortex of the monkey as seen from behind indicating the regions where fibres from different parts of the visual field terminate.
(*d*) A similar view of the right visual cortex with a segment removed. If the wall on the left of the space so formed were then removed and sectioned it would have the structure shown in (*e*).
(*e*) Vertical sagittal section of visual cortex of monkey (see *d*) showing cell layers IV and VI of the grey matter both in the outer portion and in the infolded portion. In area 17 of Brodmann layer IV is divided into IVa, IVb and IVc. In area 18 it is united into one. The optic radiation ends in layer IVc of area 17.
(*f*) The cell layers in area 17 of the visual cortex are indicated on the left of the diagram. On the right of the diagram layers I to IVb are deemed to have been removed. The shaded areas in layer IVc show how each eye is represented. Fibres from the left eye (say) end in the cells in the black zones, those from the right eye in the colourless zones (see Fig. 2.5(*b*)). The cells in this part of the cortex are arranged in vertical columns and the columns have very specific functions. For example, in one set of observations, electrodes were inserted progressively, and records taken from the cells at different depths. If the positions of the electrodes were as indicated by r, s and t the cells were found only to respond when the eye was stimulated by an illuminated bar having the orientation in the visual field indicated by the direction of the short lines opposite the positions of hypothetical cells. The sign o indicates that the cells in that layer IVc showed no such specificity. If an electrode u was inserted almost parallel to the surface and passing through layers II to IVa in a direction parallel to r, s, t the cells were found to respond only to stimuli having orientations in the visual field as indicated in the diagram below. In other words there seem to be columns of cells in the visual cortex, each one of which specialises in some oriented feature of the visual image.
(*g*) If the parts of the visual field corresponding to the cells encompassed by the grey stripes in (*f*) were also shaded black in the visual field, it would appear with a pattern essentially similar to that shown here. The fovea monopolises a disproportionately large part of the visual cortex and the periphery has only a very small representation among the cells of the cortex. Expressed differently, only a small area of cortex is involved in registering quite large peripheral fields, but small central fields occupy large areas of cortex.

In making the diagrams for this figure heavy reliance has been placed on the work of W. E. Le Gros Clark and also on the observations of D. H. Hubel & T. N. Wiesel and their Ferrier Lecture (*Proceedings of the Royal Society, Series B*, **198**, 1–59). The debt is willingly and gratefully acknowledged.

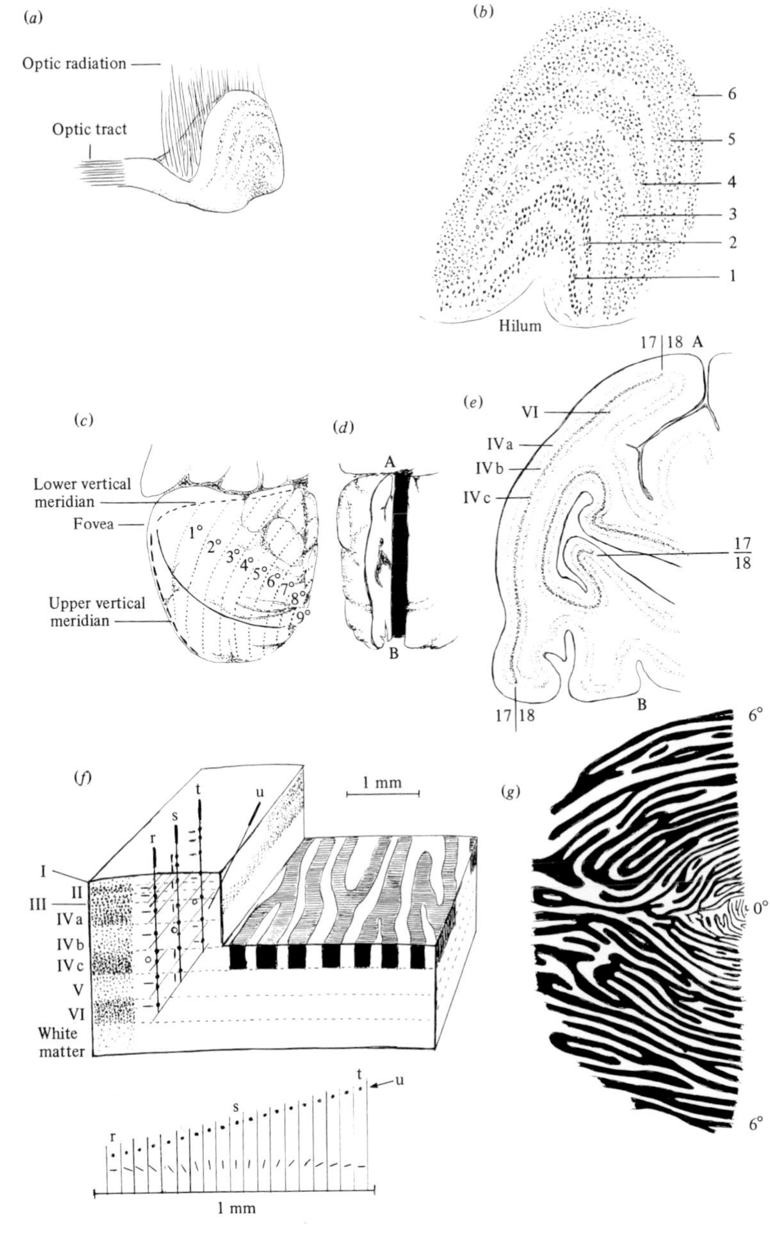
(a)
Optic radiation
Optic tract
(b)
6
5
4
3
2
1
Hilum
(c)
Lower vertical meridian
Fovea
1°
2°
3°
4°
5°
6°
7°
8°
9°
Upper vertical meridian
(d)
A
B
(e)
17|18 A
VI
IVa
IVb
IVc
17/18
B
17|18
(f)
1 mm
r
s
t
u
I
II
III
IVa
IVb
IVc
V
VI
White matter
1 mm
(g)
6°
0°
6°

CHAPTER THREE

Image formation in the eye

M. MILLODOT

3.1. OPTICAL CONSTANTS OF THE EYE

Just like those of other bodily structures, the dimensions of the human eye vary greatly among individuals. Thus average values have to be adopted, which are ironically called 'constants' of the eye. However, most of the dimensions of the various parts of the eye are normally distributed, at least in healthy eyes.

The determination of these dimensions is difficult as they must be measured *in vivo* as far as possible. Formerly, enucleated eyes were used, but because they were usually diseased and because the circulation of the blood had ceased, these eyes may have been distorted. More recently, however, most constants of the eye have been obtained on living eyes, with the exception of the refractive indices of the media.

Index of refraction

Refraction is the change in the direction of propagation when light passes from one transparent medium to another. The refractive index (n) of a medium is defined:

$$n = \frac{\text{velocity of light in vacuum}}{\text{velocity of light in the medium}} = \frac{\sin i}{\sin r}$$

where i is the angle of an incident ray (in the vacuum) and r is the angle of the refracted ray (in the medium), both angles being measured from a line perpendicular to the surface.

When light passes from one transparent medium to another, $(\sin i/\sin r) = (n'/n)$, where n and n' are the refractive indices of the first and second medium respectively. Since the refractive index of air is close to unity and that of the cornea is 1.38, substantial refraction ocurs at the outer surface of the cornea. However, when a person is swimming under water, without a mask, the cornea is bathed in water of approximately the same refractive index ($n = 1.333$) and its effective power is almost eliminated. This produces a retinal image that is vastly out of focus and vision is hazy (the eye is in fact rendered artificially hyperopic). The indices of the different media of the eye are given in Table 3.1. Since the lens is heterogeneous, its index is nominal and is based on its optical effect.

Curvature of the optical surfaces

An optical surface is defined as a boundary between two media of different refractive indices. The eye consists of four optical surfaces: the anterior and posterior surfaces of the cornea and the anterior and posterior surfaces of the crystalline lens.

The method of measuring these curvatures *in vivo* depends upon the reflection of light from objects by these surfaces. Such reflected images are called catoptric or *Purkinje–Sanson images*. In contrast, the images that are formed on the retina and give rise to vision are called dioptric images. All optical surfaces refract and reflect light but the amount reflected is only of the order of 3–4% depending on the indices of the media. You can observe this reflection from the cornea of the eye. Looking at someone else's eye you can see reflected images, of fluorescent tubes, for instance, and perhaps even of your own profile.

The sizes of these catoptric images can be measured and then the radius of curvature of the surface (r) can be assessed using a relationship between the size (y) of an object placed at a given distance (x) from the surface and the size of the image (y') formed by reflection: $r = -2\,xy'/y$. (For the derivation of this relation see Bennett & Francis (1962) in *The Eye 4*, p. 115, ed. H. Davson.) Instruments based on this principle such as ophthalmometers or keratometers, which measure the radius of curvature of the anterior surface of the cornea, have been used reliably for a century. These instruments are also used routinely to assess the astigmatism (see below) of the cornea.

The problem becomes more difficult when measuring the curvature of the other optical surfaces of the eye. Considerable ingenuity has gone into doing this using catoptric images reflected by the posterior corneal surface and the two surfaces of the lens. Modern methods of investigation have made use of photography of these images and have yielded very accurate results. Average findings are given in Table 3.1.

The four images reflected by the optical surfaces of the eye are referred to as Purkinje–Sanson images I, II, III, and IV, starting from the front surface of the cornea (I). Images I, II and III are erect

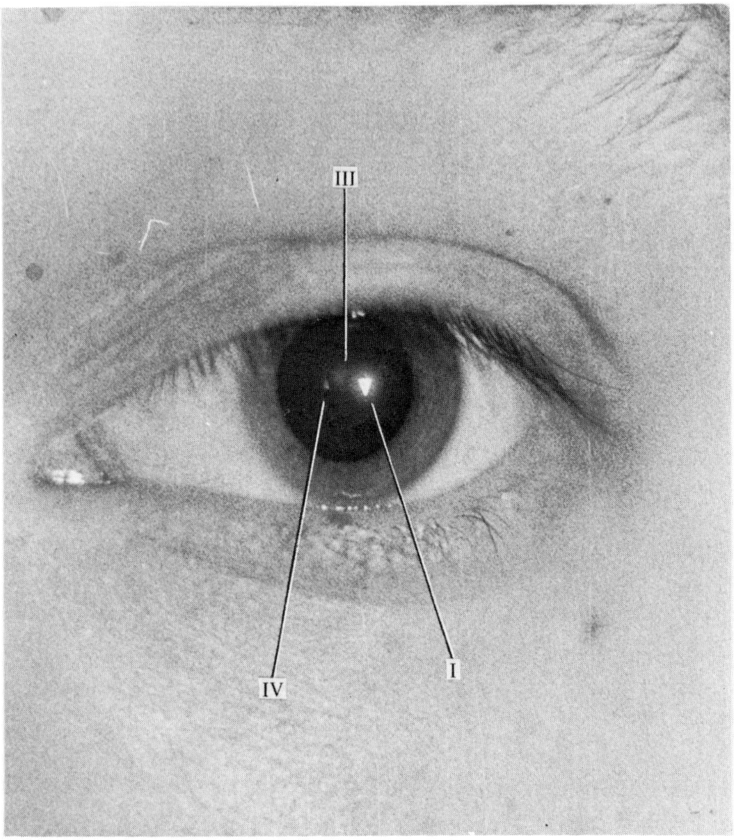

Fig. 3.1. Photographs of three of the Purkinje–Sanson images. A triangular source (point down) is seen reflected by the anterior corneal surface (I), the anterior lens surface (III), and the posterior lens surface (IV). The reflection from the posterior corneal surface (II) is very dim and is masked by I. The size of the images is related to the curvature of the surfaces, so III and IV can be used to follow the changes during accommodation. Note that IV is inverted because a real, inverted, image is formed by the concave posterior lens surface. The other images are erect.

because the surfaces upon which they are formed are convex towards the light path. Image IV is inverted (see Fig. 3.1) because the posterior surface of the lens is concave. Image I is bright but the others are very dim, especially II since the cornea and the aqueous humour have very similar refractive indices. A good deal of our knowledge of the mechanism of accommodation has been gained by measuring the III and IV Purkinje–Sanson images, since their changes show how the lens changes in shape (see p. 64).

Besides the catoptric and dioptric images there exist *entoptic images*, which are visual sensations arising from images of objects located within the eye itself. The best known of these are the 'muscae volitantes', which appear as floating specks in the field of view. They are due to shadows cast upon the retina by small particles in the vitreous body. An entoptic phenomenon that you can easily observe under special circumstances is that induced by the retinal blood vessels, which lie outside the retinal layers. Close your eye and use a small torch to throw light through the sclera by placing the torch against the eye near the outer angle of the orbit. The shadows of the vessels will now fall on the retina and, perhaps because it is an unaccustomed stimulation, they will be visible for a short period of time. If you move the torch the shadows become visible again. You will see an intricate branching pattern known as the *Purkinje tree* which corresponds to the branches of the retinal arteries and veins radiating from the optic disc. Some people have difficulty at first in seeing this effect, but once seen it is unforgettable.

Position of the optical surfaces

The separation between the various optical surfaces is also important to the dioptrics of the eye. Formerly this was determined by focussing a microscope on one surface, then the next, and so on, and recording the distance each time.

More recently, ultrasonographic techniques have been introduced to assess the relative positions of the surfaces: very high frequency sonic energy (5 to 10 megacycles) is reflected at the various optical surfaces. Knowing the speed of sound in a given medium, and the time taken, it is possible to deduce the distance. The results thus obtained are in good accord with the other methods. The locations of the various surfaces are given in Table 3.1.

Length of the eye

The length of the eye from the anterior corneal surface to the retina can also be assessed by ultrasonography. This is possible because the interface between the vitreous humour and the retina and that between the retina and the pigment epithelium reflect ultrasound as well as light. (The slight optical reflection at the former interface allows the concave surface of the fovea to be seen with the ophthalmoscope.) Another way of determining the position of the retina

Table 3.1. *Optical constants of an average adult human eye*

Structure or surface	Refractive index	Radius of curvature (mm)	Distance from anterior surface of cornea (mm)
Cornea	1.376	—	—
Aqueous humour	1.336	—	—
Lens (total)	1.41	—	—
Vitreous humour	1.336	—	—
Anterior corneal surface	—	7.8	0
Posterior corneal surface	—	6.8	0.5
Anterior lens surface	—	10.0[a]	3.6[a]
Posterior lens surface	—	−6.0	7.2
Retina	1.363	—	24.0

[a] When the eye accommodates at its maximum the anterior surface of the lens alters its curvature to about 6 mm and that surface moves forward to be approximately 3.2 mm away from the anterior surface of the cornea.

relative to the cornea is by X-rays. The retina is slightly sensitive to X-ray stimulation. A sheet of X-rays aimed at right angles to the optical axis of the eye will intersect the retina in a circle. The subject reports seeing a light circle. As the X-ray device is moved posteriorly it will eventually stimulate one small point at the posterior pole of the eye. Then the subject will report seeing a small luminous point. In this case the distance between the X-ray device and a microscope aimed at the front surface of the cornea represents the length of the eye. Ultra-sonographic and X-ray methods yield results that are in good agreement but they involve some risks of injury to the eye. The average length of the eye is shown in Table 3.1.

Dioptrics of the eye

The power of a spherical refracting surface such as those in the eye is equal to $P = (n' - n)/r$, where n and n' are the refractive indices of the object (first medium) and image (second medium) spaces respectively and r the radius of curvature in metres. The converging or diverging power (P) of a lens is expressed in a unit called the dioptre (abbrev: D), which is equal to the reciprocal of the focal length in metres. Thus the strength of a lens with a focal length (f) of 1 m is 1 D; of one with focal length of $\frac{1}{2}$ m, 2 D, and so on.

Table 3.2. *Refractive power (in D) of the optical surfaces of an average adult human eye*

Anterior surface of cornea	48.2
Posterior surface of cornea	−5.9
Anterior surface of lens	8.4
Posterior surface of lens	14.0

For example the refractive power of the anterior surface of the cornea, which is against air ($n = 1.0$), is equal to $P = (1.376 - 1.0)/(7.8 \times 10^{-3}) = 48.2\ D$. The power of the other optical surfaces of the eye can be calculated in the same fashion and the results are given in Table 3.2.

It is obvious from Table 3.2 that the major refracting surface of the eye is the anterior surface of the cornea, since it separates two media (air and cornea) differing widely in refractive indices. Thus any irregularities on this surface will have major consequences for the image formation of the eye. Oedema, which is a sequel of any lesion of the cornea and of some other ocular and systemic conditions, will alter the configuration of the cornea as well as reducing its transparency. All of these effects cause blurred vision.

The total refracting power of the eye is not equal to the simple addition of the four refractive surfaces. This is because the separation of the surfaces tends to diminish the cumulative refractive effect. Indeed the power of two refractive surfaces is $P = P_1 + P_2 - (d/n)\ P_1 P_2$ where P_1, P_2 are the powers of the two refractive surfaces, d their separation and n the refractive index of the medium separating them. If all four surfaces of the eye were against one another ($d = 0$), the total power would be 64.7 D (adding the values in Table 3.2). In reality the total power of the human eye is approximately 60 D.

There are considerable individual differences in human eyes, yet it is useful to have an optical model of a typical eye to help in understanding the dioptrics of the eye and for convenience in calculations. These optical models are called *schematic eyes* and the most widely used model was devised by the Swedish ophthalmologist, Gullstrand (1862–1930).

A simple yet often sufficiently accurate model is the *reduced eye*. It consists of a simple spherical surface assigned the average power of the eye ($P = 60\ D$) and separating air outside from a medium of refractive index $n = 1.333$ (the same as for water) inside the eye (see Fig. 3.2). The radius of curvature of this simple surface is

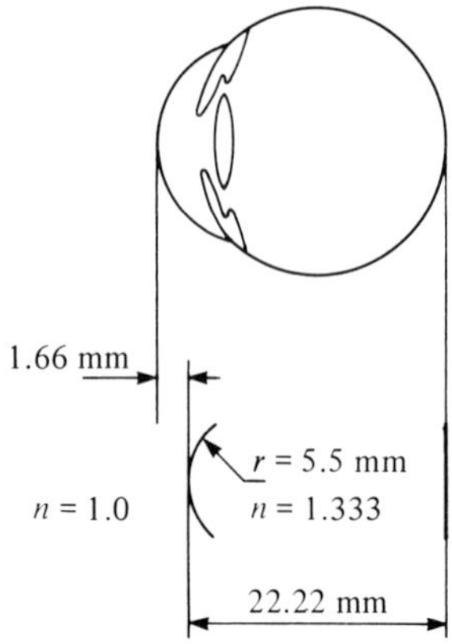

Fig. 3.2. The reduced eye (lower) consists of a single spherical refracting surface of 5.5 mm radius of curvature separating air ($n = 1.0$) and a medium with the refractive index of water ($n = 1.333$). It produces an image the same size as the average real eye with its multiple refracting surfaces, but is slightly shorter.

$r = (1.333 - 1.0)/60 = 0.0055$ m or 5.5 mm. The focal lengths (the distance between the surface and the focal points) of this eye are

anterior focal length, $f = -n/P = -1.0/60 = -0.0167$ m or -16.7 mm

posterior focal length, $f' = n'/P = 1.333/60 = 0.02222$ m or 22.22 mm.

This hypothetical simple surface is not coincident with the cornea of the eye. It occupies a place at which a simple surface has the same effect as that of the whole optical system of the eye. Therefore it is situated 1.66 mm behind the real cornea (see Fig. 3.2). It must be noted though that the refractive surface of the reduced eye has a greater curvature (and thus greater power) than the real cornea, so that it compensates for the absence of the lens.

The posterior focal point (that is where the image of an object placed at infinity is formed by the optical system) lies within the plane of the retina in the normal or *emmetropic* eye. As we shall see later (p. 70) if the retina lies either in front or behind the posterior focal point, the eye is said to be *ametropic*.

Axes of the eye

Although specialists have defined several axes of the eye (five, in fact) it is at first sufficient to consider two of them. Firstly the *optical axis*. This is the line that joins the centres of curvature of the four optical surfaces of the eye. It is a somewhat theoretical line since the eye is

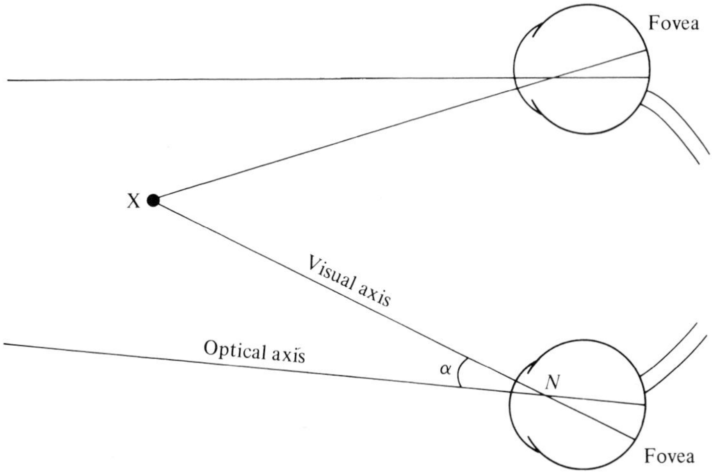

Fig. 3.3. When the *visual* axes of the eyes are converged on a fixation point X the *optical axes* (see text) are less convergent. The angle between the visual and optical axes (α) is about 5° in adults.

a biological structure and its optical surfaces are not necessarily truly spherical and their centres of curvature are not necessarily colinear. Nonetheless, the optical axis can be approximately defined as a line close to the several centres of curvature. The optical axis cuts the retina at the posterior pole of the eyeball, that is, a point somewhere between the optic disc or blind spot (where the optic nerve enters the eye and there are no photoreceptors) and the fovea (see Fig. 3.3).

The other axis is the *visual axis*. It is the line joining the fixation point (X in Fig. 3.3) to the fovea of the eye, where vision is sharpest (see Chapter 8). This line passes through the equivalent centre of curvature of all the optical surfaces of the eye called the nodal point (N) which lies on the optical axis. A light beam passing through the nodal point is undeflected.

In the reduced eye the optical and visual axes are assumed to coincide and this is convenient for the purpose of simple object–image calculations. But the real eye is such that the fovea does not lie upon the approximate optical axis of the eye. The optical axis forms an angle with the visual axis, *angle alpha* (α) which is equal, on average, to 5° in the adult. In the child this angle is greater because the retinal distance between the fovea and the posterior pole of the eye is almost the same as in an adult but the length of the eye is much smaller (about 18 mm).

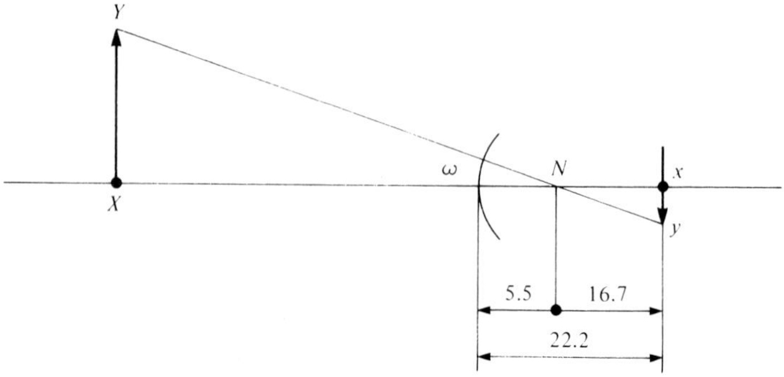

Fig. 3.4. Image formation in the reduced eye. In this simple case the ray through the nodal point N is undeviated, and the ratio of image size to object size is the ratio of *posterior nodal distance* xN to object distance XN. The construction is more complex for optical systems with multiple refracting surfaces, but the posterior nodal distance is still the main factor determining image size; it has an average value close to 16.7 mm in the human eye.

There is a clinical consequence of this fact. When a child is examined and asked to look at you, his eyes may appear to diverge, as the pupil will probably be slightly turned outwards, since it is more or less symmetrical about the optical axis (see Fig. 3.3). The child appears to be affected by a divergent squint and the uninitiated practitioner may regard this condition as abnormal.

Retinal image size

The reduced eye offers a simple model with which to show how the retinal image is constructed. Suppose we are fixating an object XY (see Fig. 3.4) at a given distance. Draw straight lines from the extremities of that object through the centre of curvature or nodal point (N) (which is defined as the point through which light rays pass undeflected) of the simple surface of the reduced eye to the retina, where an image xy is formed. It is clear from Fig. 3.4 that the image thus obtained is inverted and smaller.

The two lines crossing at N form an angle ω called the *visual angle*, which is defined as the angle subtended by the object at the nodal point. It is obviously equal to the angle xNy which is subtended by the retinal image. The size of the retinal image can be deduced from $xy/XY = xN/XN$. Since the distance between the nodal point and the retina is 16.7 mm $xy = 16.7\ XY/XN$.

For small angles, the following approximation is often adequate:

retinal image size (mm) = $17 \times \tan$ (visual angle)

A few examples may be useful.

1. *What is the visual angle subtended by an object 5.24 mm high at a distance of 6 m and what is the size of the corresponding retinal image?*

$$\tan \text{visual angle} = 5.24/(6 \times 10^3) = 873 \times 10^{-6}$$

$$\text{visual angle} = 3'$$

$$\text{retinal image} = 17 \times \tan 3' = 0.015 \text{ mm or } 15\ \mu\text{m}$$

2. *Assuming that a foveal cone has a diameter of 2.5 μm, what is the angle subtended at the nodal point?*

$$\tan \text{angle} = 0.0025/17 = 147 \times 10^{-6}$$

$$\text{angle} = 30 \text{ s of arc}$$

It is interesting to note that from this result it follows that an angle of 1 min corresponds approximately to 5 μm on the retina; that is a distance covering about two photoreceptors.

The pupil

The image-forming mechanism of the eye consists not only of the optical surfaces but also of a diaphragm, the pupil, which dilates in darkness and contracts in bright light. Typical limits for pupil diameter are 8 and 2 mm, and the sixteen-fold change in retinal illumination thereby allowed cannot alone account for the eye's ability to cope with a range of 10^{10} in average illumination (for a full account of this ability we must consider neural processes of adaptation, the bleaching of photopigment, and retinal duplexity – the presence of partially independent rod and cone systems; see Chapter 7). However, it has been shown that the rapid dilation of the pupil when the observer passes from bright illumination to dimmer does yield a substantially improved sensitivity during the early minutes of dark adaptation when the slower processes of neural adaptation are still incomplete. A second advantage of a mobile pupil is that it can act to optimise spatial resolution: when the visual scene is dimly lit, visual acuity is limited by the number of photons available and it is better to have a large pupil, whereas at higher levels of illumination the optical aberrations of the eye set a limit to resolution and then, as will become clear in later sections, it is better to have a smaller pupil.

When only one eye is exposed to light, both pupils contract; this is the consensual reflex, an important diagnostic clue in neuro-ophthalmology. The pupil also contracts when the subject is looking

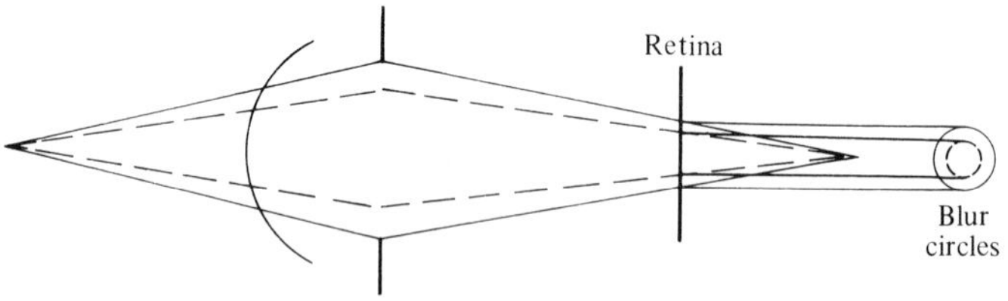

Fig. 3.5. The effect of pupil size on the blur circle formed by an out-of-focus image on the retina. For a small pupil size the blur circle is reduced and objects over a larger range of distances can be seen sharply; depth of focus is increased.

at new objects and is smaller in old age. Being under sympathetic control, it is additionally influenced by emotions such as fear (dilation) or aversion (contraction). Pupil size can be altered by the instillation of drugs: mydriatics such as homatropine dilate the pupil and miotics such as pilocarpine contract it.

Out-of-focus images

If, for some reason (e.g. error of accommodation, ametropia) the image of a point source viewed by the eye is not focussed on the retina it will form on the latter a patch of light called a *blur circle* (or diffusion circle) (see Fig. 3.5). A blur circle may result whether the image is formed in front of or behind the retina, but in the latter case the image can be moved back on to the retina, if the eye accommodates.

The size of the blur circle depends on two factors. It is clear from Fig. 3.5 that it is proportional to the diameter of the pupil. And it can be easily realised by simple drawings that it is also proportional to the distance by which the stimulus is out of focus.

If you normally wear glasses you can strikingly demonstrate for yourself the role played by the diameter of the pupil. Remove your glasses and place an object at a distance such that it appears blurred. Now view the object through a small pinhole aperture pierced in a card. The blur circle is diminished and vision is thereby much improved. In the clinic it is often necessary to assess whether the poor visual acuity of a patient is to be attributed to a need for an optical correction or to *amblyopia* (a reduced visual acuity that cannot be corrected by optical means). If when the pinhole (of about 1 mm diameter) is placed in front of the eye the patient's vision improves, it indicates that his eye is not amblyopic and an optical correction is warranted.

Depth of field of the eye

The depth of field of the eye is the distance through which a point object can be moved without appearing blurred, whilst the eye is fixating a point, and without changing accommodation. This depth of field extends in front and behind the fixation point and depends on the fact that the retina and brain cannot discern any blur until it exceeds a certain threshold value (about 0.15 *D*). Strictly speaking the *depth of focus* is the dioptric distance corresponding to the depth of field.

Pupil size is also critical and the smaller it is, the larger the depth of field. With an average pupil size of 4 mm the depth of field at infinity extends to about 3.5 m in front of the eye and the depth of field at 1 m varies from about 1.4 to about 0.8 m. With a pupil diameter of 2 mm, the depth of field at infinity extends to about 2.3 m and at 1 m it varies from about 1.8 to about 0.7 m. Thus one can understand why reading is easier with increased illumination as the pupil contracts affording a larger depth of field: the need for precise accommodation is reduced.

3.2. ABERRATIONS

The image formed by the optical system of the eye is not perfect because the eye, like most optical systems, suffers from diffraction and various aberrations. But, unlike other optical images, the image formed by the eye on the retina is not destined to be viewed by an observer. The image is information supplied to the retina for coding and transmission to the visual centres of the brain.

This concept was not understood until the nineteenth century and scientists were baffled by the fact that the retinal image is inverted. In this century scientists have been baffled by the aberrations of the eye. These are of small magnitude (for small or average sized pupils). In fact we are beginning to discover that the aberrations that Nature has not corrected may play a useful role in the visual process.

Diffraction

The image of a point object formed by a perfect optical system cannot be a point, owing to the wave nature of light. The image of a circular point of light is a pattern consisting of a central bright disc surrounded by alternate dark and bright rings of diminishing intensity. Therefore the edges of an image are never sharp. The central bright zone is called the Airy disc and its radius (r) is equal to $1.22\lambda x'/d$ where λ is the wavelength of light, x' the posterior focal length and d the diameter of the pupil. Thus even in an optically perfect eye (with 2.5 mm pupil) the Airy disc strays over several cones, although the bulk of the illumination ($> 80\%$) of the central disc falls on a single cone. For a pupil size twice as large, the Airy disc is half the size, and so on. Large pupils give rise to less diffraction, a factor beneficial to the resolving power of the eye. However, as we shall see, the other aberrations act in the reverse fashion, that is the larger the pupil the more degraded is the retinal image.

Chromatic aberration

The velocity of light in a medium varies with its wavelength; and consequently the refractive index, the power and the focal length vary accordingly. Thus the image of a source of white light consisting of a wide range of wavelengths is extended along the optical axis, as each wavelength has a corresponding focal point. As only the image of one wavelength at a time is in focus on the retina, the images belonging to the other wavelengths form blurred patches on the retina.

Fig. 2.5*a* illustrates the axial chromatic aberration of the eye. Short wavelengths (blue) are more refracted than long wavelengths (red). Actually when the eye is at rest, that is, looking at infinity, blue light is focussed well in front of the retina whilst red light is focussed just behind the retina. The dioptric extent of this elongated retinal image is about 1.5 *D*, that is, a clinically significant error. Yet this aberration goes unnoticed and does not hinder visual perception, probably because the extremes of the spectrum where the refractive error is large have poor luminous efficiency (see Fig. 5.8).

To demonstrate to oneself the existence of chromatic aberration one needs to look at a white source of light through a filter that transmits only the blue and red ends of the spectrum, such as a Kodak Wratten filter No. 35. One then sees a red point surrounded by a blue halo.

Spherical aberration

Optical surfaces of constant radius refract light rays to the same extent only if the rays are incident near the optical axis (paraxial optics). Light rays penetrating into the eye near the edge of the pupil are refracted more than the paraxial rays. Consequently the retinal image formed by such a system is a blurred patch and is said to be affected by spherical aberration.

Actual measurements have shown that the human eye is very little affected by spherical aberration. Even for large pupil diameters spherical aberration of the retinal image rarely exceeds 1 *D*. The main reason for this is the fact that the cornea is not spherical but flatter in the periphery than in the centre. Therefore peripheral rays are less refracted than they would be if the cornea were spherical.

3.3. THE OVERALL QUALITY OF THE RETINAL IMAGE

The optical aberrations of the eye and diffraction at the pupil result in an image that is not perfect. Objective measurements of its quality in the living eye can be made by the following technique. The subject looks at a special lamp that has a single straight wire as its filament. The eye forms an image of this filament on the retina, and care is taken to ensure that the image can be accurately focussed there. The retina absorbs most of the light in the image, but a small fraction is scattered and passes backwards through the lens and out of the pupil. If the optics were perfect, this light would be focussed back on to the original filament but the imperfect optics cause it to form a blurred image centred on the filament. It is actually this light scattered from the retina and refocussed by the lens that is responsible for the bright green reflection from a cat's eye when it is caught in a headlight beam; the optics of the cat's eye are poor enough for the image reformed around the headlight to spread as far as the observer's pupil, and the reflecting tapetum behind the cat's retina causes the reflected image to be much brighter than in man.

To determine image quality one inserts a semi-silvered mirror at an angle between the lamp and the eye, thus reflecting the returning light to one side where it can be scanned by a very sensitive photocell. The intensity of the light at each point in the scan is the *line-spread function* resulting from a double passage through the eye's optics. At this point use has to be made of the Fourier transform (see Chapter 1.2); one considers the line filament lamp as a source of spatial sinusoids which are demodulated by being passed *twice* through the optics of the eye; this results in the measured line spread function. Now these measurements can be expressed as a set of spatial sinusoids by performing a Fourier transform on them; knowing the composition of both input and output in terms of spatial sinusoids one can calculate the attenuation at each frequency resulting from the double passage through the optics, and the attenuation at each frequency from a single passage is simply the square root of that resulting from the double passage. This is the way the MTF (Chapter

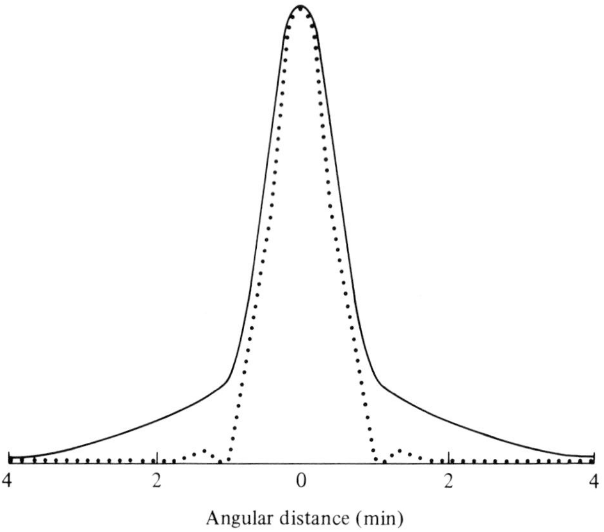

Fig. 3.6. The continuous line shows the *line spread function* of the human eye in perfect focus with a 2 mm pupil (from Campbell & Gubisch, 1966). The subject looked through an aperture at a thin, straight, electric filament and the light reflected back from the retinal image through the pupil was analysed as described in the text. Distances from the centre of the image are given in min of arc; 4.9 μm on the retina are equivalent to 1 min. Notice that although the image spreads more than 3 min (15 μm) on either side of the centre, and is more than 1 min (5 μm) wide at half height, it is not very much broader than the best possible image that can be formed with a 2 mm pupil and light of wavelength 560 nm. The dotted line shows the line spread function for such a diffraction-limited image.

1, § 1.2) was calculated. To obtain the line spread function for a single passage one simply retransforms this MTF, and the results of doing so are shown in Fig. 3.6.

The line spread function is narrowest, corresponding to optimum resolution, for pupil diameters between 1.5 and 2.5 mm. For smaller diameters the image is not only dimmer but also worse because diffraction broadens the line spread function. For larger diameters it is also broader as a result of optical aberrations, though there is an effect (the Stiles–Crawford effect) that alleviates this degradation. Measurements of effectiveness of light entering the pupil at different points show that it is less effective at the edge than at the centre, so the badly focussed light entering at the edges will be less effective.

This effect works for cones, which subserve high acuity, but for rods light entering the pupil at any point is equally effective. Thus rods reap the full benefit of the brighter image that results from pupil dilation, whereas cones 'see' a less blurred image than might be expected.

3.4. SUGGESTIONS FOR FURTHER READING

General references

Duke-Elder, S. & Abrams, D. (1970) *System of Ophthalmology*, Vol. 5. *Ophthalmic Optics and refraction.* London: Henry Kimpton.
Emsley, W. H. (1955) *Visual Optics.* London: Hatton Press.
Michaels, D. D. (1975) *Visual Optics and Refraction.* St Louis, USA: C. V. Mosby Co.

Special references

Ultrasonic measurements. Mundt, G. H. & Hughes, W. F. (1956) *American Journal of Ophthalmology*, **41**, 488.
X-ray measurements. Rushton, J. (1938) *Transactions of the Ophthalmological Society of the UK* **58**, 136.
Role of pupillary reflex. Woodhouse, J. M. & Campbell, F. W. (1975) *Vision Research* **15**, 649.
Spherical aberration. Ivanoff, A. (1956) *Journal of the Optical Society of America* **46**, 901.
Depth of field. Campbell, F. W. (1957) *Optica Acta* **4**, 157.
Line-spread function. Campbell, F. W. & Gubisch, R. W. (1966) *Journal of Physiology* **186**, 558.
Stiles–Crawford effect. Crawford, B. H. (1972) *in Handbook of Sensory Physiology*, vol. 7, pt 4, ed. D. Jameson & L. M. Hurvich. Heidelberg: Berlin: New York. Springer.

CHAPTER FOUR

Accommodation and Refraction of the eye

M. MILLODOT

4.1. ACCOMMODATION

The phenomenon of accommodation is best illustrated by making the following observations. Draw a small mark on a window with a chinagraph pencil or felt pen. Then look at a distant object through the window straight through that mark with one eye (cover the other eye with your hand to avoid confounding the experiment with double images). You will notice that the mark is blurred; then look at the mark and the distant object becomes blurred. You cannot be in focus at both distances simultaneously. In fact to view the mark on the window you must make a conspicuous ocular effort, whereas the eye seems to return naturally to focussing on the distant object. These simple observations demonstrate that the eye needs to adjust itself to see near objects clearly. This mechanism is called accommodation.

Brief history

This phenomenon has, of course, been known since ancient times although it was not always recognised to be a peripheral process. It was even attributed to a shift of attention. Scheiner in 1619 was the first to show that it depended on actual changes taking place in the eye. He made two small pinholes in a card, some 3–4 mm apart. Looking through them at say, a needle, you see it singly. If now you look at the distance or at some point closer to you than the needle, the needle will appear double. The explanation of this phenomenon is given in Fig. 4.1. However, if Scheiner's experiment proves that the eye accommodates to focus at different distances, one important question remains, that is, what structure in the eye is responsible for these changes in accommodation?

It was thought that perhaps the shape of the cornea altered with accommodation, as is in fact the case in some birds. Indeed very small changes would be significant since the cornea makes the greatest contribution to the optical power of the eye (see Table 3.2). Alternatively it was thought that perhaps the length of the eye varied

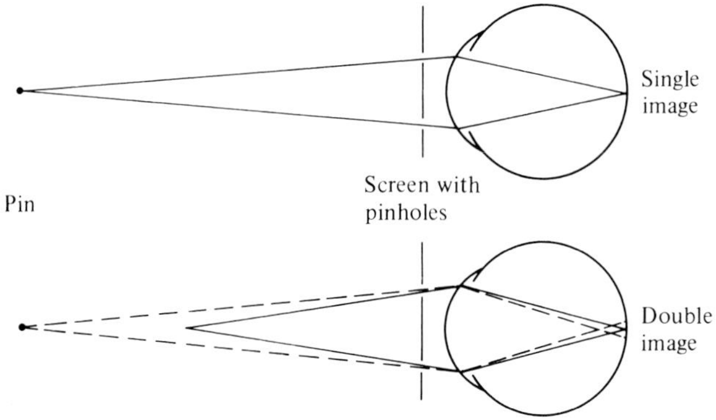

Fig. 4.1. Scheiner's experiment. A pin is viewed through two pinholes in a card held immediately in front of the cornea. An image is formed by each pinhole, but these are coincident when the eye is focussed at the distance of the pin (upper diagram), and the pin is seen clear and single. If the eye's dioptric power is increased by focussing at a closer distance (continuous lines in lower diagram) the two images coincide at a point in front of the retina and have separated again at the retina, as shown by the dashed lines, so a double image of the pin is seen.

with accommodation. These hypotheses were disproved by the audacious experiments of Thomas Young at the turn of the nineteenth century. Firstly Young immersed his eye in water, so eliminating the corneal power, since water and cornea have similar refractive indices. In these circumstances, the power of accommodation remains unaffected. Secondly, to prove that the length of the eye did not change during accommodation, Young inserted a metal hook at the outer angle of his orbit (he took advantage of his rather prominent eyes) whilst looking towards his nose, and rested it against the posterior pole of the eyeball. If the hook presses against the retina one sees a bright spot in the field of vision called a *phosphene*. However, even when the eye was maximally accommodated the hook was not displaced nor was there any difference in the appearance of the phosphene, demonstrating that no elongation of the eye had occurred. Thus the business of accommodation was attributed to the crystalline lens. It is conceivable that accommodation could take place by forward and backward movements of the lens, as is the case in some fish, but this is not the case in man.

Modifications of the eye during accommodation

The changes that occur in the eye during accommodation are revealed in various ways. Firstly by observing the eye using, for example, a biomicroscope. Secondly by observing and measuring the Purkinje–Sanson image (see Chapter 3). Thirdly by histological examination of an eye in which accommodation was induced either chemically or electrically prior to death of the animal.

The most obvious change that presents itself during accommodation is a contraction of the pupil. This contributes to the accommodative effort by reducing the diameter of the diffusion circle on the retina and thereby extending the depth of field. However, the most important modification is an increase in curvature of the front surface of the lens. This is clearly seen by merely observing the Purkinje–Sanson image corresponding to this surface (III). It becomes conspicuously smaller since the radius of curvature of the surface varies from 10.2 to 6 mm for an accommodation of 7 *D*.

At the same time the thickness of the lens increases by nearly 0.5 mm. Consequently the depth of the anterior chamber diminishes with accommodation and the edge of the iris moves slightly forward. Moreover the equatorial diameter (that is, the diameter of the lens in the vertical plane) also decreases. This has been evinced by observing people in whom part of the iris is missing. Finally the radius of curvature of the back surface of the lens varies very slightly (from 6 to 5.5 mm). All these changes are shown diagrammatically in Fig. 4.2. They take place rapidly, some 300 ms (± 80) from the time the retina has received the stimulus, although the system takes somewhat longer to adjust for a change in stimulus from near to far.

Theory of accommodation

Helmholtz, in 1855, was probably the first person to suggest that the ciliary body played an active role in accommodation. This is corroborated by examination of histological sections: myopic eyes have thinner ciliary muscle than hyperopic eyes as the latter usually exert more accommodation. It is also confirmed in observations of freshly enucleated eyes where drugs that either stimulate or paralyse accommodation had been injected prior to the death of the animal.

But if the role of the ciliary body was unquestioned, controversies developed over the manner in which this muscle contraction was supposed to induce a change in the shape of the crystalline lens. We shall not describe the numerous theories that have been suggested

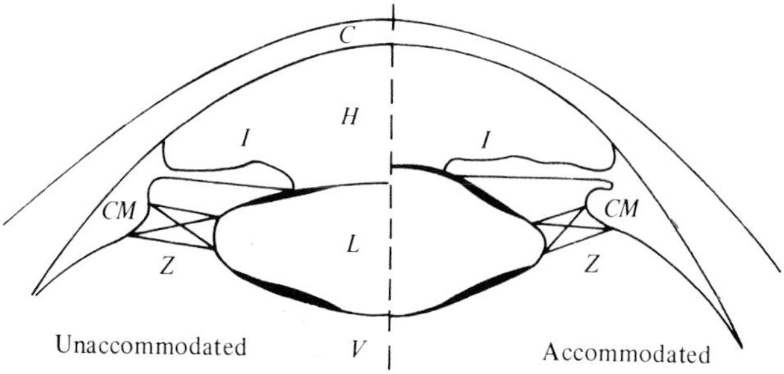

Fig. 4.2. Schematic diagram of the anterior part of the human eye unaccommodated (left half) and accommodated (right half). *C*, cornea; *H*, aqueous humour; *I*, iris; *L*, lens; *CM*, ciliary muscle; *V*, vitreous body; *Z*, zonule. See also Fig. 2.1.

in the last hundred years. We shall merely relate the presently established view, which is in fact more-or-less a confirmation of the century-old theory of Helmholtz.

When the eye is relaxed the zonule (Fig. 2.1) is tense and pulls the crystalline lens which flattens, particularly in its front surface. When the eye accommodates, the zonule is relaxed and the lens assumes a more spherical shape owing to the elasticity of the capsule and of the lens. However, it was difficult to understand how the front surface of the lens adopted its particular shape until Fincham in 1925 suggested that this peculiar form taken by the lens was due to the structure of the capsule. Indeed the capsule is not evenly thick: it is thickest and strongest near the equator and therefore the lens flattens in this area whilst it becomes more spherical near the optical axis where the capsule is thinnest and weakest. This theory is consistent with all the changes occurring during accommodation.

Stimulus to accommodation

Although some people can exercise voluntary control over their accommodation, the act of accommodating is commonly believed to be reflex in nature. Somehow the eye manages to take up the correct accommodative stance for any distance, so minimising the blur in the parts of the image falling in the central region of the fovea. In a normal environment, accommodation may be controlled by the several perceptual cues to distance (Chapter 12). Thus it is known that the mechanism controlling accommodation, like that determining our phenomenal experience, can be tricked into mistaking a change in apparent size for a change in distance. A particularly important stimulus to accommodation is an increase in convergence, the inward rotation of the eyes that serves to maintain single vision when an object is brought close to the observer (Chapter 11). However, when a single monocularly-viewed point of light is optically changed in depth, the accommodative mechanism continues to respond – and usually responds unhesitatingly in the direction required to bring the newly blurred image back into focus. In these reduced circumstances, most cues to distance are eliminated and accommodation must be controlled by properties of the blurred image itself. Chromatic aberration (Chapter 3) may provide one cue, since in over-accommodation it will be the short wavelengths that are most out of focus (see Fig. 2.5) and in under-accommodation it will be the long wavelengths. However, when chromatic aberration is eliminated by illuminating the target with monochromatic light, many subjects still respond correctly, or can learn to do so. Residual sources of information may then be spherical aberration and the direction of changes in blur during microfluctuations of accommodation. The latter are about 0.1 *D* in amplitude and 0.5 Hz in frequency with average pupil size. They are analogous to the motor tremor existing in all muscles.

Night and space myopia

When the eye is confronted by some visual stimulation situated at some distance it adjusts its focus correctly. In the absence of any objects or contours in the visual field (as when one looks into empty space or in the dark), the accommodative mechanism of the eye adopts a stance that corresponds to a certain amount of accommodation (from 0.5 to 1.5 *D*). In other words the eye becomes more powerful and its second focal point no longer coincides with the retina but is displaced towards the lens. The eye is then said to be affected by space myopia or night myopia. Actually, night myopia is not due only to accommodation: it also depends on spherical aberration and the shift in spectral sensitivity (the 'Purkinje-shift') that occurs when illumination is lowered. Clinically this condition must be kept in mind. Correction may be attempted but it is cumbersome as another pair of glasses must be substituted as soon as illumination increases. Space myopia is well known by fighter pilots and is taken into account in their training.

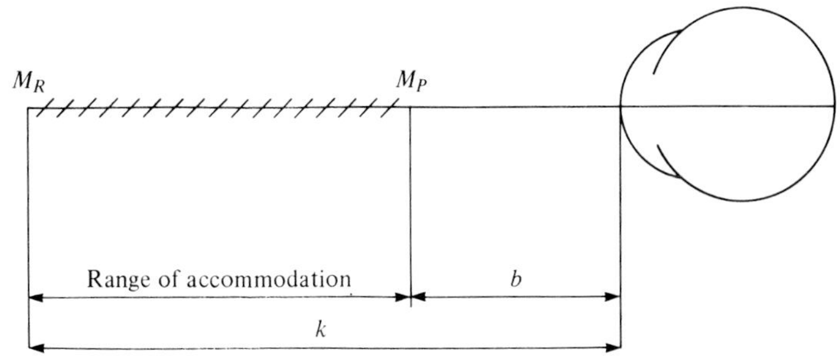

Fig. 4.3. Range of accommodation of an eye. M_R is the far point (punctum remotum) at a distance k metres; it would be at infinity in an emmetropic eye. M_p is the near point (punctum proximum) at a distance b; this recedes with age. The distance between them is the range of accommodation, and is best expressed as the amplitude of accommodation in dioptres, namely $1/k - 1/b$ (note that k and b are, by convention, both negative since they are in front of the cornea). This is simply the additional power in dioptres required to change focus from far to near point. It is plotted as a function of age in Fig. 4.4. A similar diagram would be applicable to short or far-sighted individuals wearing lenses to correct their ametropia.

Amplitude and range of accommodation

The furthest distance away at which an object can be seen clearly is called the *far point* of the eye (or *punctum remotum*). This point is represented as M_R in Fig. 4.3. To see at such distance the eye must be fully relaxed. And the fact that this object is seen clearly means that its image is formed exactly on the retina. If the far point of the eye is situated at infinity, the eye is said to be *emmetropic*. Otherwise it is *ametropic*. We shall return to this point later (see §4.2).

When maximum accommodation is exerted (i.e. the refracting power of the eye is increased to its maximum), the nearest point at which an object can be seen clearly is called the *near point* of the eye (or *punctum proximum*). This point is represented as M_P in Fig. 4.3.

The distance between the far point and the near point is called the *range of accommodation*. The maximum amount of accommodation that can be exerted is called *amplitude of accommodation* (A). It is always positive and equal to A (in dioptres) $= K - B$ where $K = 1/k$ and $B = 1/b$. k and b are the distances between the eye and the far and near point, respectively (see Fig. 4.3). K and B represent the amount of dioptres corresponding to objects situated at M_R and at M_P. If the far point is at infinity (emmetropic eye), $K = 1/\infty = 0$.

If M_R is at 1 m, $K = 1/1 = 1\ D$; at 0.25 m, $K = 4\ D$, etc. A problem might be helpful here.

1. *Suppose the far point of an eye is at 1 m (we shall see that this is the case of a myope) and the near point 0.12 m away. What is the range and amplitude of accommodation of this eye?*

The range of accommodation is $100 - 12 = 88$ cm or 0.88 m. The amplitude of accommodation is (the distances are negative because they are anterior relative to the cornea) $A = 1/(-1) - 1/(-0.12) = -1 + 8.33 = 7.33\ D$

The far point of an emmetropic eye (or corrected ametrope) is at infinity. Thus the main clinical measurement of accommodation consists of measuring the near point of the eye. The most common technique is the 'push-up' method. A card made up of small print is brought towards the patient's eye whilst he fixates the smallest letter that he can see distinctly. The distance at which the letter gets blurred is the near point. So if a person wearing his spectacles sees the test-type blurred at a distance of 25 cm his amplitude of accommodation is 4 *D*.

Age factors

As we grow older our eyes inevitably change. Our crystalline lenses become harder and less elastic. This leads to a loss of the amplitude of accommodation since the lens has a lesser tendency to become more convex. Consequently the near point of the eye recedes. This is illustrated in Fig. 4.4.

It is interesting to note in Fig. 4.4 that in fact we lose our accommodation from a very early stage in life. This fact may give some credence to J. J. Rousseau's view that we start dying the moment we are born.

When the accommodative ability of the eye is so reduced that the near point of the eye retreats further than the normal required distance (e.g. that required for reading a book), the eye is said to have become *presbyopic*. This condition occurs in most people between the ages of 42 and 48 years. Young people whose amplitude of accommodation is inordinately low for their age ought to be suspected of some systemic condition.

Cataract

With age the lens usually loses some of its transparency and becomes somewhat yellowish. If the loss of transparency is such that vision is noticeably impaired the eye is said to have a cataract. Other

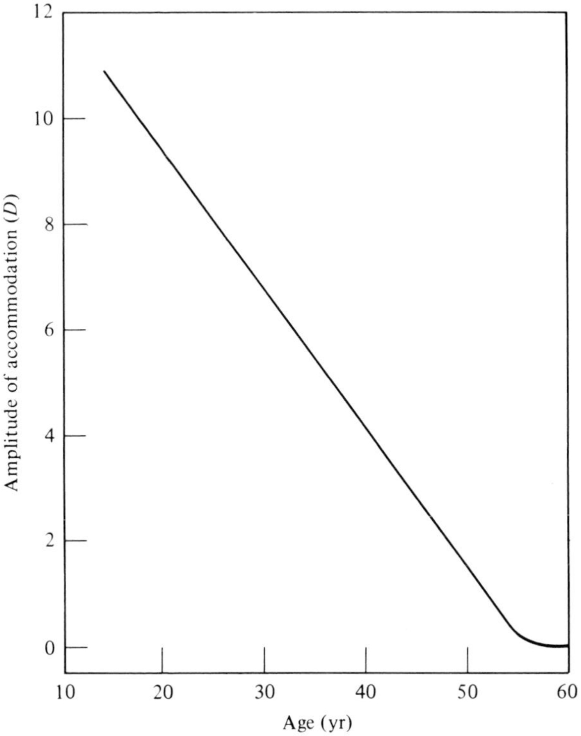

Fig. 4.4. Relationship between age and the range of accommodation. (After R. F. Fisher.) The function shown is based both on subjective measurements and on physical measurements of lenses taken from cadavers.

circumstances may also give rise to cataract: trauma to the eye, excess radiation such as infra-red, ultra-violet or X-rays, some diet deficiencies or uncontrolled diabetes. The principal treatment of cataract is to remove the crystalline lens, thus making the eye *aphakic*. We shall come back later to the optical correction of this condition (see §4.2).

4.2. REFRACTION OF THE EYE

Classification

An eye that, with its accommodation relaxed, produces an image on its retina of an object situated at infinity, is said to be *emmetropic*. It must be kept in mind, though, that emmetropia is a necessary but not sufficient condition for clear vision. Whether the patient sees distinctly depends upon the integrity of the retina and of the neurophysiological stages of the visual process, the subject of subsequent chapters.

However, if the retina is not situated in the plane of the eye's focal point (the focal point is the point of convergence of light rays emerging from an optical system when the incident rays are parallel) we have a condition called *ametropia*. In this condition the far point of the eye does not lie at infinity.

There exist two types of spherical ametropia, depending on whether the retina is situated in front of or behind the focal point of the eye. These are *myopia* (or short-sightedness) and *hyperopia* or hypermetropia (or long-sightedness). Later we shall discuss another ametropia called *astigmatism*. In passing, it is worth noting that the lack of harmony of the various components of the optics of the eye needs to be only very slight to cause an ametropia. Indeed an error of 0.3 mm in the length of the eye or a variation of 0.2 mm in the radius of curvature of the cornea gives rise to 1 *D* of ametropia. It is surprising, therefore, that about half of the population is emmetropic.

Myopia

Myopia is a condition in which the image of a distant object is formed, not on the retina, but in front of it (see Fig. 4.5). This may occur for either of two reasons. (1) The eye is too long whilst the focal length is normal (axial type of ametropia). This is the more common case. (2) The power of the eye is too great (refractive type of ametropia), a condition that gives rise to a shorter focal length whilst the length of the eye is normal. Some cataracts provide examples of the latter: the refractive index of the lens increases and thus the power of the eye increases and the patient becomes myopic (or less hypermetropic if that were the case).

As illustrated in Fig. 4.5 the image of a distant point on the retina of a myope is blurred. As the object is moved closer to the eye its image moves closer to the retina: the point in space where the object first has an image sharply formed on the retina is the *far point* of the eye. Myopes may spontaneously complain of blurred vision, but usually they are discovered at school when they have difficulty in reading the blackboard.

The reciprocal of the distance in metres between the cornea and the far point represents the amount of myopia in dioptres. This is

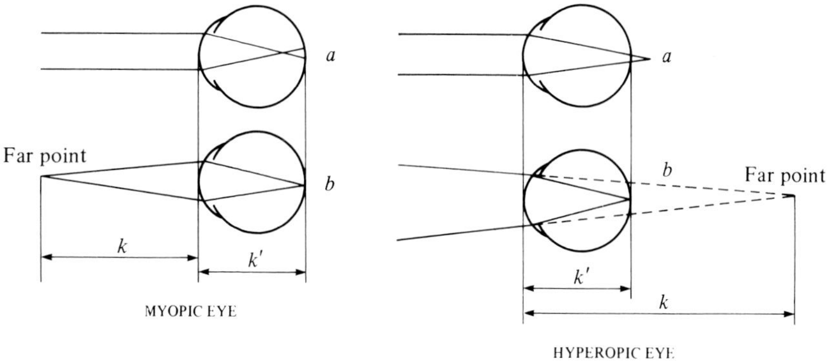

Fig. 4.5. Left: a myopic eye with relaxed accommodation; (*a*) looking at infinity; (*b*) looking at its far point.
Right: a hyperopic eye with relaxed accommodation; (*a*) looking at infinity; (*b*) with a convergent beam of light focussed at its far point.

called the *ocular refraction* or *refractive error* of the eye given as $K = 1/k$ where k (in metres) is the distance from the cornea to the far point. For example, if the far point of a myope is situated 25 cm in front of the eye, his ocular refraction $K = 1/-0.25 = -4\ D$.

Myopia usually develops around puberty and increases somewhat with further growth. It stays more or less constant thereafter. There is a strong hereditary influence in the development of myopia. Environmental factors may also contribute but the evidence supporting this hypothesis is uncertain. The incidence of myopia in the general population varies according to age, race and geography. We can consider that about one person in five is myopic (of over 0.5 D) but this number doubles among Chinese and Japanese. There also exists a type of myopia called progressive myopia, which is considered pathological since it is accompanied by some damage to the eye structures.

Hyperopia

In this condition the image of a distant object is focussed behind the retina, as illustrated in Fig. 4.5. This can be attributed either to an eye that is too short compared to the normal, or to an abnormally low refractive power of the eye (as occurs when there is a decrease in sugar concentration of the blood).

The far point of a hyperope is located behind the eye (see Fig. 4.5). In other words an object ought to be placed there so that its image be formed on the retina. This is obviously impossible (anatomically, all that is behind the eye is orbital fat!) and the far point of a hyperope is not real, it is virtual.

However, to compensate for this unlucky optical situation hyperopes accommodate continuously thereby increasing the refracting power of the eye and bringing the focal point back on to the retina. Therefore uncorrected hyperopes, unlike myopes, see objects at all distances clearly (provided they have enough accommodation) but at the expense of eye-strain and sometimes headaches. This is especially so after prolonged close work because of the constant high demand on accommodation.

The ocular refraction (K) of a hyperope is equal to $1/k$, where k (in metres) is the distance from the cornea to the far point. For example, if the far point of a hyperope is situated 50 cm behind the cornea $K = 1/50 \times 10^{-2} = 2\ D$.

Hyperopia usually develops soon after birth and strong evidence seems to indicate that genetic factors play an important role in determining this ametropia. One may consider that about one person in three is hyperopic in Western countries.

Correction of ametropia

The fundamental principle of the correction of ametropia is simple. It suffices to find an ophthalmic lens or a contact lens of a power such that its focal point coincides with the far point of the eye. An object at infinity will thus be focussed ultimately on the retina after passing through the correcting lens.

The power of the required correcting lens is $P = 1/SM_R$, where SM_R is the distance from the lens to the far point (see Fig. 4.6). P is called the *spectacle refraction.* It differs from the ocular refraction (which is relative to the cornea) by an amount dependent upon the separation between the eye and the spectacle which is about 12 to 14 mm. If the eye is corrected by a contact lens placed against the cornea the spectacle refraction is equal to the ocular refraction.

The optical correction of a myope consists of a negative lens since this lens has a virtual focal point in front of the eye (SM_R is negative). This is illustrated in Fig. 4.6. The optical correction of a hyperope consists of a positive lens since such a lens has its focal point behind the eye (SM_R is positive) as shown in Fig. 4.6.

The difference between spectacle and ocular refraction is usually small but it becomes quite significant in cases of large ametropias. For example, suppose an eye has a far point situated 10 cm behind

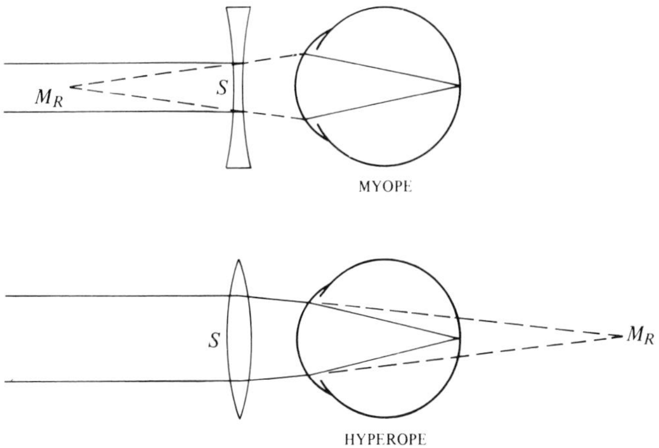

Fig. 4.6. A myope (top) requires a diverging, concave lens ($-^{ve}$ power) in order to provide the diverging beam that is focussed on his retina when accommodation is relaxed (see Fig. 4.5, left, *b*). Similarly a hyperope (lower) requires a converging, convex lens ($+^{ve}$ power) to provide a converging beam (see Fig. 4.5, right, *b*).

the cornea; the ocular refraction is $K = 1/k = 1/0.10 = 10\ D$. Assuming that this patient wears spectacles 14 mm from his corneas his spectacle refraction is $P = 1/SM_R = 1/114 \times 10^{-3} = 8.77\ D$. In this case the difference is equal to 1.23 *D*. This fact is important in optical correction by contact lenses: hyperopes need contact lenses with more power (1.23 *D* in this case) than glasses; it is the reverse for myopes. Similarly the power of the correction changes as the spectacles are moved away from the eye. You sometimes see people getting some temporary relief by moving their spectacles further down their nose, an indication that the correction has become incorrect.

For most purposes an ametropic eye with an appropriate correction (either in the form of spectacles or contact lenses) functions more or less like an emmetropic eye. However, corrected ametropes do not exert exactly the same amount of accommodation and the size of their retinal image is slightly different.

Astigmatism

In the ideal eye the refractive surfaces are spherical with equal curvatures along all meridians. However, most human eyes do not fulfil these criteria, and the cornea in particular may have slightly

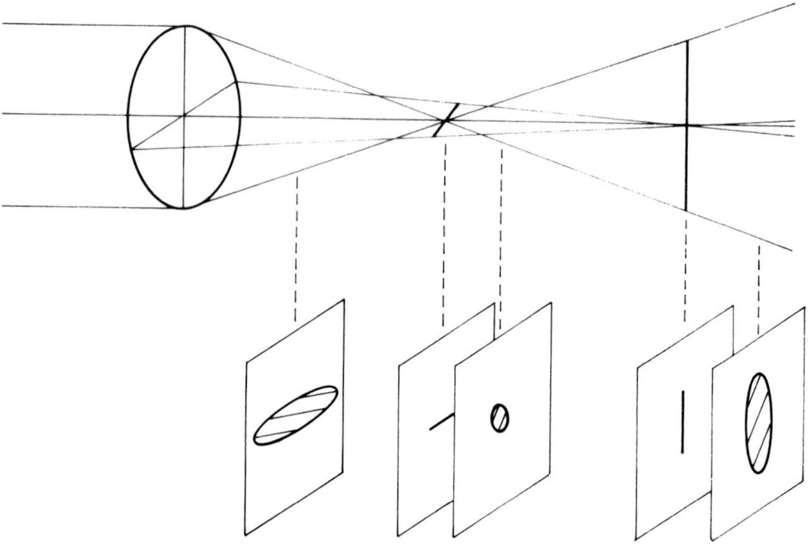

Fig. 4.7. Astigmatic pencil of light emerging from a toric lens. In planes at successively greater distances from the lens the beam forms a line, then a disc, then a line at right angles to the first. The aberration is corrected by a cylindrical lens of appropriate power and orientation.

different curvatures in the various meridians. As a result, light rays will be refracted more in one meridian (that of the greatest curvature) than in the other at right angles to it (that of least curvature). The image of a distant object will no longer be simple. This defect is called *astigmatism*, a term derived from the Greek roots *a* ('without') and *stigma* ('a point') and meaning the condition in which an optical system does not form a single point focus. Fig. 4.7 shows how an astigmatic lens refracts the light from a distant point source, the lens being assumed to have its greatest power in the vertical meridian. The bundle of transmitted rays does not have a disc-shaped cross-section converging to a point, as is the normal case. At one point the cross-section of the bundle forms a horizontal line (corresponding to one focus) and at a greater distance from the lens it forms a vertical line (corresponding to the other focus); at one intermediate point the cross-section is disc-shaped and at all others it is elliptical, the long axis depending on the nearest focal line. The amount of astigmatism is defined in terms of the dioptric distance between the two focal lines.

Astigmatism is an optical anomaly that is superimposed on the

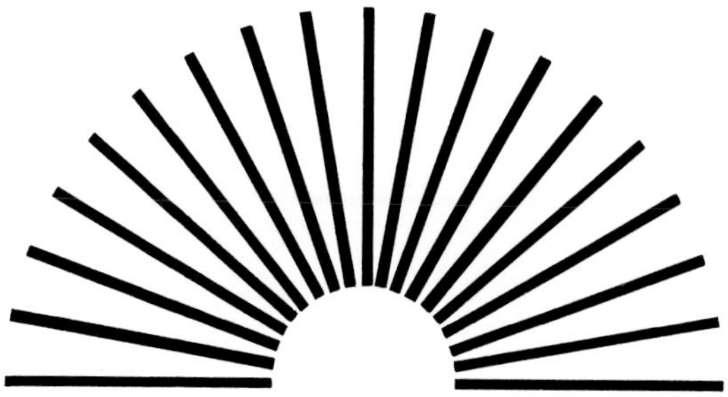

Fig. 4.8. Chart for testing astigmatism. If the eye is astigmatic, some lines will be out of focus whatever the state of accommodation.

spherical ametropia, which may be present. Yet, it is usually only half as bad as an equal amount of spherical ametropia. Indeed the eye can see relatively clearly the contours in an object parallel to the focal line focussed on (or near) the retina. If, for example, the vertical focal line is focussed on the retina the eye will discern clearly all vertical features of an object and thereby recognition will be easier than if the whole object were blurred uniformly. It is for this reason that the standard subjective test for astigmatism is a chart composed of black lines orientated in different directions (see Fig. 4.8). A patient who has astigmatism will see some lines more sharply and others, at right angles, more blurred. The directions thus found indicate the two meridians of astigmatism. In clinical practice corneal astigmatism is measured accurately with a keratometer or ophthalmometer which give either the radii of curvature or the powers in the two meridians.

Correction of Astigmatism

To correct astigmatism we must secure a lens such that it forms the image of a distant object at the two far points of the astigmatic eye, one for each principal meridian. Such an optical correction will have different curvatures in different meridians, but its astigmatism will, of course, be complementary to that of the eye. A lens with different powers in its two principal meridians is a spherocylindrical lens or toric lens and a lens with power in only one meridian is called a cylindrical lens.

Since the cornea is the main site of astigmatism in the eye, hard contact lenses offer a valuable means for such corrections, besides being inconspicuous. As the tears fill the space between the lens and the cornea the contact lens effectively eliminates the natural cornea.

The main refracting surface becomes that formed by the front surface of the contact lens which needs to be spherical only and of an appropriate curvature needed to correct the remaining spherical ametropia, if any.

Astigmatism that can be corrected by a spherocylindrical lens of appropriate power and angle is called *regular astigmatism*, but sometimes the distortion of the cornea is more complex and its optical aberrations cannot be remedied by any spherocylindrical lens. This condition is called *irregular astigmatism* and is typically the result of trauma or of keratoconus or of distortion following a cataract operation. In such cases, contact lenses are especially valuable.

It is absolutely essential that spherical ametropia and astigmatism be corrected as early in life as possible. Failure to do so may lead to irreversible amblyopia (see Chapter 20).

Aphakia

When the crystalline lens is absent, the eye is said to be *aphakic*. This usually occurs as a result of the operation for cataract (see §4.2). Extremely rare are cases in which the lens was dislocated in such a way that it is no longer situated in the optical path of light. Aphakes cannot see clearly at any distance, and are of course, unable to accommodate. Therefore they require two corrections, one for distance and one for near.

The power of an aphakic eye is reduced by about a third and its optical characteristics are those of the remaining cornea (see Chapter 3). As a result the aphakic eye is hyperopic, or is less myopic if it was highly myopic before the operation. In fact it could be emmetropic if it were between $-13\ D$ and $-11.5\ D$ prior to the operation. The correcting lens consists of a strong positive lens with very curved surfaces. The visual acuity of a corrected aphake is usually quite good, about 6/6 (or 20/20). Correction by contact lenses provides increased visual field, improved cosmetic appearance and slightly better vision, since the cornea is usually quite astigmatic as a result of the operation, a condition that is corrected more adequately with hard contact lenses.

Furthermore, contact lenses produce a smaller retinal image than spectacles and this is a distinct advantage in unilateral aphakia. In this case the image size in the normal eye is different from that in the corrected aphakic eye (a condition called *aniseikonia*) and the brain cannot fuse the two images. This difference can amount to 25%

in a pair of eyes in which the aphakic eye (which was originally emmetropic) is now corrected by a +11 D lens placed 12.5 mm in front of the cornea and the other eye is emmetropic. If the aphakic eye is corrected with a contact lens the aniseikonia would amount only to 6%. Binocular vision might not be easy but is, in these circumstances, possible.

4.3. MEASUREMENT OF THE REFRACTION OF THE EYE

The clinical measurement of the eye's refractive error is often referred to elliptically as simply 'refraction'. Clinical refraction was codified by Donders in the middle of the last century. The importance of refraction need not be stressed, as clear and comfortable vision is certainly one of the prerequisites of a better and more useful life in the modern world.

Refraction is determined in two ways, called subjective and objective according to whether or not the patient is required to judge what he sees. The objective method is considered to be only a guide to the subjective measurement, which is regarded as the last court of appeal.

The subjective measurement of visual performance is a psychophysical procedure of the type discussed in Chapter 7. However, the psychophysics employed in the clinic is a far cry from that used with trained observers in the laboratory. Indeed it is an art to extract the desired information out of patients with the minimum amount of testing under variable conditions and fluctuating attention. This is one of the main reasons why refraction is almost always assessed objectively as well as subjectively. Besides, objective methods are sometimes the only possible means of refracting the eye, particularly when dealing with malingering, illiterate or psychotic patients.

Retinoscopy

Although Sir William Bowman in 1859 noted it without appreciating its clinical value, this method of objective refraction is credited to a French ophthalmologist Cuignet who rediscovered it, almost inadvertently, in 1873.

Retinoscopy utilises the light reflected by the interior back surface of the eye. The retina plays the role of an object (R) the image of which is formed at the far point (M_R) by the dioptrics of the subject's eye (*see* Fig. 4.9). One is merely left with finding this far point. The illumination of the retina is very simple (*see* Fig. 4.9*a*). It consists

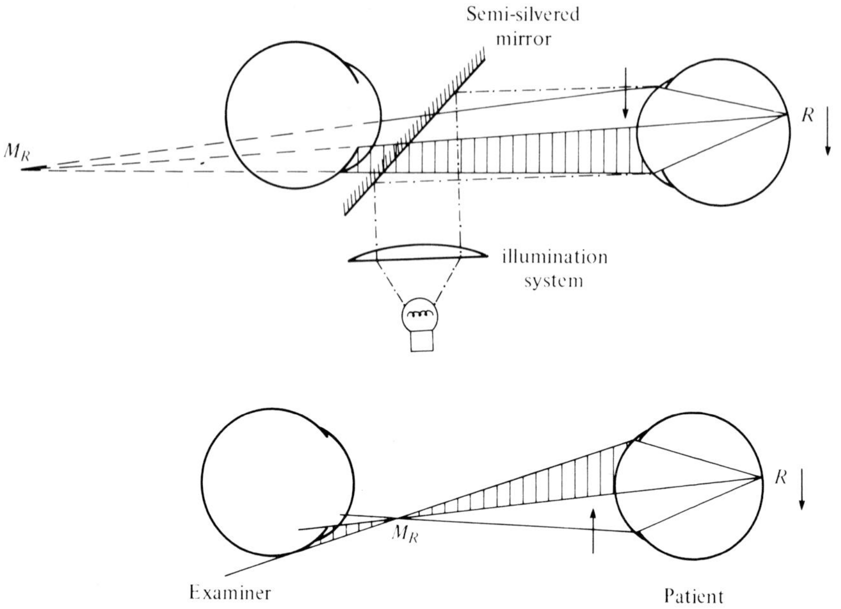

Fig. 4.9. Basic principle of retinoscopy. The illumination system is shown only in (*a*) so as to retain simplicity in diagram (*b*). The hatched areas appear dark to the examiner and the vertical arrows represent movement of the light beam.

of a source of light (usually a small battery-operated bulb), a condensing lens, which gathers light emanating from the bulb, and a semi-silvered mirror (or a mirror with a small aperture) which reflects light into the patient's eye. The mirror arrangement lets the examiner view the patient's pupil along the same axis as the path of light to and from the fundus.

If the examiner views the subject's pupil other than from the far point, he will see more or less of it illuminated depending upon the fraction of the reflected beam of light that penetrates his own pupil; but if the patient's far point happens to be situated in the plane of the examiner's pupil (this is called the neutral point), the latter will see the patient's pupil as either completely bright or completely dark because the nearly punctate image will either lie within the examiner's pupil or outside it. Therefore, to use the retinoscope the examiner tilts it to and fro, so moving the position of *R* on the retina, and he observes the change in the illuminated area of the pupil (called the reflex). If the patient's far point is as drawn in Fig. 4.9*a*, as *R* moves downwards, the examiner will see the reflex gradually extend over the pupil from top to bottom. Thus, when the far point lies behind the examiner's eye the reflex moves in the same direction as R and consequently in the same direction as the tilt of the retinoscope. Positive lenses can be placed in front of the patient's eye until the examiner reaches the neutral point. The patient's refractive state can then be calculated as described below. If the patient's far point is as drawn in Fig. 4.9*b*, as R moves downwards, the examiner will see the reflex gradually extend over the pupil from bottom to top. Thus when the far point lies between the examiner and the patient the reflex moves in the direction opposite to the tilt of the retinoscope. Negative lenses are placed in front of the patient's eye until the examiner reaches the neutral point.

The final estimate of the patient's ametropia takes into account the lens that was placed in front of the patient and the distance separating the patient and examiner. If, for example, the examiner is 50 cm from the patient, $-2\ D$ is added to whatever lens is in front of the patient. If there is no such lens the patient is simply $-2\ D$ myopic. If a $+5\ D$ was needed to reach the neutral point the patient is $+3\ D$ hyperopic, etc.

Ophthalmoscopy

Although Charles Babbage had devised a method of observing the fundus of human eyes, it is to Helmholtz in 1851 that the ophthalmoscope is attributed, as he improved and thoroughly described the technique and its application. Being able to view the internal structures of the living human eye was rightly hailed as a revolution in medical science in the nineteenth century. The ophthalmoscope has never ceased to be the most invaluable companion of the eye practitioner and even of the general practitioner. It is used to assess the state of health of the eye as well as some other systemic conditions (e.g. diabetes, arteriosclerosis or intra-cranial pressure). However, we describe this instrument here as it can be used to assess, grossly, the refraction of the eye.

Like retinoscopy, ophthalmoscopy makes use of the minute amount of light reflected by the fundus of the eye (about 1/10000 of the light that enters the eye). Thus one must illuminate the fundus in such a way that the examiner's eye is located along the path of the incoming light without being in its way. This is achieved by the use of a semi-silvered mirror (or by reflecting the ingoing light from the tip of a small prism), as shown in Fig. 4.10. If the patient is

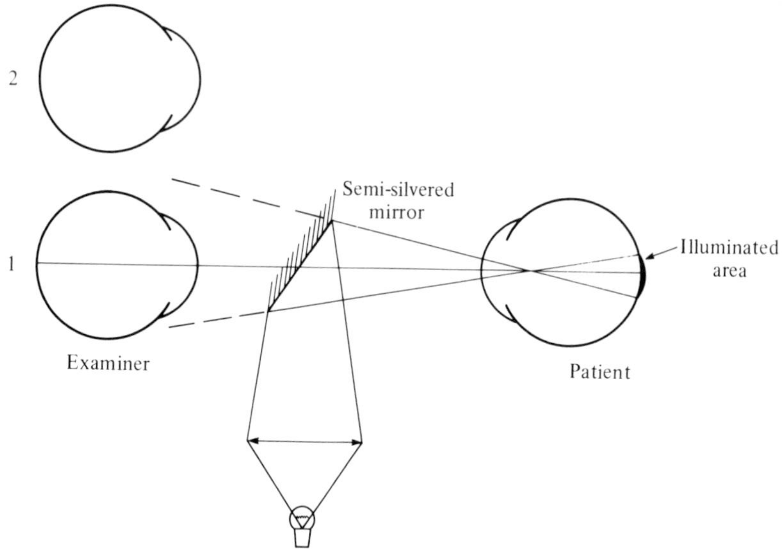

Fig. 4.10. Principle of the ophthalmoscope. In position 1 the examiner views the illuminated retina, but this is not possible in position 2.

emmetropic his retina will be imaged at infinity and will appear clearly to the examiner (assuming the latter is emmetropic and unaccommodated). If, on the other hand the patient's retina appears blurred, the examiner will place lenses in front of his own eye until he sees it clearly, that is when the image of the patient's retina will be formed at infinity through his eyes' optics plus the correcting lens.

As the ophthalmoscopic examination is carried out with the examiner's eye situated as close as possible to the patient's eye the correcting lens is almost situated in the spectacle plane of the patient. The dioptric power of that lens gives an approximate indication of the patient's spectacle refraction. This method of objective refraction is far simpler than retinoscopy but it is much less accurate. It depends on the examiner's ability to relax his accommodation and his appraisal of the sharpness of the patient's fundus, both of which are difficult to achieve consistently and accurately.

4.4. SUGGESTIONS FOR FURTHER READING

General references

Bennet, A. G. & Francis, J. L. (1962) In *The Eye*, Vol. 4. London: Academic Press.

Borish, I. (1970) *Clinical Refraction.* Chicago: Professional Press.
Duke-Elder, S. & Abrams, D. (1970) *System of Ophthalmology*, Vol. 5. London: Henry Kimpton.
Helmholtz, H. (1924) *Treatise on Physiological Optics*, translated by J. P. C. Southall. New York: Dover (Reprint 1962).
Le Grand, Y. (1965) *Optique Physiologique*, Vol. 1. Paris: Masson.
Michaels, D. D. (1975) *Visual Optics and Refraction.* St Louis, USA: C. V. Mosby Co.

Special references

General review of accommodation. Toates, F. M. (1972) *Physiological Reviews*, **52**, 828–63.
Stimulus to accommodation. Fincham, E. F. (1951) *British Journal of Ophthalmology*, **35**, 381–93.
Influence of apparent distance on accommodation. Ittelson, W. H. & Ames, A. (1950) *Journal of Psychology*, **30**, 43–63.
Reaction time of accommodation. Campbell, F. W. & Westheimer, G. (1960) *Journal of Physiology*, **151**, 285–95.
Accommodation in coloured light; effects of learning. Charman, W. N. & Tucker, J. (1978) *Journal of the Optical Society of America*, **68**, 459–70.
Evidence for night and space myopia in the absence of any objects or contours in the visual field. Leibowitz, H. W. & Owens, D. A. (1978) *Documenta Ophthalmologica*, **46**, 133–47.
Changes in the lens with age. Fisher, R. F. (1977) *Journal of Physiology*, **270**, 51–74.
Fisher, R. F. & Pettet, B. E. (1973) *Journal of Physiology*, **234**, 443–7.
Millodot, M. & Newton, I. (1981). *British Journal of Ophthalmology*, **65**, 294–8.
Permanent neural consequences of uncorrected astigmatism. Freeman, R. D., Mitchell, D. E. & Millodot, M. (1972) *Science*, **175**, 1384–6.

CHAPTER FIVE

The biochemical aspects of vision

A. KNOWLES

The vertebrate eye is a highly specialised structure for the formation of an image of the outside world upon the retina, and the conversion of this image into neural signals suitable for transmission to the brain. These functions have widely differing structural and metabolic requirements: image formation is the function of the cornea, aqueous humour, lens and vitreous, which will be collectively termed the *pre-retinal media.* These provide the refractive properties of the eye, and so high refractive index and good transmission of light are essential qualities (see Chapter 3). In contrast, the other major components of the eye – the retina and pigment epithelium – must absorb light efficiently; the retina in order to obtain the maximum response from light entering the eye and the pigment epithelium in order to absorb stray light which would otherwise reduce the quality of the retinal image.

The pre-retinal media are relatively inert and have low metabolic rates and turnover of cellular material, while the retina is highly vascular, consumes energy very fast and has a high rate of protein replacement. This chapter is accordingly divided into two parts.

5.1. THE PRE-RETINAL MEDIA

The cornea

The cornea functions both as a window and as the principal refractive surface of the eye, and so it must combine optical clarity with structural stability. It must also contain the pressure within the eye – the *intraocular pressure* – and therefore has strictly controlled permeability. The structure of the cornea is shown in Fig. 5.1, and consists of a thick central layer, the *stroma*, sandwiched between semi-permeable membranes, and enclosed by the epithelium on its external surface and the endothelium on the inner surface. The stroma is composed of collagen fibrils with mucopolysaccharides filling the interstitial surfaces. These compounds are important in the regulation of the water content of the tissue, and hence its

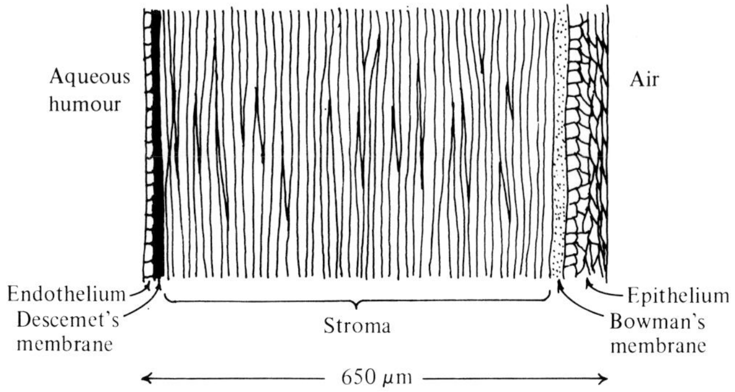

Fig. 5.1. Diagrammatic section of the human cornea. The stroma consists of about 200 layers, each composed of bundles of collagen fibrils. The fibrils in each layer lie parallel to one another, while successive layers are orientated so that their fibrils are roughly perpendicular.

transparency. A structure composed of collagen fibrils would be expected to be opaque because of the scattering of light at the numerous interfaces, and it was thought that the exceptional transparency of the cornea was due to a regular quasi-crystalline arrangement of the fibrils. However, Benedek has now proposed that the tissue is transparent simply because the fibres and the spaces between them are too small to scatter visible light. If a piece of cornea is immersed in water it will swell and become opaque, and changes in hydration after death or on the administration of some drugs will also cause clouding. Damage to the epithelium by abrasion or radiation will cause zones of local hydration which are quickly repaired, but deeper lesions lead to a permanent loss of transparency. These observations all suggest that an increase in the spacing of the fibrils, or their disruption, will increase the light scatter. Consequently the corneal membranes have active pumps that maintain the water content of the stroma to an optimum 75–80%, but the mechanism of the pumps is not yet established. It was thought that active transport of sodium ions across the corneal membranes from the stroma to the tears and to the aqueous humour would lead to a regulation of water content by osmosis, but it is now established that sodium ions are pumped *into* the stroma across the endothelium. Nevertheless the sodium-ion content of the stroma is critical, and it seems that sodium bound to the interstitial mucopolysaccharides regulates their degree of hydration.

The cornea is, of necessity, non-vascular and there are three routes

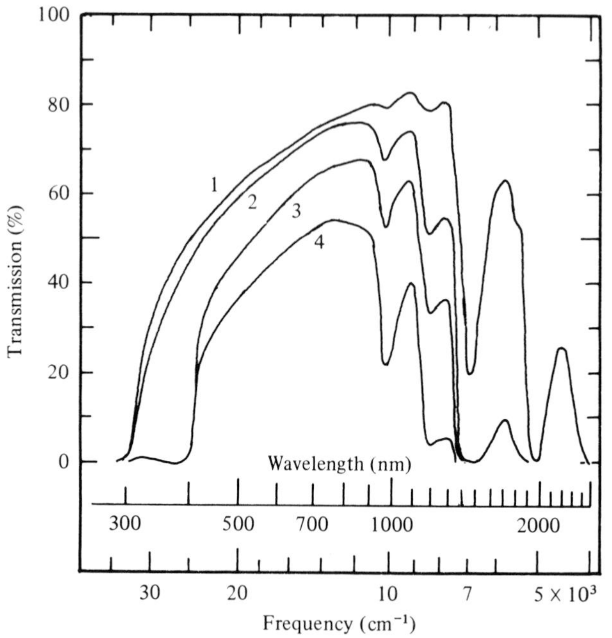

Fig. 5.2. Cumulative transmission spectra of the pre-retinal media. 1: transmission of cornea; 2: cornea + aqueous humour; 3: cornea + aqueous + lens; 4: cornea + aqueous + lens + vitreous. The curves are based on the light absorbed by the media together with losses by scattering and reflection. (From Boettner, E. A. & Wolter, J. R. (1962) *Investigative Ophthalmology*, **1**, 776–83.)

by which metabolites can enter and leave it: from the capillaries of the limbus, through the epithelium from the tears, and through the endothelium from the aqueous humour. Of these, the last is by far the most important and accounts for the major supplies of glucose and oxygen. Larger molecules such as plasma proteins probably enter through the limbus. Aerobic glycolysis is the principal mechanism of energy production.

Fig. 5.2 shows the transmission of the cornea. This falls rapidly in the ultraviolet (UV) region because of absorption by the proteins, while all the constituent molecules, including water, contribute to the infrared cut-off. Although the cornea appears transparent to visible light, reflection at the surfaces and scatter reduce its maximum transmission well below 100%.

Aqueous humour

The chamber formed between the cornea and the lens is filled with a watery liquid aptly called the aqueous humour. This has several important functions: (i) supplying nutrients to the cornea and lens; (ii) acting as the hydraulic agent that maintains a positive pressure within the eye, thus preserving the shape of the eye and in particular the cornea; and (iii) acting as a medium of low refractive index compared with the lens, and so maximising the refractive power of the latter.

The aqueous humour is generated in the posterior chamber, the small annular volume bounded by the iris and the lens, and is secreted by the epithelium of the ciliary body. Several mechanisms are thought to operate in its formation. The intraocular pressure is about 16 mm Hg, and is thus less than the arterial pressure in the capillaries of the epithelium. It therefore seems that there is an 'ultrafiltration' process by which water and small molecules pass into the aqueous humour through the membranes of the ciliary epithelium. There is also evidence of active transport of sodium, chloride and other ions into the aqueous humour which also causes the passage of water by osmosis.

The aqueous humour flows from the posterior chamber through the iris and into the anterior chamber; from there it is returned to the venous system through the *trabecular meshwork* which is located in the angle between the cornea and the iris (Fig. 2.1*d*). The flow is quite rapid so that the entire volume is replaced every 2–3 h. The rate of flow out of the anterior chamber, and hence the intraocular pressure, is controlled by the resistance of the trabecular meshwork. Reduction in the rate of outflow is the primary disorder in glaucoma, the increased pressure of the aqueous humour then causing excessive pressure upon the vitreous which in turn restricts the capillary flow in the retina and results in blindness.

The aqueous humour has a major nutritional role for both cornea and lens, and has been shown to be the principal source of amino acids for protein synthesis in the lens. The lens has a high ascorbate content and this is reflected in the ascorbate concentration of the aqueous humour, which is several times higher than that of blood plasma.

Lens

The lens is the most rigid structure in the eye, although the human lens is relatively soft compared with most other species. It is built up from long ribbon-like cells that run from the anterior pole through the equator to the posterior pole. New cells are added to the periphery of the lens throughout life so that the oldest cells form a central core – the nucleus. The component proteins are called crystallins and can be classified according to their solubility in water, the peripheral cells having a high content of soluble protein, while that of the nuclear cells is relatively insoluble. This is probably due to increased cross-linking as the cells age, and this part of the lens is correspondingly denser and has a higher refractive index. No protein synthesis can be detected in the nuclear cells, while the peripheral ones show both synthesis and a high rate of protein turnover.

The lens absorbs strongly in the near-UV region, and the tail of this absorption band extends into the visible region giving the lens a pale yellow colour. The effect of this absorption upon the transmission of the eye can be seen in Fig. 5.2, which demonstrates that it is the lens that determines the UV cut-off of the eye and thus the short-wave limit of the visible spectrum. It is generally supposed that this restriction of the wavelength range is beneficial, for if shorter wavelengths reached the retina, chromatic aberration would reduce the sharpness of the retinal image (see Chapter 3).

One of the compounds contributing to the absorption is the glycoside of 3-hydroxykyurenine – an unusual metabolite of tryptophan. The yellow colour increases with age, particularly in cases of nuclear cataract, when a dark brown protein species forms in the nuclear cells.

The transparency of the lens is dependent upon the condition of its constituent cells and the state of the protein molecules within them. Loss of transparency is called *cataract* and results in a loss of image quality and, in extreme cases, in blindness. Metabolic disorders affecting the water or salt concentration of the cells can cause an immediate effect that is generally reversible, though the slow aggregation of the proteins can lead to a permanent cataract if the aggregates become large enough to scatter visible light. Cataracts can be generated by dietary deficiencies and by defects in carbohydrate metabolism. Artificially raising the blood-sugar level in animals leads to cataract, and although human diabetes does not necessarily cause loss of transparency, diabetics are more prone to senile cataract. Some drugs and a variety of toxic agents cause cataract, as can both infra-red and ionising radiation. The cornea will transmit beyond 1250 nm, and so near-infra-red radiation can reach the lens and cause local heating; this results in 'glass-blower's cataract'. Ionising radiation affects only the developing cells at the periphery, and opacification of the lens takes place slowly as the affected cells develop.

The supply of nutrients to the lens and the removal of waste products is primarily carried out by the aqueous humour, and there appears to be active transport of ions and amino acids across the surface. There is an uptake of ascorbate ions from the aqueous humour, though their function in the lens is not yet established. Vitamin-C deficiency does not necessarily cause changes in the lens.

The lens proteins are remarkable in their action as organ-specific auto-antigens which can elicit an immune response in other cells in the body. This may be a consequence of the mechanism of lens development for although new cells are being continually added, none are shed, and so the lens proteins are contained within the capsule and do not come into contact with the antibody-forming cells of the rest of the body. Removal of the lens from one eye of a patient suffering from cataract may cause enough cells to be released to sensitise him to his own lens protein, especially if the lens is removed from the capsule within the eye. If the lens of the second eye is injured at a later date or if the lens capsule becomes permeable in old age, antibodies will gain access to the second lens and a severe inflammatory reaction with loss of vision results.

Sympathetic ophthalmia is another condition where injury to one eye can cause damage to the other eye by an immunological mechanism. If one eye is damaged by a penetrating wound the patient may become sensitised to antigens from the uveal tract (the iris, the ciliary body and the choroid) or the retina. This allergic reaction can cause ophthalmitis in the affected eye and may subsequently affect the second eye in a similar way. This 'sympathetic' reaction can thus lead to severely impaired vision in both eyes. It is to prevent sympathetic ophthalmia that severely damaged eyes are removed, and minor injuries treated with corticosteroids.

Vitreous

The vitreous occupies the major part of the volume of the eye, and is a semi-solid gel with a high water content. The structure is an open framework of collagen fibres permeated by a relatively fluid solution of hyaluronic acid and other compounds. This fluid is termed the *vitreous humour* by some authors to distinguish it from the structure as a whole. The collagen fibres are not uniformly distributed throughout the vitreous, but are concentrated around the periphery and converge on points of attachment to the base of the ciliary body and at the periphery of the retina. The low refractive index of the vitreous humour, like that of the aqueous humour, serves to maximise the refractive power of the lens.

Water and nutrients enter the vitreous from the ciliary body, and while there appears to be a flow of vitreous humour back towards the retina in some animals, the gel of the human vitreous seems too rigid to permit an appreciable flow. The vitreous has a phagocytic activity which helps to maintain the clarity of the vitreous if 'floaters' – blood or cells from the surrounding tissues – enter it.

The macula

Although it is a part of the retina, the *macula lutea* (yellow spot) falls within our definition of pre-retinal media since it is an inert light filter. Its function seems to be to prevent short-wave light from reaching the central area of the retina, and it completely covers the fovea – the region that has the highest acuity. The macula absorption varies between individuals, but on average exceeds 50% of the incident light below 495 nm. The purpose of the macular pigment seems to be to enhance the acuity of the fovea by further reducing the effect of chromatic aberration.

5.2. THE RETINA AND PIGMENT EPITHELIUM

The receptor cells

Having traversed the vitreous, light travelling through the eye must then pass through the neural layers of the retina before reaching the *receptor cells* (see Fig. 2.2). The latter are elongated cells divided into two sections connected by a narrow cilium which are termed the *inner segment* (i.e. the proximal part) and *outer segment*. The inner segment contains the cell nucleus, mitochondria, and other components typical of a mammalian cell, while the outer segment is a structure specialised for the optimum absorption of light (Fig. 5.3). Detailed studies by electron microscopy have revealed that the arrangement of the membranes in rod and cone receptors is different: the cone outer segment consists of a single membrane folded back and forth

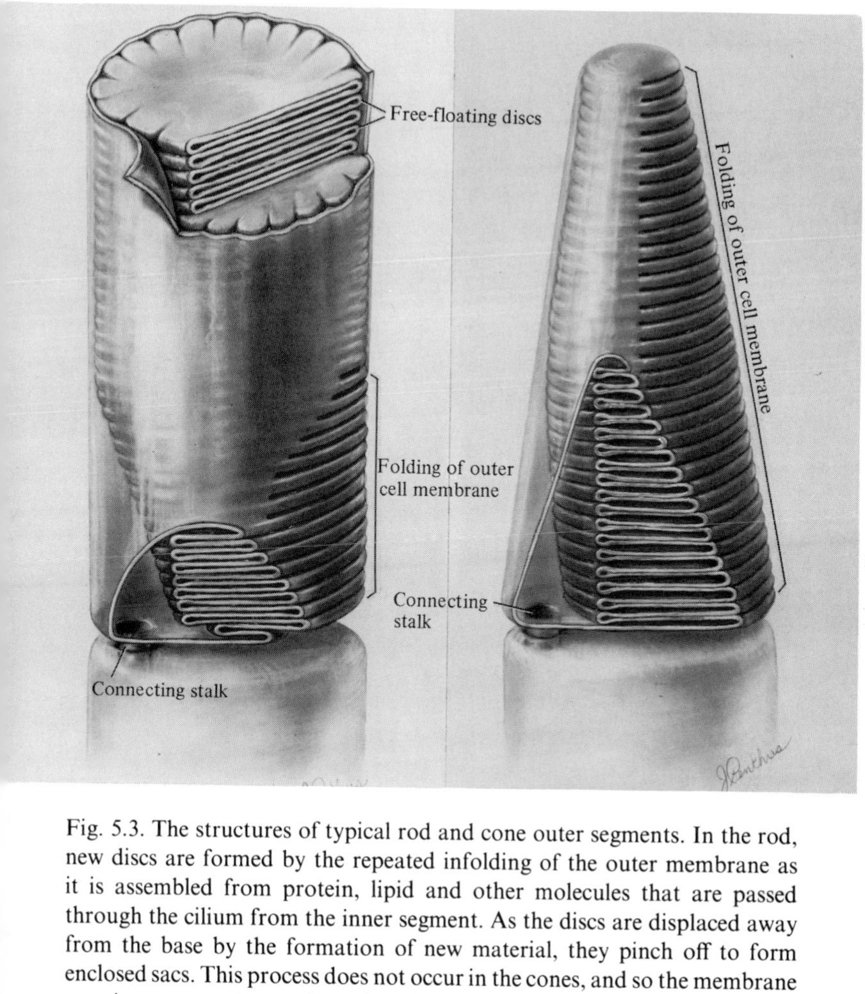

Fig. 5.3. The structures of typical rod and cone outer segments. In the rod, new discs are formed by the repeated infolding of the outer membrane as it is assembled from protein, lipid and other molecules that are passed through the cilium from the inner segment. As the discs are displaced away from the base by the formation of new material, they pinch off to form enclosed sacs. This process does not occur in the cones, and so the membrane remains as a continuous invaginated sheet. (From Young, R. W. *Scientific American*, (1970), **223**, pt. 4, 80–91.)

forming a concertina-like structure that encloses the intracellular space. The rod outer segment is more complex, the membrane layers being joined together in pairs to form flattened sacs that are called *discs*. The stack of discs is enclosed by the outer cell membrane, and so the outer segment is divided into two kinds of enclosed volume, the *intradisc space* and the *interdisc space*.

In both types of cell, the membranes are composed of protein molecules embedded in a double layer of closely-packed lipid molecules. It is the protein molecules that absorb light and are thus the key to the visual transduction process. Since they absorb light they appear coloured when viewed in white light, and are therefore termed *visual pigments*. Careful comparison of the fraction of light absorbed by pigment molecules in an intact retina with their absorption when the membranes have been broken down by treatment with a solubilising agent shows their absorption to be greater in the intact membrane. This suggests that the molecules are orientated in the membrane to maximise their light absorption, and in fact, this alignment increases the absorption of the molecules by a factor of 1.4 over their absorption when randomly distributed in solution. The development of a number of techniques for the study of proteins in membranes, such as X-ray crystallography, freeze-etch electron microscopy, and so on, has led to a model for the rod disc membrane which is illustrated in Fig. 5.4. In this, the protein is an elongated structure that is held with its long axis perpendicular to the membrane surface – the orientation required for the maximum absorption of light. The lipid bilayer is relatively fluid and so the pigment molecules are free to rotate about their long axes and to diffuse sideways.

The distal ends of the outer segments are embedded in the pigment epithelium, so-called because of the black melanin granules contained within it. The pigment epithelium serves as a light-trap absorbing light that leaks from the sides or ends of the outer segments, thus preventing the light from entering neighbouring receptors and blurring the image. The pigment epithelium has another important role in maintaining the metabolism of the receptor cells: it serves as a route for the transport of nutrients from the blood stream to the receptor, and carries out the dispersal of discarded receptor membranes.

Visual pigments

The major component of the visual pigment molecule is a protein of molecular weight about 40000. However the isolated protein is colourless, that is, it does not absorb visible light, and the colour results from union with a much smaller molecule called *retinal*, which is a polyene with an aldehyde functional group. The complex formed between the protein and retinal is called a 'rhodopsin' but since this term has various usages, the general classification of *visual pigment* is to be preferred. The visual pigments have high thermal stability –

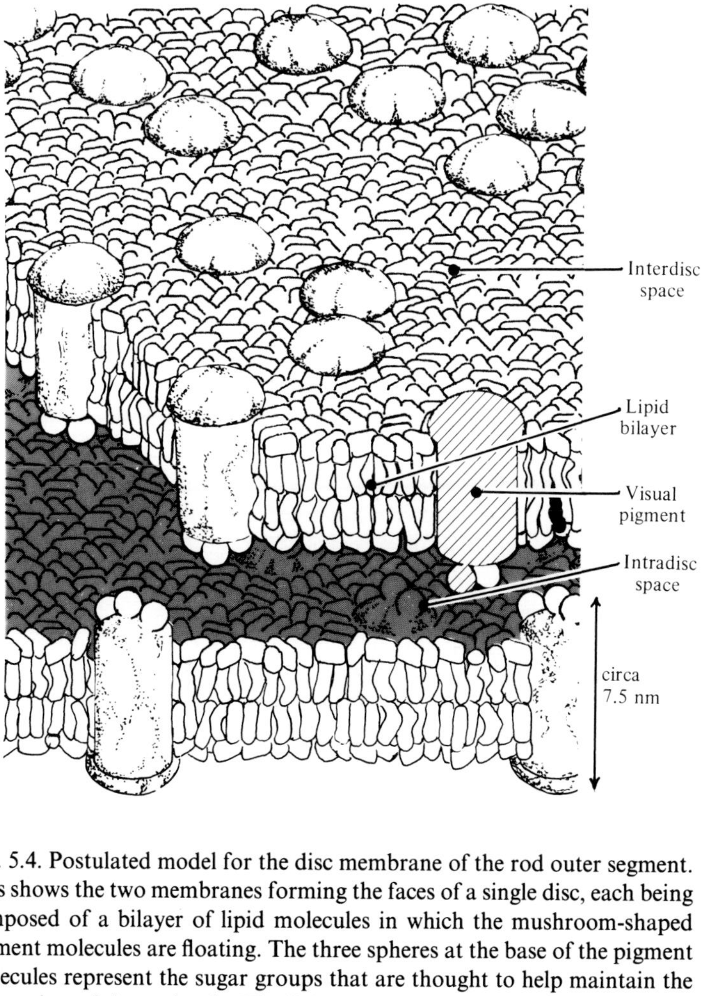

Fig. 5.4. Postulated model for the disc membrane of the rod outer segment. This shows the two membranes forming the faces of a single disc, each being composed of a bilayer of lipid molecules in which the mushroom-shaped pigment molecules are floating. The three spheres at the base of the pigment molecules represent the sugar groups that are thought to help maintain the orientation of the molecule. The lipid head-groups are close-packed in the membrane surfaces while their fatty acid tails form the hydrophobic core of the membrane. (E. A. Dratz, G. P. Miljanich & S. Schwartz, *personal communication*.)

this is essential not only because they are structural components of the receptor membrane, but because spontaneous breakdown of the molecule could cause the receptor to generate spurious signals. On the other hand, to be good receptor molecules, they must readily undergo some specific change on the absorption of light.

It was recognised 100 years ago that the retina of a dark-adapted eye looks pink when first exposed to light and then bleaches to a pale yellow colour; this is due to the photobleaching of the rod pigment. It was subsequently shown that free retinal can be extracted from the bleached pigment, demonstrating that the loss of colour is due to a weakening of the bond between protein and retinal. Wald later showed that this is probably the result of a change of shape of the chromophoric group. Polyenes can have a number of *stereoisomers*, that is, molecules with the same composition of atoms can have different physical properties owing to different spatial arrangements of the bonds between their atoms. Interconversion of such stereoisomers upon the absorption of light or UV radiation is a typical reaction of polyenes.

Wald showed that one stereoisomer of retinal, 11-*cis*-retinal, is found in native visual pigments, while photobleached pigments contain a different form, termed all-*trans*-retinal. The currently accepted mechanism of the photobleaching process is illustrated in Fig. 5.5: In the native pigment the curved 11-*cis*-retinal molecule fits into a specially shaped binding site in the protein. Absorption of a photon causes it to change to the linear all-*trans* isomer, which then springs out of the binding site. The interaction between chromophoric group and protein is thereby lost and the molecule loses its colour.

Regeneration of visual pigments. Colour can be restored to the photobleached pigment simply by the addition of fresh 11-*cis*-retinal. This will spontaneously enter the binding site to give a pigment indistinguishable from the original, thus confirming that it is indeed the 11-*cis* stereoisomer that is present in the native pigment. The presence of the 11-*cis*-retinal molecule in the binding site also confers stability on the protein, for it is readily degraded while the pigment is in the bleached state. This is borne out by studies by Dowling and Wald on retinol (Vitamin A)-deficient rats. Retinol is the normal dietary source of retinal and after prolonged retinol deprivation, the retinal content of the eyes of the rats was found to fall. There was a corresponding drop in the content of visual pigment, and also a breakdown of the receptor membranes, for without a supply of

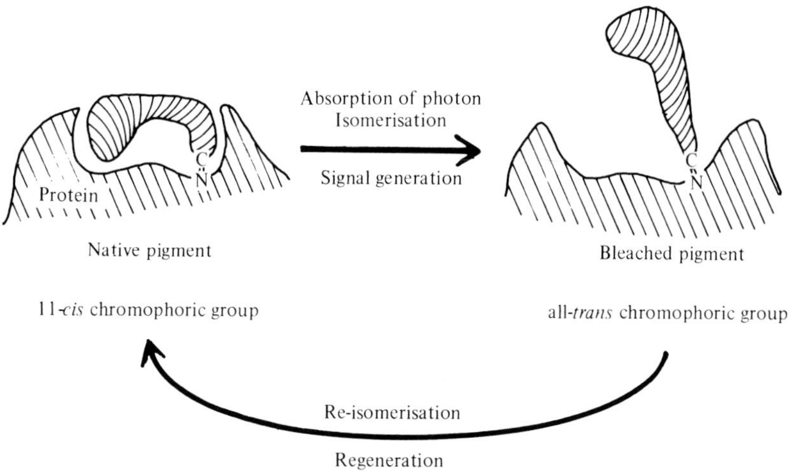

Fig. 5.5. Diagram showing how the light-induced isomerisation of the chromophoric group could cause a conformational change in the visual pigment protein. (Based on Wald, G. (1968) *Nature*, **219**, 800–7.)

retinal, bleached pigment molecules cannot regenerate and are liable to decompose.

It appears that there is to some extent a cycling of retinal within the retina:

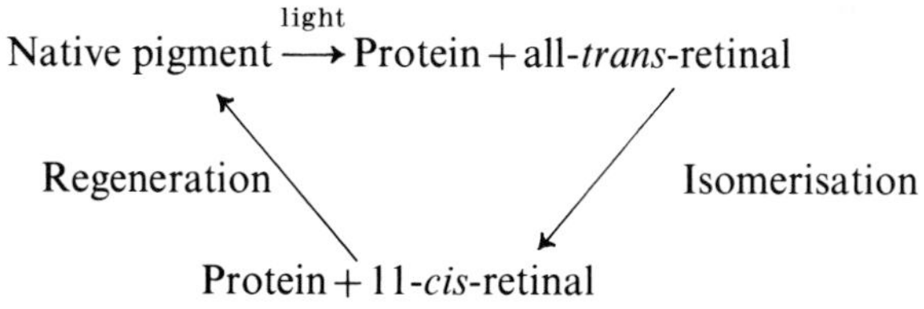

Despite a great deal of study, this process is not properly understood: the extent of recycling, the site and mechanism of isomerisation, and whether the retinal released from bleached pigment has to be converted to retinol before it can be re-used are questions that have still to be answered.

Renewal of receptor membranes. The technique of autoradiography permits individual protein molecules labelled with radioactive amino acids to be located in the receptor membrane. Richard Young and his co-workers have used this technique to study protein replacement in the receptors of a number of species, and have shown that it is a rapid, continuous process. In the case of rod outer segments, complete discs are formed very frequently; for example, in the Rhesus monkey a new disc is formed every 20 min, and since there are about 900 discs in the outer segment, the whole receptor is replaced every 12 days. Fig. 5.6 shows autoradiograms of a frog rod and cone 5 days after a single dose of radioactive amino acid. The newly synthesised protein at first formed a compact band at the base of the rod, and this has been displaced upwards by subsequent disc formation. In contrast, the new material in the cone is randomly distributed for the new material apparently diffuses throughout its continuous membrane.

While disc formation is a continuous process, the old material is shed from the distal ends of the rods in blocks of 10–20 discs, and this is a diurnal process taking place in the morning soon after wakening. These blocks of discs are called *phagosomes* and are slowly digested by the pigment epithelium, their components being returned to the bloodstream. The rate of replacement of rod material by the renewal process is so rapid that the rate of pigment renewal approaches the rate of its normal turnover by bleaching and regeneration in the visual process.

The renewal process may be important in retinal dystrophy. An inherited form seen in some strains of rat is characterised by a mass of undigested phagosomes that choke the pigment epithelium. It has been suggested that the basic disorder is a failure of the digestion process leading to a build-up of discarded material and the eventual disruption of receptors.

Light absorption by visual pigments

Light absorption is the essential property of the receptor molecule and, indeed, most of our knowledge of the visual pigments has been derived from this property. The absorption process can be visualised as the interaction of a quantum of visible light – a photon – with the electrons of a molecule, and results in the elevation of one electron into a discrete state of higher energy; the whole molecule is then said to be in an *excited state*. Whether or not a photon is absorbed is dependent upon the correspondence of its energy to the magnitude of this energy step. The energy of the photon is determined by its wavelength, while the size of the energy step is related to the composition of the molecule; hence the wavelength of maximum absorption varies between different types of molecule. A molecule

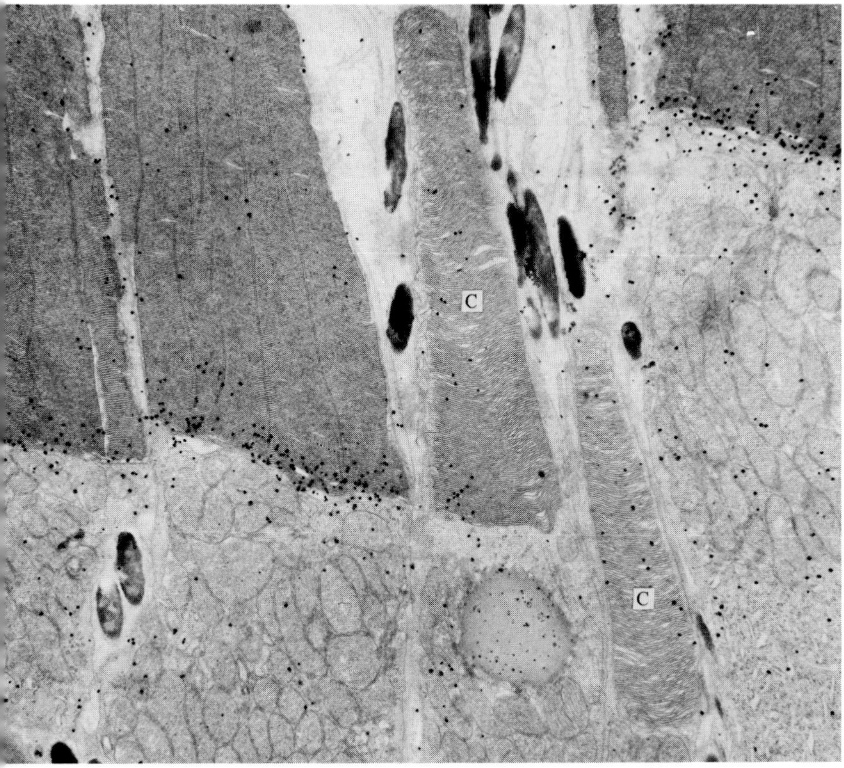

Fig. 5.6. Autoradiogram of cone and rod outer segments of a frog made 2 h after the injection of radioactive leucine. The black dots show the position of the resulting radioactive protein. In the rods these form a band at the site of disc formation at the base of the outer segment, while in the cones (c) they are diffusely distributed. (From Young, R. W. (1967). *Investigative Ophthalmology*, **15**, 700–25.)

can thus be characterised by the probability of its absorbing a photon of given wavelength, and a plot of this probability against wavelength is called an *absorption spectrum*; typical absorption spectra are given in Fig. 5.7. Curve 1 is the absorption spectrum of a solution of free retinal. The wavelength of the absorption peak (λ_{max}) is 390 nm, which is in the UV region. The tail of the curve extends into the visible; thus the solution absorbs violet light and so appears pale yellow. Combination with the visual pigment protein displaces the λ_{max} to 500 nm (Curve 2), which means that the molecule absorbs green light most strongly, and looks red in solution. Photobleaching

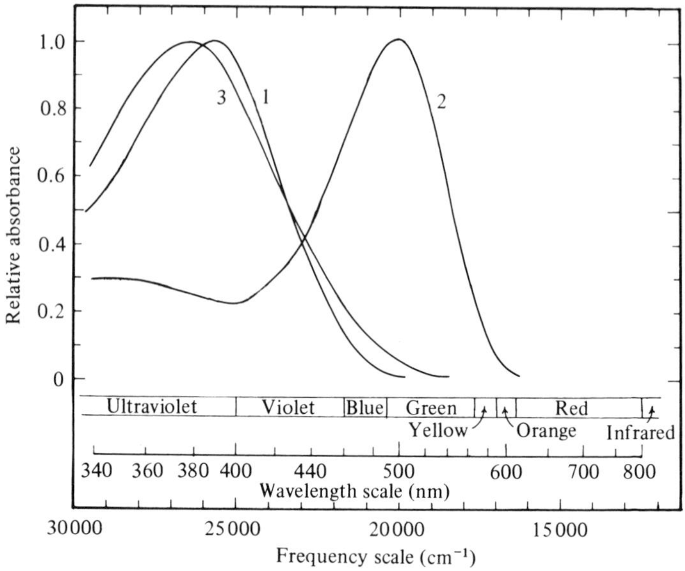

Fig. 5.7. Absorption spectra of: (1) retinal; (2) visual pigment from bovine rods; (3) the pigment bleached by exposure to light. The abscissa is linear in frequency (cm^{-1}) while a supplementary wavelength scale is given together with the subjective names of the various frequency bands.

the pigment displaces the peak back to the UV region (Curve 3), the spectrum of the photo-product resembling that of retinal.

The abscissa of Fig. 5.7 is provided with two scales, wavelength (λ) in units of nanometres (nm, i.e. 10^{-9} m), and frequency ($\bar{\nu}$) in units of 'wavenumbers', which is the number of waves per cm. The two quantities are related by $\bar{\nu} = 10^7/\lambda$ cm^{-1} and in the figure, the curve is plotted on a linear frequency scale. The reason for this is that the absorption spectra of the visual pigments from a large number of species have been found to have similar shapes if plotted on a frequency basis. This is very useful when comparing the relative absorptions of pigments from different species, or of different receptors in the same retina. However this rule is only a first approximation, and recent measurements on pigments with λ_{max} values ranging from 345 nm to 570 nm show that long-wave pigments have spectra narrower than the norm, while short-wave pigments have spectra that are broader than expected. Although several of the absorption spectra presented in this volume are linear in frequency, we shall follow current practice in giving the λ_{max} values of pigments in nanometres, and presenting the spectra with their highest wavelength (lowest frequency) at the right-hand side of the plot.

Quantum efficiency

The absorption of light and the photobleaching process can be studied more easily with pigments that have been extracted from the retina, and such experiments have shown that visual pigments follow established photochemical principles. Excitation of the molecule with visible light generates the same excited state irrespective of the wavelength, that is, the energy of the photon. The excited molecule is a reactive entity and the course of its reaction is also independent of the wavelength of light that generates it. Since any individual receptor cell is filled with identical pigment molecules, the response of the cell will be the summed response of the individual molecules, and must also be independent of wavelength. This means that an individual receptor cannot distinguish the wavelength of the incident light; this is the basis of the 'Principle of Univariance' which will be discussed in Chapter 9.

The principal reaction of the excited pigment molecule is the isomerisation of the chromophoric group, which leads to bleaching. If every quantum absorbed by the molecule resulted in bleaching, the *quantum efficiency* of the photobleaching reaction would be said to be unity. However measurements on isolated visual pigments have shown that their quantum efficiency, i.e. (Number of molecules bleached)/(Number of quanta absorbed), is 0.66. Thus about one-third of the excited molecules undergo a second kind of reaction, which is a simple deactivation process, the electronic excitation being converted to thermal energy and the molecule returning to its original condition. In general, the quantum efficiency of a reaction varies with wavelength, but the visual pigments are remarkable in that their quantum efficiency of bleaching – and therefore isomerisation – does not vary appreciably throughout the visible region.

As far as we know, these principles can be applied to the pigment molecule in the receptor outer segment. The chances of a single photon causing an isomerisation will therefore be determined by two factors: (i) the probability of its absorption which is determined by the absorption spectrum of the pigment and the length of the receptor, and (ii) the quantum efficiency of photobleaching. For light passing along a human rod outer segment, the fraction absorbed at the wavelength of maximum absorption (498 nm) is about 40%. Combining this with the quantum efficiency, the probability that a

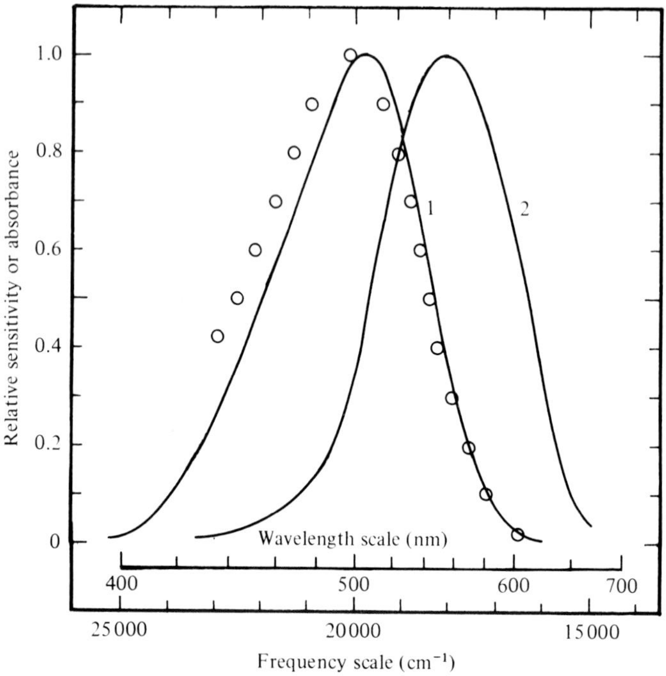

Fig. 5.8. Human spectral sensitivity curves. (1) scotopic sensitivity; (2) photopic sensitivity; Circles: Absorption spectrum of human rod pigment.

quantum passing down the rod will cause an isomerisation is $0.66 \times 40 = 26\%$. This figure applies only to 498 nm light; the probability is less at longer and shorter wavelengths in a manner corresponding – to a first approximation – to the absorption spectrum. The factors that limit the overall sensitivity of the system will be discussed in Chapter 7.

Pigment absorption and spectral sensitivity. Since the quantum efficiency does not vary with wavelength, the relative response of a receptor to different wavelengths is determined solely by the absorption spectrum of its pigment. This can be tested fairly readily for the human rod system, for at low light levels the cones do not interfere, probably because they have a high level of spontaneous activity which is not changed significantly by weak lights. Curve 1 of Fig. 5.8 shows the relative sensitivity of the eye under dim-light conditions, that is, the *scotopic sensitivity curve*. The absorption spectrum of the human rod pigment is shown as circles and it can be seen that while the curves are similar, they do not fit exactly. This is in part because the short-wave sensitivity of the eye is lower than expected from the absorption curve because of the absorption of shorter wavelengths by the pre-retinal media. A better fit can be obtained with the sensitivity curves of subjects whose lenses have been surgically removed.

Under bright conditions, the rod signals reach their maximum value (called 'saturation'), and the light becomes effective in raising the cone signals above their spontaneous rate. The resulting *photopic sensitivity curve* does not correspond to the absorption spectrum of any particular class of cone: it depends primarily on the combined responses of cones with peak sensitivities at 530 nm and 560 nm (see Chapter 9). Depending on the experimental technique used to derive the photopic sensitivity curve, the way in which the signals are combined may apparently vary and the violet-sensitive cones (λ_{max} = 420 nm) may sometimes make a small contribution. Thus the photopic curve does not have an absolutely fixed form.

The limits of the visible spectrum given in Fig. 5.7 are in no way absolute cut-off wavelengths and vary with the state of adaptation of the eye and the intensity of the source. The retina is able to respond to wavelengths outside this range, and the UV limit is due to absorption by the lens while the apparent long-wave limit is due to the falling absorption spectra of the visual pigments: these probably extend beyond 1000 nm, although by this wavelength the absorption is extremely small. A fully dark-adapted eye can see intense sources beyond 800 nm, though a point is reached at which the sensation of heat from the skin overwhelms the sensation of light from the eye.

Signal generation

The absorption of light, the isomerisation of the chromophoric group and the conformational change in the pigment molecule are reasonably well-established concepts. The next step is to explain how a single molecular event amongst the 10^7–10^8 pigment molecules in the outer segment of a rod can change the release of transmitter at the remote base of the cell. An early proposal based on the measurements of potential described in Chapter 6 was that absorption of photons released Ca^{2+} ions which migrated to the outer membrane of the cell where they blocked the entry of Na^+ ions. The discovery of a light activated enzyme system in the rod outer segment now leads to the following well-supported alternative explanation.

Absorption of photons initiates a cascade of reactions that activate a phosphodiesterase; this hydrolyses cyclic guanosine monophosphate (c-GMP) to 5′-GMP; the system amplifies, so that a single photon absorption results in the hydrolysis of as many as 10^5 molecules of c-GMP. Cyclic GMP holds open the sodium channels, so its reduction allows the channels to close, preventing sodium entry and causing hyperpolarisation of the whole cell and a reduction in transmitter release. The most recent evidence suggests, however, that there is also a role for Ca^{2+} ions. Calcium normally enters along with Na^+ through the sodium channels, but calcium is continuously extruded from the outer segments by a Na/Ca exchange pump; during exposure to light calcium entry is blocked along with sodium, but calcium continues to be extruded so its concentration in the cell declines. This decline in calcium speeds up an enzyme, guanate cyclase, whose action restores the c-GMP concentration. This negative feedback is thought to underlie light adaptation in the photoreceptors.

Neurotransmitters in the retina

The neural layers of the retina contain a number of different types of cell connected with one another and with the receptor cells by a variety of different kinds of synaptic contact (see Fig. 2.4). Some of these have a structure and electrical behaviour similar to those in the brain, while others are peculiar to the retina and consequently a variety of transmitter substances are involved.

There is good evidence that acetylcholine, dopamine and γ-amino-butyric acid (GABA) are retinal neurotransmitter substances, and are probably released by different sub-populations of amacrine cells. There is as yet no clue to the transmitter released from bipolar cells, but recent evidence suggests that the transmitter of the receptor synapse is aspartate. Other neurotransmitters such as taurine, glutamate, glycine and 5-hydroxytryptamine (serotonin) have been suggested as transmitter candidates.

5.3. SUGGESTIONS FOR FURTHER READING

General reviews

Adler's Physiology of the Eye. 7th edn (1981) ed. R. A. Moses. St Louis: C. V. Mosby Co.

Knowles, A. & Dartnall, H. J. A. (1977) *The Photobiology of Vision.* In *The Eye*, 2nd edn, vol. 2b, ed. H. Davson. New York and London: Academic Press.

Fein, A. & Szuts, E. Z. (1982) *Photoreceptors: Their Role in Vision.* Cambridge: Cambridge University Press.

Scientific Foundations of Ophthalmology (1977) ed. E. S. Perkins & D. W. Hill. London: Heinemann Medical Books.

Special references

Receptor structure and renewal. Young, R. W. (1978) *Vision Research*, **18**, 573–8.

Corneal structure and metabolism Klintworth, G. K. (1977) *American Journal of Pathology*, **89**, 719–808.

Transparency of lens and cornea. Benedek, G. B. (1971) *Applied Optics*, **10**, 459–73.

Lens biochemistry. Bloemendal, H. (ed.) (1981) *Molecular and cellular biology of the eye lens.* New York: Wiley.

Lens crystallins and cataract. Zigman, S. (1971) *Science*, **171**, 807–9; Stevens, V. J. *et al.* (1978) *Proceedings of the National Academy of Sciences, USA*, **75**, 2918–22.

Scotopic and photopic sensitivity curves. Wyszecki, G. & Stiles, W. S. (1967) *Color Science*, New York: Wiley.

Rhodopsin molecular structure. Hargrave, P. A. *et al.* (1983) *Biophysics of Structure and Mechanism*, **9**, 235–44.

Quantum efficiency of rhodopsin bleaching. Dartnall, H. J. A. (1968) *Vision Research*, **8**, 339–58.

Neurotransmitters in the retina. Thomas, T. N. & Redburn, D. A. (1979) *Experimental Eye Research*, **28**, 55–61; Lam, D. M. K. *et al.* (1980) *Neurochemistry*, **1**, 183–91.

Transduction mechanism. Pugh, E. N. & Cobbs, W. H. (1986) *Vision Research,* **26**, 1613–43. Stryer, L. (1986) *Annual Review of Neuroscience,* **9**, 87–119.

Light adaptation and calcium. E. Pugh & J. Altman (1988) *Nature*, **334**, 16–17.

CHAPTER SIX

Physiology of the retina

H. B. BARLOW

Much of the classical research on human visual performance was intended to elucidate physiological mechanisms of the retina, and in many cases indirect conclusions drawn from such research have been strikingly confirmed by more recent direct experimental investigations. But this direct approach has also revealed important new facts that were previously unsuspected, and the aim of this chapter is to give a brief description of retinal physiology emphasising these new aspects. Details that fit in well with previous knowledge will be found elsewhere, mainly in Chapter 2 for anatomy, Chapter 5 for biochemistry, and Chapters 7 to 9 where psychophysical measurements of performance are covered systematically.

6.1. DYNAMIC RANGE OF RETINAL INPUT AND OUTPUT

The retina is a thin sheet of photoreceptors and nerve cells lining the back of the eye where the image is formed (Figs. 2.2, 2.3, 2.4). It is obvious that its functional role is to encode the image falling on the retina as a pattern of nerve impulses in order to transmit the picture up the optic nerve to the brain, but a few words about the natural difficulties of this task may enable us to understand better the more detailed account that follows.

The total number of quanta that enter the pupil and form the retinal image varies from about 100 s^{-1} on a dark night where the light is just sufficient to help us orient ourselves, up to some 10^{14} s^{-1} on a sunny beach in summer when the light is uncomfortably strong. Nerve fibres are wretched channels to signal a quantity that varies over this range because their impulse frequency can only go from zero to about 1000 s^{-1}, so only a dozen or so distinguishable levels of activity are possible within the fraction of a second allowable if information is to be provided promptly. An electronic engineer uses differential inputs and zero offsets, or special devices such as automatic gain controls and contrast controls, to ensure that the meaningful aspects of the image are preserved in the electrical signal.

The very restricted dynamic range of nerve fibres compared with electrical transmission channels suggests strongly that equivalent functions are necessary in the retina. It is worthwhile bearing in mind the problem of matching the dynamic range of visual signals to the limited capacity of nerve fibres when reading about retinal physiology.

6.2. PHOTORECEPTORS: RODS AND CONES

The photoreceptor cells form the outermost layer of the retina, furthest from the vitreous and lens, and they are depicted in Figs. 2.3 and 2.4. Their outer segments, which contain the photosensitive pigment, have been discussed in Chapter 5.

Of the two main classes of receptor, rods are far more numerous in the human eye (over one hundred million in each) and also more sensitive; they are distributed widely over the retina but are absent in a region subtending about 1° in the centre of the fovea where powers of visual discrimination are greatest. They contain *rhodopsin* which is the photosensitive pigment that has been used for the vast majority of biochemical studies (see Chapter 5); its peak absorption occurs for blue-green light of wavelength 500 nm. Individual rods are linked electrically with their neighbours so that each of them signals the average illumination over a small region of the retina that overlaps the regions of its neighbours. Their response is relatively slow, which enables the results of quantum absorptions to accumulate for about 100 ms.

Rods are responsible for vision under conditions of poor illumination, roughly from bright moonlight downwards. The properties of *scotopic* vision, as it is called, result mainly from the facts about rods mentioned above. Cones do not respond at these low levels, so one is totally dependent upon the rods, but because of their interconnections they only provide poor spatial resolution compared to the cones, and because of their slow response they can only detect flicker up to about 12 Hz. All rods have the same spectral sensitivity, so there is no colour vision, and because they are absent from the central fovea and are most densely packed about 3 mm away from it, scotopic vision is actually best at 10°–20° eccentricity from the point of regard; at low light levels you can often see something better by looking slightly to one side of it, rather than directly at it.

There are far fewer cones, only about six million in each human eye, but they are much more important in ordinary daylight or good artificial lighting. They are very tightly packed in the central fovea, where there are 150000 mm^{-2}, and in this region each cone probably has an individual ganglion cell and nerve fibre connecting it to the LGN and thence to the cortex. However it does not follow from this that there are no lateral connections through horizontal cells and amacrines giving the spatial and chromatic opponent aspects of the ganglion cell receptive fields described below. The time course of the cone response to a flash of light is quicker than for rods. There are three different types containing photosensitive pigments with different peak sensitivities; a red-sensitive one peaking at 560 nm, a green-sensitive one peaking at 530 nm, and a blue-sensitive one peaking at 420 nm. These blue-sensitive cones are curious in having substantially different properties from the other two types (see Chapter 9).

Vision using cones is called *photopic*, and the properties of such vision might be deduced from those of cone receptors. Spatial resolution is high, especially in the fovea, temporal resolution is improved by their quick response (see Fig. 8.12), and flicker can be seen up to about 55 Hz. Because their individual spectral sensitivities differ, colours as well as intensities can be signalled by the cone system.

The evidence for much that has been described so far comes from direct and rather simple observations. For instance the rod–cone 'duplicity' of retinal function was first enunciated by the German microscopist Max Schultze in 1866 on the basis of comparative studies of retinal histology in many animals; he found that the more nocturnal the species, the greater the preponderance of rods, and he concluded that rods were receptors specialised for vision at low luminances. The demonstration that psychophysical measurements of human visual functions usually show distinguishable scotopic and photopic regions corresponding to rod and cone functions was mainly accomplished by Selig Hecht in New York some 40 years ago; some measurements of this type will be found in the next section. The basic reason for the duplicity of the visual system is not altogether clear, but the hypothesis that cone pigments are less stable than rhodopsin, and therefore have a higher level of unavoidable intrinsic noise, would make sense of many of the structural and functional differences (see Chapter 7).

6.3. THE ACTION OF LIGHT ON PHOTORECEPTORS

In the last decade or so the biophysical mechanisms whereby the absorption of light activates the system have been intensively studied by measuring the flow of electrical currents around the receptors in light and darkness, and by measuring intracellular potential changes with very fine glass microelectrodes. Fig. 6.1 shows how the rods are thought to react to the absorption of light. It is found that, in darkness, a strong, steady current flows out of the inner segments of the rods towards their outer segments, and then in through the outer segment membranes to return to the inner segment intracellularly. The inner segment contains many mitochondria, and the resting current that flows in darkness is thought to result from a metabolically driven sodium ion pump similar to that of other cells. The expelled sodium ions are replaced by potassium ions to which the membranes are relatively freely permeable, and the excess internal concentration of potassium ions is accompanied by the well-known negative intracellular resting potential, which would be close to the electrochemical equilibrium potential for potassium ions if there was no resting current of sodium ions. However, unlike nerve cells, the membrane of the outer segment of rods in its unexcited state in darkness is quite freely permeable to sodium ions, so these pass down their large electrochemical gradient and cause the current shown flowing round the oval circuit in Fig. 6.1. This current short circuits the resting potential and reduces it from its usual value of more than -60 mV to about -25 to -30 mV. Now when light is absorbed this short-circuit current is found to be reduced, with the result that the negative intracellular potential increases towards the value typical for other neurons in their resting state. Records of intracellular responses to three different intensities of flash are shown in Fig. 6.1. Notice that most sensory receptors are *depolarised* when stimulated, whereas photoreceptors become *hyperpolarised* (their negative intracellular potential becomes more negative) when light is absorbed.

This hypothesis about the mechanism of rod activation was initially based on measurements of the potential difference between a pair of extracellular electrodes with their tips at slightly different depths in the retina. The flow of extracellular current was derived from these and has been confirmed by the much more sensitive method shown in Fig. 6.2: the outer segment of a rod is gently sucked into a glass micropipette so that all the current entering it can be

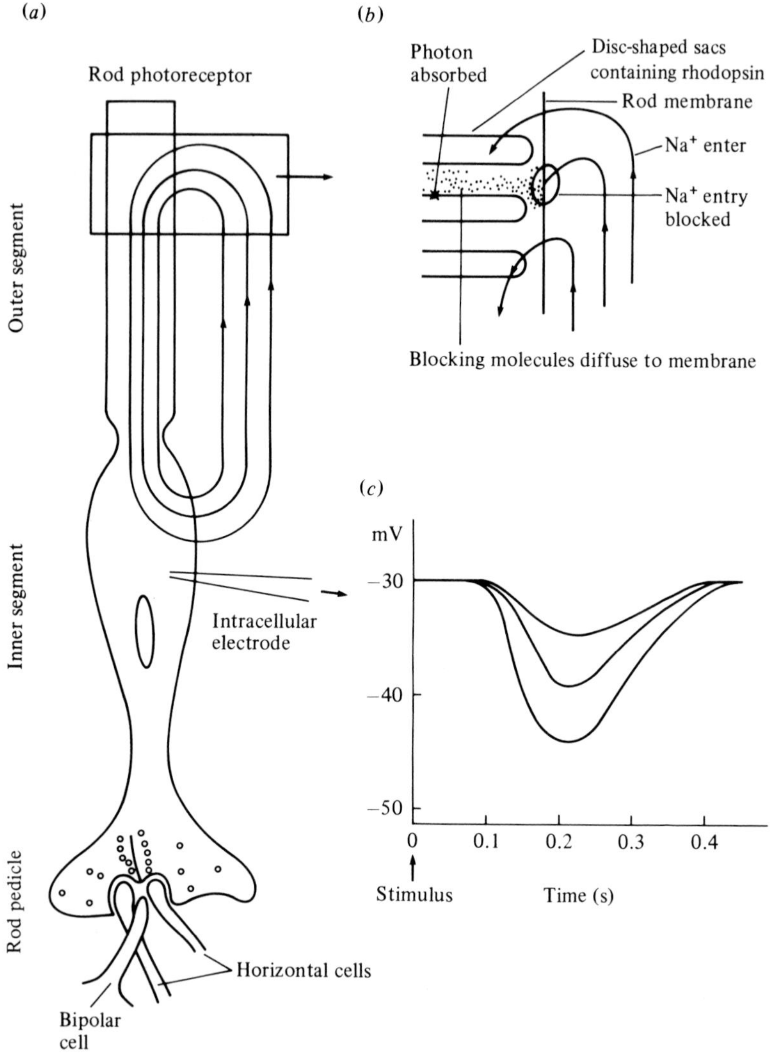

Fig. 6.1. (*a*) Diagram showing how the absorption of light is thought to excite a rod. A metabolic pump in the inner segment extrudes sodium ions; these are free to enter the outer segment, whose membrane has a high conductance for sodium ions in its resting state in darkness. (*b*) When photons are absorbed this high conductance is transiently decreased, perhaps as a result of 'blocking molecules' released at the site of absorption of the photons, though other mechanisms have been proposed (see Chapter 5). The reduction in the entry of sodium ions causes the interior to become more negative leading to the waveforms (*c*) recorded through an intracellular

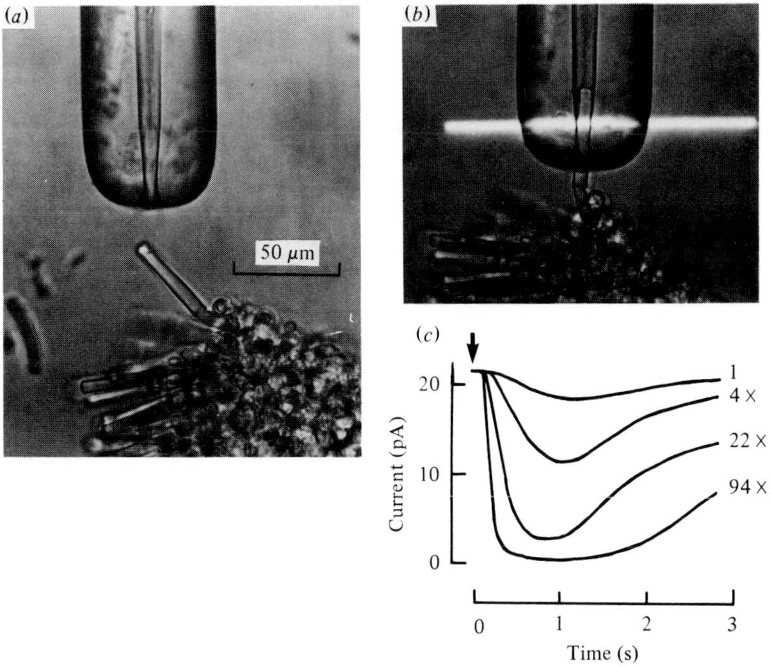

Fig. 6.2. The experiment shown here confirms that the hyperpolarising responses of rods result from blockage of the flow of current into their outer segments. (*a*) The outer segment of a toad rod, which is much larger than a mammalian rod (see 50 μm scale), is sucked into a glass micropipette so that it forms a tight seal at the tip. Almost all the current entering the outer segment has to be supplied through the pipette and can thus be measured. (*b*) When stimulated by a flash of light placed as shown at top right and delivered at the time indicated by the arrow in (*c*), the current is reduced from its resting value of about 20 pA. The reduction is graded with intensity as shown by the four responses illustrated; the strongest intensity is 94 times the weakest, and the entering current is reduced almost to zero at the flattened peak of this response. Much smaller responses than the smallest shown here can be detected, and in fact the sensitivity of the method is such that the responses to the absorption of single photons can be detected (see Fig. 7.8). (After Baylor, Lamb & Yau.)

electrode for three different intensities of flash stimulus. Note that the response to a light stimulus is a graded hyperpolarisation; most sensory receptors respond to stimulation by depolarising. The responses of mammalian receptors at 37 °C are faster than those shown here. (After Penn & Hagins and others.)

measured as it passes into the pipette (except for a small fraction leaking through the seal between the outer segment and the pipette tip). When flashes of light are delivered to the part of the outer segment above the seal there is a reduction of this entering current, and the time courses of these responses for four intensities of flash are shown in Fig. 6.2 (see also Fig. 7.8). Views on the mechanism whereby light blocks the entering current of sodium ions have changed radically over the past decade and are outlined at the end of Chapter 5: in darkness c-GMP holds the sodium channels open, and light decreases c-GMP concentration by activating the hydrolytic enzyme phosphodiesterase through a cascade of other substances. Calcium ions, previously thought to act as an intracellular transmitter closing the sodium channels, are now thought to decrease in concentration during light and thereby mediate light adaptation.

The increased intracellular negativity resulting from the blockage of the current entering outer segments spreads electrotonically throughout the receptor cell and reaches the inner segment and pedicle, which synapses with bipolar cells and horizontal cells. Intracellular recordings from receptors in non-mammals have revealed another unsuspected detail. The rods do not function in isolation, but as already mentioned they are linked to each other by electrical synapses. The result of this arrangement is to pool the excitation reaching a number of receptors, so that the change in intracellular potential of each rod does not depend solely upon the number of quanta absorbed in that particular rod, but rather upon the average number absorbed in the local population of rods, thereby reducing the 'noisiness' of the messages transmitted and making better use of the dynamic range available at the receptor–bipolar synapses.

The pipette technique has been successfully applied to primate and human rods and cones, thereby confirming that they behave in most respects like the non-mammalian receptors upon which most of the pioneering work had been done. There are some important differences, for instance in the fact that the rods of primates do not adapt like those of toads and salamanders, and the method has made possible greatly increased accuracy in the determination of spectral sensitivities.

6.4 HORIZONTAL AND BIPOLAR CELLS

When the photoreceptors absorb more light the hyperpolarisation of their terminals is thought to decrease the release of a chemical transmitter whose identity is so far uncertain. The action on the postsynaptic cells is variable; the intracellular potential of horizontal cells always moves in the same direction as the receptors (at least in the case of those connected to rods) and therefore the chemical transmitter must have a depolarising action on them. Some of the bipolar cells also hyperpolarise when light is absorbed by the underlying receptors, but other bipolars are depolarised under the same circumstances; for these bipolars the receptor transmitter must cause hyperpolarisation. Fig. 6.3 shows these responses together with a diagram of the synaptic organisation of a vertebrate retina. This figure also indicates the action of the horizontal cells. They are connected to receptors over a much larger area than that within which rods are connected to each other, and they are hyperpolarised when light is absorbed. Their action is to modulate the connection between rods and bipolars, so they carry a signal from rods at a distance from a bipolar cell which opposes that of the receptors immediately underlying it; if one set hyperpolarises the bipolar, the other set depolarises, and vice versa.

There are two particularly interesting points about the organisation and function of the retina described above. The first is that it provides an example of nerve cells interacting with each other by means of graded depolarisations and hyperpolarisations, rather than through all-or-none impulses arriving at synaptic terminals. Previous to this work on the retina it had usually been taken for granted that the input to a synapse was an impulse, and that the output from the postsynaptic neuron was also in the form of impulses propagated down its axon, though it was recognised that the intracellular potential of the postsynaptic neuron represented the graded combination of all its synaptic inputs. Now it is clear that neurons can also interact by graded potentials, and examples are being found elsewhere in nervous systems. Interactions over distances greater than a few millimetres must be mediated by impulses, but over shorter distances graded interaction may prove to be the rule rather than the exception.

The second point of interest is that the role of the horizontal cells appears to be to mediate *lateral inhibition*, which is found very commonly in all sensory pathways. Its action in the spatial domain may be likened to that of the adaptation of sense organs in the temporal domain; both of these mechanisms emphasise change and discontinuity at the expense of uniformity in space or constancy in time. In vision, lateral inhibition plays a part in bringing about simultaneous contrast effects and also the reduction in sensitivity to low spatial frequencies illustrated in Fig. 8.1 and described in that chapter. Though it has only recently been demonstrated that hori-

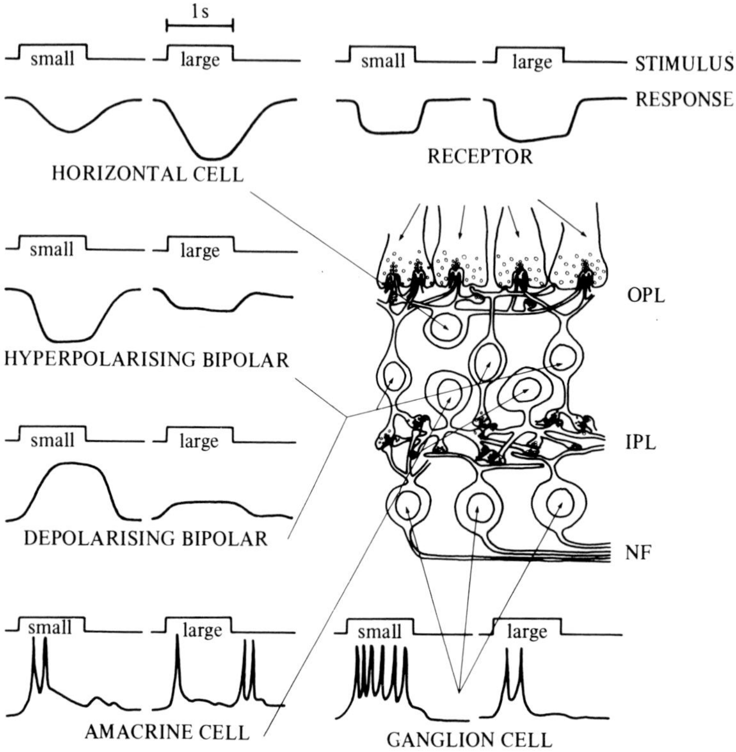

Fig. 6.3. Diagram of the synaptic interconnections in the vertebrate retina based on electron microscopy, and examples of the potentials that can be recorded intracellularly from the different types of cell. For each cell, responses are shown for two types of stimuli, each lasting 1 s as shown by the stimulus traces. The first stimulus is a small spot of light focussed on the centre of the region of retina to which the cell responds. The second is for a large spot which stimulates a wider region of retina. For a receptor the large spot produces a slightly larger hyperpolarisation (downward deflection), because of the lateral interconnections with other receptors, and also because of any stray light that may be present. For horizontal cells this effect is even bigger, because they pick up from an extensive population of rods. But for both types of bipolar cell the large spot produces a smaller response than the small spot; this is because the horizontal cells oppose the action of receptors on bipolars, this taking place in the complex synaptic interconnections occurring in the outer plexiform layer (OPL). The responses so far have all been graded hyper- or depolarisations, but the amacrines and ganglion cells show spikes. The former are characteristically bursts at 'on' and 'off'. The ganglion cell shown here gives a well-sustained burst of spikes

zontal cells are responsible, studies of visual performance showed more than a hundred years ago that such a mechanism must exist.

Some horizontal cells are connected with cones, and they are often connected to two different types of cone in an opponent fashion so that red light causes hyperpolarisation while green light depolarises, or vice versa. The details of such interconnections are not known in man, or any other mammal, but they have been studied in fish.

6.5. AMACRINE AND GANGLION CELLS

So far we have traced the results of light absorption in the receptors only through the outer synaptic, or plexiform, layer to the second order neuron, the bipolar cell. These connect to *amacrine* and *ganglion cells* in the inner plexiform layer (Fig. 2.4), and it is only at this level that impulses are recorded. The ganglion cells generate the streams of action potentials that pass up the optic nerve fibres to the brain. The role of amacrine cells, which also generate spikes, is uncertain but they are probably concerned with the pattern selectivity of ganglion cells. Electron microscopy and intracellular recording have given a preliminary view of these mechanisms, but the account is so far incomplete.

Recordings from optic-nerve fibres have provided some examples of what the mechanisms achieve, and a brief account has been given in Chapter 1. The first point to realise is the great diversity of types of ganglion cell. It was known that this must be the case from the histological studies of the retina by Raman y Cajal in the last century, but it is only in the last few decades that a corresponding diversity of physiological function has been uncovered. Some ganglion cells are amazingly selective in what they will respond to, and only send impulses centrally when a highly specific pattern of stimulation is present in the image falling on the receptors to which they connect (see Fig. 1.6).

It is not certain what ganglion cells the retina of man contains, but the common types found in monkey are as follows. The majority have

which would be transmitted up the optic nerve fibres (NF) to the LGN. Note that there are many other types of ganglion cells (see text and Figs. 1.6 and 1.7); the very complex interactions in the inner plexiform layer (IPL) are probably involved in achieving selectivity for particular trigger features, such as direction of motion. Although the retina is simple compared with many parts of the nervous system, it should not be thought that this diagram covers all types of retinal neuron. (After Dowling, Werblin and others.)

small receptive fields whose centre is predominantly sensitive either to red or to green, with their surrounds sensitive to the complementary colour. Furthermore these surrounds respond at the opposite phase of illumination to the centre and counteract excitation of the centre. Thus if, for instance, the centre is excited by red light when turned on, a response will be elicited from the surround when green light is turned off. Since there are four possible centre types (red, green, and 'on' or 'off' for each) and four surround types, there are sixteen (4 × 4) possible arrangements, but only four of these are found, namely those for which the surround is opposite to the centre both in spectral type (red or green) and in the phase to which it responds. The mechanism through which this double antagonism is achieved is not certain, but it probably involves the horizontal cells whose contribution to centre-surround antagonism has already been mentioned.

Ganglion cells of this type seem to signal a combination of spatial and colour information. Take a red, on-centre unit with its green, off-surround, for instance. It will not respond to a large, uniform white spot of light, for this excites both centre and surround and these annul each other. But if the large spot is tinted red, the 'on' mechanism will predominate and a response will occur when it is turned on, whereas if it is tinted green a response will occur when it is turned off. However if a white spot is small so that it only fills the centre, then it will elicit responses at 'on' like a large red spot, whereas a white annulus will behave like a uniform green spot and elicit a response at 'off'. The ambiguity in the meaning of impulses from one cell is decreased when the responses of other types are taken into account, and this must be performed centrally by the organised combination of signals from the four different types of unit.

Some colour opponent cells are encountered which lack the spatial opponent feature. The receptive fields of this type are larger than for the other type, and the fields for the two opponent colour mechanisms are the same size. The consequence is that such units cannot signal fine spatial detail, but respond to changes of tint of the light.

In addition to the red–green opponent units, blue–yellow units are found. For these the blue mechanism is derived from the blue-sensitive cones, just as red and green mechanisms are derived from their respective cones in the commoner type of colour-opponent unit. The yellow mechanism, however, appears to be formed by combining signals from red and green cones.

Most of the colour-opponent units have spatial opponency, but as mentioned some do not. There are also units with spatial opponency but no colour opponency. These have rather large receptive fields with an 'on' centre and 'off' surround, or vice versa. They have another characteristic in that they tend to give brief, poorly sustained, responses to continued stimulation and they are probably analogous to the 'transient' or Y-type cells of cats (Chapters 1 and 8). The exact significance of this system is not understood; possibly they should be regarded as elements that specialise in high temporal resolution and signalling movement while other units with small receptive fields specialise in high spatial resolution.

One of the lessons neurophysiologists have learned is that electrodes do not sample from all neural types in an unbiassed fashion. We know that the types so far described occur in monkey retina, but it is probably not a complete list of the types likely to be present, so it is quite possible that the human retina also has ganglion cells selective for direction of motion and other specific spatiotemporal patterns.

The foregoing brief account of the retina tells us something about its mechanism, but it does not tell us how well it works. The best way to find out what information the retina successfully extracts from the image and transmits to the brain is to measure the overall performance of the visual system, and the next chapters are devoted to this problem.

6.6. SUGGESTIONS FOR FURTHER READING

General references

Dowling, J. E. (1987). *The Retina.* Harvard University Press. (An account of the histology and organisation of the retina by the person who discovered much of it.)

Baylor, D. A. (1987). Photoreceptor signals and vision. Proctor lecture. *Investigative Ophthalmology and Visual Science,* **28**, 34–49. (Nice summary of the photoreceptor work.)

Rodieck, R. W. (1988) In *Comparative Primate Biology*, Vol. 4, pp. 203–78, ed. H. D. Stecklis & J. Erwin. New York: Alan R. Liss (Detailed review of the primate retina.)

Special references

Automatic control systems in retina. Werblin, F. W. (1973) *Scientific American*, **228**, no. 1, 70.

Rod and cone function. Hecht, S. (1937) *Physiological Reviews*, **17**, 239.

Electrical response of photoreceptors. Lamb, T. D. (1984) *Recent advances in Physiology*, ed. P. F. Baker, **10**, 29–65. Edinburgh: Churchill Livingstone.

Types of ganglion cell. De Monasterio, F. M. & Gouras, P. (1975) *Journal of Physiology*, **251**, 167. Shapley, R. & Perry, V. H. (1986) *Trends in Neuroscience*, **9**, 229–35.

CHAPTER SEVEN

Psychophysical measurements of visual performance

H. B. BARLOW AND J. D. MOLLON

When you use your eyes in everyday life what you 'see' is, in one sense, the physical reality of the world around you: the table, book, cat, and so on. But for the sensory physiologist what you 'see' is something very different: it is the pattern of nerve activity, occurring at some level in the brain, that enables you to recognise the table and place your cup on it, to read the print in the book, or to reach out and stroke the cat. What level is involved in this conscious perception is quite unknown, and it is probably over-simple even to talk about it as 'a level in the brain', but what is abundantly clear from the chapters in this book is that a very intricate, very cunningly evolved, sequence of steps intervenes between the physical reality in the world about us and the patterns of nervous activity by which we experience the sight of that reality. The next sections describe the results of measurements that have been made to explore how well this sequence of steps works. How much light do we need to see? How sharp is the picture given by our eyes? How quickly can changes be followed? What is the range of our colour discriminations?

The direct, quantitative study of our sensory performance is called *psychophysics*. Psychophysical measurements of vision are important for two practical purposes: first, one must know how well the normal eye works in order to assess whether an individual is suffering from defective vision; and second, one needs to be able to give the engineer practical guidance about many everyday problems. How much light is needed to read a traffic sign, and how big must the letters be to read it from a certain distance; is the flickering of a light visible when it is very dim; how would a particular type of fluorescent lighting affect the anaesthetist's ability to detect cyanosis in an operating theatre?

But psychophysical measurements are also indispensable theoretically; and indeed they have historically been the primary source of our knowledge of sensory mechanisms. By systematically manipulating the input to the system and by restricting the range of permitted responses, the psychophysicist can learn much about the logic of the mechanisms within the black box without lifting its lid. Thus psychophysical experiments on the mixture of colours (Chapter 9) allowed the three types of cone photoreceptor to be postulated as 'hypothetical constructs' many years before it was possible to identify them by direct measurements, rather as the gene existed as a hypothetical construct before it could be identified with a particular segment of a DNA molecule.

Despite the glamorous successes of single-unit electrophysiology in the last three decades, psychophysics remains a powerful and complementary source of information. The reason may best be understood by considering the limitations of single-unit electrophysiology. Firstly, the electrophysiologist can never be sure that the electrical response he records is part of the process of seeing: it is easy enough to observe changes in a complex, reactive tissue like the retina when it is excited by light, but it is quite a different matter to establish that the changes you are recording are functional links between physical reality and our perception of it. Secondly, the electrophysiologist can at best record from only a small number of units concurrently and cannot know whether the activity of a given cell has an absolute meaning for later stages of the system or whether its meaning depends on the state of other members of a set of cells. Thirdly, differences between types of cell in morphology and in vulnerability to anaesthetics mean that electrophysiology often yields a badly biassed sample of the full population of cells. The late William Rushton, a distinguished physiologist whose own psychophysical experiments were as elegant as his humour was wry, once likened the electrophysiologist to a person who wishes to discover the foreign policy of a central European country and who therefore crosses the frontier on a dark night, enters a border town and asks random passers-by for their opinion. Psychophysical measurements are important in telling the electrophysiologist what his mechanisms must achieve, how the stimulus should be manipulated and what ought to be the stimulus–response properties of a particular mechanism. The sections on vision that follow are organised according to the *functions* of the visual system: we draw on both psychophysics and electrophysiology (and also anatomy) as suits our local purpose.

The vividness of our visual sensations, and the fact that they almost always prove to be reliable guides to action, tend to make us believe we understand vision and do not need to be told how it works. But although one can gain some knowledge of perception simply by paying attention to one's own subjective experiences as they occur in a varying world, this source of knowledge is treacherous and cannot be very penetrating. A well-functioning television set usually reproduces faithfully what is before the cameras in the studio, but you are at the mercy of the studio engineers, who can easily deceive you if they wish, and anyway you cannot tell much about the working of television simply by looking at the picture.

The link between the receptors and the brain is at least as complicated as television and one must remember that in this case there is an additional step. The television picture is simply transmitted from studio to receiver, but our visual system not only transmits the picture on the retina, it also interprets it; our conscious perception is only the final product of these processes. The interpretative process involves memory and inference, and is consequently much harder to understand than transmission, but since our perceptions depend on these hazily understood operations we have to very careful how we use them to try to analyse sense organs. Figs. 7.1 and 7.2 reinforce this point.

7.1. MATCHES AND THRESHOLDS

The trick that has served well to show up the properties of the peripheral links in the sensory chain is to ask only the simplest possible questions of these more complex interpretative mechanisms. Thus the experimental subject might be asked to judge whether two stimuli can be distinguished from each other or not; if they appear the same in all respects they are said to *match*, and much valuable information, particularly about colour vision, has been obtained by finding what sets of lights, differing in their physical composition, are nonetheless indistinguishable to the observer's eye. Incidentally, simplifying the task of interpretation is a useful trick for practical purposes, because it leads to more accurate and reliable judgements: if you doubt whether a screw is the right size you will compare it directly with one known to be correct. In clinical medicine there are many such opportunities to improve the reliability of one's observations by substituting an easy direct comparison for a difficult absolute judgement.

Observations of *threshold* are only one step more complicated than matches. In this case the value of a physical parameter of the stimulus, such as its intensity, is adjusted until the subject can distinguish whether the stimulus is present or not with a specified rate of success over a number of trials. More complicated judgements

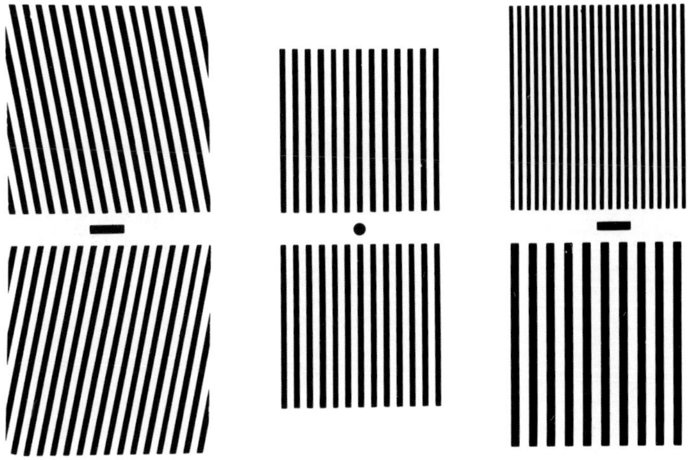

Fig. 7.1. Place the figure at a distance of about 50 cm and fix your eye on the black strip between the tilted gratings (on the left) for at least 30 s. Then transfer your gaze to the round dot between the vertical gratings. Each vertical grating should then briefly appear tilted in a direction opposite to that of the corresponding adapting grating. This effect shows that half a minute's experience of tilted lines can seriously disturb your ability to judge orientation.

Now try looking at the black strip between the right-hand patterns for a minute or so. (Move your eye about a little so that you don't fixate the same point on the strip throughout this adaptation period.) When you now look at the central spot, the lines of the upper grating should appear more coarsely spaced and those of the lower grating, more finely spaced. (A related after-effect is demonstrated in Fig. 8.8.)

The explanation of both these 'negative after-effects' is discussed in Chapter 12. Do the effects spread outside the region of visual field exposed to the adapting stimulus? If one eye only is exposed, do you experience the after-effect when viewing with the other eye? Where in the visual pathway might it occur? (After Blakemore.)

have proved useful, for instance measuring the physical intensities of two lights that appear 'equally bright', even though they differ in colour. Although such judgements can often be made reliably and repeatedly, there is a serious difficulty in interpreting them, for there is a linguistic or verbal element involved in asking people to judge a quality such as brightness in the presence of another quality such as colour that is also variable. This verbal element tends to be even stronger if we ask for more complex judgements, such as expressing

Fig. 7.2. At first you may see no sign of human occupation in this innocent island scene. However its title is 'St Helena', so look again. If you still have no luck, look for Napoleon in the space between the trees on the left. Once you have seen him, for how long can you look at the picture without him obtruding into your gaze? This illustrates first that mental 'set' is important in what you perceive, for without hints one can easily miss Napoleon altogether. Second, notice how Napoleon and the trees are alternative constructs for that part of the figure; you cannot see both simultaneously, and what you see tends to flip from one interpretation to the other. Gestalt psychologists would say that the trees are either 'figure', in which case you see them as such, or 'ground', in which case the space between them forms the figure. Finally you probably would not interpret the space between the trees as Napoleon unless you had previously seen a picture of him in this posture, so you are being influenced by memories from some time in the past. More factors influence perception than we intuitively believe.

brightness by rating it on a 10-point scale. Such procedures necessarily involve interpretative mechanisms, and although they may have their uses, the following sections rely mainly on matches and thresholds, for these give the most reliable information about the early steps.

7.2. FACTORS THAT DETERMINE VISUAL SENSITIVITY

It might seem a simple matter to make a light dimmer and dimmer until it can no longer be seen, and then to record its physical intensity, but many different factors would influence the result and these must be brought under control before a meaningful and consistent figure is obtained. These factors will be described first before discussing what ultimately limits the eye's performance.

Dark adaptation

If one comes into a dimly lit room after being outside in bright sunlight everything appears dark at first, but after a few minutes one's eyes adapt to the darkness and one is often surprised how much can be seen. Recovery is complete in a few minutes after moderate levels of light adaptation, but a recovery curve lasting more than 30 min, as shown in Fig. 7.3, will be obtained if the pre-exposure has been very strong. Note that the recovery occurs in two stages, the first attributable to cones, the second to rods, and note also that threshold drops by a factor of more than 10000 (4 log units) during about half an hour.

There are many causes of these changes. The pupil expands, but this occurs rapidly and can only account for a tenfold change at most. There is also a resynthesis of the photosensitive visual pigments that were bleached during pre-exposure, but the gain in sensitivity is not a simple result of the increased proportion of light absorbed by the increased concentration of pigment. This is very clearly shown in the case of the rods, where it is found that more than a hundredfold drop in threshold is associated with the resynthesis of the final 10% of the rhodopsin, from 90% to 100% of its full concentration. The major factors that increase sensitivity appear to lie in the little-understood intracellular transmitter mechanism that causes changes in conductance of the rod membrane. However there are also interesting changes in the neural connections in the retina.

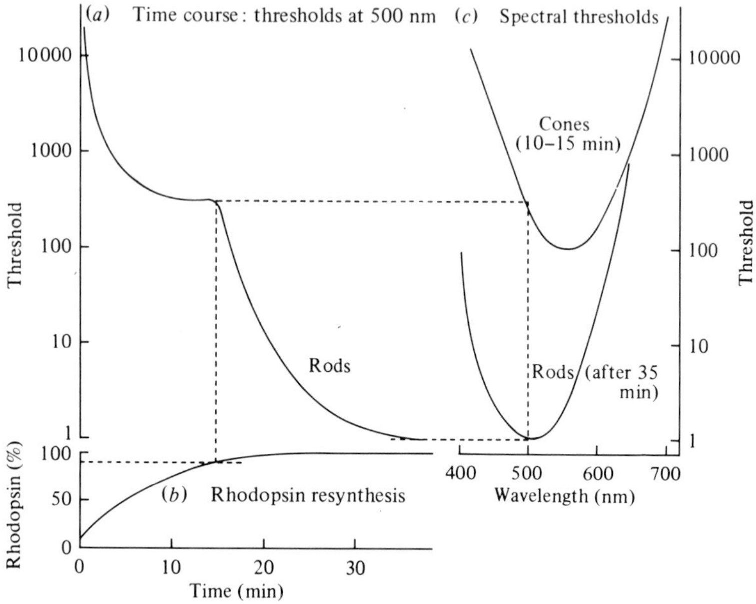

Fig. 7.3. Dark-adaptation curves, the resynthesis of rhodopsin, and spectral sensitivity curves showing the Purkinje shift. (*a*). The drop of log threshold with time following adaptation to a light that bleached a high proportion of rhodopsin in the rods. (*b*). The time course of resynthesis of rhodopsin on the same time scale; notice that less than 10% of the rhodopsin escaped the initial bleach, but 90% had been regenerated after 15 min, when the break occurred in the dark-adaptation curve. (*c*). Two sets of thresholds plotted against wavelength. The upper curve was obtained between 10 and 15 min after starting dark adaptation, and it represents the spectral sensitivity of the cone system. The lower curve was obtained after full dark adaptation and corresponds to the rods. Note that the peak of sensitivity moves (Purkinje shift). These curves represent similar data to V_λ and V'_λ plotted as relative sensitivity in Fig. 5.8, though the latter are obtained by different methods. The fact that the hundred fold drop in rod threshold occurs while the final 10% of the rhodopsin is being resynthesised shows that the change in the threshold is not a simple matter of increased light absorption in the photopigment. (After Rushton and others.)

Spectral sensitivity

If the threshold is measured using monochromatic lights of different wavelengths, results such as those shown in Fig. 7.3 are obtained. Measurements made on the 'cone plateau', within 15 min of intense light adaptation, yield a curve with a minimum threshold near 555 nm wavelength, corresponding to a yellow-green light. If the measurements are repeated after complete dark adaptation, the threshold is much lower and the minimum has shifted to 505 nm, in the blue-green part of the spectrum. This change of the position of maximum sensitivity is called the 'Purkinje shift', after the Bohemian physiologist of the last century who gave his name to many optical effects and anatomical structures: these include the images formed by reflections at the lens surfaces, and the appearance of the shadows of the retinal blood vessels, described in Chapter 3, and also the conducting fibres of the heart, and the neurons named after him in the cerebellum.

The spectral sensitivity curves shown in Fig. 7.3 are of great practical and theoretical importance. They are usually plotted as sensitivity (defined as 1/threshold energy), and they show the relative effectiveness of lights of different wavelengths in stimulating the photopic (cone) and scotopic (rod) systems. They form the basis for calculating the photopic and scotopic *luminance* of a light, for the degree to which a light of physical energy E_λ at wavelength λ nm excites the eye is given by the product $E_\lambda V_\lambda$ in photopic vision and $E_\lambda V'_\lambda$ in scotopic vision, where V_λ and V'_λ are the sensitivities at λ nm relative to the peak of the photopic and scotopic sensitivity curves respectively. This product must then be integrated over the whole spectrum ($\int E_\lambda V_\lambda d\lambda$) to assess the luminance of a light containing energy at several wavelengths or continuously distributed over the spectrum.

The curves used for the standardisation of V_λ and V'_λ were mainly obtained by photometric methods different from the threshold determinations suggested in Fig. 7.3, and they differ slightly from these. The shape of V'_λ is theoretically important because it depends primarily upon the absorption spectrum of rhodopsin, and the agreement between the two provides the key evidence implicating this pigment in visual excitation. For cones the situation is more complicated because the photopic sensitivity curve is some kind of average of the sensitivity curves of the red-, green-, and blue-sensitive receptors.

Background luminance

The apparent brightness of a light is much affected by the background against which it is seen, as one knows from the dim appearance of the moon by day, and the invisibility of stars. Fig. 7.4 shows an experiment in which special conditions were chosen to discourage the cones and thus ensure that rods were being utilised over a wide range of luminance levels. The subject saw a large field whose luminance I could be varied over a range of 10^6; this value is plotted horizontally on a logarithmic scale. To this was added a patch of light whose luminance was adjusted to threshold, ΔI, and this value is plotted vertically on a logarithmic scale. Over the middle range the line has a slope very near to 1, which implies that $\Delta I \propto I$. This is an example of Weber's law, which is familiar in many sensory judgements; it should be thought of as an approximate empirical generalisation rather than a basic law of operation of the nervous system, and even where it holds quite accurately its basis is not well understood. In the present case, desensitisation from the background light occurs either in the receptors, or in transmission to bipolars.

Notice that Weber's law breaks down both at low levels and at high. The former may result from spontaneous events mimicking the absorption of light; this is sometimes called 'dark light', since the level of such intrinsic noise can conveniently be expressed as a luminance. The steepening at the upper end is called 'saturation', and may result from the current that enters the outer segments of rods being completely blocked, or from another step in the excitation process being driven nearly to completion. At the very top the curve stops rising so steeply; this is where cones finally enter under these conditions, which were selected to keep them out of the way until a very high background was reached.

Eccentricity

Under photopic conditions the greatest sensitivity is almost always found at the fovea, which is automatically directed to what we wish to see best. But under scotopic conditions this is not so; lowest thresholds are obtained by giving the subject a dim point to fixate his gaze on, and then arranging for the test light to be exposed 10° to 20° in the periphery. The direction selected must not be temporal, for the blind spot is located 15° in the temporal field.

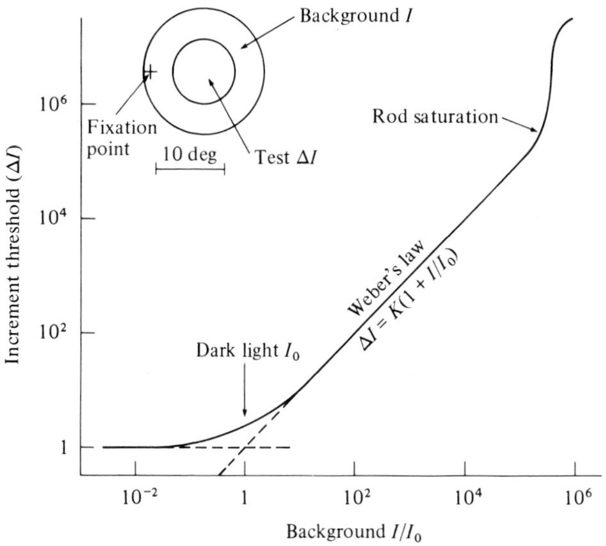

Fig. 7.4. The effect of background luminance on the detection of an added stimulus. The subject fixated on the cross at top left and saw the large uniform background in the right visual field. In the centre of the background a patch of light was added, and adjusted to threshold. Background luminance I is plotted horizontally, and increment threshold ΔI vertically, both on logarithmic scales. To ensure that the threshold was dependent on rods a blue-green light was chosen for the stimulus, together with peripheral location in the visual field, a large stimulus, and long duration exposure. In addition this light was sent into the periphery of the subject's pupil; because of the Stiles–Crawford effect (see p. 130) this light was relatively ineffective for cones. To ensure that the background field had more effect on cones and thus desensitised them more, its colour was orange. Below a certain value of the background, termed the *dark light*, the curve runs almost horizontally, indicating that ΔI is not affected by I. Above this value it rises, and over a range covering a 10000-fold increase in I the line has a slope close to 1, indicating that ΔI is directly proportional to I (Weber's law). Above this the values of ΔI rise more steeply, indicating *rod saturation*, and at higher levels still the values rise less steeply because cones are at last becoming operative. (After Aguilar & Stiles.)

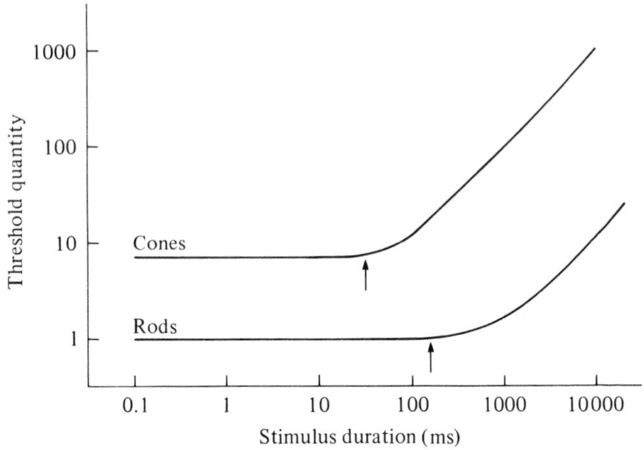

Fig. 7.5. The effect of stimulus duration on visual threshold. Stimuli of variable duration were adjusted in luminance until they could just be seen. This threshold was then multiplied by the duration, and also by the stimulus area, which was held constant, to obtain the threshold quantity of light, and this is plotted vertically against the duration horizontally, both on logarithmic scales. Notice that threshold quantity is independent of duration up to a certain value termed the *critical duration* or *summation time*, and thereafter rises. The summation time for rods is longer than that for cones, so the difference between their thresholds is greater for long-lasting stimuli. The value of the summation time is not a constant but varies with the background upon which test stimuli are superimposed, as well as the system being used.

Pupil diameter

As the pupil expands it admits more light, so under natural conditions the weakest lights can be seen when the pupil is large, provided this does not degrade the image quality to the point where the object's visibility is lost by blurring. In the laboratory it is convenient to use an artificial pupil smaller than the natural pupil, for then the amount of light entering the eye can be calculated without having to measure the natural pupil. The threshold *intensity* of light will be higher than with the natural pupil, but the *quantity* entering the eye will be the same.

Area and duration of test stimulus

Fig. 7.5 shows how the threshold varies with the duration of a stimulus flash. The threshold quantity of light is plotted vertically; this is the product of the threshold intensity, the duration (plotted

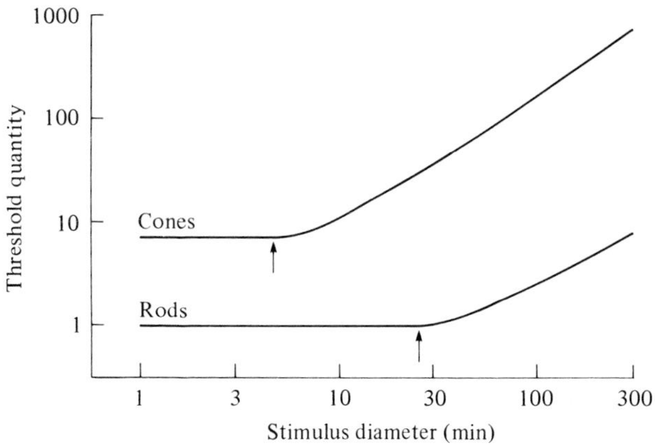

Fig. 7.6. The effect of the size of a test stimulus on threshold. The experiment was done as for Fig. 7.5, but the stimulus diameter was varied and its duration held constant. As before, the threshold quantity is unaffected until the stimuli exceed a certain size called the *critical area* or *summation area*, and thereafter it rises. The summation area is larger for peripheral rods than for foveal cones, so the difference between their thresholds is increased by using large stimuli. As with summation times, summation areas vary with background luminance as well as with eccentricity and the system, rods or cones, which is employed for the threshold tested.

horizontally) and the area of the test stimulus, which is kept constant in this experiment. Two curves are shown, the lower for rods in the periphery where they are most sensitive, the upper for cones in the fovea. For short durations the threshold quantity does not change, which means that the threshold intensity required is proportional to 1/duration. This is referred to as Bloch's law, and it holds for many photochemical reactions (the Bunsen–Roscoe law). However for durations beyond about 30 ms for the cones and 200 ms for the rods the threshold quantity increases, and at long durations it increases in direct proportion to the duration; at this point the threshold intensity is constant and unaffected by the duration of the stimulus. The point where the curves rise is called the critical duration or summation time. Notice that the threshold for foveal cones is about 7 times that of peripheral rods for short duration stimuli, but rises to almost a hundredfold for long durations because the long summation time for rods gives them a further advantage.

Fig. 7.6 shows the very similar relations that are obtained if threshold is determined as a function of the angle subtended by the stimulus. Here again the quantity of light required at threshold is constant, provided that the diameter is less than about 5 min for foveal cones and 30 min for rods. This relation implies that threshold intensity is inversely proportional to stimulus area, and is sometimes called Ricco's law after its discoverer. It occurs because the retina and visual pathways have the capacity to pool excitatory effects over a certain area, which is referred to as the 'summation area'. Summation may result from the interconnections between the receptors, from convergence of connections from receptors, through bipolar cells, on to retinal ganglion cells, or possibly from more central interactions. As with summation time, summation areas are reduced in the presence of background lights; in this case it is probably because lateral inhibitory mechanisms are more potent in the light-adapted state.

Summation time and area are reciprocally related to temporal and spatial resolution; a system which integrates over a long time or large area gains sensitivity for persisting and extended targets, but it is unable to specify exactly when or where they occur, so it loses temporal and spatial resolution. In both respects the resolution of cones is better than that of rods, but it would be a mistake to regard summation time and area as fixed and invariable constants for the two systems; they vary with adaptation level, retinal eccentricity, and other characteristics of the stimuli used to measure them.

Reliability of response

It is not altogether easy to decide whether a very weak light can be seen or not, for very weak sensations have a dubious character to them. Fig. 7.7 shows the responses of a subject to lights of six different intensities presented in random order. Some he definitely saw and the percentages of these are marked as circles. Others he was a little uncertain about; if these are included among the 'seen' responses the crosses are obtained. This curve is shifted to the left compared with that for the definitely seen responses, so the effect of including 'uncertains' is to lower the threshold, taking for this the mid-point of the curve corresponding to the intensity detected on 50% of trials. However this gain in sensitivity is achieved by sacrificing reliability; the subject was right to call these sensations uncertain, for he gave a small proportion of 'uncertain' responses to test trials which were actually of zero intensity, whereas he did not claim he definitely saw any of the blank stimuli. The allowable proportion of such false-positive responses is a factor determining the value of the threshold that will be obtained.

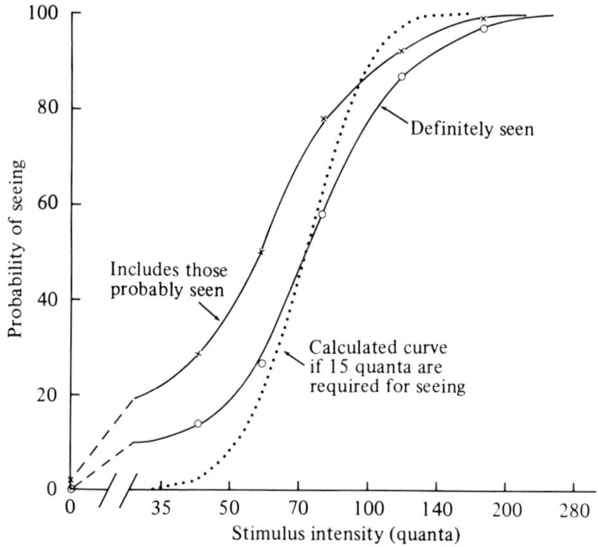

Fig. 7.7. Flashes of light of five different intensities were delivered in irregular order until 100 of each had been delivered. In addition 300 blanks of zero intensity were mixed in. The subject could respond with a definite 'Yes', or could indicate 'Probably yes' if he was not certain. The circles indicate the frequency-of-seeing curve for yes, the crosses the result when probables were added to yes responses. Notice that including probables lowers the threshold, defined as the intensity seen in 50% of trials, from 73 to 59 quanta. But a price was paid for this greater sensitivity because there were three 'Probably yes' replies among the responses to the 300 blanks of zero intensity. Unavoidable fluctuations in the numbers of quanta absorbed from these weak flashes play a part in the uncertainty of seeing near the threshold, but they are not the only cause of variability in responses. The number of photo-isomerisations resulting from 73 quanta entering the pupil is thought to be about 15 on average. The dotted curve shows the predicted frequency-of-seeing curve if the subject had said 'Yes' whenever 15 or more were absorbed, 'No' otherwise. This hypothesis predicts less variability in response than was observed, so there are other intrinsic sources of response variability, or noise. (After Barlow.)

Absolute threshold

The results shown in Fig. 7.7 are the ones we have been aiming towards, for they were obtained under the conditions that yield the lowest possible figure for the quantity of light that yields a visual sensation. To summarise the previous sections, the subject had been dark-adapted for more than 45 min; light of wavelength 500 nm, at the peak of scotopic sensitivity, was used; there was no background light or other light to elevate the threshold; the test target lay 15° eccentric from the fixation point in the nasal field; an artificial pupil was used of known diameter, smaller than the natural pupil and well within it; the stimulus duration and area were below the upper limits of temporal and spatial summation; and on this occasion the subject used two criteria of acceptability, one of which gave no false responses with 300 blanks, the other three false positives, corresponding to 1%.

The result shows that a stimulus causing an average of 73 quanta to enter the pupil was definitely seen on 50% of trials. The exact result of course varies a bit from person to person but we may take 100 quanta of light of 500 nm as a typical figure for the human absolute threshold. Note first that this represents a very small amount of energy compared with the quantities we customarily handle; for instance the work available from 1 g falling 1 cm is more than 10^{12} times the absolute threshold. The eye is considerably more sensitive than photographic emulsions, and photoelecric devices have only recently surpassed it. However the ear is more sensitive and can detect an even smaller amount of energy.

We can follow further the quanta that enter the eye to cause a sensation of light. Some fail to reach the retina, and some are lost by passing through the retina without being absorbed. Then, of those absorbed, some fail to cause isomerisation of the rhodopsin molecule, which is thought to be the step that initiates excitation. The number of effective absorptions resulting from a just-visible stimulus is thus probably 10 to 15, though this figure cannot be given with great confidence or precision.

When one is dealing with such a small number of events one must expect considerable random fluctuation from trial to trial, and the amount of this variation is readily calculated. If the average number is 15 events, the numbers on individual trials will follow a Poisson distribution* that will have a standard deviation of $\pm\sqrt{15}$; hence

* If the events that *actually* occur are only a small fraction of the events that *might* occur, the numbers of such events in successive trials follow the Poisson distribution $P(n/a) = e^{-a}a^n/n!$ where a is the mean number of events, n the number on a particular trial, and $P(n/a)$ the probability of obtaining that number at that mean rate. The number of events that might occur in the present case is equal to the number of molecules of photopigment in the region of retina considered, and it is clear that only a small fraction of these are isomerised in this type of experiment, so the conditions hold for a Poisson distribution to be followed.

The calculation of the chance of a rod receiving a multiple hit when 15 molecules on average are isomerised in 1000 rods is done by obtaining from the above it will not be at all unusual for the number on a trial to drop below 10 or rise above 20. The question naturally arises whether this accounts for the fact that a stimulus delivering on average between 50 and 150 quanta to the cornea is sometimes seen, sometimes not seen, as shown in Fig. 7.7. In other words, do statistical fluctuations in quantal absorption cause the variability in response at threshold?

If the condition for seeing is that 15 or more quanta are absorbed one can calculate the expected shape of the frequency-of-seeing curve, and this is plotted in Fig. 7.7 as a dotted line. It is steeper than the observed curve, so one must conclude that quantum fluctuations are not the only factor causing variable responsiveness, though they must of course contribute to it. Other contributory factors are likely to be the excitatory events that occur in the absence of all light ('dark light', see Fig. 7.4), the 'noise' of synaptic transmission in the retina and elsewhere, and the inefficient use of the messages from the eye on the part of the central nervous system.

In following the weak excitatory process a step further one must realise that the 15 isomerisations do not occur in a single rod but are shared between many; how many will depend upon the size of the stimulus, but they could certainly be shared among 1000 rods without the threshold requirement being increased, and one can then calculate the chance of any rod receiving 2 or more isomerisations (see footnote). This turns out to be small (about 11%), so one can reasonably conclude that a rod does not require the absorption of more than a single quantum in order to initiate a signal. Fig. 7.8 shows that this is indeed so; the outer segment of a single rod from a toad retina was sucked into a fine pipette by the technique shown in Fig. 6.2, and the current being drawn into it by the inner segment was measured. A weak flash of light was delivered at regular intervals, and on some trials a change in entering current was recorded. The frequencies of occurrence of small bumps, and missing bumps, fitted well the expectations for a Poisson distribution (see footnote) and they can confidently be identified as the responses to the isomerisation of a single rhodopsin molecule, or none. Thus these neuro-

expression the probabilities of zero and one hit in a given rod; this is raised to power 1000 for the probability that all the rods absorb 0 or 1; then the probability that one or more rods absorb 2 or more quanta is the only remaining possibility and is 1 minus that figure:

$$P = 1-(e^{-0.015}+0.015\,e^{-0.015})^{1000} = 0.1054.$$

The calculation of the probability of obtaining a small bump or missing bump in the experiment illustrated in Fig. 7.8 is simply obtained from the Poisson expression for $n = 1$ or 0, and $a = 0.53$ in this particular case.

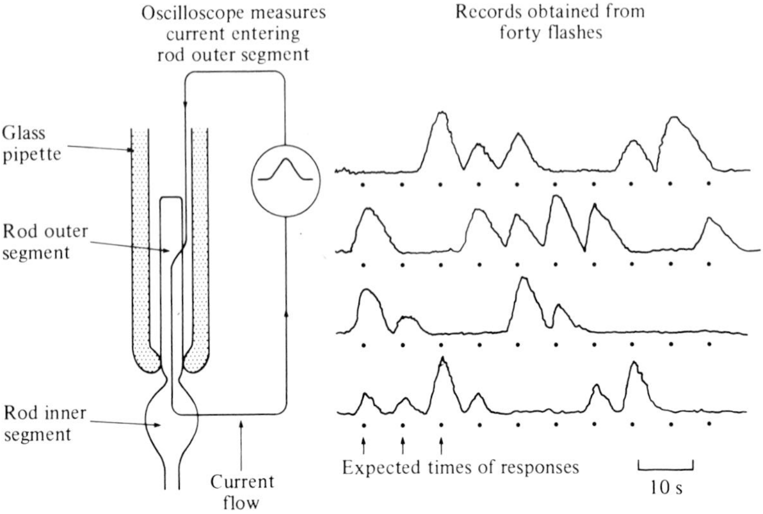

Fig. 7.8. Responses of a rod to the absorption of single photons. The outer segment of a toad rod was sucked into a micropipette by the technique illustrated in Fig. 6.2, and the current entering was measured while a weak flash of light causing an average of 0.53 isomerisations was delivered repeatedly. The records show 40 successive responses. As would be expected for a Poisson distribution with mean 0.53, isomerisations often failed to occur and there was no response. The small-sized bumps correspond to a single isomerisation and occurred at the expected frequency. Large bumps correspond to more isomerisations, but the sizes for more than two cannot be reliably resolved from each other. The very slow, transient, reductions of photocurrent are characteristic of toad rods at 20 °C, but mammalian rods at 37 °C would be much faster. (After Baylor, Lamb & Yau.)

physiological experiments confirm predictions made many years earlier, and they have also confirmed the speculation that the threshold for seeing is greater than the single quantum required by an individual rod because thermal isomerisations of rhodopsin occur at a slow but measurable rate. These would cause frequent false impressions of light if fewer than about 10 isomerisations caused a sensation.

7.3. SENSITIVITY OF CONES

It is worth summarising here the reasons why cones do not function at the low luminance levels that rods do.

(1) The cones are shorter and do not absorb such a high proportion

of the light that enters them. This, however, is partially offset by the fact that each appears to have an optical arrangement that concentrates the light from the pupil on to the pigment in the outer segment. This arrangement is responsible for what is known as the Stiles–Crawford effect: light entering the pupil's centre is, for cones but not for rods, a more effective stimulus than light entering the periphery. It is also responsible for the paradoxical fact that cone photopigments in situ show greater photosensitivity than does rhodopsin in rods; that is, for a given quantal flux, a higher proportion of cone pigment molecules are photochemically changed.

(2) The cone system has a smaller summation area and shorter summation time than rods (see Figs. 7.4 and 7.5).

(3) The electrical response to photon absorption is less.

(4) Whereas single photon absorptions produce very significant and readily detected changes in rods (Fig. 7.7), in cones many photons are required to produce a significant change detectable against the background noise. It is an attractive hypothesis that this high intrinsic noise level is the most important difference between the two types of receptor. The shift to longer wavelengths requires a decrease in the energy barrier that prevents thermal isomerisation occurring in darkness, and although it is only a decrease of about 10% this would be expected to increase the thermal isomerisation rate by a very large factor, about 4000 times. That would be enough by itself to prevent the cones giving useful signals at low light levels, so their unresponsiveness to single photons and the other differences listed above should perhaps be regarded as adaptations to overcome the handicap of a higher intrinsic noise level.

7.4. SUGGESTIONS FOR FURTHER READING

General references

Hochberg, J. E. (1978) *Perception.* New Jersey: Prentice-Hall Inc. (Perception from a psychologist's viewpoint with clear account of Gestalt approach.)

Graham, C. H. (1965) *Vision and Visual Perception.* New York: John Wiley. (A good source for the work on visual psychophysics before 1964.)

Ripps, H. & Weale, R. A. (1976) *On human vision.* In *The eye*, Vol. 2A, *Visual Function in Man*, ed. H. Davson. London: Academic Press.

MacLeod, D. I. A. (1978) Visual Sensitivity. *Annual Review of Psychology*, **29**, 613.

Special references

Tilt and other after-effects. Blakemore, C. (1973) In *Illusion in Nature and Art*, pp. 8–47. London: Duckworth.

Dark and light adaptation. Barlow, H. B. (1971) In *Handbook of Sensory Physiology*, Vol. 7, pt 4, ed. D. Jameson & L. M. Hurvich. Heidelberg: Berlin: New York. Springer. Lamb, T. D. (1988) In *Night Vision*, ed. K. Nordby & R. Hess. Cambridge University Press.

Rhodopsin concentration and threshold. Rushton, W. A. H. (1961) *Journal of Physiology*, **156**, 166.

Spatial and temporal summation. Barlow, H. B. (1958) *Journal of Physiology*, **141**, 337.

Weber's Law and saturation of rods. Aguilar, M. & Stiles, W. S. (1954) *Optica Acta*, **1**, 59.

Reliability, false positives, thermal isomerisation and threshold. Barlow, H. B. (1956) *Journal of the Optical Society of America*, **46**, 634. Baylor, D. A., Matthews, G. & Yau, K-W. (1980) *Journal of Physiology* **309**, 591–621.

Rod responses to single photons. Baylor, D. A., Lamb, T. D. & Yau, K-W. (1979) *Journal of Physiology*, **288**, 613.

Thermal stability of photopigments and Purkinje shift. Barlow, H. B. (1957) *Nature*, **179**, 255.

CHAPTER EIGHT

Spatial and temporal resolution and analysis

J. M. WOODHOUSE AND H. B. BARLOW

Our eyes enable us to recognise and discriminate the shapes of objects around us. We have seen how an image of such objects is formed on the retina and have gone into the factors that determine how much light is required to see. In the first part of this chapter we consider what is known of the mechanisms responsible for spatial resolution and analysis. The first requirement is that the eye should be able to respond separately to different parts of the image; these are the problems of resolution. But the visual system achieves form recognition by combining information obtained from different parts of the image, and we go on to consider the first steps of this process. It must be emphasised that there is a big difference in the level of our knowledge about these two stages; helped by the analogy with physical instruments, we have a good understanding of the eye's resolution, but we hardly begin to know how it proceeds to the analysis and recognition of the objects around us.

Just as the eye responds separately to light entering from different parts of the visual field, so it also responds quickly to light and thus separates events in time. We include problems of temporal resolution in this chapter, and also the problem of detecting movement, which is the simplest example of temporal pattern analysis.

8.1. ACUITY AND CONTRAST SENSITIVITY

The most complete way to represent the eye's resolving power is by determining its *contrast sensitivity function*. The subject looks at sinusoidal gratings (see Chapter 1 and Fig. 1.3), which are usually generated electronically on an oscilloscope screen. For each spatial frequency the contrast (see Fig. 1.3) is adjusted until the subject can just barely see that there is a grating present, as opposed to a uniformly illuminated surface. This is the contrast threshold and for a spatial frequency of 20 cycles deg^{-1} it would have a value of about 0.5% or 0.005. Contrast sensitivity is the reciprocal, 200, and this is what is plotted against frequency as the continuous curve in Fig. 8.1.

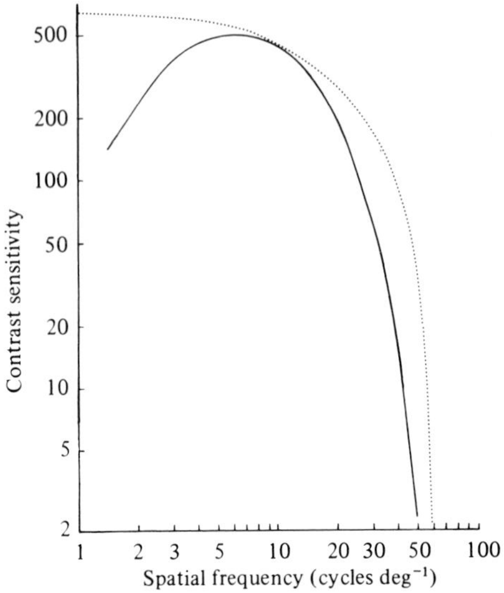

Fig. 8.1. The contrast sensitivity curve (CSF) of the human eye with a 2.5 mm pupil and high screen luminance is shown by the solid line. Contrast is defined as $(I_{max} - I_{min})/(I_{max} + I_{min})$, and contrast sensitivity is the reciprocal of threshold contrast, the contrast at which the subject can just discriminate the grating from a uniform screen. The dotted line shows the CSF expected if optical demodulation is the only factor to limit contrast sensitivity. It was calculated from the modulation transfer function of the human eye (see Fig. 1.4), assuming that the retina and brain could detect the same modulation (nearly 0.002) at all frequencies. This assumption is clearly wrong; other factors cause additional losses of sensitivity both for high frequencies and for low. Lateral inhibition may account for the loss at low frequencies.

Notice that the scales in this figure are logarithmic. The peak of the curve – maximum sensitivity – lies at 3–5 cycles deg^{-1}. (Here contrast sensitivity is about 500; this corresponds to a contrast threshold of 0.2%.) Sensitivity falls for the lower and for the higher spatial frequencies. The high-frequency cut may be extrapolated to a contrast sensitivity of 1, corresponding to a contrast threshold of 100%. The frequency that can be detected at this contrast represents the observer's acuity, or limit of resolution, for a grating stimulus. In the case of a normal observer the limit lies between 50 and 60 cycles deg^{-1} (equivalent to a bar width between 37 and 30 s of arc). The dependence of contrast threshold on spatial frequency is vividly demonstrated in Fig. 8.2, which shows a series of bars of sinusoidal

Fig. 8.2. The contrast of the sinusoidal grating increases from top to bottom, and the spatial frequency increases from left to right. The contour separating the visible from the invisible bars has approximately the shape of the CSF (Fig. 8.1).

luminance profile increasing in spatial frequency from left to right and decreasing in contrast from bottom to top. The contrast threshold is lowest for the middle range of spatial frequencies. (If the figure is held at a viewing distance of 1 m the centre bars are at a spatial frequency of about 4 cycle deg^{-1}.) The contrast has to be greater before the bars at higher and lower spatial frequencies may be discriminated.

The contrast sensitivity curve is closely related to the modulation transfer function of the eye's optics which was described in Chapter 1 (p. 19) and illustrated in Fig. 1.4. High spatial frequencies are demodulated by the optics, so that the image of a 20 cycle deg^{-1} grating of full (100%) contrast has a contrast of only 40% on the retina. To what extent are the eye's different contrast sensitivities at

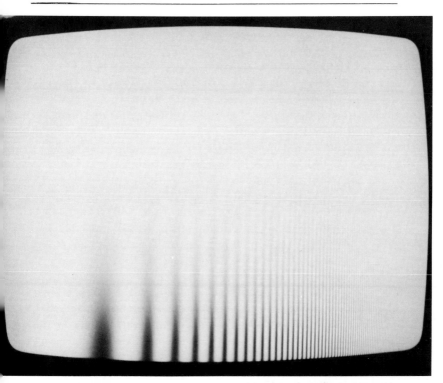

different spatial frequencies caused by variable degrees of physical degradation of the image by the optics? To answer this question the dotted curve in Fig. 8.1 was calculated and shows what the contrast sensitivity would be if optical losses were the only ones. It was assumed that the retina and central pathways are equally sensitive to all spatial frequencies and have the contrast sensitivity (nearly 500) which makes the two curves touch near 10 cycles deg^{-1}. It will be seen that there is an additional loss of contrast sensitivity at both high and low spatial frequencies that cannot be attributed to optics and must be caused by retinal and central mechanisms.

The additional loss at high frequencies is not fully understood. The spacing of the receptors could limit resolution but in the centre of the fovea they are placed at intervals close to 2.0 μm, which corresponds to 25 s of arc. Information theory tells us that an image with spatial frequencies up to 50 cycles deg^{-1} needs to be sampled at intervals of 1/2F(max) i.e. 1/100 deg or 36 s. There are both red- and green-sensitive cones in the fovea, so the separation between cones of like class is $\sqrt{2} \times 25 = 35$ s. One must conclude that a well-balanced design has been achieved by evolution in this respect. Other factors that may cause loss of high spatial frequencies are the finite size of the receptors, the leakage of light between them and possible neural interconnections, the blurring effects of residual eye movements, and incomplete use by the nervous system of information available at the receptor level.

The decline in contrast sensitivity at low frequencies is not explainable on optical grounds and results, at least in part, from lateral inhibition in the neural pathways. When the wavelength of the sinusoidal grating is long compared with the receptive field size, the centre and surround are nearly equally stimulated and tend to cancel each other out, thus reducing sensitivity. The situation is complicated, however, by the fact that receptive fields vary in size, and hence in the frequencies they prefer.

Pathological conditions may lead to a loss of contrast sensitivity; of course much the commonest cause of decreased contrast sensitivity at high spatial frequencies is failure or inability to focus the image accurately (see Chapter 4). For many purposes it is unnecessary or too time-consuming to determine a complete sensitivity function and it is enough to know if the upper limit of the curve lies in its usual position or is shifted to the left. For the clinical testing of visual acuity a Snellen letter chart is commonly employed. This comprises a series of dark letters of decreasing size on a bright background. The smallest letters that are correctly identified give a measure of the acuity. The design of the chart is based on the assumption that the 'normal' limit of resolution is a gap subtending 1 min of arc. The letters are constructed with the gaps between individual components of the letters subtending 1 min at specified viewing distances; each complete letter then subtends 5 min of arc. The acuity is recorded as a fraction, the numerator being the testing distance (normally 6 m or 20 ft) and the denominator the distance at which the gaps would subtend 1 min of arc. Thus 6/6 (20/20) represents 'normal' acuity and 6/18 (20/60) represents an acuity one third the 'normal' value. Acuities of 6/5 and even 6/3, better than normal, are not uncommon. Current vision standards for driving licences in the UK demand identification of a number plate at a distance of 75 ft – corresponding to an acuity of 6/14.

The use of the Snellen chart of course requires knowledge of the alphabet, and a second disadvantage is that all letters are not equally easy to identify. A simple test that meets these criticisms is the Landolt C chart, which comprises letter Cs at different orientations. The patient is required to indicate the position of the gap in each letter. Again the limiting resolution for a normal observer is reached when the angle subtended by the gap is of the order of 1 min of arc, though $\frac{1}{2}$ min can often be resolved.

Although acuity tests of the type just described are likely to continue to provide the most efficient clinical measure of spatial vision, there do exist occasional patients for whom high-frequency loss is slight but for whom relatively large losses occur at middle or low spatial frequencies. Since much of the information that we use in visual tasks lies at middle or low spatial frequencies, the contrast sensitivity function gives a fuller account of an individual's capacity to process the visual scene.

Some tests of acuity yield results far below 1 min of arc, but these are not really tests of resolution, the ability to separate parts of the visual image. An example of such a test is the visibility of a thin wire, which can be seen against the bright sky when its diameter subtends an angle of only 0.5 sec of arc; however its image on the retina then consists of a dim streak that is much more than a hundred times as broad as the geometrical image would be. Fig. 8.3 shows other examples. For *vernier acuity* two lines lying end-to-end are adjusted to be collinear. This task can be done with a precision of about 5 s of arc, but what it tests is the ability of the visual system to judge the position of objects in the visual field, not the ability to separate

(*a*) (*b*) (*c*) (*d*) (*e*)

Fig. 8.3. Examples of judgements of relative position that can be made with a precision much finer than the separation of foveal receptors. The tasks are as follows: (*a*) aligning the two halves of a vernier; (*b*) adjusting the curvature of a line to zero; (*c*) aligning the three dots; (*d*) aligning the two dots with the true vertical; (*e*) setting the dot to the centre of the ring. A convenient measure of precision is given by the standard deviation of the settings, which would be under 5 s of arc in all cases; foveal receptors are separated by 25 s. In some cases it has been shown that these tasks can be performed nearly as well when the image is moving over the retina at a speed of 3 deg s^{-1}. (From Westheimer & McKee.)

or resolve them. Dots can be aligned, or the curvature of a line detected, with about the same accuracy. Tests of *stereo acuity*, in which the slightly different spacing of objects seen by the two eyes is interpreted in terms of depth, gives an even lower figure, down to about 2 sec of arc. The visual system must do these tasks by some process of averaging or interpolation, using signals from several of the quite widely spaced observation points corresponding to the cones. It may appear remarkable that a positional accuracy of 2 sec of arc can be achieved when the cones of the retinal mosaic are spaced apart by at least ten times that distance, but one should recall two facts. First, one can specify the mean of a set of measurements with greater accuracy than that of the individual measurements. Second, though the cones and their pathways to the brain are quite widely spaced, corresponding to a rather coarse grain or texture in the transmitted picture, the structure in which these pathways terminate, the visual cortex, has an extremely fine grain, for there are several hundred times as many cells in area 17 of visual cortex as there are pathways coming in from the eye through the lateral geniculate nucleus.

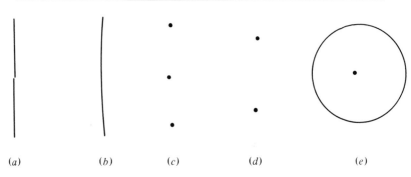

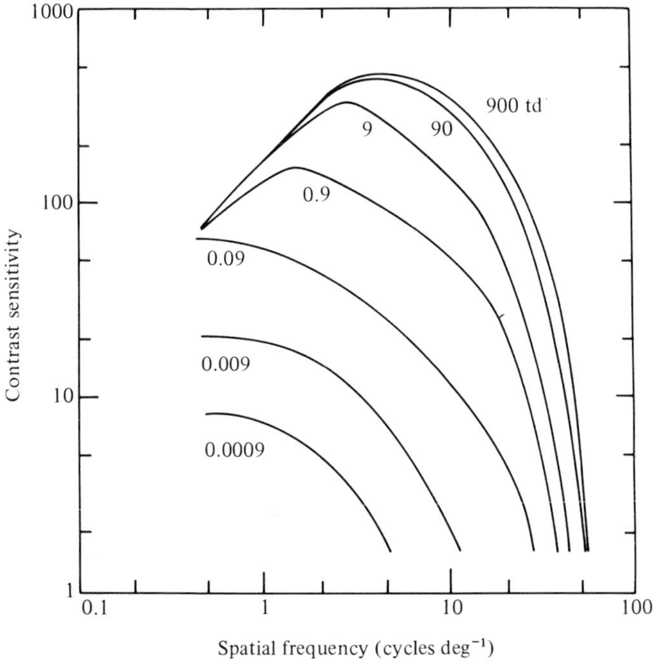

Fig. 8.4. Spatial contrast sensitivity curves at seven different retinal illuminance levels between 0.0009 and 900 trolands. The subject viewed the gratings through a 2 mm diameter artificial pupil. The wavelength of the light was 525 nm. Notice the loss of sensitivity for medium and high frequencies as the retinal illumination is decreased. (Adapted from Van Nes & Bouman, 1967.)

Some important factors that affect acuity and the contrast sensitivity function

The values given earlier for resolution limits and for contrast sensitivity apply only to optimal conditions of testing. We must now examine these conditions and the way in which they affect the results attained.

The shape of the contrast sensitivity function varies with illumination. Dinıming sinusoidal gratings reduces contrast sensitivity to medium and high spatial frequencies, so the curves become less peaked and shift downwards and to the left, as shown in Fig. 8.4. The same results have been replotted in Fig. 8.5 to show how the spatial frequency that can be detected at a given contrast rises with

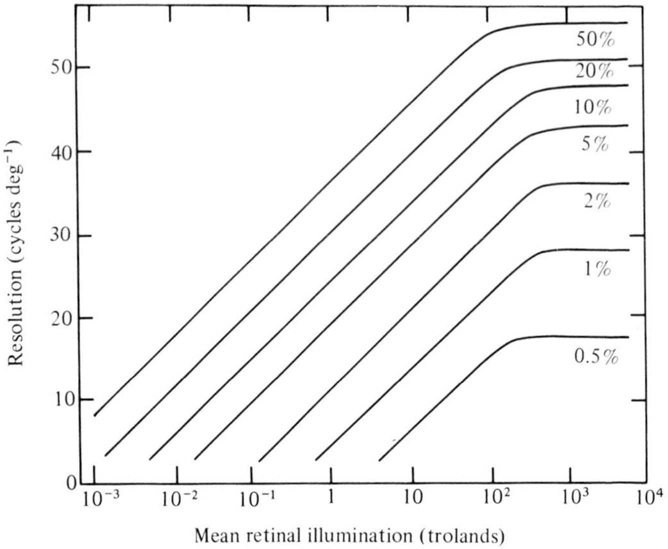

Fig. 8.5. The results shown in Fig. 8.4 were replotted to obtain these curves. They show the highest spatial frequency that could be detected at seven contrasts between 0.5% and 50%, plotted against mean retinal illuminance in trolands. Resolution is given here in cycles deg^{-1}; division by 60 gives a figure comparable to *acuity*, defined as 1/(minimum resolvable angle in minutes). Notice that performance is not improved by increasing retinal illumination above about 500 trolands. (Adapted from Van Nes & Bouman, 1967.)

retinal illumination. The data are those found for a constant pupil size of 2.0 mm – optically the best where the effects of optical aberrations are minimised. The highest line is for 50% contrast and is very close to the highest spatial frequency that can be resolved at a particular illumination level. The curve apparently reaches a plateau as illumination increases, and no further improvement occurs beyond about 1000 trolands. At extremely low scotopic levels resolution is limited by the sensitivity of the eye and, as we have seen, quantal fluctuations in light absorbed are relevant at these levels.

Figs. 8.4 and 8.5 were obtained by looking directly at the gratings, so they refer to foveal vision. If a peripheral region of retina is tested, the contrast sensitivity curve is shifted down and to the left. Since acuity is highest when the image falls on the fovea, voluntary eye movements act to place an object of interest on this central part of the retina. Acuity falls if the image moves as little as 10 min of arc

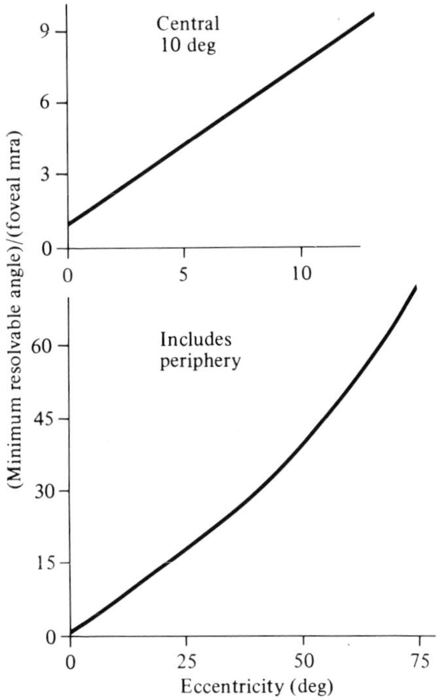

Fig. 8.6. The minimum resolvable angle (mra) increases nearly linearly with eccentricity for the central 25°, and increases more rapidly after that. Note that scales are enlarged fivefold in the upper figure The mra can be measured in many ways, and the rise with eccentricity always occurs at about the same rate proportionally; hence the ordinate chosen is the ratio of mra at a certain eccentricity to that found with the same test at the fovea, and it is valid to a first approximation for all acuity measures. Note that mra is increased more than twofold, and acuity is therefore less than half, at an eccentricity of only 2°.

from central 'fixation'. Fig. 8.6 shows that the increase in the minimum resolvable angle is nearly directly proportional to eccentricity up to 25°. Thereafter resolution deteriorates even more rapidly.

Though it is true that the optical quality of the image deteriorates slightly with eccentricity this is not the main cause of the deterioration of acuity. Instead it is a feature of the neural organisation of the retina: there are fewer ganglion cells in the periphery and they have larger receptive fields, so poor peripheral acuity should be regarded as a means of simplifying the visual information provided to the brain. The mobility of the eyes of course allows a selected region to be transmitted without any such simplification.

At scotopic levels, when vision is mediated by the rod system, the situation is rather different. There are few rods in the central 1 deg, so that under these conditions objects cannot be perceived if fixated directly. (The reader may verify this for himself by observing a dim star or luminous watch face in darkness: the object is visible only if one fixates slightly to one side.) Highest acuity is then recorded for an area of the retina several degrees from fixation.

It is interesting to notice that acuity varies with the orientation of the stimulus. For vertical and horizontal gratings the limiting resolution is equally good, but for oblique directions the spacing of bars has to be increased by about 20%. Resolution for one orientation can be selectively impaired by astigmatism, and the deficit in oblique directions was at one time attributed to optical defects. This explanation has been ruled out by performing very careful astigmatic correction, and also by a technique in which a diffraction grating is formed on the retina from two spots of coherent light, usually from a laser, formed in the plane of the pupil. Such diffraction patterns are less disturbed by optical aberrations than normal images but the deficit in acuity at oblique orientations persists. It must therefore be a result of receptor spacing in the retinal mosaic, or of the neural organisation of the pathways.

8.2. ANALYSIS OF THE IMAGE BY SPATIAL FREQUENCY FILTERING

The problems of resolution and acuity in vision were greatly clarified by treating the optics as a spatial frequency filter with characteristics given by its modulation transfer function (see Fig. 1.4). This treatment already required the technique of generating sinusoidal gratings and the concept of Fourier transforms (see Chapter 1, p. 15) so it was natural to go ahead and look at the retina and brain in the same way. The results of this approach suggest that a segregation of neural activity according to spatial frequency occurs and that different spatial frequencies excite different nerve cells. This has been confirmed by recording from neurons of retina, lateral geniculate, and cortex in cats and monkeys while the eye is being stimulated by sinusoidal test gratings.

Psychophysics of frequency filtering

Both the main psychophysical effects suggesting that different spatial frequencies are handled separately in the brain can be easily demonstrated. Fig. 8.7 shows a square wave grating at the top, a sinusoidal grating of the same frequency in the middle, and a sinusoidal grating of 3 × the frequency at the bottom. The contrasts of the two sinusoidal gratings are chosen to correspond with those that are present in the Fourier series representing the square wave at the top, which has a contrast of 50%; thus the middle grating has a contrast of 63.5% ($50 \times 4/\pi$) and the bottom a contrast of 21.2% (63.5/3) (see Chapter 1, pp. 15–17).

Prop up Fig. 8.7 in a good light and walk back a distance of about 10 m. The top two gratings should still be visible, but the lowest one will appear a uniform grey; at the greater distance its visual angle has been decreased and its spatial frequency increased to a point where 21% contrast is below threshold, so that it cannot be seen. Now compare the top square wave with the middle sinusoid; they will be indistinguishable because your eye is failing to detect the third, fifth and higher harmonics, which are the spatial frequencies that make a square wave differ from a sine wave. Approach closer, to the point where you can just see that the square wave is different from the sine wave: does this distance coincide with the point where the isolated third harmonic becomes visible? It should be pretty close.

What this observation shows is that the detectability of the 3rd harmonic is the same in the presence or absence of the fundamental. Now in discussing Weber's law (Chapters 1 and 7) it was pointed out that a background of excitation in a sensory pathway usually raises the threshold for detecting an added stimulus, so the lack of interaction is surprising and suggests that the fundamental and 3rd harmonic are being carried in separate pathways; these are sometimes called 'frequency channels'.

We have seen that the presence of the fundamental has little influence on the detectability of the harmonic. What about the converse? Do the harmonics contribute to the detectability of the square wave? The answer is no: for all spatial frequencies above 3 cycles deg^{-1} the contrast sensitivity for a square wave is $4/\pi$ times the contrast sensitivity for a sinusoid, and that is because the contrast of the fundamental in a square wave is $4/\pi$ times the contrast of the square wave itself (see Fig. 1.2 in Chapter 1). For frequencies below 3 cycles deg^{-1} complications arise because the eye may become more

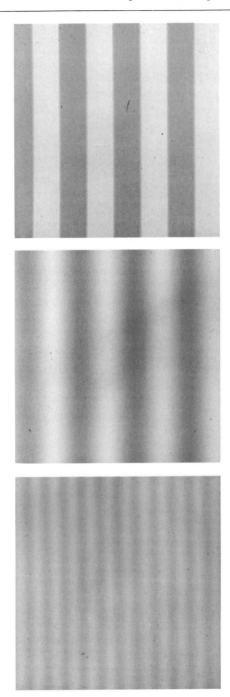

sensitive to the third harmonic and thus detect it before it detects the fundamental.

The second main observation suggesting that spatial frequencies are separated in the brain is shown in Fig. 8.8. Look at the top left grating for 60 s, then at the central one; it should remain invisible for a few seconds after you first look at it. This effect does not take place if you preadapt to a grating of substantially different frequency or orientation. Clearly the fact that different spatial frequencies and orientations cause different adaptation effects suggests that they activate different neurons in the brain. An estimate of the bandwidth and range of orientations to which neurons adapt can be made by measuring threshold elevation at frequencies and orientations near to that of the adapting grating. The band of frequencies to which a neuron responds appear to be quite broad – more than an octave – and the range of orientations about $\pm 15°$.

Neurophysiology of spatial frequency filtering

The neurophysiological basis for these effects became clear from measurements of responses of neurons in the visual pathways to sinusoidal gratings. Fig. 8.9 shows the contrast sensitivity curve for the cat. This can be obtained by behavioural tests, and also by using cortical evoked potentials. It is a broad curve like that for humans, but shifted to lower frequencies by a factor of about 10; the peak contrast sensitivity of 200, corresponding to a contrast threshold of

Fig. 8.7. Demonstration of independent sensitivity to fundamental (F) and third harmonic (3F) of a square wave. The sine wave in the middle has the contrast and frequency of the fundamental of the square wave at the top, and the sine wave at the bottom has the contrast and frequency of its third harmonic, though difficulties of reproduction make the contrasts only approximately correct. By moving away one reaches a critical distance at which the square wave becomes indistinguishable from the sine wave, and the third harmonic becomes invisible. This should happen at the same distance if the visual system detects the two components in the square wave independently, but the critical distances would be different if the presence of the fundamental hindered or facilitated the detection of the third harmonic. Note that higher harmonics (5F, 7F, etc) are also present in the square wave, but they would become invisible at shorter distances because they are of lower contrast and higher frequency, so the third harmonic is the important one for distinguishing sine from square wave. Experiment supports independent detection of F and 3F, thus suggesting 'channels' for different frequency bands in the visual system. (After Campbell & Robson.)

Fig. 8.8. Demonstration of selective adaptability of frequency channels. Gaze at the top left grating for about a minute, allowing your eyes to fixate points within the circle but not outside it. Then transfer your gaze to the central patch, fixating on the cross at the middle of it. Initially the faint grating that is normally visible will not be seen, but it will become visible in about 5 to 10 s. Repeat the experiment by fixating on the other three gratings: you should obtain negative results for these, indicating that the 'channel' for the central grating is not adapted by exposure to gratings of the wrong orientation or spatial frequency, but only by the grating of the same orientation and spatial frequency. (Adapted from Blakemore & Campbell.)

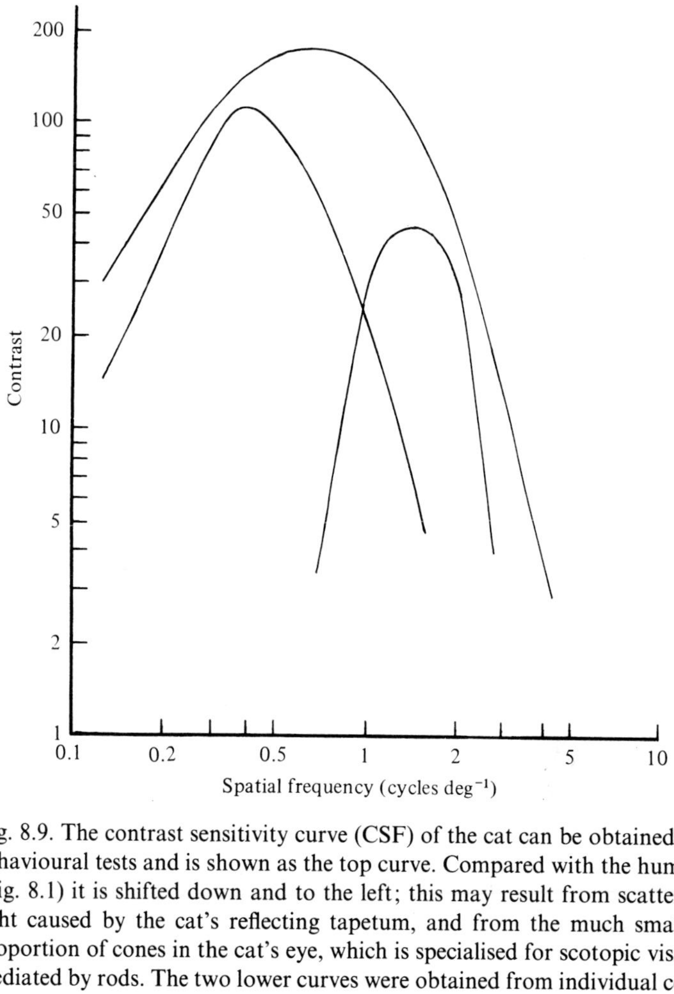

Fig. 8.9. The contrast sensitivity curve (CSF) of the cat can be obtained by behavioural tests and is shown as the top curve. Compared with the human (Fig. 8.1) it is shifted down and to the left; this may result from scattered light caused by the cat's reflecting tapetum, and from the much smaller proportion of cones in the cat's eye, which is specialised for scotopic vision mediated by rods. The two lower curves were obtained from individual cells in area 17 (striate cortex). It is presumed that the CSF represents the upper envelope of many such cells.

0.5%, is also substantially less than that of humans. Both these differences are probably connected with specialisations of the cat's eye for nocturnal vision. First, its retina has far fewer cones, and it is these that give humans good acuity and high contrast sensitivity; and second the cat has a reflecting tapetum which must increase the amount of scattered light, thereby lowering image contrast.

Within the broad curve are two much narrower curves. These were

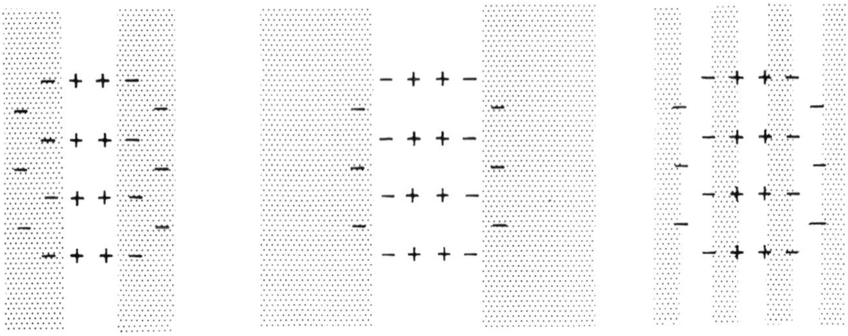

Fig. 8.10. The explanation for simple cortical neurons acting as spatial bandpass filters is shown here. At the left the bright bar of a grating is shown fitting the central 'on' zone of a simple cell receptive field while the neighbouring dark bars (stippled) lie on the flanking 'off' zones; both 'on' and 'off' zones are excited, so the response is maximal. In the centre a good fit is no longer obtained because the spatial frequency of the grating is too low, whereas at the right its spatial frequency is too high. This simple explanation might lead one to expect that the experiment would work as well with square wave gratings as with sinusoids, but this is not the case: the upper harmonics of a square wave (see Fig. 1.2) cause misleading responses to low frequency gratings. Complex cells may also have bandpass characteristics, but it is harder to explain them because 'on' and 'off' zones of their receptive fields do not interact in a simple way.

obtained by testing two different simple cells of the cat's primary visual cortex (area 17) with sinusoidal gratings. For each of them the minimum contrast required to elicit impulses was measured, and these results are plotted as contrast sensitivity curves. The overall contrast sensitivity is thought to be formed as the envelope of several such tuning curves, each covering only a portion of the total range of spatial frequencies to which the visual system responds.

These simple cells were originally described by Hubel and Wiesel, who emphasised the fact that they had strong orientational preferences, and also that they varied over a wide range in size. However until they were directly tested with spatial sinusoids it was not realised how well they fitted the idea that filtering for spatial frequencies occurs. Fig. 8.10 illustrates graphically that a grating of correct period 'fits' a simple cell nicely, with a bright bar on the 'on' centre and the neighbouring black bars on the 'off' regions. This delivers the maximum excitation to the cell, and a grating of half or double the frequency cannot excite it nearly as much. The excitation will clearly also be reduced if the orientation is inappropriate.

The use of sinusoidal grating stimuli has brought out other interesting facts about the neuronal organisation of the visual system. Cortical neurons responding to the highest spatial frequencies are only found close to the visual axis, in agreement with the fact that visual acuity is high there and declines rapidly with eccentricity. However in each region of the cortex neurons are found responding optimally to a range of frequencies, so that the two neurons illustrated in Fig. 8.9 could easily have been recorded from the same part of the cortex.

The technique has also been applied to neurons in the retina and LGN. These are unselective for orientation, as expected from their concentrically organised receptive fields. They also respond to broader ranges of spatial frequencies, so the one- to two-octave bandwidth of the cortical neurons is achieved by progressive narrowing.

Contrast sensitivity curves are easier to interpret when a system is linear (see footnote p. 21), so those who recorded from neurons in the visual pathway were interested in the question of linearity. This led to the unexpected discovery that some cells in the cat retina (called Y-type) behave in a highly non-linear way when tested with sinusoidal gratings, while others (X-type) behave nearly linearly. It is probably the linear type that feeds the simple cells of the cortex, and it is only in this X system that receptive-field plotting and contrast-sensitivity curves give results that are concordant in the simple way suggested in Fig. 8.10. The linear, X, and non-linear, Y, properties are strongly correlated with others, such as how well sustained the discharge is, the conduction velocity of the axons, and the size of the cell body and of the receptive fields. We shall return to them in the next section.

Interpretations

The results of testing the visual system with sinusoidal gratings have been interpreted in very different ways by different people. One school leapt to the conclusion that a full scale Fourier analysis was performed and that pattern recognition was done on this transform of the whole visual image. This view was aided by the analogy with hearing, where frequency analysis is certainly very important and forms the basis for pattern recognition in speech and music, but there are serious objections to the view that a Fourier transform of the whole visual image is made. First, all evidence points to frequency tuning being very much broader in vision than in hearing, and such broadly-tuned spatial frequency channels could not provide the detail that we know is utilised in vision. Second the tuned elements that have been recorded from have quite small receptive fields and hence retain a large measure of positional selectivity, whereas a component of a global Fourier transform should be sensitive to the same spatial frequency and orientation over the whole visual field. Third, the advocates of Fourier transforms in vision have not explained why or how it would aid the performance of pattern recognition; the transform would contain all the information in the original image, and it is not clear what would have been gained by performing it.

A current interpretation of the role of frequency channels in vision is that local, but not global, Fourier analysis is performed. On this view patches of the visual image are analysed into about half a dozen different spatial frequency bands at about twenty different orientations. There would be several thousand such patches in the whole visual field, and they would subtend a fraction of a degree at the fovea, and several degrees in the periphery. The result of the analysis would correspond to coefficients of some hundred sinusoids and cosinusoids of differing frequency and orientation for each patch, and the range of frequencies covered would vary with the eccentricity and size of the patch. This scheme is a tentative one, but it is consistent with much of the psychophysical and neurophysiological evidence. However at the moment it is not easy to see how it helps the brain provide us with the picture we see. One can speculate that frequency analysis confers an advantage in vision analogous to that for hearing. There it enables a sound pressure fluctuating up to 20000 times s^{-1} to be accurately represented by a number of slowly varying quantities, the amplitudes of the different frequency components; such quantities, fluctuating only about 100 times per second, can be more readily transmitted by nerves and responded to by the brain. In vision, the comparable advantage may be that the spatial frequency components represent a description of 'texture' which applies to the whole of each patch. This would be a step beyond a point-by-point description, just as the cochlea goes beyond a moment-by-moment transduction of sound pressure. But several thousand such 'patch descriptions' would be required, and if this is the correct interpretation it can be no more than the first step leading to the representation of the world we are aware of and make use of. It is also important to remember that the brain characteristically conducts analyses in parallel, as the example of the rabbit retina showed (Chapter 1, p. 23). As well as performing a spatial frequency analysis we can be sure that the visual brain pays much attention to motion, colour, and binocular disparity.

To conclude this section, consider a paradox whose solution may depend on the frequency-selective channels we have seen discussing. Suppose you look at a grating (such as one in Fig. 8.7) from close at hand and then retreat away from it, keeping it in sharp focus all the time. If it has high contrast, it will look deep black and white initially and if it has low contrast it will look dark grey and light grey. Moreover, these appearances will not change much as you retreat – the high contrast one will still appear black and white and the low contrast one two shades of grey. Now look at Fig. 1.4, which shows what happens to the image contrast as the frequency of a grating with 100% contrast is changed; the image is progressively demodulated, until ultimately, at about 60 cycles deg^{-1}, it has zero contrast. The constant apparent contrast of the gratings seen at varied viewing distances is, in a sense, an illusion; the real contrast of the retinal image is not constant but changes. Thus the apparent constancy of contrast is a nice example of an 'illusion' that makes vision more, not less, veridical; what we see corresponds more closely to the contrast of the object than it would if we did not have the mechanism responsible for this effect.

Fig. 8.11 shows an experiment analysing this observation in more detail. The contrasts of gratings of various frequencies were adjusted to match a standard at 5 cycles deg^{-1} at various contrasts. For very low contrasts, close to the threshold contrast of the standard grating, all the other gratings were set at contrasts close to their own contrast thresholds. But when the contrast of the standard was raised to, say, 30%, the matching grating was not increased by the same factor, but by a smaller one. The resulting setting was close to a true contrast match in the object, and was therefore far from a true contrast match in the image.

Apparently the visual system has the capacity to correct for the frequency-dependent demodulation caused by the optics, provided that the contrast is substantially above threshold. The degree of compensation is frequency-dependent, and it is hard to imagine that it could be applied without the segregation of different spatial frequency bands in the central neural pathways. A similar phenomenon has long been known to occur when sounds of different frequency are matched for loudness.

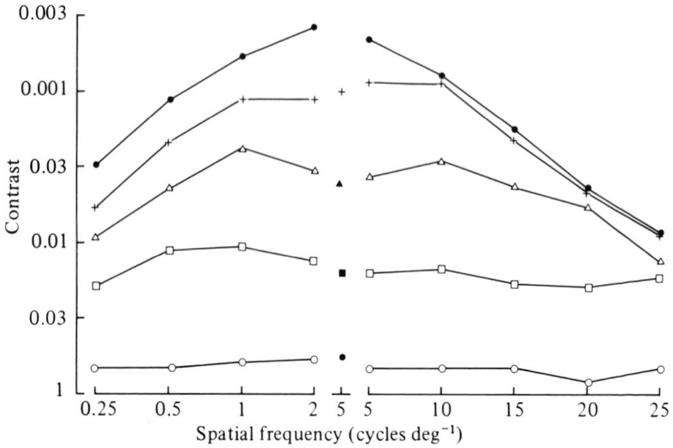

Fig. 8.11. Constancy of apparent contrast at different spatial frequencies. The cross and filled symbols in the centre show the values at which a 5 cycle deg^{-1} grating was set. A grating of different frequency was then adjusted in contrast until it was judged to have the same contrast as the 5 cycle deg^{-1} grating. For the highest contrast of 0.56, this matching is done quite accurately, so that the contrast of all gratings was set to a value close to 0.56. This is done in spite of the fact that the retinal image of a 25 cycle deg^{-1} grating with contrast of 0.56 only has a contrast of about 0.17; apparently when matching a contrasting grating the visual system can compensate for optical losses. The dots at top show the contrast sensitivity (1/contrast threshold) of this subject. When the 5 cycle deg^{-1} grating was set at 0.001, the matching gratings were set to contrasts that are a uniform distance from the contrast threshold, and higher than 0.001 at all frequencies. Therefore the compensation mechanism does not operate at low contrasts. Notice that the loss of contrast at low spatial frequencies, which is attributed to lateral inhibition, is also compensated at high contrasts. (After Georgeson & Sullivan, 1975.)

8.3. TEMPORAL FACTORS IN VISION

The effect of light on most photographic films is permanent, so the total amount of change in the silver grains depends upon the total amount of light that has ever fallen on them; in other words, a photographic film can integrate perfectly in time. Clearly the eye operates differently, for it supplies us with a moment-by-moment record of the visual scene, but nonetheless it does integrate for a limited period; this is demonstrated by the fact that the quantity of

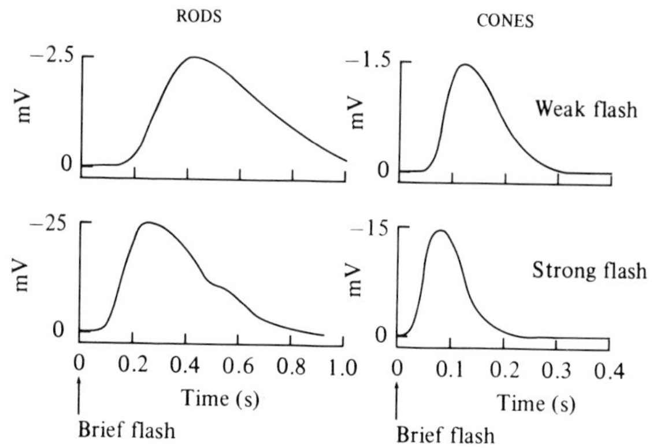

Fig. 8.12. Responses recorded intracellularly from a rod (left) and a cone (right) of the turtle to weak (top) and strong (bottom) flashes of light delivered at $t = 0$. Note that the time scales differ and cones respond more than twice as fast as rods. Brightening the flash (about 100 times in these examples) makes the responses peak earlier. Human receptors at 37 °C are probably about three times as fast as these, which were recorded at 20 °C, but even so the responses are slow compared with those of many sensory receptors. (From Baylor & Hodgkin.)

light required at threshold is independent of the duration of the stimulus, provided this is less than the summation time (see Fig. 7.5).

This summation time can extend to several tenths of a second under some conditions, and as pointed out in Chapter 1, long summation times are desirable to obtain high sensitivity. On the other hand, short summation times are needed for good temporal resolution and to obtain speedy responses to changing illumination. Speedy response is especially necessary for the vision of movement, and seeing movement is obviously important for predators and prey, as well as sportsmen. This chapter deals with these temporal aspects of vision.

Fig. 8.12 shows the electrical responses of photoreceptor cells to brief flashes of light delivered at time zero. The responses start after a latency of up to 200 ms for a weak flash on a rod, the peaks occur at times from 400 ms down to 70 ms for a strong flash on a cone, and the total duration may last up to a second. For human receptors at 37 °C the responses are thought to be about three times as fast as these turtle photoreceptors recorded at 20 °C, but even so it will be seen that a brief flash initiates a process that peaks after a considerable delay and persists a long time; a fast-bowled cricket ball travels over 1 m during the 25 ms it takes a brightly illuminated cone to reach its peak response.

The effects of these long-drawn-out responses are in some ways like the effects of optical aberrations in the spatial domain, for it becomes impossible to distinguish when a stimulus occurred, just as blurring mars spatial resolution. However, in other ways the optical spread is very different, for the aberrations spread out the light, and thereby diminish its intensity at each point. The long duration of the response, on the other hand, is thought to result from the persistence of the 'blocking molecules' released by an absorbed quantum (see Fig. 6.2) and their continued action increases the total amount of entering current that is blocked; thus it improves sensitivity. It will be seen that cones respond more briefly, and that both rods and cones respond more briefly to an intense flash than to a weak one. These are presumably adaptations to avoid the disadvantages of prolonged summation, and though it is not shown in Fig. 8.12, responses occurring to weak incremental stimuli added to a steady background light are also speeded up.

The retina is thus a very refined mechanism compared with a photographic film, for it combines a sensitive, long-latency, slow-exposure monochromatic system (the rods) with a less sensitive, shorter-latency, brief-exposure, colour system (the cones). Furthermore each system automatically speeds up its responses as illumination rises and high sensitivity is no longer necessary.

Just as the spatial contrast sensitivity function (Fig. 8.1) shows how the visual system handles spatial sinusoids so the *temporal contrast sensitivity function* (Fig. 8.13) shows how it handles temporally fluctuating lights. A subject looks at a source of light fluctuating sinusoidally in luminance at a particular frequency, and adjusts the contrast of the fluctuation until he barely detects that the light is flickering. The frequency is then changed, and he adjusts contrast to threshold again. As before, contrast is defined as $I_{max}-I_{min}/I_{max}+I_{min}$. Fig. 8.13 shows a series of curves of contrast sensitivity, the reciprocal of contrast threshold, plotted on logarithmic axes; the different curves correspond to different mean luminances of the flickering light.

The first point to note is that the maximum frequency the eye can detect as flickering is 60 Hz; above that frequency a light is seen as steady, even when its contrast is 100% (that is, when it is varying in luminance between zero and twice the mean). The lower curves corresponding to lower mean luminances have lower high-frequency cut-off points, and this cut-off point is called the *critical fusion*

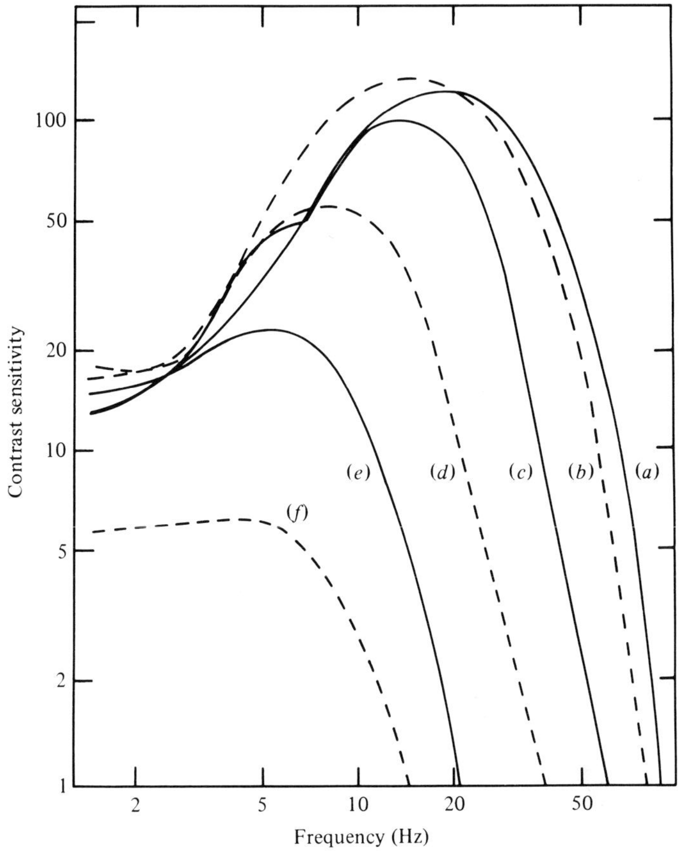

Fig. 8.13. Temporal contrast sensitivity curves for a human subject at various mean luminance levels. At high luminances, contrast sensitivity peaks between 10 and 20 Hz, but as the mean luminance is decreased our sensitivity for rapid flicker is decreased progressively. Between 2 and 5 Hz, mean luminance has little effect on contrast sensitivity until it is reduced to the lowest level. (*a*), 9300 trolands; (*b*), 850 trolands; (*c*), 77 trolands; (*d*), 7.1 trolands; (*e*), 0.65 trolands; (*f*), 0.06 trolands. (One troland is the retinal illuminance when viewing a photopic luminance of 1 cd m^{-2} through a pupil of 1 mm^2.)

frequency. Fig. 8.14 shows how it varies with mean luminance, and it will be seen that this curve breaks into two limbs: the lower one has a plateau at 15 Hz, and this is the highest flicker rate that rods normally detect.

Another point to note in Fig. 8.13 is that contrast sensitivity peaks

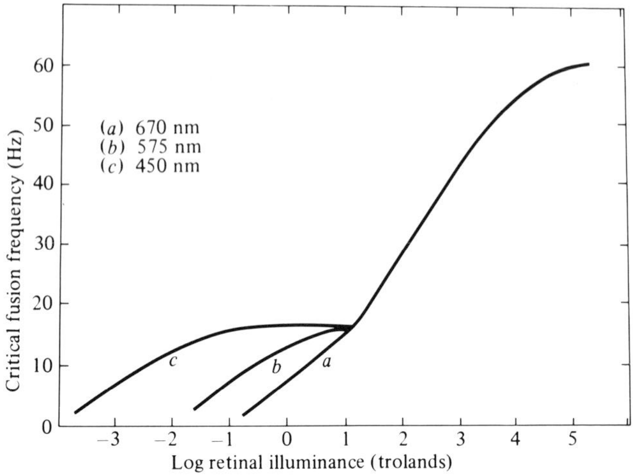

Fig. 8.14. Critical fusion frequency (cff) is the highest frequency at which a subject can just detect flicker in a field alternating between zero and twice the mean luminance. The cones show a plateau at about 60 Hz. Below a mean luminance of 10 trolands rods are operative and cff is higher for blue or green light. They appear to show a plateau at about 15 Hz, but it is possible to devise conditions where rods detect flicker at much higher rates. Note that the choice of photopic trolands as the unit for the abscissa in this figure equalises the cff for cones at the three wavelengths, but does not do so for rods, owing to the Purkinje shift.

at 10 Hz under the conditions of these measurements. As mean luminance is lowered, contrast sensitivity is reduced most at medium and high frequencies, so the peak disappears.

A natural question to ask is whether the detailed shape of the human temporal contrast sensitivity function can be explained by the shape of the impulse response curve of the receptors; as with the line-spread function and the modulation transfer function, one might expect the one to be the Fourier transform of the other. The shape of the temporal impulse response may account for the decline of sensitivity at high temporal frequencies, but it could not account for the decline at low frequencies: to do this one would have to take into account a desensitisation phenomenon in the photoreceptors that is thought to be responsible for their non-linear behaviour with strong stimuli. It is also quite possible that neural mechanisms central to the receptors have an influence on contrast sensitivity.

Another natural question to ask is whether there is segregation according to temporal frequency, as there appears to be according to spatial frequency (see pp. 142–52). Is the temporal contrast sensitivity function the envelope of several mechanisms, each tuned to a separate temporal frequency? Tests of this hypothesis have not given such impressive results as the equivalent experiments in the spatial domain, but they do support the notion that there is a division into a rapid, transient, system, probably fed by Y-type cells of the retina (see pp. 161–2 and below), and a slower, sustained, system fed by X-type cells. The different results in spatial and temporal domains are well illustrated by the following experiments performed by Robson and Watson.

The usual measure taken to define sensory performance is the threshold – the dividing line between seeing something and seeing nothing. But it seems obvious that, when one sees something, one can usually say more than just 'I have seen something': one can also describe *what* one has seen. This can be put on a more satisfactory basis than a verbal description by delivering a variety of stimuli at just-suprathreshold levels, and finding if a subject can correctly classify them according to some prearranged scheme. Checking the correctness of the responses is important because, as with ghostly shapes on a dark night, the imagination will readily supply what the senses have failed to provide, and it is the senses not the imagination that one wants to investigate.

In this experiment a whole range of stimuli were available consisting of spatial sinusoids of various spatial frequencies flickering at various temporal rates. They appeared briefly in a restricted region of the visual field, and each was adjusted to be in the range close above threshold. Tests were then performed to see how well a subject could discriminate between stimuli that differed either in spatial frequency or in temporal frequency. The results showed good performance in the spatial domain, consistent with the idea that there are spatial bandpass filters of about one octave bandwidth that enable the system to classify a just-visible sinusoid instantly into one of about eight categories. Subjects were not so good in the temporal domain, however, and could only classify them successfully into two categories; these are thought to correspond to the transient or Y, and the sustained or X, pathways.

8.4. SEEING MOVEMENT

The detection of motion is one of the simplest types of pattern recognition that the visual system performs, and it is clearly one of the most important. Many animals feed on moving prey, or escape moving predators, so efficient detection of motion has obvious survival value for that reason. It is also very important in maintaining one's orientation and balance, as can be shown by experiments in which a person is surrounded by a moveable cage or drum: if this is slowly rotated while he is seated on a stationary chair he will experience a very compelling illusion that the chair is rotating with him on it, while the cage or drum is assumed to be a stationary, stable, reference and is seen as such. Likewise if the cage is displaced or tilted while he is standing he will involuntarily move to hold himself in a stationary relation to it, and will consequently tend to lose his balance and fall.

For the reactions just described the middle to far periphery of the field of vision can be shown to be especially important, and it is sometimes said that the peripheral field of vision is specialised for detecting motion. Though movement in the periphery is certainly important in the above reactions, and may have special attention-grabbing value in other circumstances, measurements of the minimum extent or velocity of motion that can be detected do not indicate best performance in the periphery; lowest values are obtained on the visual axis, and higher values are required in the periphery. The minimum angle of displacement that can be detected is another measure that confirms the excellent accuracy of positional judgements performed by the visual system, for the lowest results are of the order of 10 s of arc, comparable to vernier acuity.

The pattern of fixational pauses separated by rapid saccadic eye movements, together with vestibular reflexes and ocular reflexes compensating for body movements (see Chapter 16), seem designed to provide the retina with an image that is as steady as possible for as much of the time as possible. Ocular following reflexes obviously also serve this purpose. Nevertheless the image is never completely steady, and the question arises how well the eye performs when the image moves, for one might expect this to cause serious blurring. It would be very hard to avoid blurred photographs with a camera as unstable as the eye using the relatively long exposure times that would correspond to the slow impulse responses of Fig. 8.12. Experiments have shown that visual performance resists image movement remarkably well, for the accuracy of positional judgements is hardly affected by image drift at velocities up to 3 deg s^{-1}. If one takes the breadth of the human cone impulse response as 33 ms (less than half that of the turtle cone shown in Fig. 8.12), one finds that an image drifting at 3 deg s^{-1} will have moved across a dozen cones during this time: that is six times the minimum resolvable angle and almost 100 times the positional accuracy the system is capable of.

Why do we not notice the image-blur caused by movement? One suggested answer is that image motion is taken account of and compensated for at a very early level in the processing of the visual image in the cerebral cortex. Our ability to interpret the sequence of frozen images presented at the cinema in terms of smoothly moving, sharply focussed, objects may well be the consequence of these processes but the mechanisms are not understood in any detail.

Physiology of motion detection

From the time when physiological activity was first recorded from visual pathways it was clear that moving stimuli were particularly effective. Since then some of the means by which movement is analysed have been discovered, and ganglion cells in the rabbit's retina selective both for direction and velocity of motion were described in Chapter 1 (see Figs. 1.5, 1.6). For each small region of the visual field a different population of ganglion cells will be active, depending on the speed and direction of movement of objects in that region. Conversely, which population is active of course tells the rabbit about the direction and velocity of movement of objects in each part of its field of vision. Directionally selective cells are found in the visual cortex of cat and monkey, and also in the mid-brain visual centre, the superior colliculus.

A strong hint about the mechanism for analysing movement is given by the psychophysical observation that a compelling sensation of motion results from seeing a stimulus in just two positions at just two moments; the continuum of positions at a continuum of times that occurs when seeing real motion is not necessary. This phenomenon has been studied for over a century and the conditions for the optimum appearance of movement are well known, but the simple fact that two positions at two times give a convincing impression of movement is perhaps the most important result. First it brings out the fact that detection of motion requires resolution in both space and time; if either the positions or the times could not be distinguished, motion could not be detected. And second, it brings out the fact that the key operation required physiologically is the discrimination of sequence; if one could not tell that stimulation at one place occurred before the other, one could not even decide that movement had occurred, and one certainly would not know its direction. Finally, it is impressive that the nervous system instantly interprets a single step in space (δs) and time (δt) as motion (ds/dt), for this is in a sense the minimum possible cue.

A good many lines of evidence obtained in neurophysiological

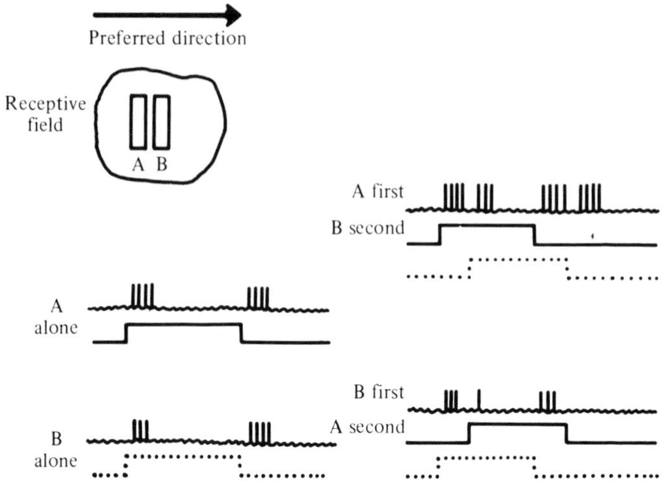

Fig. 8.15. Experiment showing inhibitory mechanism of directional selectivity. At top left two rectangular patches of light, A and B, are shown within the outline of the receptive field of a directionally selective ganglion cell in the retina of a rabbit. Continuous movements of patterns of light from left to right (corresponding to sequence A first, B second) cause a vigorous response, whereas continuous movement from right to left (sequence B first, A second) cause little or no response. The left-hand traces indicate the 'on' and 'off' responses to A alone and B alone. The lower trace to the right shows the response to the sequence B first, A second; whereas both on and off responses to B are unaffected, the responses to A are reduced compared with A alone. Apparently an inhibitory influence from B prevents the normal excitatory influences from A reaching the ganglion cell, and this inhibition, spreading in only one direction, is presumably the mechanism for the different responses to patterns moving continuously in the different directions. At top right the responses to the sequence A first, B second are shown; both responses are unaffected. Occasionally some increase of the second response is shown for the preferred sequence, but the inhibitory effects in the non-preferred (null) sequence are more prominent. (After Barlow & Levick.)

experiments suggest that this sequence discrimination depends upon inhibition preventing responses to the sequence that is ineffective in causing a response, rather than upon detecting the conjunction of sequential excitations that occurs for motion in the direction that does cause a response. Fig. 8.15 shows one type of experiment on directionally selective cells of the rabbit's retina that supports this. The detailed synaptic mechanism is not known, but it is thought to depend upon the interactions of amacrine cells and ganglion cell dendrites in the inner plexiform layers of the retina (see Figs. 2.4 and 6.3). A similar inhibitory mechanism for directional selectivity probably holds in cat visual cortex, and it may well turn out that selective inhibition is the most important means of achieving pattern detection elsewhere in the nervous system.

The Y system

The problems resulting from the long-lasting impulse response of the photoreceptors (see Fig. 8.12) are obviously most serious for fast-moving objects, yet if one recalls the sight of a kitten playing one realises that the feline nervous system must make rather a speciality of quick responses to quick movements. It has been suggested that the Y system subserves this, and a brief summary of the X and Y systems will be formed around this notion, even though it cannot be regarded as firmly established.

The Y-type ganglion cells of the cat retina were originally distinguished from the X because the excitatory and inhibitory regions of their receptive fields interacted differently. For X cells, it is possible to balance a stimulus light falling on the 'on' or 'plus' (see Fig. 1.6) region of the receptive field by a light placed simultaneously on the 'off' or 'minus' region. For a Y cell this is not possible: in whatever ways one adjusts the two lights in area, intensity, or position, there will always be a response, and the best balance that can be achieved results in equal responses at both 'on' and 'off'. This seemingly not very important difference was found to be associated with others. The Y cells characteristically give more transient, less sustained, responses than X cells, and the two types are named 'transient' and 'sustained' by some authors. The Y cells are also capable of giving brief, high-frequency, responses to rapidly moving objects, which fits in with the notion we are pursuing. For a given eccentricity from the visual axis, Y receptive fields are about twice the diameter of the fields of X cells; furthermore their axons are larger and hence conduct impulses more rapidly. Their central destinations are also different, for whereas the X type go only to the LGN and thence primarily to the simple-type cells of area 17, Y type go to the superior colliculus as well as the LGN, and from the LGN they relay to area 18 (in the cat) as well as area 17. In the cortex they form a major input to the 'complex' type neuron, though they do not provide the only input to these cells. There are about 1/10 as many Y cells as X cells, but like the X cells there are more of them near the visual axis than in the peripheral retina. Their relative rarity has aided the positive identification of Y cells with one particular histological type of ganglion cell, namely the so-called alpha cells with large dendritic trees and cell bodies.

There is one other curious property that is much more prominent in Y cells than X cells; this is their sensitivity to movements of the image occurring well away from the receptive field proper. This unexpected response, called the *periphery* or *shift* effect, has an interesting consequence when one records from the Y pathway in the LGN of a cat that is executing the normal pattern of saccadic movements separated by fixational pauses (see Chapter 11). During a saccade the image shifts over the retina and this causes a vigorous burst in every Y cell, regardless whether it is an 'on' centre or 'off' centre. In contrast, cells of the X pathway in the LGN pause during the saccade, probably as a result of inhibition from Y cells. Each X cell then assumes a firing rate during the succeeding fixational pause that is related as expected to the local contrasts in the image overlying its receptive field. The different properties of X and Y cells do not seem very important when listed separately, but acting together they lead to radically different performance when the eye is executing its normal pattern of movements.

The signals from X cells give an account of the pattern of luminance in the visual field, but what role do the Y cells play? They give bursts of impulses during saccades, but during the ensuing fixational pauses their response rates rapidly decline and can add little information about the pattern of light and shade. Consider, however, what will happen if something, such as a mouse or a bird, moves in the visual field: to this the appropriate Y cells will respond vigorously and promptly. Perhaps, then, this is the system that signals moving objects and subserves the rapid, visually-guided, responses of the playful kitten or marauding cat.

Some of what we know about cats probably carries over to primates, where the primary division that can be made is into the *magnocellular* and *parvocellular* systems, these taking their names from the large and small celled layers of the LGN. There appear to be two distinct pathways from retina right through to the parastriate cortical areas beyond V1. It is probable that the magnocellular pathway is more concerned with motion, while the parvocellular subserves colour vision, but it is not yet clear how high resolution spatial vision is handled.

8.5. SUGGESTIONS FOR FURTHER READING

General references

Graham, C. H. (ed.) (1965) *Vision and Visual Perception.* New York: John Wiley. (A good source for the work on visual psychophysics before 1964.)

Jameson, D. & Hurvich, L. M. (eds.) (1972) *Handbook of Sensory Physiology*, Vol. 7, pt 4, *Visual Psychophysics.* Heidelberg: Springer. (A massive handbook of which Chapters 1, 7, 11 are particularly relevant to the contents of this chapter.)

Robson, J. G. (1980) *Neural images: the physiological basis of spatial vision.* In *Visual Coding and Adaptability*, ed. C. S. Harris, pp. 177–214. Hillsborough, N.J.: L. Erlbaum and Associates. (A general account of the use of Fourier theory in vision.)

Special references

Contrast sensitivity, Fourier analysis and vision. Campbell, F. W. & Robson, J. G. (1968) *Journal of Physiology*, **197**, 551.

Optics of human eye. Campbell, F. W. & Gubisch, R. W. (1966) *Journal of Physiology*, **186**, 558.

Vernier acuity and interpolation. Barlow, H. B. (1979) *Nature*, **279**, 189.

Variation of contrast sensitivity with luminance. Van Nes, F. L. & Bouman, M. A. (1967) *Journal of the Optical Society of America*, **57**, 401–6.

Grating adaptation. Blakemore, C. & Campbell, F. W. (1969) *Journal of Physiology* **203**, 237.

Spatial frequency channels in vision. Braddick, O., Campbell, F. W. & Atkinson, J. (1978) In *Handbook of Sensory Physiology VIII, Perception*, ed. R. Held, H. W. Leibowitz & H-L. Teuber. Heidelberg: Springer.

Responses of neurons to spatial sinusoids. Maffei, L. & Fiorentini, A. (1973) *Vision Research*, **13**, 1255. Shapley, R. & Lennie, P. (1985) *Annual Review of Neuroscience*, **8**, 547–83.

Cortical neurons as band-pass spatial frequency filters. Movshon, J. A., Thompson, I. D. & Tolhurst, D. J. (1978) *Journal of Physiology* **283**, 53. DeValois, R. L., Albrecht, D. G. & Thorell, L. G. (1978) In *Frontiers in Visual Science*, ed. S. J. Cool & E. L. Smith. Heidelberg: Springer.

Contrast constancy. Georgeson, M. A. & Sullivan, G. D. (1975) *Journal of Physiology* **252**, 627.

Responses of photoreceptors to flashes. Baylor, D. A. & Hodgkin, A. L. (1973) *Journal of Physiology*, **234**, 163.

Flicker sensitivity of the eye. Kelly, D. H. (1972) In *Handbook of Sensory Physiology*, Vol. 7, pt 4, ed. D. Jameson & L. M. Hurvich. Heidelberg: Springer.

X and Y ganglion cells, linear and non-linear, in cat retina. Enroth-Cugell, C. & Robson, J. G. (1966) *Journal of Physiology*, **187**, 517.

Sustained or X and transient or Y ganglion cells. Lennie, P. (1980) *Vision Research*, **20**, 561.

Distinguishing flickering spatial sinusoids. Robson, J. G. & Watson, A. B. (1981) *Vision Research*, **21**, 1115.
Apparent rotation induced in drum. Brandt, Th., Dichgans, J. & Koenig, E. (1973) *Experimental Brain Research*, **16**, 476.
High acuity with moving images. Westheimer, G. & McKee, S. P. (1975) *Journal of the Optical Society of America*, **65**, 847.
Absence of blurring by motion. Burr, D. C. (1980) *Nature*, **284**, 164.
Anatomical identification of Y cells. Cleland, B. G., Levick, W. R. & Wässle, H. (1975) *Journal of Physiology*, **248**, 151.
Eye movements, X and Y cells. Noda, H. (1975) *Journal of Physiology*, **250**, 579.
Mechanism of pattern selectivity in retina. Barlow, H. B. & Levick, W. R. (1965) *Journal of Physiology*, **178**, 477.
Magno- and parvo-cellular pathways in primates. Movshon, J. A. (1989) Chapter **8** in *Images and Understanding*, ed. H. B. Barlow, C. Blakemore & E. M. Weston-Smith. Cambridge University Press.

CHAPTER NINE

Colour vision and colour blindness

J. D. MOLLON

9.1. TRICHROMACY

We are all colour blind. Despite the endless variety of hues that we experience, the colour vision of normal observers differs only in degree from that of those we label 'colour blind'. Mixtures of wavelengths that are physically very different may look identical to us. Our eyes cannot analyse the spectral composition of lights in the way that a spectroscope can.

This fundamental limitation of our vision is expressed in the fact of *trichromacy*. Suppose we illuminate the left half of a bipartite field with three wavelengths, say a red, a green and a blue (λ_1, λ_2, λ_3 in Fig. 9.1*a*). (The precise wavelengths of these 'primaries' do not matter, although it is convenient to space them out along the spectrum and it is crucial that no one of them can be matched by a mixture of the other two.) Suppose further that we illuminate the right side of the field with a fourth, 'comparison' light (*L* in Fig. 9.1*a*), which may be any mixture of wavelengths or may itself be monochromatic. If now we ask a normal observer to inspect the bipartite field foveally, he will be able, by adjusting the intensities of the three primaries on the left, to match most lights that we choose to place on the right; and if, where necessary, we allow ourselves the additional operation of adding one of our three primaries to the comparison light that is to be matched (Fig. 9.1*b*), then it will *always* be possible to effect a match. Thus, with just three numbers we can specify the mixture that our comparison light will match (provided only that we sometimes allow one of the numbers to be negative, in the sense that we transfer the corresponding primary to the other side of the match). It is true that the recent or concurrent presence in the visual field of other coloured lights will affect the hue of both sides of the match; but the match will continue to hold.

Although it is often said that trichromacy was first stated in 1757 by the Russian poet and scientist, Lomonosov, the fact was well known (in a simplified form) in the first half of the eighteenth century.

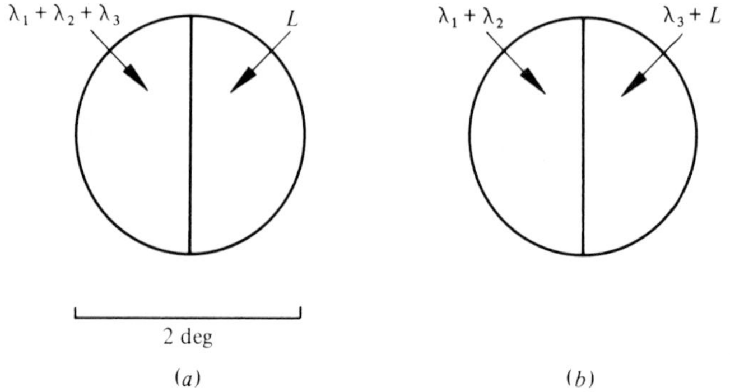

Fig. 9.1. The two possible arrangements for a trichromatic match. For explanation see text.

For in 1719 a mezzotint engraver, J. C. Le Blon, introduced in London his technique of 'printing paintings' by a three-colour process, and he shortly afterwards set down the principles in his curious essay, *Il Coloritto* (Fig. 9.2*a*). Some of his three-colour prints, it is said, were so good that, varnished, they were sold by unscrupulous dealers as oil paintings; and one impressive varnished portrait of George I in the British Museum collection lends credence to this tale. But Le Blon had a taste for high living and little talent for business and he died in poverty in Paris in 1741.

The fact of trichromacy became a commonplace in the second half of the eighteenth century, but it was almost universally regarded as a fact of physics rather than as a fact of human physiology. The confusion lingered among leading scientists until well into the nineteenth century and it is still common among non-specialists. An intermediate historical stage is represented by the prescient London tradesman, G. Palmer, who in 1777 suggested that there were three kinds of light and three corresponding types of retinal fibre (Fig. 9.2*b*).

9.2. THE PRINCIPLE OF UNIVARIANCE

As Thomas Young first clearly suggested in 1801 (Fig. 9.2*c*), trichromacy is strictly a fact of physiology; and we now know that it has its basis in the existence of just three classes of retinal cone, with peak sensitivities at different points in the spectrum (see Fig. 9.4). Any single cone, or any single class of cone, is in itself colour blind. That is to say, its response is governed by what the late

(*a*)

Of *Preliminaries.*

COLORITTO, or the *Harmony* of Colouring, is the *Art* of *Mixing* COLOURS, in order to repreſent naturally, in all Degrees of *painted* Light and Shade, the ſame FLESH, or the Colour of any other Object, that is repreſented in the true or *pure* Light.

PAINTING can repreſent all *viſible* Objects with three Colours, *Yellow*, *Red*, and *Blue*; for all other Colours can be compos'd of theſe *Three*, which I call *Primitive*; for Example,

Yellow and *Red* } make an *Orange Colour*.

Red and *Blue* } make a *Purple* and *Violet Colour*.

Blue and *Yellow* } make a *Green Colour*.

And a *Mixture* of thoſe *Three* Original Colours makes a *Black*, and all *other* Colours whatſoever; as I have demonſtrated by my Invention of *Printing* Pictures *and* Figures *with their* natural *Colours*.

I am only ſpeaking of *Material* Colours, or thoſe uſed by *Painters*; for a *Mixture* of *all* the primitive *impalpable* Colours, that cannot be felt, will not produce *Black*, but the very Contrary, *White*; as the Great Sir ISAAC NEWTON has demonſtrated in his Opticks.

White, is a Concentering, or an *Exceſs* of Lights.
Black, is a deep Hiding, or *Privation* of Lights.

(*b*)

Of VISION.

PRINCIPLES.

1. *The ſuperficies of the retina is compounded of particles of three different kinds, analagous to the three rays of light; and each of theſe particles is moved by his own ray.*

2. *The complete and uniform motion of theſe particles produces the ſenſation of white: this motion is the moſt tireſome for the eye, and may be ſtrong enough to hurt, or even deſtroy, its organization.*

3. *The abſolute want of motion in theſe particles, whether by the interception of light, or by the aſpect of a black body, produces the ſenſation of darkneſs; and this ſenſation is the perfect quietneſs of the eye.*

4. *The motion of theſe particles by diſcompoſed rays, whether by coloured bodies, or by priſmatic refractions, produces the ſenſation of colours.*

5. *Any uniform motion of theſe particles by rays not diſcompoſed, but only decreaſed, from the white to the black, produces only ſenſations of more or leſs white; but none of colours.*

6. *Theſe particles may be moved by the rays which are not analagous to them, when the intenſeneſs of theſe rays exceeds their proportion.*

(*c*)

as it is almost impossible to conceive each sensitive point of the retina to contain an infinite number of particles, each capable of vibrating in perfect unison with every possible undulation, it becomes necessary to suppose the number limited, for instance, to the three principal colours, red, yellow, and blue, of which the undulations are related in magnitude nearly as the numbers 8, 7, and 6; and that each of the particles is capable of being put in motion less or more forcibly, by undulations differing less or more from a perfect unison; for instance, the undulations of green light being nearly in the ratio of $6\frac{1}{2}$, will affect equally the particles in unison with yellow and blue, and produce the same effect as a light composed of those two species: and each sensitive filament of the nerve may consist of three portions, one for each principal colour.

Fig. 9.2. Three stages in the understanding of trichromacy. (*a*) The opening page of Le Blon's *Il Coloritto* (*c.* 1722). Notice that Le Blon distinguishes between the mixture of lights and the mixture of pigments, and gives rules for the latter. (A pigment gains its colour by selectively absorbing part of the spectrum. Trichromacy will hold for mixtures of pigments provided they do not interact chemically.) Le Blon assumes that trichromacy has a physical basis. (*b*) A remarkable passage from G. Palmer's *Theory of Colours and Vision* (1777). Palmer postulates both physical and physiological trichromacy. He also passes under the names Girod de Chantilly and Giros von Gentilly, and both his identity and his nationality have been the subject of speculation; but he can almost certainly be identified with G. Palmer (1740–1795) a London glass-seller, who specialized in coloured glass and lived at 118 St Martin's Lane (for references, see §9.10 below). (*c*) From Thomas Young's Bakerian Lecture of 1801. Young assumes that light varies continuously in wavelength while human vision is trichromatic. His contribution was not a sudden insight but the judicious conclusion of a man who had carefully read the eighteenth-century literature.

W. A. H. Rushton called the *Principle of Univariance*: although the stimulus may vary in intensity and in wavelength, the response of a cone is thought to vary along only one dimension – the extent of its electrical hyperpolarisation (Chapter 6). Once an individual photon has been absorbed, all information about its energy level* is lost, although as wavelength is varied the *probability* of a given photon's being absorbed will vary according to the spectral sensitivity curve† of the photopigment that the cone contains. Thus a change in intensity or a change in wavelength may produce the same change in the electrical polarisation of the cone. The Principle of Univariance means that it is mistaken to say 'cones discriminate colour'; only by comparing the outputs of the individually colour-blind cones is the visual system able to discriminate wavelength.

We have noted that in a colour-matching experiment one of the primaries may have to be given a negative value. Does this mean that the Principle of Univariance is wrong and that cones are bivariant, giving responses of different sign to steady stimuli from different parts of the spectrum? No. The need to transfer one of the primaries to the other side of the match arises when the test light consists of a very narrow band of wavelengths (and thus has a very 'saturated' – strongly coloured – appearance). Such a light is producing large differences in the absorption rates of the three classes of cone and there will be cases where we cannot reproduce these large differences by any mixture of our primaries. Essentially this is because none of our primaries will stimulate only a single class of cone (see Fig. 9.4). By mixing one of the primaries with the test stimulus the differences in absorption rates are reduced and a match can be achieved.

9.3. SPECTRAL SENSITIVITY FUNCTIONS

Ideally, to specify the set of lights that will match our test light, we should like to take as our three variables the rates of absorption of photons in the three classes of cone, rather than the intensities of our three primaries. We should not then have to trouble with negative values. However, the spectral sensitivities of the individual classes of cone cannot be uniquely derived from the results of colour-matching experiments alone, and it has not proved straightforward to establish these sensitivities by other methods.

* The energy of a photon is proportional to its frequency, and thus inversely proportional to wavelength in a given medium.

† The *spectral sensitivity curve* is the function relating sensitivity to wavelength. See, for example, Fig. 9.4 and the discussion of such functions in Chapter 1.

Until recently almost all the methods of estimating the spectral sensitivities have been psychophysical. In most procedures the principle has been to ensure that the observer's response depended on only one class of cone. This has been achieved by one or a combination of several devices. The simplest is to use colour-blind observers. Another is to adapt the other classes of cone with monochromatic fields that have little effect on the class of cone whose sensitivity is being measured. A third is to select retinal positions or stimulus parameters that favour a particular class of cones. As an example, we shall review the psychophysical method introduced by W. S. Stiles, in part because it instructively illustrates some basic principles and in part because the results it yields are in first-order agreement with those recently obtained from direct measurements of individual primate cones. Better than any other, the work of Stiles demonstrates the power and limitations of the psychophysical approach (see Chapter 7).

Two-colour increment-threshold procedure

In Stiles' psychophysical method, a monochromatic test-flash is presented in the centre of a larger, concentric, adapting field (Fig. 9.3*a*) and the threshold for detecting the test-flash is measured. Typically the test-flash is square with sides subtending 1 deg of visual angle and it is delivered to the centre of the fovea; its duration is 200 ms. The wavelength of the test-flash is conventionally represented by λ and that of the field by μ.

Central to the understanding of Stiles' method is the Principle of Univariance, which he explicitly assumes. A second important assumption is that the adaptive states of the individual cone systems are essentially independent; that is to say, the sensitivity of any particular class of cones depends only on the rate at which photons are absorbed from the background field by that class of cones and is independent of the state of other classes of cone. We shall refer to this as the *Principle of Adaptive Independence*. Stiles' own results show that it is correct to a first approximation. Always a cautious man, Stiles himself did not speak of isolating 'classes of cone' but merely of isolating 'cone mechanisms' or 'π mechanisms': a π mechanism is simply a hypothetical association of receptors and accompanying neural elements, and its defining characteristic is obedience to the Principles of Univariance and Adaptive Independence.

The adapting field in Stiles' experiments serves two distinct purposes. Firstly, its wavelength (μ) can be chosen so that it

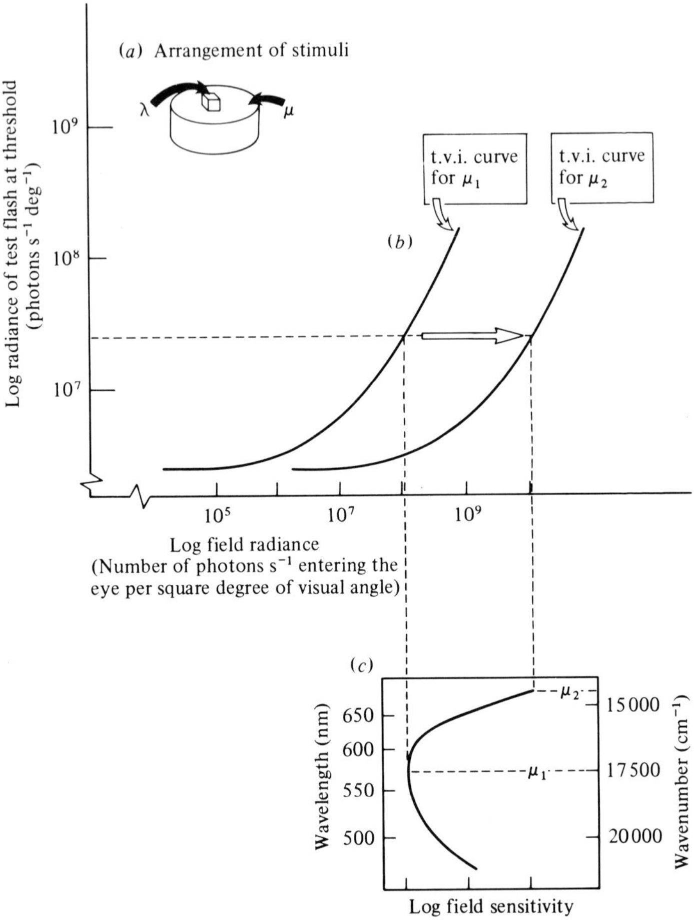

Fig. 9.3. Derivation of field sensitivity by the two-colour method of Stiles. (*a*) Spatial arrangement of test-flash and adapting field. (*b*) t.v.i. curves for the same test wavelength but two different field wavelengths, μ_1 and μ_2. The horizontal broken line shows the criterion elevation of threshold (1 log unit) usually used by Stiles for deriving the 'field sensitivity' of a π mechanism. (*c*) Derivation of field sensitivity. When μ_2 is used instead of μ_1 the t.v.i. curve is displaced along the abscissa of (*b*) by 2 log units; this displacement corresponds directly to a 10^2 difference in sensitivity in plot (*c*). The abscissa of plot (*c*) is wavenumber rather than wavelength; Stiles favoured this way of plotting because the corneal spectral sensitivities of the different cone mechanisms are then more similar in shape and because frequency (which is directly proportional to wavenumber) is independent of the medium. The sensitivity derived in the figure is in fact that of Stiles' long-wavelength mechanism π_5.

differentially adapts those cone mechanisms that are not being examined while having little effect, relatively, on the mechanism that is favoured by the wavelength (λ) of the test-flash. But secondly, by manipulating μ (within a range compatible with the first purpose), Stiles obtains an estimate of the spectral sensitivity of the mechanism that has been isolated, an estimate that is independent of the measure he can obtain by varying λ. Fig. 9.3*b* illustrates the principle. The primary experimental result is a plot showing how the threshold for detecting the test flash increases as a function of the log intensity of the background. Stiles calls such a function a *t.v.i.* ('threshold vs intensity') curve. If μ is changed this function is found to be displaced horizontally along the abscissa without distortion, provided only that the same cone mechanism remains isolated. This is one of Stiles' two Displacement Rules and it must follow from the Principles of Univariance and Adaptive Independence. The idea can be readily understood by reference to Fig. 9.3*b*, where t.v.i. curves are shown for two adapting wavelengths, μ_1 and μ_2. If we could plot on the abscissa of Fig. 9.3*b* the actual number of photons absorbed from each adapting field by the cone mechanism that has been isolated, then the two curves would be superimposed, since the wavelength of the photons, once they are absorbed, does not matter and since the effect of the field on other classes of cone is assumed to be irrelevant. But in fact we know only the number of photons delivered to the observer's eye and we find that the curve moves laterally as we vary μ. It is these shifts that give us a measure of spectral sensitivity, for they show how the proportion of photons absorbed is varying as we alter the background wavelength. To raise the threshold by a fixed amount we need 100 times as much light at μ_2 as at μ_1 and that means that our mechanism is 100 times ('2 $\log_{10}$ units') less sensitive to μ_2. Sensitivity (in this case 'field' sensitivity) is defined simply as the reciprocal of the intensity required to produce a given effect. The derivation is shown graphically in Fig. 9.3*c*.

If now we hold μ constant and vary λ (the wavelength of the test-flash), we find that the entire t.v.i. curve is shifted vertically along the ordinate without distortion. This is Stiles' other Displacment Rule and by an exactly analogous procedure he derives a second measure of spectral sensitivity, 'test sensitivity'. A crucial result is that the two independent measures, 'test' and 'field' sensitivity, give closely similar results in those parts of the spectrum where both can be obtained for the same cone mechanism. This is the evidence for the Principle of Adaptive Independence; if the sensitivity of one class of cone were controlled not only by its own rate of quantum catch but also by photons absorbed in other cones, its field sensitivity curve would be distorted relative to its test sensitivity.

Sooner or later as μ or λ is varied the appropriate Displacement Rule will fail. In Stiles' scheme this means that a new mechanism has taken over detection of the test flash. The three main π mechanisms so derived have peak sensitivities, measured at the cornea, of approximately 440 nm (π_3), 540 nm (π_4) and 570 nm (π_5). Only marginal modifications need to be made to these three mechanisms for them to satisfy one property of cones that is theoretically critical: when standard and comparison lights are phenomenally matched in a colour-mixing experiment, the calculated absorptions in π_3, π_4 and π_5 also come very close to being matched.

However, Stiles found more failures of the Displacement Rules than were compatible with three independent cone systems. For example, there appeared to be three blue-sensitive mechanisms, π_1, π_2 and π_3, which all have their peak sensitivity near 440 nm but differ in the extent of their sensitivity to long-wavelength fields. It is nowadays thought that this multiplicity represents a limited failure of the Principle of Adaptive Independence rather than the existence of three separate blue-sensitive cone systems (see §9.5, 9.6).

Microspectrophotometry

Direct measurements of the pigments in individual primate cones can be made by 'microspectrophotometry', a procedure in which a tiny monochromatic beam is passed through individual receptors in freshly dissected tissue. Typically the measuring beam must be less than 2 μm broad if it is to be confined within the outer segment of a single receptor. Using an infra-red converter to avoid bleaching the pigment, the experimenter lines the cell up in the beam; and then the light absorbed at each visible wavelength is measured. Since light will be absorbed by material other than the photopigment, the experimenter either passes a reference beam through adjacent tissue and uses this as a baseline or else takes the difference between absorption in the receptor before bleaching and absorption after bleaching. The former procedure yields an *absorption spectrum* (e.g. Fig. 5.7), the latter a *difference spectrum*; neither is necessarily the same as the *action spectrum*, the actual effectiveness of different wavelengths in producing a given physiological or psychophysical effect. One problem for microspectrophotometry is that the photoproducts of bleaching may not themselves be transparent. Such photoproducts may contribute significantly to absorption at short wavelengths (see Chapter 5). They may thus distort difference spectra and, if they are present before measurements begin, they may contaminate absorption spectra too.

The microspectrophotometrist must decide whether to pass his measuring beam axially or transversely through the outer segments of individual receptors. In the earliest microspectrophotometric work a fragment of retina was mounted with the receptors on end and the measurements were axial. This arrangement was favoured because the optical density of the pigment was thought to be low and so the maximum path-length was sought. The disadvantage of axial measurement is that the focussed measuring beam must be confined within a cylinder 30 μm long and 1–2 μm in diameter; in practice light is likely to leak into adjacent cones and rods, and the absorption spectrum will thereby be distorted. More recently it has been realised that the optical density of the pigment is high and thus good records can be obtained with the 2-μm path-length available in transverse measurements. The advantage of transverse measurements is that the individual receptors, lying on their sides, can be dispersed so that they are well separated; and thus the experimenter can be confident that his beam passes through only a single cell.

In recent work by Bowmaker, Dartnall and Mollon, transverse measurements were obtained for large samples of macaque and human cones. For macaques the peak sensitivities of the cones lie at approximately 420 nm, 535 nm and 565 nm; for man, the corresponding values are approximately 420 nm, 530 nm and 560 nm. In Fig. 9.4 we show the mean absorbance spectra for seven human eyes. Notice that the so-called 'red-sensitive' and 'blue-sensitive' cones have their peak sensitivities in the yellow-green and violet respectively.

How well do these microspectrophotometric measurements agree with Stiles' psychophysical estimates? The comparison of absorption spectra with psychophysical action spectra is not straightforward. First, we have to correct the psychophysical data for absorption by the media of the eye and by the macular pigment (Chapter 5). Second, we have to allow for the fact that *in vivo*, where the light beam is axial, the pigment in the anterior part of the outer segment will significantly screen the pigment in the posterior part; this 'self-screening', being maximal at the wavelength of peak sensitivity of the pigment, has the result of broadening the absorption spectrum for the cell as a whole. Third, we must allow for any wavelength-dependent optical funnelling that may occur for an axial beam *in vivo*.

The first of these corrections, and its attendant uncertainties, can

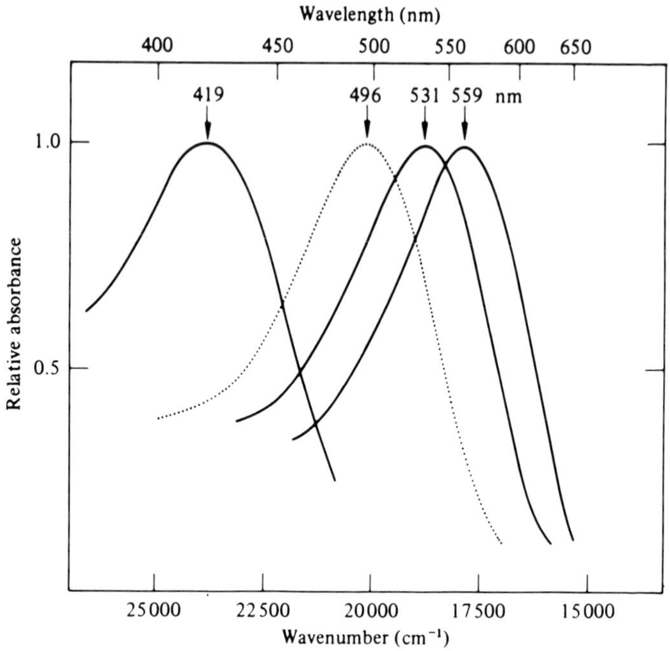

Fig. 9.4. The absorbance spectra of the four photopigments of the normal human retina. The solid curves are for the three kinds of cone, the dashed curve for the rods. The quantity plotted is *absorbance*, i.e. log (intensity of incident light/intensity of transmitted light), expressed as a percentage of its maximum value; when 'normalised' in this way, absorbance spectra have the useful property that their shape is independent of the concentration of the pigment.

These curves are mean results obtained microspectrophotometrically from seven human eyes and are reproduced here by kind permission of Professor H. J. A. Dartnall and Dr J. Bowmaker.

be side-stepped by the comparison of Fig. 9.5: here the ratio of the sensitivities of Stiles' psychophysical mechanisms, π_4 and π_5, is compared at each wavelength with the ratio of the sensitivities of 'green' and 'red' human cones. It is plausible (though not completely safe) to assume that absorption by the media and the macular pigment will affect 'green' and 'red' cones equally and so will not distort results plotted in this way. The agreement is good, although the ratio π_5/π_4 does not show the concavity at short wavelengths that is seen in the microspectrophotometric function and is required by other psychophysical data.

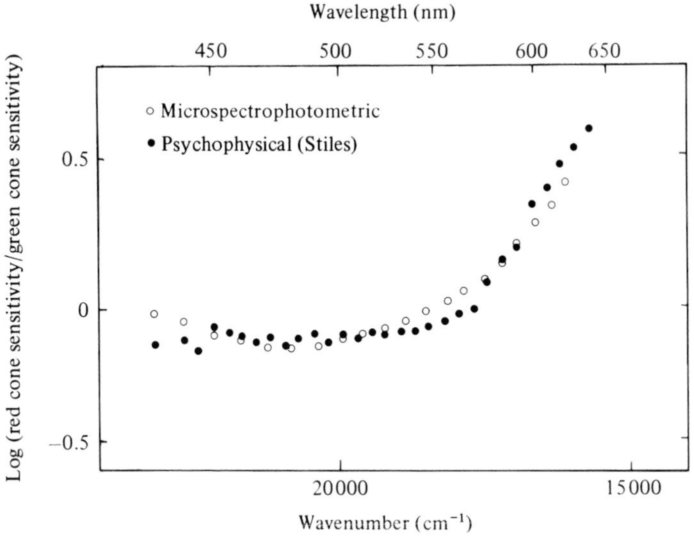

Fig. 9.5. A comparison of microspectrophotometric and psychophysical estimates of the sensitivities of long- and middle-wavelength cones. The psychophysical data are derived from field sensitivities for π_4 and π_5 as tabulated by Wyszecki & Stiles (1967). To obtain the microspectrophotometric estimates, the data of Fig. 9.4 have been used to calculate the axial *absorptance* (fraction of light absorbed or $(I_{inc} - I_{trans})/I_{inc}$) for a foveal cone 30 μm long and having an absorbance of 0.015/μm. What is plotted in each case in this figure is the logarithm of the ratio of the sensitivity of long- and middle-wavelength cones. By plotting *ratios* of sensitivities in this way we can compare the psychophysical and microspectrophotometric data without making assumptions about absorption by the ocular media.

9.4. NEURAL ANALYSIS

If the three classes of cone individually obey the Principle of Univariance (see §9.2), there cannot be colour discrimination without neural machinery to compare the outputs of different classes of cone. The necessity of this comparison has traditionally been obscured by a controversy between proponents of the 'trichromatic' (three-cone) theory of colour vision and proponents of the rival 'opponent-process' theory of Hering, a controversy that has now become redundant.

The way in which the comparison is made between absorptions in different classes of cone is closely analogous to the way a spatial comparison is made between absorptions in adjacent retinal regions

(Chapter 6). Thus, many retinal ganglion cells and LGN cells in macaque monkeys appear to receive antagonistic inputs from different types of cone: they show an increase in firing rate when their receptive fields are stimulated with one part of the spectrum and a decrease (below any spontaneous level of activity) when stimulated by another part of the spectrum. Such cells are called 'colour opponent'. In detailed studies of retinal ganglion cells, chromatic adaptation has been used to separate different types of cone input to a given cell, and test sensitivity has then been measured by an attenuated form of Stiles' technique (see §9.3). Almost all possible combinations of excitatory and inhibitory inputs from the three classes of cone have been demonstrated in different ganglion cells. Most common are cells receiving opposed inputs from 'red' and 'green' cones, but many cells are 'trichromatic', having inputs from all three types of cone, two being opposed to the remaining one. Inputs from 'blue' cones have been seen only in 'trichromatic' cells and appear to be almost always excitatory: the existence of 'blue-off' units is questionable.

Most colour-opponent cells also show spatial opponency: one type of cone provides the input to the centre of the receptive field while one or both of the other types provides input to the concentric surround (see Chapter 6). As a class the colour-opponent cells are most common near the centre of the visual field and have small receptive fields; the field centres probably correspond to only a few cones at the most.

Other cells in the retina and lateral geniculate are chromatically *non-opponent*: their response is of the same sign whatever the wavelength of the stimulus. According to P. Gouras and his collaborators, such non-opponent ganglion cells have larger receptive fields and their responses are transient, whereas sustained responses characterise colour-opponent cells. Whether, as has often been suggested, the non-opponent cells signal brightness is an open question.

9.5. THE ANOMALIES OF THE BLUE MECHANISM

There is nothing grossly odd about the blue-sensitive cones when (in the course of microspectrophotometry) they are seen by light microscopy, but our vision is definitely odd when it depends on the blue cones alone. The psychophysical blue mechanism, defined by Stiles' procedure, differs in a number of properties from the red- and green-sensitive mechanisms. For example, the blue mechanism is disproportionately vulnerable to diseases that affect the retina, such as retinitis pigmentosa and diabetes mellitus, although conversely it is much less frequently subject to genetic anomalies (see §9.9). Even in its normal state its absolute sensitivity is low: its quantum efficiency at the wavelength of its peak sensitivity is 100 times poorer than that of π_4 and π_5. In part this is accounted for by the increased pre-receptoral absorption at short wavelengths (Chapter 5), but these increased transmission losses do not explain why *differential* sensitivity is poorer for the blue mechanism: the Weber fraction $\left(\frac{\Delta I}{I}\right)$, the percentage increment in intensity that can just be detected (Chapter 7), is about 9% compared with a value of < 2% for the red and green mechanisms. These are the Weber fractions for Stiles' standard 1-deg, 200-ms flash, but if the target is made very small or very brief the difference in sensitivities is exaggerated and our vision becomes *tritanopic*, resembling that of the rare individuals who congenitally lack the blue-sensitive cones (see §9.9). The reason that the Weber fractions of the blue- and green-sensitive mechanisms differ by only a factor of about 4.5 when the target is large and long (as in Stiles' standard conditions) is that larger space and time constants provide some compensation for the basic insensitivity of the blue system. Thus the blue-sensitive mechanism shows much more extensive spatial integration than the red and green mechanisms, in that Ricco's law (Chapter 7) holds over a larger area; and its spatial resolution is correspondingly poor. Similarly, its time constants are found to be larger, in that Bloch's law (Chapter 7) holds over a greater interval.

The basic insensitivity of the blue mechanism and several of the associated properties could be explained (directly or indirectly) if the blue cones were relatively rare. This possibility is also suggested by the microspectrophotometric results (see above), although the reader will realise that the latter could be distorted by a sampling bias (if, say, the outer segments of blue cones differentially adhered to the pigment epithelium as the retina was lifted away). Psychophysical experiments suggest that blue cones are almost completely absent in an area at the very centre of the fovea corresponding to the central 20 min of the visual field.

A second class of anomalies are associated with the light and dark adaptation of the blue mechanism. For example, strange losses of sensitivity to short-wave targets may occur at the onset and offset

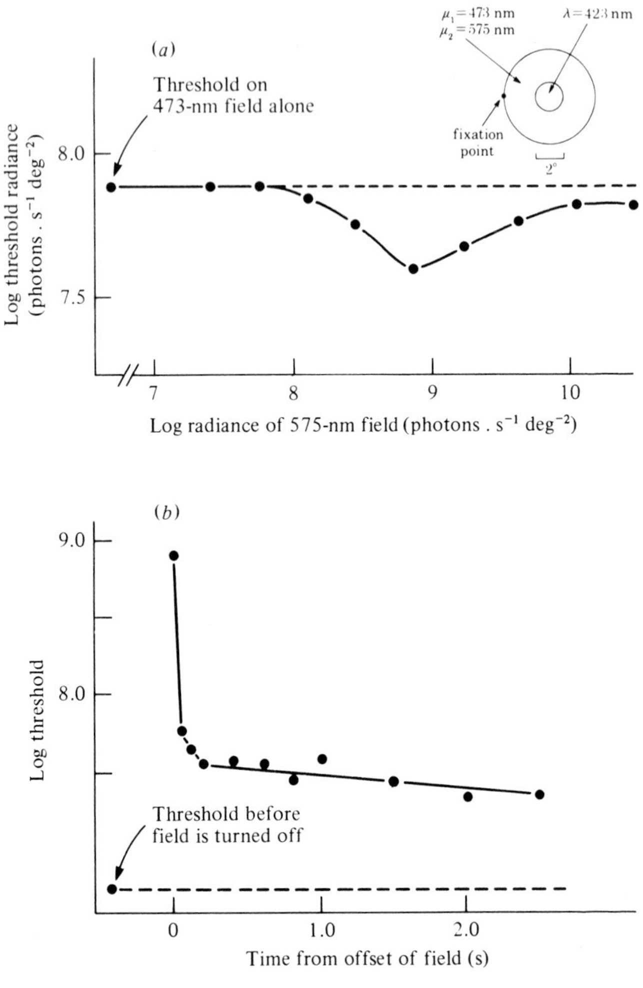

Fig. 9.6 (*a*) Combinative euchromatopsia. The ordinate represents the log threshold intensity for detecting a short-wavelength test-flash (423 nm) presented on a composite steady background consisting of (i) a blue field of fixed intensity and (ii) a complementary yellow field of the intensity indicated on the abscissa. The duration of the test-flash was 200 ms and it subtended 2 deg of visual angle. (From Polden & Mollon, 1980, Fig. 9.) (*b*) Transient tritanopia. The data points show the threshold for a short-wavelength (445-nm) test flash just before, and for 2.5 s after, a yellow adapting field is turned off. (For details of conditions see Mollon & Polden, 1977.)

of long-wave fields (Fig. 9.6*b*), even though such fields produce negligible absorptions in the short-wave cones themselves. These phenomena represent failures of the Principle of Adaptive Independence and we have already hinted (§9.3) that such failures account for why Stiles was led to postulate three blue-sensitive π mechanisms. In explanation, it is thought that signals from the short-wave cones are transmitted only by colour-opponent channels of the visual system and that the adaptational anomalies reflect not the behaviour of the blue cones themselves but that of the chromatically opponent channels to which their signals are confined (see below). The idea that the short-wave cones make little or no contribution to non-opponent channels is compatible with known electrophysiology and with the fact that changes in the intensity of the short-wave primary in colour-matching experiments can produce large changes in hue while having negligible effects on brightness.

9.6. PSYCHOPHYSICAL DEMONSTRATIONS OF OPPONENT PROCESSES

The adaptational 'anomalies' of the blue mechanism are no longer in fact thought to be curiosities peculiar to the short-wave system. Analogues of them can be found for signals originating in the red- or green-sensitive cones provided that detection depends on chromatically opponent channels. The trouble is that these cones enjoy access to a variety of post-receptoral channels, some opponent, some non-opponent; and only under a limited range of stimulus conditions are the opponent channels the more sensitive and thus available for study in psychophysical experiments.

Shown in Fig. 9.6 are two adaptational phenomena that illustrate properties of opponent processes in the visual system. In both cases the targets were of short wavelength, but analogues of both phenomena have been convincingly demonstrated in conditions where detection depends on signals originating in the long-wave cones.

Combinative euchromatopsia

The first phenomenon (Fig. 9.6*a*) is a steady-state one, called 'combinative euchromatopsia'. The term refers to the *enhanced sensitivity to hue differences* ('euchromatopsia') produced by combining certain adapting fields. The spatial arrangement of stimuli is shown inset in the figure. The threshold is first measured for a violet

($\lambda = 423$ nm) test-flash delivered on a blue ($\mu_1 = 473$ nm) adapting field. This gives the left-most point in the figure. The blue field remains fixed in intensity for the rest of the experiment, but now increasing increments of yellow light ($\mu_2 = 575$ nm) are added to the adapting field and it is the intensity of this added light that is plotted on the abscissa. Paradoxically, the threshold falls as the yellow light is added; and it passes through a minimum when the composite adapting field appears white. This striking failure of Weber's law is thought to occur because opponent channels are maximally sensitive to input perturbations when in the middle of their operating range. Strongly coloured fields (in this case blue or yellow) polarise the 'opponent channel', driving it to one or other extreme, where it is less sensitive.

Combinative euchromatopsia is not merely an oddity of experiments on increment thresholds but reflects a basic property of colour vision. Classical measurements of the observer's ability to discriminate one colour from another of the same luminance show that the threshold (if expressed in terms of the change required in receptoral absorptions) is smallest when the absorptions in different classes of cone are similar: when, in other words, the discriminanda are close to white, and when opponent channels are unlikely to be polarised.

Transient tritanopia

If the eye is adapted for a few minutes to yellow or red light, and the adapting field is then turned off, our sensitivity to blue or violet flashes does not recover according to the normal dark-adaptation curve. Instead the threshold actually rises and for several seconds remains above the level of that obtained when the adapting field was present. This 'transient tritanopia' is shown in Fig. 9.6*b*. Significantly, the phenonemon cannot be obtained with a white adapting field. It is thought to arise because a restoring force acts to oppose the maintained polarisation of any opponent channel and so bring the channel back closer to the centre of its operating range. When the adapting field is suddenly taken away, the restoring force continues to act unopposed for some seconds, the channel is thereby polarised in the opposite direction, and sensitivity is transiently lost.

9.7. CENTRAL ANALYSIS OF COLOUR

In area 17 of the primate cortex (the primary visual projection area) colour opponent cells are common, although not as frequent as in the lateral geniculate nucleus. At this stage a new characteristic is seen: many of the colour-opponent cells show a spatial centre-surround antagonism that is derived from the same cone mechanism. Thus, for example, a cell might give an on-response to a long-wavelength spot in the middle of its field and an off-response to a long-wavelength annulus, while giving an off-response (or no response at all) to short-wavelength stimuli of any spatial characteristic. Such cells may draw their (antagonistic) centre and surround inputs from two or more LGN units of the same opponent-colour type.

A question of some interest is that of the extent to which colour is analysed independently of other attributes of the visual image (see also Chapter 12). We can at the start exclude the two extreme possibilities. It would be no good having a system in which the properties of an object were not dissociated at all, a system in which, for example, there were single units specific to chartreuse-coloured Volkswagens moving left at a distance of three metres. The main difficulty is not the numerical one of the number of cells required but our need to explain how the system recognises a Volkswagen for what it is, independently of its accidental features, such as hue, position and direction of movement. (Sometimes, granted, hue may be one of the defining characteristics of an object.) But equally, we cannot postulate complete independence of analysis, since a person looking at an intricate visual scene can quickly and accurately report which colour belongs to which object. He or she can tell us that the Volkswagen is yellow and the Mini Metro, scarlet.

Given these preliminaries, how do electrophysiological findings bear on our question? In area 17 of the primate cortex, colour-specificity has been found to be associated with the several types of receptive field identified by Hubel and Wiesel as 'non-oriented', 'simple', 'complex' and 'hypercomplex' (see Chapter 1). Thus a cell might be found that gave its maximum response only to a red bar of a particular orientation moving in a particular direction. But opponent-colour characteristics have been most often found in association with 'non-oriented' or 'concentric' receptive fields, cells with such properties being particularly frequent in layer 4B of the striate cortex, close to the input from the optic tract. Does this show that spatial and chromatic analyses are progressively dissociated in the visual cortex? The reader should draw this conclusion with caution, since a cell that will respond only when a large number of attributes are narrowly specified is thereby less likely to be sampled in an electrophysiological experiment, especially if it has no spontaneous activity.

Bearing on our question in a distinct way are the reports of the

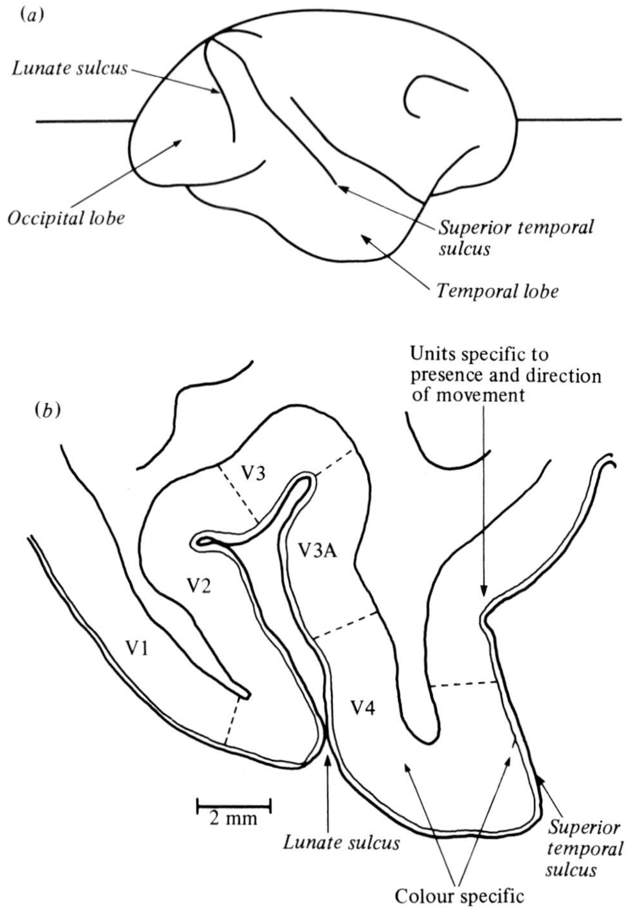

Fig. 9.7. (*a*) A lateral view of the brain of the rhesus monkey, showing the positions of the lunate sulcus and the superior temporal sulcus. (*b*) A horizontal section through the prestriate cortex. The section has been made at the level of the horizontal line in (*a*). After Zeki (1977). *Proc. Roy. Soc. B*, *197*, 195.

London anatomist, S. M. Zeki, who has described two adjacent areas in the prestriate cortex of the rhesus monkey that appear to be specialised for the analysis of colour. The first of these, V4, lies on the anterior bank of the lunate sulcus (Fig. 9.7) and the second on the lateral part of the posterior bank of the superior temporal sulcus. Over 56% of cells sampled in these two areas were colour-opponent and some had very narrow action spectra. Some colour-opponent cells had an orientational preference, but most were not selective for either orientation or direction of motion. In contrast to the striate cortex (VI), where opponent-colour cells are seldom driven by both eyes, most of the colour-specific cells of V4 are binocularly driven, the colour-preference and receptive-field properties being similar for the two eyes. Conversely, in other prestriate areas (V2, V3 and V3A) opponent-colour cells are rare while orientational preferences are common. Zeki also reports that in the posterior bank of the superior temporal sulcus, medial to the colour area, there is a region where cells are specialised for detecting the presence and direction of movement but are not concerned with colour.

Further evidence bearing on our question is provided by some very odd perceptual phenomena known as *contingent after-effects.* The reader may experience one of these (the McCollough effect) by observing, for about 4 min, the red and green pattern printed on the cover of this book. Notice that red areas are associated with leftward tilted bars and green areas with rightward tilted bars. If afterwards the black and white pattern (Fig. 9.8) is observed, it will appear pale green where the bars are tilted leftward and pink where they are tilted rightward. The illusory colours are 'contingent', in that they are seen only when a second stimulus dimension has a particular value. The most widely favoured explanation of such effects attributes them to the presence in our visual system of neurons specific both to colour and to some other attribute. We are to suppose that our perception of the colour of a black and white grating depends on the relative activity in a population of neurons all specific to the same orientation but differing in the colour to which they most strongly respond. If we selectively adapt those neurons specific to red and to rightwards tilt, these units will not make their normal contribution when a black and white grating is observed and so the grating will appear pale green.

The effect demonstrated here is only one of a large number of contingent after-effects. There are after-effects of colour that are contingent on spatial frequency or direction of movement, and there are after-effects of orientation and movement that are contingent on colour. In each case the after-effect is attributed to the presence in our visual system of multiply specific neurons. However, in several characteristics these strange phenomena resemble Pavlovian conditioning: contingent after-effects may persist for hours, days and even months; they show spontaneous recovery, reappearing after they have once disappeared; and the rate of their decay depends on the frequency of test trials. We might suppose that the unconditioned response of the visual system to an excess of redness in the world is to turn down the gain in the red-sensitive channels; a correlated visual attribute (in our example, orientation) then becomes the conditioned stimulus for this response and is analogous to the bell in Pavlov's paradigm.

Pathology also provides evidence relevant to our question of the independence of analyses and this evidence will be discussed below.

9.8. COLOUR CONSTANCY

'When a considerable part of the field of vision is occupied by coloured light, it appears to the eye either white, or less coloured than it is in reality; so that when a room is illuminated either by the yellow light of a candle, or by the red light of a fire, a sheet of writing paper still appears to retain its whiteness.' (Thomas Young, *Lectures on Natural Philosophy*, 1807.) In this passage the author of the trichromatic theory makes it clear that the appearance of a particular local coloured surface in our world does not depend *only* on the rates of absorption in the three classes of receptor in the corresponding retinal area. The important phenomenon that Young describes has since become known as 'colour constancy': objects in our world appear to retain a constant hue despite large changes in the spectral composition of the illuminant and thus in the local spectral flux that reaches our eye from a given surface. Many readers will at some time have made the error of using 'daylight' colour film to photograph an indoor scene lit by tungsten light: when our photograph is returned from processing it is little more than a chiaroscuro study in yellow and browns, a very poor representation of what we saw. The reason is that tungsten light is dominated by long wavelengths, and cameras, unlike the eye, are not yet able to compensate automatically for the change in the spectral composition of the illuminant.

The power of colour constancy is prettily illustrated in a classical experiment by Ewald Hering (1834–1918). Hering took two pieces of matte paper, one brown, one ultramarine, and he placed one on each of two faces of the right-angled wooden prism of a Bouguer photometer (Fig. 9.9). By means of a mirror the brown paper was illuminated by the light of the white sky, while the ultramarine paper was illuminated by the artificial light of a gas flame. Looking

Fig. 9.8. A figure for demonstrating the McCollough effect. For instructions, turn to the back cover of the book.

monocularly through the vertical tube of the photometer, the observer saw a bipartite field, rather like that shown in our Fig. 9.1. At a suitably chosen intensity of the gas flame, the ultramarine paper appeared brown and perfectly matched the brown paper illuminated by daylight. The match was possible because the artificial gas-light was weak in short wavelengths, and under these conditions we may assume that the two surfaces produced the same ratios of absorptions in the three classes of retinal cones. Now Hering closed his window shutters, illuminated the whole room with gas-light and took both

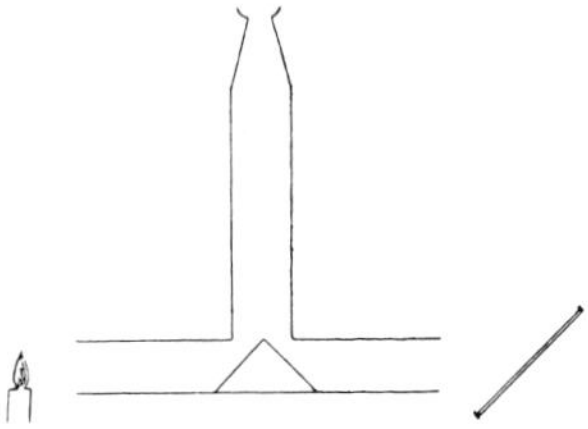

Fig. 9.9. Left: Ewald Hering in his later years. Right: an arrangement used by Hering to demonstrate colour constancy.

papers out of the photometer. Instantaneously the ultramarine paper ceased to look brown and appeared as blue as in daylight, even though it was illuminated by the same artificial light as before. What this experiment shows is that stimulation of an area of retina with the very same spectral flux may yield quite different sensations (brown or blue, in this case) according to what we reveal to the eye about the spectral fluxes from other regions, and thus, indirectly, what we reveal about the spectral composition of the illuminant.

A very grand version of Hering's experiment has been described in a *Scientific American* article (see References) by Edwin Land, inventor of polaroid and of the Land camera. But it is essentially the same experiment; and we still do not know how the visual system actually achieves colour constancy. Land himself, however, has made one specific proposal. In his 'retinex' theory, he supposes that each of three cone systems independently extracts the spatial pattern of illumination *as seen by that system*, scaling each local signal according to the total range of illuminations that it finds over a larger region. The latter scaling gives a pattern of lightnesses that is specific to a particular class of cone. So a blue patch, say, in the external array will be represented as a dark area in the pattern extracted by the long-wave cone system and as a light area in the pattern extracted by the short-wave cone system. At a later stage in the visual system, a comparison is made of the three separate lightnesses signalled for the same local area by the three cone systems, and this comparison gives the colour of the corresponding surface. Since changes in the overall amount of, say, long-wave light in the image will not alter the pattern of lightnesses extracted by an individual cone mechanism, the retinex theory does yield colour constancy.

It is interesting to notice that both Stiles and Land, ostensibly by very different routes, are led to suggest that sensitivity is adjusted independently for each class of cone signal before interactions occur between signals from different classes. Perhaps colour constancy will prove to be the functional advantage of the adaptive independence discussed in §9.3.

9.9. COLOUR BLINDNESS

We began this chapter by defining the colour blindness of the normal eye in terms of the number of variables needed to achieve all possible colour matches, and it is in similar terms that the colour-deficient are best classified. About 2% of the Caucasian male population are *dichromats*, who require only two variables in a colour-matching experiment and for whom there is a band of wavelengths in the spectrum that matches white. Dichromacy is usually inherited, as a recessive, sex-linked characteristic, and occurs in less than 0·03% of women. The most probable explanation of dichromacy is that one of the three classes of cone is absent, a suggestion first made in 1781 by G. Palmer (whose anticipation of the trichromatic theory we met above, Fig. 9.2*b*). Those in whom the red-sensitive cones are thought to be absent are called *protanopes*, those thought to lack the green-sensitive cones, *deuteranopes*. Protanopes and deuteranopes are about equally common, but cases of congenital *tritanopia* (absence of the blue mechanism) are rare, and the condition is not sex-linked.

A second major class of colour-deficient observers are the *anomalous trichromats* who, as the term suggests, require three variables to make all possible colour matches but who make matches different from those of normals. In practice they are identified by one particular match, the Rayleigh equation, in which red and green lights are mixed to match a monochromatic yellow. (The clinical instrument that allows this match to be made is called an *anomaloscope*.) If the anomalous observer requires more red than normal in his matching field, he is termed *protanomalous*; if he requires too much green, he is termed *deuteranomalous*. It is thought that in such observers an anomalous pigment is substituted for one of the normal pigments, replacing the red pigment in the case of protanomaly and replacing the green in the case of deuteranomaly. The anomalous pigment may not be the same for all anomalous trichromats but its peak sensitivity almost always seems to lie between the peaks of the normal red and green pigments. Thus the two long-wavelength pigments of the anomalous observer will have closely overlapping absorption spectra and consequently there will be no wavelength that produces a large difference in their absorption rates; so wavelength-discrimination in this spectral region will usually be poorer than that of normals. Congenital *tritanomaly*, an analogous anomaly of the blue mechanism, is said to be very rare, although it should be remarked that an equivalent shift (< 20 nm) in the peak sensitivity of the blue-sensitive pigment would probably have little effect on colour discrimination, since the peak of the normal blue pigment is so far removed from that of the green-sensitive pigment.

A truly colour-blind, or *monochromatic*, observer can match all lights by adjusting the intensity of a single wavelength. Monochromacy (or *achromatopsia*) is rare and has several forms. In one type the patient appears to retain only the rods: he or she is dazzled by moderately bright light and shows other characteristics that might be expected in exclusively rod vision (as an exercise the reader could deduce these from information given in earlier chapters). In another type of congenital monochromacy, a single class of cones appears to be present.

Our understanding of the common forms of inherited colour deficiency has been advanced by the work of Jeremy Nathans and his colleagues, who have isolated and sequenced the genes that code for the protein parts of the three cone pigments. Two of their findings are especially significant. First, the genes that code for the red-sensitive and green-sensitive pigments lie very close together on the q-arm of the X-chromosome. Secondly, 96% of the inferred amino-acid sequences are identical for these two pigments, whereas the blue-sensitive pigment shows only a 43% identity with the other two. Thus the genes for the red- and green-sensitive pigments appear to have evolved relatively recently (in evolutionary terms) as a result of a duplication of an ancestral gene. And their similarity and juxtaposition render them vulnerable to the genetic phenomenon of 'unequal crossing-over': when the two X-chromosomes align themselves at the stage of meiosis, the gene for the red-sensitive pigment may appose itself to the gene for the green-sensitive pigment; and if now there occurs a crossing-over (an exchange of a strand of DNA from one chromosome to the other) and if the breakpoint lies in the duplicated non-coding region between genes, then one chromosome can end up with two green genes and the other with none. A man who inherits the latter X-chromosome will be a deuteranope. If, on the other hand, the breakpoint occurs within the misaligned genes, then hybrid genes will be formed that consist of part of the green gene and part of the red. A man who inherits an X-chromosome carrying a hybrid gene may be either a dichromat or an anomalous trichromat according to the exact position of the breakpoint and according to what normal genes he also inherits.

Acquired colour deficiency is a very common symptom both in specifically ophthalmic disorders, such as glaucoma and retinitis pigmentosa, and in systemic disorders, such as diabetes mellitus and alcoholism. Anoxia and many toxins produce deficiencies of colour vision; a famous example is the early deterioration of colour vision in 'tobacco amblyopia'. Acquired deficiencies often do not resemble the common forms of genetic colour deficiency. The discrimination of wavelength may be impaired in all parts of the spectrum and the patient may have difficulty in identifying desaturated colours; and an impairment of the blue mechanism is very commonly an early symptom, as can be revealed by a clinical version of the Stiles' technique.

Disorders of colour vision may also be of central origin and such cases may throw interesting light on the question, raised above, of the extent to which colour is processed independently of other attributes of the retinal image. In rare cases the ability to match, sort or identify colours may be entirely lost without any measurable impairment of spatial acuity. One such patient studied by the present writer was shown (by means of the Stiles procedure) to retain three functional cone mechanisms with normal spectral sensitivities, each of which could independently control his behaviour in a threshold task. The existence of such a case nicely brings out the implications of the Principle of Univariance (§9.2): the patient can respond to signals from any of the three classes of colour-blind cone, but lacks the machinery to compare the signals from different classes of cone. Central achromatopsia of this kind has been associated with bilateral lesions of the inferior part of the occipital cortex and it is possible that this region of the human neocortex contains the homologue of area V4, which appears to be specialised for colour in the rhesus monkey. But certainly such cases point to some functional independence between the analysis of colour and the analysis of other dimensions of the stimulus.

9.10. SUGGESTIONS FOR FURTHER READING

General references

Boynton, R. M. (1979) *Human color vision*. New York: Holt–Rinehart–Winston. (A well-balanced, clear and comprehensive textbook.)

Rushton, W. A. H. (1972) *Journal of Physiology*, **220**, 1–31*P* (A personal essay by one of the masters of this field.)

Lennie, P. & D'Zmura, M. (1988) *CRC Critical Reviews in Neurobiology*, **3**, 333–400.

Cornsweet, T. N. (1970) *Visual Perception*. Pp. 135ff. New York: Academic Press. (A particularly clear introduction.)

Mollon, J. D. (1982) *Annual Review of Psychology*, **33**, 41.

Wyszecki, G. & Stiles, W. S. (1967) *Color Science*. New York: Wiley. (An authoritative reference work on the measurement of colour.)

Specific references

Trichromacy and colour matching. Wright, W. D. (1969) *The Measurement of Colour*. 4th Edn. New York: Van Nostrand Reinhold.

Early statements of trichromacy. Mortimer, C. (1731) *Philosophical Transactions of the Royal Society*, **37**, 101. (Contemporary account of the additive and subtractive enterprises of J. C. Le Blon.) Walls, G. (1956) *Journal of the History of Medicine*, **11**, 66. (A splendidly entertaining account of an unsuccessful search for the true identity of G. Palmer); Mollon, J. D. (1985) *Actes du 5eme Congres de l'Association International de la Couleur*, **1**, 23. Paris: AIC (The identity of G. Palmer revealed.)

Two-colour increment-threshold procedure. The classic papers of W. S. Stiles, many of which are peculiarly difficult to obtain, have been collected together in Stiles, W. S. (1978) *Mechanisms of colour vision*. London: Academic Press; but Stiles' own papers are a little austere and the student should first read the review of his work by Marriott, F. H. C. (1976) in *The Eye*, vol. 2A, ed. H. Davson. London: Academic Press. A more advanced review is that by Enoch, J. M. (1972) in *Handbook of Sensory Physiology*, vol. VII, 4, ed. D. Jameson & L. M. Hurvich. Berlin: Springer Verlag.

Microspectrophotometry. Bowmaker, J. K. *et al.* (1980) *Journal of Physiology*, **298**, 131; Dartnall, H. J. A. *et al.* (1983) *Proceedings of the Royal Society, Series B*, **220**, 115–30.

Action spectra of receptors. Schnapf, J. L., Kraft, T. W. & Baylor, D. A. (1987) *Nature*, **325**, 439.

Electrophysiology. De Monasterio, F. M. & Gouras, P. (1975) *Journal of Physiology*, **251**, 167. Lennie, P. (1984) *Trends in Neurosciences*, **7**, 243. Derrington, A. M., Krauskopf, J. & Lennie, P. (1984) *Journal of Physiology*, **357**, 241.

Anomalies of blue mechanism. Pugh, E. N. & Mollon, J. D. (1979) *Vision Research*, **19**, 293. Mollon, J. D. & Polden, P. G. (1980) *Nature*, **286**, 59.

Combinative euchromatopsia. Polden, P. G. & Mollon, J. D. (1980) *Proceedings of the Royal Society, Series B*, **210**, 235. Wandell, B. A. & Pugh, E. N. (1980) *Vision Research*, **20**, 625.

Transient tritanopia. Mollon, J. D. & Polden, P. G. (1977) *Philosophical Transactions of the Royal Society, Series B*, **278**, 207. Augenstein, E. J. & Pugh, E. N. (1977) *Journal of Physiology*, **272**, 247.

Independence of analysis of colour and other attributes. Mollon, J. D. (1977) in *The Perceptual World*, ed. K. von Fieandt & I. K. Moustgaard. London: Academic Press. Livingstone, M. S. & Hubel, D. H. (1984) *J. Neuroscience*, **4**, 309; (1988) *Science*, **240**, 740. Hubel, D. H. & Livingstone, M. S. (1987) *Journal of Neuroscience*, **7**, 3378. Zeki, S. & Shipp, S. (1988) *Nature*, **335**, 311. DeYoe, E. A. & van Essen, D. C. (1988) *Trends in Neuroscience*, 11, 219. Tootell, R. B. H. *et al.* (1988) *Journal of Neuroscience*, **8**, 1569; 1594.

Colour-contingent after-effects. Stromeyer, C. S. (1978) in *Handbook of Sensory Physiology*, vol. 8, ed. R. Held & H. Leibowitz. Berlin: Springer-Verlag. Barlow, H. B. (1989) In *Vision: Coding and Efficiency*, ed. C. Blakemore. Cambridge University Press.

Colour constancy. Hering, E. (1964) *Outlines of a theory of the light sense*, trans. L. M. Hurvich & D. Jameson. Cambridge, Mass: Harvard University Press. Land, E. (1977) *Scientific American*, **237**, no. 6, 108. A number of papers on colour constancy appeared in the *Journal of the Optical Society of America* for October, 1986.

Colour blindness. Pokorny, J., Smith, V. C., Verriest, G. & Pinckers, A. J. L. G. (1979) *Congenital and acquired color vision defects.* New York: Grune & Stratton. Mollon, J. D. *et al.* (1984) in *Colour Vision Deficiencies VII.* The Hague: W. Junk.

Molecular biology of visual pigments. Nathans, J. *et al.* (1986) *Science*, **232**, 193–202; 203–10. For introductory accounts, see Nathans, J. (1989) *Scientific American*, **260**, No. 2 (February) pp. 29–35, or Piantanida, T. (1988) *Trends in Genetics*, **4**, 319–23.

Cerebral achromatopsia. Meadows, J. C. (1974) *Brain*, **97**, 615. Mollon, J. D., Newcombe, F. W., Polden, P. G. & Ratcliff, G. (1980) in *Colour Vision Deficiencies V*, ed. G. Verriest. Bristol: Hilger.

CHAPTER TEN

Binocular vision

O. J. BRADDICK

10.1. BINOCULAR CORRESPONDENCE AND DISPARITY

Human beings have two frontally set eyes which look at heavily overlapping regions of the visual world from two positions about 6 cm apart. This arrangement sacrifices the panoramic visual field of many lower animals; its compensating advantage is *stereoscopic vision* or stereopsis, one of the ways we obtain an awareness of the distance and solidity of objects. Fig. 2.5 shows how, in the visual system of primates and some other mammals, fibres originating in each of the two eyes divide at the optic chiasma, so that each half of the visual cortex (and of the superior colliculus) receives signals from only one half of the visual field, but from the same half-field in each eye. (In many lower animals, the crossing at the chiasma is complete, so that each eye projects entirely to the opposite half of the brain, and any interaction between the two eyes must depend on commissures within the brain.)

Fig. 10.1 illustrates the geometry of binocular vision. The two eyes are converged so that both fixate object *A*. An object, such as *B*, at the same distance as *A* and not too far from it, will form images at *corresponding points* with the same coordinates on both retinae. For *C*, which is closer than *A*, this is not so. The angular separations of the images *A* and *C* in the two eyes are different, and the difference $(\alpha_R - \alpha_L)$, is known as the disparity of the images of *C*. For an object behind the plane of fixation, the direction of disparity would be the reverse of that shown: the right-eye image would then appear further to the right than the left-eye image. The amount of disparity depends on how far the object is from the plane of fixation.

A pair of pictures like those shown at the bottom of Fig. 10.1 can be presented to an observer in a stereoscope, an instrument that shows one picture to each eye. When the stimulation of the two eyes by an arrangement of objects in space is mimicked in this way, there is a striking sense of the third dimension: the objects depicted appear to float in space. This is the evidence that our visual system is indeed

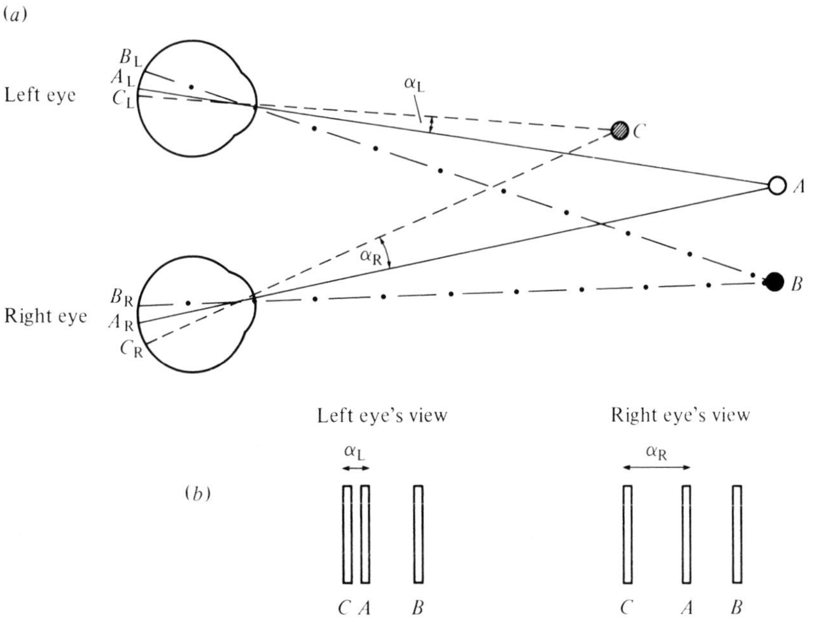

Fig. 10.1. (*a*) Projections of three objects on to the two retinae. A_L B_L C_L are the positions of the retinal images of *A B C* in the left eye, and A_R $B_R C_R$ the positions of their retinal images in the right. Since *A* is fixated, A_L and A_R are the central foveas of each eye. The angular difference $(\alpha_R - \alpha_L)$ is the disparity of the images of *C*. *A* and *B* have zero disparity. (*b*) The spatial relationship of *A*, *B* and *C* as viewed by each eye, showing the disparity between the two eyes' views. This pair of pictures, displayed in a stereoscope, would give the same retinal patterns as the objects in space and would lead to the same perception of stereoscopic depth.

sensitive to binocular disparity, and uses it as a source of information about relative distances.

10.2. VERGENCE EYE MOVEMENTS AND STEREOPSIS

In Fig. 10.1, the observer might change his fixation from *A* to *C*. *C* would then be imaged on corresponding points, and *A* and *B* on disparate points. These changes in the absolute disparities of *A*, *B*, and *C* would not, however, affect their *relative disparities*, and in fact the perceived depth relations would remain the same.

A and *C* differ in direction as well as distance. The change in fixation would require both a *vergence* change (a term covering both convergence and divergence of the eyes) and a sideways shift. These two aspects of the eye movement appear to be controlled quite independently: the sideways shift is a fast *saccade* (see Chapter 11) while the vergence change is much slower, and the overall eye movement is a simple sum of these two components.

Stereoscopic depth perception is possible in brief flashes and with stabilised images (see Chapter 11), for which vergence movements cannot alter disparities on the retinae. Stereopsis, therefore, does not depend on a 'rangefinder' action of the eyes converging successively on each target. The visual system does not simply detect when images fall on corresponding points, but is sensitive to the extent and direction of disparities at any moment. Indeed, this information is used not only for stereopsis but also to control vergence changes, since the initial velocity of vergence movements has been shown to be proportional to the disparity that they are acting to annul.

Although vergence movements are not responsible for stereopsis, they may help to set its scale. A given disparity, expressed in angular terms, does not correspond to a fixed difference in depth: if you redraw Fig. 10.1 with *A* at three times the distance, say, the separation of *A* and *C* will have to be much greater (actually nine times greater) to produce the same disparity ($\alpha_R - \alpha_L$). Our distance judgements reflect this, i.e. disparities are interpreted as perceived depth with a variable scaling factor that depends on the distance of fixation.

10.3. THE LIMITS OF STEREOSCOPIC VISION

Stereoscopic depth discrimination is exquisitely sensitive: under optimal conditions, disparities as small as 2 sec of arc can be detected. This would correspond, for instance, to a depth difference of less than 0.05 mm at 50 cm, or 4 mm at 5 m. This very high *stereoacuity* is an example of the high positional accuracy that the visual system also demonstrates in tasks such as vernier acuity (see Chapter 8 and Fig. 8.3), implying that retinal spatial information is transmitted for stereoscopic comparisons with all the precision available.

Perceived depth does not increase indefinitely with disparity. For disparities of more than about 1 deg, only a qualitative sense of 'in front' or 'behind' is obtained. It is not yet certain whether this involves a distinct mechanism from quantitative stereopsis with small disparities; there may be a continuum of decreasing precision with increasing disparity.

Most real scenes will produce disparities much greater than 1 deg. However, it should be remembered that they are usually viewed in a series of fixations at different distances; and also that the limits quoted apply to foveal vision. As with other spatial parameters of vision, the limits increase progressively towards the periphery of the visual field.

10.4. SINGLE VISION AND RIVALRY

If disparate images in the two eyes are inspected separately, they will appear in different visual directions. One might therefore expect that when seen with both eyes together, *diplopia* (double vision with the images side by side) might occur. In fact, for small disparities (less than about 10 min of arc in central foveal vision – a limit called Panum's area) the separate visual directions associated with non-corresponding points are lost and the object seen in a single, fused, visual direction. This single vision, and stereopsis, must both depend on the visual system attributing a pair of disparate images to a single external object; so it is plausible to suppose that both depend on the same mechanism. However, single vision and stereopsis are not necessarily associated. For moderate disparities, stereoscopic depth perception can coexist with visibly double images.

When the two eyes' images are grossly different, fused single vision cannot occur; instead, one eye's image prevails in any particular region of the visual field. Which eye is dominant varies across the field and with time; e.g. if the two eyes see gratings at right angles, a continually fluctuating basket-weave pattern is seen. In this *binocular rivalry*, lines in the two directions are never actually seen to cross.

In everyday life, we are rarely aware of diplopia even with large disparities, except in extreme cases such as when a finger is held up close to the nose. Partly this is because large disparities will tend to occur in peripheral vision (where Panum's areas are large), partly because the rivalry process tends to suppress one of the diplopic images, and partly because, even when two images are visible, one of them is often simply ignored.

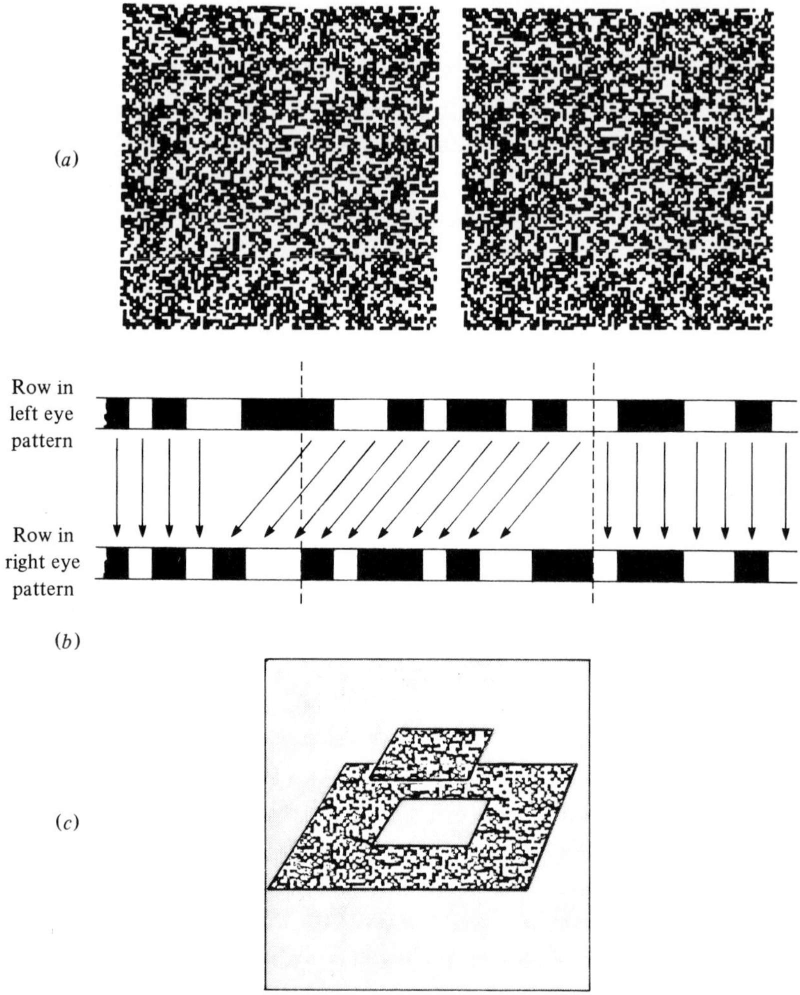

Fig. 10.2. (*a*) A random-dot stereoscopic pair. (*b*) The way in which a single row of each pattern is generated: outside the central region the dots in corresponding rows of the two patterns are identical random sequences; within the central region the sequence in the right eye's pattern is shifted a constant number of positions to the left. The gap on right of centre is filled with further random dots. In practice the sequences are generated by a computer program. (*c*) The appearance of (*a*) when viewed stereoscopically: the disparity of the dots in the central region causes that region to appear standing out from the surround. A reverse disparity can produce the appearance of a 'hole' in the central region.

10.5. THE 'PAIRING PROBLEM' IN STEREOPSIS AND RANDOM-DOT STEREOGRAMS

The mechanism that detects disparity must be able to identify correctly the pairs of images whose positions on the two retinae have to be compared. This problem may not seem very important when dealing with the isolated simple figures often used in stereoscopic experiments, but it becomes much more acute with the densely packed detail that may appear in real scenes.

Such densely packed detail is presented in an experimental form in 'random dot stereograms' (Fig. 10.2), often called 'Julesz patterns' after their inventor. Each pattern of the pair is homogeneously random; an overall shape exists only in the relationship between the two, i.e. the disparity of a certain region of the dots. The ability to see this form, as a region standing out in depth from the rest, shows unambiguously the ability of the visual system to detect disparity, and the patterns provide an elegant test of individuals' stereoscopic ability. It has been suggested that in real scenes, stereopsis helps us to detect objects by similarly picking them out from the clutter of their background.

If you consider any particular white dot in the disparate region, it will have a white 'partner' with the proper disparity. But since the patterns are random, there is a 50% probability that any other nearby dot will be white, by chance. Why does the depth perceived derive from the correct partners and not from any of these chance pairings? Given the randomness, there are no large-scale shapes that are consciously recognised as being identical for each eye; the effectiveness of the patterns shows that this kind of monocular recognition is not necessary for stereopsis.

Somehow, the visual system must do more than detect the disparity of individual dot pairs considered in isolation. Julesz argues that the pairing is of individual dots, but that the system can 'try out' many such pairings at the same time. The pairings selected are those that are consistent over a wide area, giving the perception of a coherent form in depth. An alternative possibility is that what are paired are not dots, but clusters of dots forming 'micropatterns'; the likelihood that identically shaped clusters would appear by chance is much less than for individual dots. Some of Julesz's demonstrations show stereopsis even when many of these micropatterns are disrupted. However, it is not possible to disrupt all possible micropatterns, and the next section will describe physiological evidence that single cells do respond to the disparity of particular shapes.

10.6. BINOCULAR INTERACTION IN SINGLE NEURONS

What physiological machinery extracts the information that is used in stereoscopic vision? The earliest site in the visual pathway where signals from the two eyes have the chance to interact is the lateral geniculate nucleus (LGN) (see Fig. 2.5). LGN cells, however, are found in alternating laminae which receive input from the left and right eye respectively. A left-eye LGN cell may show inhibitory influence from the right eye, but there is no evidence for cells that can be excited by both eyes.

Most LGN cells project to layer IVc of the striate cortex (Fig. 2.6), and there the signals from the two eyes are still separate; the cortex is divided into alternating bands which receive left-eye and right-eye input exclusively.* But intracortical connections apparently fan out from these bands in such a way that cells in the layers above and below show increasingly balanced responses to inputs from either eye. Cells in area 18, one of the regions to which the striate cortex projects, also mostly show binocular inputs.

When each eye is stimulated independently, the receptive fields of these binocular neurons appear similar in orientation but differ in exact location. When both eyes are stimulated together, however, they may show a much more vigorous response than would be predicted from their monocular properties, and in many neurons this occurs only if the stimulus has a particular disparity (which may be zero, i.e. a response occurs only when corresponding points are stimulated). Disparities other than this optimal value may lead to inhibition, i.e. a response less than that elicited by either eye alone. These neurons can therefore be described as 'disparity detectors' and presumably their activity is responsible for stereoscopic discriminations of depth. It is less clear what might be the neural correlates of binocular single vision and of rivalry.

Note that the two eyes' stimuli must be similar in orientation (and probably in other properties too) to activate these disparity detectors. This makes the 'pairing problem' discussed in the last section less severe than if the neurons responded to any pair of stimuli having the optimal disparity.

* Called ocular dominance columns (see Fig. 2.6*g*). This account is derived from the macaque monkey; the cat's cortex is broadly similar but differs in important details. Man is presumably more like monkey than like cat.

10.7. DEFECTIVE STEREOSCOPIC VISION

The organisation of binocular input to cortical neurons described in the last section is not fixed and immutable. Experiments to be described in Chapter 20 have shown that the visual experience an animal receives during a critical period in early life is important in determining how fibres from the lateral geniculate nucleus, carrying signals from the two eyes, form connections to cortical neurons. The neural connections required for stereoscopic vision must be extremely precise. The modifiability of the cortex in the critical period may reflect the fact that such precision places impossible demands on genetically directed developmental mechanisms; instead the connections of cortical cells are adjusted to reflect the correlation of the two eyes' inputs that is experienced during the critical period.

Stereopsis also demands great precision in the binocular co-ordination of eye movements. Strabismus ('squint'), in which the visual axes of the two eyes are not properly aligned, is one of the commonest functional disorders of vision, suggesting that this precision is not easily attained.

About two to five percent of people have no stereoscopic vision, even though each eye may function well separately. They are presumably cases in whom, because of strabismus or some other reason, these developmental processes have failed. A much higher proportion may fail to detect very fine or very large disparities, and there is some evidence that some individuals specifically fail to detect disparities of objects beyond the fixation point, while others fail for nearer objects.

10.8. SUGGESTIONS FOR FURTHER READING

General references

Ogle, K. N. (1950) *Researches in binocular vision.* Philadelphia: W. B. Saunders. (A full account of the classical geometry and psychophysics of stereoscopic vision.)

Julesz B. (1971) *Foundations of cyclopean perception.* Chicago: University of Chicago Press. (A copiously illustrated account of work with random-dot stereograms, as well as a more general review.)

Poggio, G. F. & Poggio, T. (1984) *Annual Review of Neuroscience*, **7**, 379. (An up-to-date review of computational and physiological approaches, with extensive references. Some background necessary.)

Bishop, P. O. & Pettigrew, J. D. (1986) *Vision Research*, **26**, 1587. (Brief, historically treated, review of the electrophysiology of stereopsis.)

Special references

Dynamics of vergence eye movements. Rashbass, C. & Westheimer, G. (1961) *Journal of Physiology*, **159**, 360.

The 'pairing problem'. Julesz, B. (1978) Chapter 7 in *Handbook of Sensory Physiology*, Vol. VIII, *Perception*, ed. R. Held, H. W. Leibowitz & H-L. Teuber. Berlin: Springer-Verlag. (Gives a more up-to-date account of his work and theories than the general reference above.) Sec. 3.3 in Marr, D. (1982) *Vision*. San Francisco: Freeman. (Discusses theoretical considerations, and presents a quite different model from that of Julesz.)

Binocular cortical organization in primates, and its modifiability. Hubel, D. H. & Wiesel, T. N. (1977) *Proceedings of the Royal Society, Series* B, **198**, 1.

Disparity selectivity in cortical neurons. Fischer, B. & Poggio, G. F. (1979) *Proceedings of the Royal Society, Series* B, **204**, 409 (monkey). Ferster, D. (1981) *Journal of Physiology*, **311**, 623 (cat).

Anomalies of stereo vision. Jones, R. (1977) *Journal of Physiology*, **264**, 621.

CHAPTER ELEVEN

Eye movements and strabismus

MATHEW ALPERN

11.1. REGIONAL DIFFERENCES IN VISUAL ACUITY

Colour is one of the hobbies of the human retina. It has a fascination for us perhaps not unrelated to the beauty it brings to our world. But the important *business* of the eye has nothing to do with colour. What eyes do better than any other sense organ is resolve two objects in space. This resolution – the smallest angular separation at which two points, say, can still be distinguished as two – has been discussed in Chapter 8. The reader will recall that the reciprocal of the minimum separable angle is called visual acuity.

As a camera, the eye is optically not all that bad. But the retina devotes many of its neurons and its limited nerve connections to its hobbies, i.e. processing information about colour, and about light intensity, which have nothing at all to do *per se* with its major business, the resolution of objects. In squeezing all this neural processing into limited space, the human eye has evolved in such a way that its visual acuity is not uniformly high for each part of the retina. Instead, acuity is good only at the very centre of the visual field, i.e. around the fixation point; as the angle of eccentricity gets larger, the visual acuity falls precipitously and it becomes very poor indeed only a few degrees from the point of fixation. This is seen in Fig. 11.1.

A photograph with the optical quality of an image produced by such a system would be worthless. The figure shows that it would have very great detail for a very small area (of about 20 min of arc or less) in its very centre; away from this centre, detail is quickly lost and only a few degrees from the fixation point that picture would be seriously blurred. A system of this kind makes no sense at all for an organ whose business it is to be better than all others at localising the precise positions of objects in space. But attach to it six very quickly acting muscles (the fastest in the human body), and this region of high sensitivity can be effectively pointed in very rapid sequences at one object after another in succession. In this way the

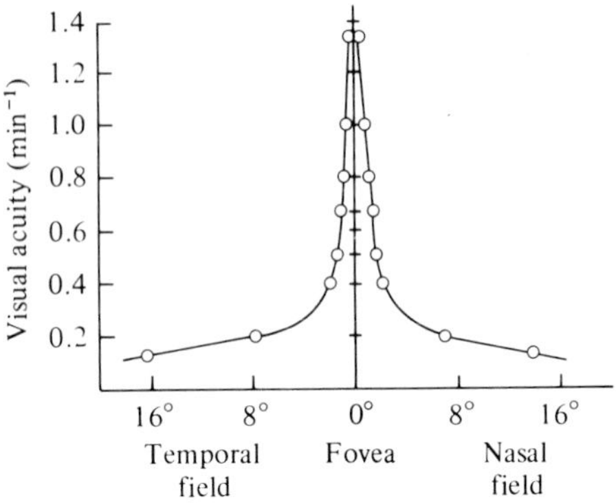

Fig. 11.1. Visual acuity for different angles of oculocentric localisation in the horizontal meridian of the normal eye. (Adapted from Alpern, 1969.) Note: the data of this figure are essentially the same as those of Fig. 8.6, but in that figure minimum discriminable angle is plotted whereas here it is the reciprocal of that value (= 'visual acuity'). Notice how the same data can be plotted in different ways to suit different arguments!

details of every part of the visual field can be quickly scanned by the retina's (spatially) most sensitive centre in a matter of moments.

Such is the importance of eye movements for human vision. But the description leaves a host of unanswered questions, some even unanswerable. Why is it we are aware neither of the very poor peripheral resolution, nor of the movements of the eyes that do so much to overcome this deficiency? From successive maps, how do we piece together an impression of a stable and solid visual space which, if its details are not in every place uniformly high, is not in any place obviously poor? For the moment we must put such bothersome questions to one side with the hope that laboratory research will some day resolve the mystery.

11.2. REFLEX EYE MOVEMENTS

Eye movements are to be found among all vertebrates; indeed many backboneless animals have evolved analogous effector mechanisms. The late Gordon Walls suggested that initially the eye movements had the function of maintaining a stable image of the world as the animal moved about. Reflex movements of this kind are driven by sense organs in periarticular tissues of the head and neck and by the otolith organs of the vestibular apparatus (see Chapter 14). The latter are sensitive to linear accelerations of the head. Since the earth's gravitational field is itself such an acceleration, these sense organs detect changes in head position within the earth's gravitational field and drive the eyes accordingly. These eye movements persist in man although modulated and overlayed by more complex, visually-guided eye movements. They are clinically useful in differentiating paralysis of extraocular muscles, where they are lost, from supranuclear damage to the oculomotor control system, in which vestibular-ocular reflexes remain intact. Like most old, established and traditional institutions they are very conservative, resistant to change. They utilise whatever power is at their disposal to maintain the *status quo ante*, the way things were before the change. If the head is moved down, these reflexes move the eyes up, if the head is moved to the left, the eyes move to the right and so on.

The hair cells of the crista ampullaris of the six semicircular canals play a similar role in response to angular accelerations of the head. The inertia of the endolymph performs a mechanical integration so that in the physiological range the neurophysiological input is the angular velocity of the head. A second integration carried out by the nervous system produces the information necessary for compensatory eye rotations. Such movements are also useful to ophthalmologists in differential diagnosis; furthermore they help otologists to evaluate the integrity of the vestibular apparatus.

The vestibulo-ocular reflexes are described in more detail in Chapter 14.

11.3. TWO FUNDAMENTAL PRINCIPLES

All these eye movements, and those shortly to be examined, are guided by two fundamental principles set down many years ago.

Sherrington cut the left lateral rectus muscle* of an otherwise normal monkey and thus produced a left convergent strabismus. Stimulating a part of the brain (the right frontal eye fields) that he knew from previous work would move normal eyes conjugately to the left, he found again that the left eye moved to the left, but now only up to the midline. Since the severed left lateral rectus could no longer produce this effect, he inferred that it must be due to the inhibition of the activity of the left medial rectus. He concluded that normal eye movements always occur with *reciprocal innervation to antagonistic muscles*: when the agonistic muscle is excited, its

* The anatomical arrangement of the six external eye muscles is shown in Fig. 2.1.

antagonist is inhibited and each of these events contributes to the resulting displacement of the globe.

The second principle of eye movements is due to the nineteenth-century physiologist Ewald Hering. Every muscle of one human eye has a mate muscle (the two are called a yoked muscle pair) of the other eye which moves it in the identical way. Hering's law states that the two eyes move together because *yoked muscle pairs are always equally innervated.* For example, if you move your right eye to the right, the left eye also moves to the right and in almost perfect synchrony by exactly the same amount. (See, for example, the eye movements in Fig. 11.2 to be discussed below.) This is so even if one of the two eyes is covered throughout the movement. Hering's law of equal innervation is as valid for intraocular, as it is for extraocular, muscles. For example, flashing a light into one eye causes both pupils to constrict, then dilate by exactly the same amount and in perfect harmony.

Visually guided eye movements are of two kinds: *versions* and *vergences.*

11.4. VERSION EYE MOVEMENTS

Conjugate (i.e. version) movements are those in which there is no change in the angle of intersection of the lines of sight of the two eyes during the movement. The two eyes move together to the right, left, up or down in some combination. There are two kinds.

Saccadic eye movements

These are very fast, the fastest of any movement of the entire body. To study them, instruct a patient to look direcıly at any light which appears in his visual field. Turn on one such light and, after fixation is assured, turn it off while turning on a second in a different position. Fig. 11.2 shows a record of the resulting movements. The top trace is the time base. At the point *S* the stimulus was changed, and after a short latent period, the two eyes move synchronously, very quickly achieving an extremely high velocity that can be as large as 1000 deg s^{-1}. The whole movement is over in about 50 ms and even large movements are completed less than a tenth of a second after they begin. In the illustrated example the end of the movement found both eyes precisely on the new target, but that will not always be so, particularly in the case of very large movements. If, at the end of the movement, the eyes are not on target, what happens? A second

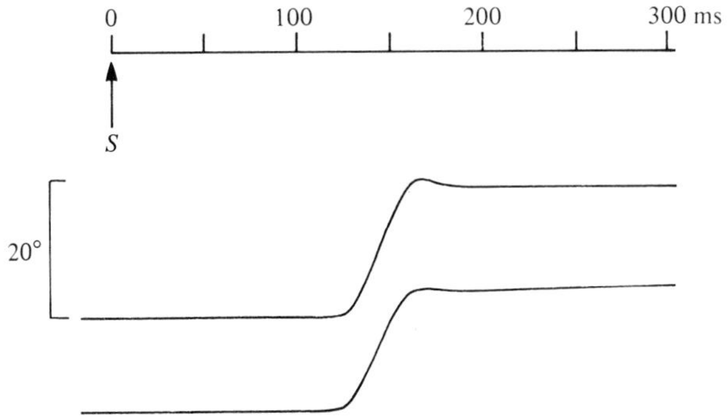

Fig. 11.2. Typical record of a 20 deg saccadic eye movement, the stimulus for which was introduced at *S*. The two traces are for left and right eyes. (Adapted from Westheimer, 1954.)

saccade occurs; but only after another latent period almost as long as the initial one. Why this second delay? Evidently errors in eye position are only intermittently sampled; a saccade follows at a short interval after a position error is first detected. Independent evidence is consistent with this sampled data model of eye position error, but this explanation cannot be the full answer. A second saccade often occurs even when the saccadic change in eye position takes place in complete darkness.

Smooth pursuit eye movements

These are evoked by errors in eye velocity. Unlike the case for the saccadic system, errors in the eye velocity are continuously monitored; hence the latent period for smooth pursuit (125 ms) is shorter than for saccades. (The latent period for the saccade in Fig. 11.2 is less by about 75 ms than it would have been if the subject had no prior knowledge of the position and moment of onset of the stimulus.) In Fig. 11.3*a* the observer was instructed to follow a target moving with constant velocity when it suddenly appears at an expected place. About 125 ms after the target has appeared the eye starts to move with a velocity that matches that of the target, but the delay has caused an error in eye position that can only be nulled by a saccade which occurs after a considerably longer latent period. After this saccade the eye continues to track as long as the target moves (and 125 ms thereafter). Because the target movement – and the consequent smooth pursuit – is slow, one readily distinguishes the

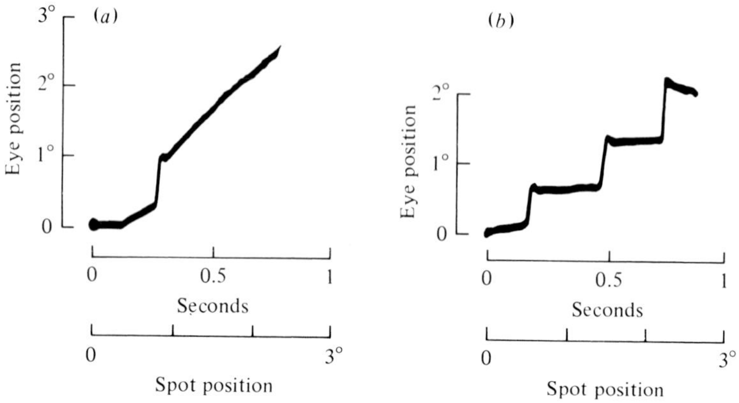

Fig. 11.3. Records of eye movements of a subject fixating a spot moving with a constant velocity of about 3 deg s^{-1}. (*a*) Before, and (*b*) after, intravenous injection of 100 mgm 'thiopental'. (Adapted from Rashbass, 1959.)

two varieties of version movements in this record. The picture doesn't change much even if the target velocity is slower than that shown in the figure. But as the target velocity increases, the eyes find it progressively more difficult and there is an upper limit (10–20° s^{-1}) to the target velocity the smooth pursuit system can follow. At higher velocities eye movement records consist largely of saccadic movements attempting to correct errors in eye position every 200 or 250 ms.

The functional role of saccadic and smooth pursuit eye movements

We have seen that the two types of version movements may go on together in ordinary tracking. At the level of the final common path (say the VI nerve nucleus) each eye position corresponds to a fixed firing pattern of every motor neuron independent of the kind of movement evoked to bring it to that position. But at supranuclear levels, the nerve pathways for smooth pursuit and saccadic movements are quite distinct and various clinical conditions can destroy one leaving the other intact. For example, in a condition known as congenital oculomotor apraxia, the neurology of which is unknown, there is a complete paralysis of saccadic eye movements while all other varieties are perfectly normal. More common is the opposite: under the influence of small doses of barbiturates or somewhat larger doses of ethanol, and in some degenerative diseases such as amyotonic dystrophy, and also occasionally in multiple sclerosis, the smooth pursuit system is destroying leaving the saccadic and other varieties of eye movement intact. Tracking targets moving with a slow constant velocity under these conditions elicits eye-movement records similar to those of normal people when the target velocity exceeds the limit of $20° s^{-1}$. The eyes make a series of successive saccades following one after the other every 200 or 250 ms. Fig. 11.3 shows two records of a normal person tracking the same target moving in the same way with the same velocity. The difference between the records is that the right-hand one was obtained shortly after intravenous injection of 100 mg of thiopental, a barbiturate hypnotic for which the normal dose to induce anaesthesia is 150 to 300 mg.

Two of the functions of version movements are obvious: they first shift the retinal images over each retina to cause the region of current interest to lie on the two foveae, and when this has been achieved they hold the image steady on the retina by reflex movements compensating for rotations of the head, and also by pursuit movements if these are necessary. As pointed out in Chapter 8 the photoreceptors respond rather slowly with an integration time of many milliseconds, so that resolution is impaired when the image moves, just as a photograph is blurred by movement.

If one pursued that analogy one might expect that, if all movements of a visual image over the retina could be prevented, one might see the image more sharply and clearly than usual, but that is not at all the result obtained. It is possible to stabilise an image on the retina either by mounting a small test object and supplementary lens on a contact lens that moves with the eye, or by making an optical system for projecting an image on the retina that includes a mirror mounted on a contact lens in the optical path. When stabilisation is achieved the image is not seen more clearly than usual, but instead it fades away almost completely within a few seconds of being turned on; thereafter what is seen are *changes* in the image. One can observe signs of this fading during prolonged voluntary fixation: this is called the Troxler effect and it works best for low-contrast images in the peripheral field of view which presumably give signals too weak to be revived by the small residual eye movements. Also, as described in Chapter 3, the retinal blood vessels can be made visible by shining a bright light through the sclera, thus causing the shadows to fall in an unaccustomed place, but like a stabilised image they disappear rapidly unless the light source is shifted. A good idea of the subjective appearance of stabilised images and the way they fade can be gained by inspecting the after-images that result from photographic flash guns; they can be made visible by changing the luminance of the background against which they are seen, but if this luminance is held constant they fade within a second or so, and any detailed pattern they contain disappears even more rapidly.

These results from stabilised-image experiments suggested a third function for eye-movements, that of maintaining the visibility of stationary objects. Records of eye-movements taken with more sensitive methods than that used for Fig. 11.2 show several classes of fine movement even when a subject tries to keep his eyes as still and accurately fixated as possible. The first are 'micro-saccades' that occur every second or so and shift the gaze by 5 or 10 min of arc. One is not consciously aware of making these movements and they are hard to suppress, though some people claim to be able to do so. A second class are slow drifts in eye position occurring at a rate of a few minutes (of arc) per second (of time) in the intervals between micro-saccades. Finally there are oscillations, sometimes called 'micronystagmus', at rates of about 50 s^{-1} and amplitudes up to about 1 min of arc, together with slower oscillations that may be of larger amplitude. These oscillations are probably best regarded as residual noise, for it is not an easy engineering task to hold steady an instrument with such a low moment of inertia as the eye, especially when it must also be moved to a specified position at a velocity up to 1000 deg s^{-1}. All classes of movement must help to preserve continuous vision, but the unremitting sequence of saccades at intervals of 200 or 2000 ms would probably be sufficient by themselves to prevent fading.

11.5. VERGENCE EYE MOVEMENTS

Disjunctive (i.e. vergence) eye movements are those in which the angle of intersection of the lines of sight of the two eyes changes during the movement. There are two basic types. Each has a very slow velocity rarely exceeding 10° per second or so. Vergence movements evolved as the eyes migrated from the side to the front of the head and the visual fields of the two eyes more and more overlapped.

Fusional movements

They are a direct consequence of the inability of nature to ensure that after so migrating the lines of sight of the two eyes will perfectly intersect at the same object of regard. These movements automatically correct for any such misalignment. The stimulus for such movements is a difference (sometimes called a *disparity**) in position of the retinal image in one eye and in the other. The two eyes automatically move to correct this disparity, moving their lines of sight to intersect at the same object of regard. If the disparity is large, it will produce double vision – diplopia. (You can demonstrate this by gently pressing on one of your eyes through the lower lid while keeping both eyelids open.) But disparities in the retinal images of the two eyes do not need to be large enough to produce diplopia in order to elicit fusional movements. Fusional movements may be vertical, horizontal, and – though this is controversial – cyclofusional (that is, rotations around the respective lines of sight of the two eyes). Disparities are continuously monitored but the latent period of a fusional movement is still relatively long (about 200 ms).

Accommodative vergence

A second kind of vergence movement occurs only in the horizontal plane. This is accommodative convergence. Looking from far to near causes the automatic increase in the refractive power of the two eyes called *accommodation* (Chapter 4), it causes the pupils of both eyes to become smaller (i.e. *miosis*), and it brings into play a certain amount of convergence of the lines of sight called accommodative convergence. This triad of responses is sometimes referred to as the *near reflex*. The amount of accommodative convergence associated with a unit change in accommodation is a characteristic that varies widely in the population. But for a given individual it is a characteristic that does not vary with age and is unaffected by most drugs (ethanol excepted), learning or eye exercises. Because these individual differences are so large, the accommodative vergence in looking from far to near rarely brings the lines of sight on to the target; the disparity of over- or under-convergence must be made up by a horizontal fusional movement.

* The term disparity is also used to define an important cue for stereoscopic depth perception. Horizontal fusional eye movements can only eliminate disparities at one distance. Objects whose images possess residual horizontal disparities are seen closer or further away than this distance. The latter disparities are called *oculocentric* (i.e. retinotopic) disparities to distinguish them from the disparities which stimulate fusional movements (egocentric disparities) (see also Chapter 10).

11.6. STRABISMUS

Unfortunately, there are limitations on the size of fusional movements; it sometimes happens that the requirements imposed by accommodative convergence are greater than these limitations and the lines of sight fail to intersect at the point of regard. The patient then has a strabismus (or squint) – a misalignment of the two eyes such that the line of sight of one of them, when it is used alone, is not the same as when both are open and the other, non-squinting eye determines the orientation of both.

Inadequate fusional vergence can account for strabismus even when accommodation plays no role, and strabismus can result from a host of pathological conditions including trauma, brain tumours and many other causes including some that paralyse extra-ocular muscles. But an important clinical variety of strabismus is associated with the accommodation brought into play by uncorrected hyperopia. In this condition the eyes accommodate to see distant objects clearly despite the error in refraction. The associated near reflex automatically brings about accommodative convergence. If the hyperopia is small, or the characteristic amount of accommodative convergence associated with each unit change in accommodation is small, the misalignment is easily corrected by the available fusional divergence, but if these combinations are such that the amplitude of the latter is exceeded, a strabismus results. This is one defect which, if detected early enough, can be completely cured. But if it is not detected early, the nervous system develops abnormal sensory adaptations to deal with the mismatch of the retinal images that result. These abnormal adaptations quickly become established as part of the observer's way of using his two eyes and are not any longer easily got rid of. Once this happens the child has developed an incurable debilitating defect, and in treating very young children one must be constantly alert so that when strabismus occurs, the patient is promptly treated by an ophthalmologist. (See also the discussion of monocular deprivation in Chapter 20.)

11.7. CONCLUSION

Nothing has been said about the neural organisation that underlies these different varieties of eye movements. The discussion has analysed each movement in isolation but outside the laboratory and clinic all varieties of eye movement (reflex and visually evoked, conjugate and disjunctive) go on simultaneously and the person most concerned is completely oblivious of their occurrence. Precisely because the types of movements are few and the nerve pathways and anatomical connections well known, relating these movement patterns to neural events at the cellular level in alert behaving primates is one of the most exciting chapters in modern neural biology. But such work is not yet capable of reduction to the simple organising principles that would justify space in this chapter. This may not be the case, however, in the next edition of this book.

11.8. SUGGESTIONS FOR FURTHER READING

General reviews

Alpern, M. (1969) *Eye movements and accommodation* in *Muscular Mechanisms*, vol. 3 of *The Eye*, ed. H. Davson. Academic Press: New York & London.

Carpenter, R. H. S. (1977) *Movements of the eyes*. London: Pion.

Special references

Saccadic eye movements. Westheimer, G. (1954) *AMA Archives of Ophthalmology*, **52**, 710–24.

Barbiturate nystagmus. Rashbass, C. (1959) *Nature*, **183**, 897–8.

Disjunctive eye movements. Rashbass, C. & Westheimer, G. (1961) *Journal of Physiology*, **159**, 339–60.

Eye movement control in primates. Robinson, D. A. (1968) *Science*, **161**, 1219–24.

CHAPTER TWELVE

Higher functions in vision

O. J. BRADDICK AND J. ATKINSON

The neural message in the visual pathway can be thought of as a description in code of the pattern of light received by the eyes. At the levels considered in the preceding chapters, this description is an analysis of the pattern into many elements. However, the functional purpose of vision is not to give us information about patterns of light, but information about objects in space. To fulfil this purpose, the coded message must be transformed by processes that combine information from many of these elements. These processes are much less well understood than the initial analysis, not only in the sense that we cannot yet identify the detailed mechanisms, but also because our theoretical understanding is not deep enough to be sure that we are asking the right questions. For instance, for some stages of the process plausible models have been proposed in terms of the properties of single neurons; however, it may well be that for other stages a knowledge of how individual neurons were responding would not be helpful for understanding how the process as a whole was working.

Many demonstrations of higher perceptual functions take the form of 'illusions', in which what we perceive differs strikingly from the real object or event we are witnessing. It is worth emphasising that, although illusions appear mysterious, the fundamental mystery is how we perceive objects and events at all. If we had any adequate account of the mechanisms that enable us to judge the straightness of a line – a judgement that is usually exquisitely accurate – we should probably understand without difficulty the special cases when lines that are really straight appear curved. This chapter will not deal with visual illusions as a topic by themselves, but examples will be used when they throw light on the mechanisms of normal perception.

The chapter begins by considering how pieces of information about particular features or dimensions of the image are extracted from the signals in the visual pathway. Then we shall discuss how these separate items can be organised into visual objects in visual space. The latter requires us to step beyond even speculative physiology, into theories and results from experimental psychology and from 'artificial intelligence', the attempt to devise computer programs that can emulate human performance in analysing and recognising patterns.

12.1. NEURAL ENCODING OF STIMULUS PROPERTIES

'*Feature detectors*'

Earlier chapters of this book have discussed examples of sensory neurons that respond only if the stimulus meets some highly specific requirements. For example, visual cortical neurons in several species respond only to an edge with a particular orientation, and often only if it is moving in one direction rather than in the opposite one. Such findings have led to the idea that neurons may be 'feature detectors'. The activity of such a neuron could signal the presence of a feature that defined, or partially defined, an object or event that was important to the organism. For instance, in the retina of the frog there is a class of ganglion cells that respond especially well to small moving objects. The same stimuli elicit very effectively a behavioural feeding response (jumping and snapping); so cells of this class have been dubbed 'fly detectors'.

Information from the distribution of activity among detectors

In general, however, the simple level of activity in a particular class of neuron will not be enough to signal the required information. For example, the activity of an 'orientation detector' might change because of a change in the orientation of an edge, or a change in the contrast of a fixed edge. According to a widely-held model, a stimulus dimension (i.e. some way in which stimuli may vary, such as orientation) can be encoded by the distribution of activity among a number of neural channels with different, although overlapping, sensitivities.

Evidence from after-effects

Experiments on *selective adaptation* have been taken to indicate how such distributions determine human perception. One example is the *motion after-effect*. This can be observed by fixating a point on a continuously moving surface such as a waterfall for, say, 60 s; if you then look at a stationary part of the scene, you will see movement in the reverse direction. Apparently, distinct neural channels respond

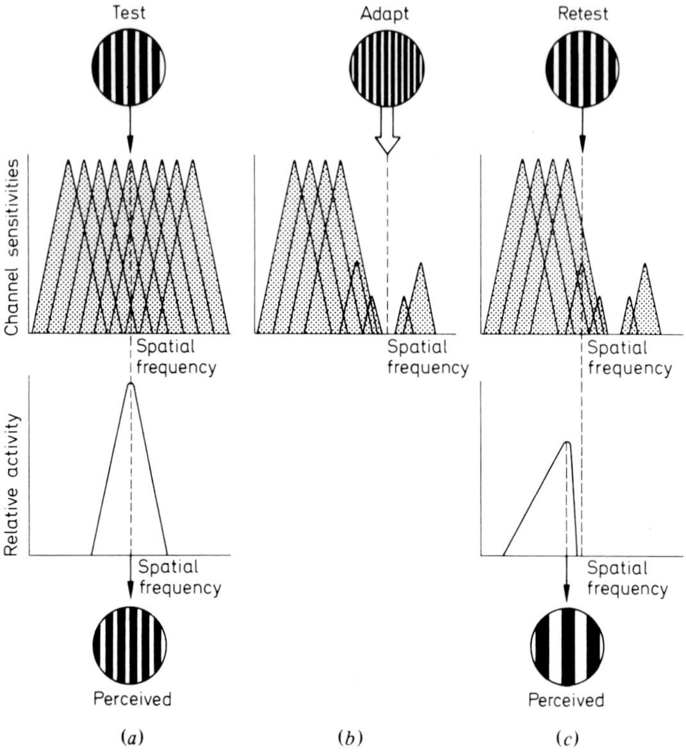

Fig. 12.1. A model for the perception of spatial frequency, and why it changes after inspection of the right-hand adapting pattern of Fig. 7.1. The upper row of graphs represents the sensitivities of a set of channels, i.e. how the response of each channel varies with spatial frequency. Note that the spatial frequency ranges of different channels overlap. (The exact shape of these curves should not be taken literally; nor should the way they have been drawn as equal in height.) The lower graphs represent the distribution of activity in these channels produced by the stimulus shown: a point on the abscissa corresponds to the channel centred on a particular spatial frequency. (*a*) The response of the system in an unadapted state: the perceived frequency is assumed to be that of the most active channel, and corresponds to the true frequency of the stimulus: (*b*) the effects of adapting the system with a grating stimulus of higher spatial frequency than the test grating. The adaptation is assumed to depress the sensitivity of each channel by an amount depending on how strongly it is stimulated; (*c*) the response of the adapted channels to the same stimulus used in (*a*). Since the channels centred on frequencies higher than the test are more adapted than those lower than

to opposite directions of motion. Their activity is normally in balance for stationary stimuli, but it may be unbalanced if one of the channels is adapted or fatigued by prolonged stimulation. The perception of motion results when one channel is more active than its opposed 'partner'. This is one of the few cases where the proposed neural origin of the perceptual after-effect has been observed directly, in the adaptation of activity in opposed directionally-selective neurons of the rabbit's retina (see Fig. 1.6).

Movement is an example of a dimension encoded by the human perceptual system. Selective adaptation is one experimental technique that has been used to identify such a dimension; another related type is *selective masking*, exemplified by Stiles' experiments on cone mechanisms discussed in Chapter 9. *Spatial frequency* is another important example of a visual stimulus dimension. Chapter 8 has introduced the idea of neural channels responding to different spatial frequencies in the image; one line of evidence for these channels came from selective adaptation experiments which showed that the contrast thresholds for different spatial frequencies could be raised independently (see Fig. 8.8). Such experiments do not themselves prove that the distribution of activity in these channels determines our perception of size or spatial periodicity in the way suggested by Fig. 12.1*a*. However, adaptation affects the perceived spatial frequency, as well as contrast: in Fig. 7.1 test gratings of a lower spatial frequency than the adapting grating appear lower still in frequency, while those of higher frequency are apparently increased in frequency. Fig. 12.1*b*, *c* shows how this *apparent spatial frequency shift* could be explained if adaptation depresses the sensitivity of certain channels, thereby shifting the peak in the pattern of activity produced by the test grating.

There are important differences between the movement and spatial-frequency after-effects discussed above. The grating used to produce frequency-specific adaptation does not appear to change in spatial frequency itself (although its apparent contrast declines with adaptation) while the motion used to induce the motion after-effect

---

the test, the distribution of activity is skewed towards lower frequencies, resulting in a lower perceived spatial frequency. Note that, if the test frequency had been the same as the adapting frequency, the depression would have been symmetrical about the test frequency. In this case the distribution of activity, while depressed, would not be skewed and the perceived frequency would be unaltered.

does itself appear to slow as adaptation proceeds. On the account given, this reflects differences in the way the pattern of neural activity encodes the two stimulus dimensions. In the case of motion, a higher velocity is signalled by a higher level of activity; different spatial frequencies, however, are signalled by the peak activity occurring in different channels, and the level of that peak signals not spatial frequency but (perhaps) contrast.

The explanation of perceptual after-effects in terms of the fatigue of neural detectors is not the only possible one, and is unlikely to explain all of the great variety of effects that have been discovered. An alternative view is that each sensory dimension has a working range and mid- or neutral-point, and these are continuously adjusted or 'normalised' to handle the range of stimuli currently being received. The explanations may be contrasted for the case of the tilt after-effect, a shift in apparent orientation as a result of adaptation to tilted lines (Fig. 7.1). From the fatigue of detectors it would be expected that the effect would be restricted, like the analogous spatial frequency effect, to stimuli close to the adapting value on the dimension of orientation. However, if the whole horizontal–vertical frame of reference has been re-normalised, changes in apparent tilt for lines near the horizontal could result from adaptation to near-vertical stimuli, and vice versa. This effect has indeed been reported, but it is not as strong as for orientations close to the adapting stimulus.

Are different dimensions processed separately?

Electrophysiological and psychophysical evidence implies that the same retinal image is encoded in terms of many different dimensions – contrast, orientation, size or spatial frequency, colour, stereoscopic disparity, direction and speed of motion, and so on. How are these different analyses related to each other? Neurons in the visual system generally respond selectively on several dimensions – e.g. both orientation and spatial frequency. Indeed, the spatial-frequency shift effect is greatest if the adapting and test gratings are in the same orientation. Doubly selective neurons of this kind, for orientation and colour, have also been proposed to explain the *McCollough effect*, in which the subject is adapted by alternately viewing a red and a green grating of opposite orientations. This procedure leaves no net after-effect of colour for a blank screen, but black-and-white test gratings appear greenish or pinkish according to their orientation. (see Chapter 9).

This line of argument suggests that a given neuron may carry information about a number of different stimulus dimensions, and be part of a number of different analyses of distribution of the kind described in the preceding section. On the other hand, there is evidence that different dimensions may be encoded by quite separate neural mechanisms. Electrophysiological studies of cortex outside the primary visual receiving area (area 17) have shown an area where cells show a special selectivity for colour (see Chapter 9), and another where motion is the main dimension determining each cell's response. A larger scale division that has been proposed, and supported by the effects of brain lesions in some species, is between a cortical system that transmits the information necessary to identify the shapes of objects, and a subcortical (collicular) system that transmits the information needed to localise an object and orient the animal to it in space. The independence of different dimensions is also sometimes apparent in perception. The motion after-effect, for instance, may lead to a perceptual paradox: objects may appear to be in quite rapid motion for several seconds, but their positions appear to remain unchanged. If motion and position were processed by independent systems, they might easily become dissociated in this way; the relationship between them in the outside world does not hold in their internal representations. Indeed, in so far as different dimensions are signalled separately, we need an explanation of how the different kinds of information about a single object are associated together at all in perception.

12.2. HOW DO WE RECOGNISE OBJECTS?

So far we have discussed the encoding of features or dimensions of the retinal image. However, these do not necessarily correspond very closely to the information we need about *objects*. First, a very diverse set of objects may be, for us, members of a single significant category. Consider, for example, how much variation there can be among things that we would immediately recognise as examples of the category 'dog' or 'hand-printed letter B'. Secondly, an unvarying object can produce a wide range of different retinal images when it is viewed from different angles, different distances, or in different lighting conditions. The standard name for the first problem is 'pattern recognition' and for the second 'perceptual constancy'.

Pattern recognition

To recognise that an object belongs to a significant category, we must be comparing a description derived from the sensory input with some stored definition or prototype. The question is, in what language or code are the two compared? The most simple-minded view is that the stored definition is a 'template' – a prototype image that can be compared more or less directly with the sensory input. The variety among 'dogs' and 'letter Bs' makes this untenable for most perceptual categories of any importance. No specified pattern of light will serve to define a dog, nor will extracted features, or values of simple stimulus dimensions of the kind discussed earlier in this chapter. However, features of a more elaborate kind might do; it is plausible that the feature 'tail' enters into the perceptual definition of a dog, for instance. Such features might be defined in terms of the presence of features of a more elementary kind. This argument has led theorists to a hierarchical model of recognition; each level of a hierarchy describes the stimulus in terms of features, which are themselves the building blocks for the definition of features at the next level up. At the topmost level the description would be in terms of the categories being recognised.

The hierarchical model has been attractive partly because visual neurophysiology seemed to support it: Hubel and Wiesel proposed that the simple cells they discovered in visual cortex (see Chapter 1) served as the input to complex cells, which encoded orientations in a more generalised way, and that these in turn were the input to hypercomplex cells, which responded to more elaborate features such as 'stopped edges'. However, this scheme of how cortical cells interconnect is by no means established; for instance, there is evidence that simple and complex cells receive parallel inputs from the lateral geniculate nucleus rather than, or as well as, one feeding the other. Furthermore, if the scheme were correct, it would still represent only the very lowest levels of the analysis required to test the definition of a significant category such as 'dog'. It may be that hierarchies of feature-recognising cells exist, culminating in cells whose activity corresponds to the recognition of a particular category of objects; but at present the test of the hierarchical scheme must be how well it fits the facts of perception rather than those of neurophysiology.

There are three related problems that the hierarchical model fails to deal with (apart from the essential, but specific, problems of producing a watertight definition of a dog, or anything else, in these terms). First, just how do features, at any level, enter into definitions? A definition might simply be a list of the features required for a particular category (or higher level feature). This in turn raises the question, how does the system make the decision when such requirements are only partially satisfied? Most of the considerable mathematical effort that has gone into pattern categorisation by computer has been devoted to this problem. Despite some successes of this approach to specific applied problems, it is doubtful that lists of features are adequate for defining perceptual categories in general. A dog must not just have a tail, but it must be at the right end. The definition involves a relationship between features, which is more difficult to represent than the simple presence or absence of a feature.

Secondly, the output of a categorisation process is simply whether or not the object is a member of the category – the response of the topmost level of a hierarchy. However, the detailed information used in arriving at this decision is not lost in the process of recognition. So, although dachshunds and Great Danes must both satisfy the definition of 'dog', they do so by a different combination of features and can be recognised as distinct from one another. In fact, it is uncertain what should be considered the top of the hierarchy – consider the recognition of a dog as an example of the category 'animal'.

Thirdly, we are not usually trying to categorise the whole of the visual input under a single heading. Rather, we want to treat it as a scene which may contain a number of categorised objects. This implies that dividing up the overall pattern into objects that are distinct from each other and from the background is an essential preliminary to recognition. 'Perceptual segregation' and 'scene analysis' are stages in this process that are discussed below.

These three problems are connected: they all suggest that perception must involve a richer form of description than just the presence or absence of certain features, or of a certain category of object. The nervous system must arrive at a representation that describes something of the spatial structure of the scene – the relationship between features and their assignment to separate objects. Furthermore, a variety of detailed information, not just the highest level of categorisation, is accessible in perception.

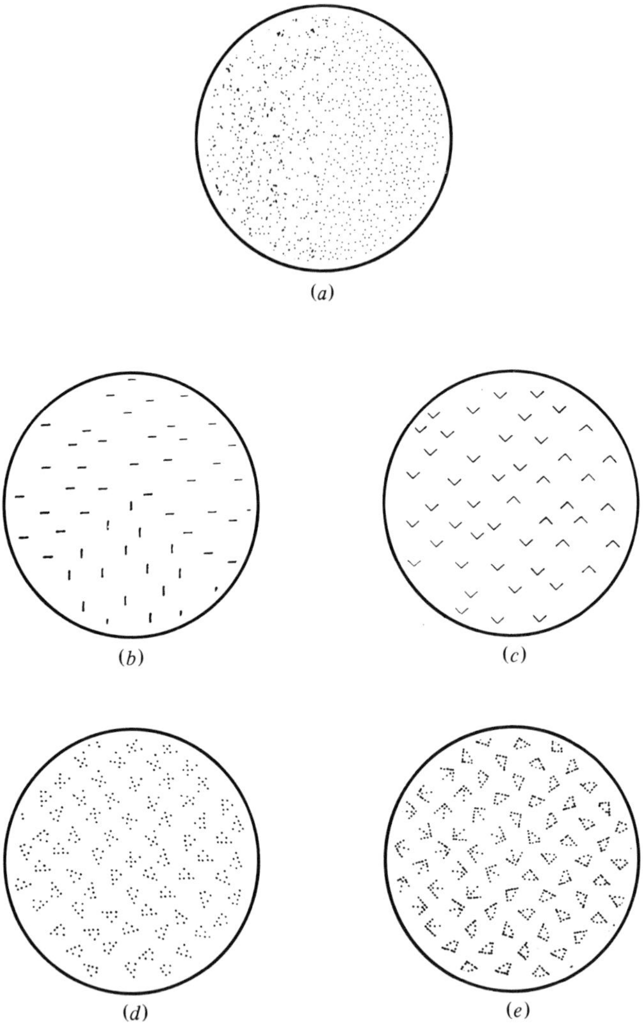

Fig. 12.2. Perceptual segregation caused by differences in texture. (*a*) The average number of dots per unit area is the same all over the circle, but the two halves differ in 'second-order statistics', i.e. in one half the distances between dots are fairly uniform, while in the other half they are more variable and the dots appear to clump together. The boundary between the halves is immediately visible; (*b*), an immediately visible boundary also occurs around a quadrant whose texture elements differ in orientation from those in the rest of the circle; (*c*), the texture elements in the right-hand quadrant are discriminably different from the rest (they are inverted) but this can only

Perceptual segregation

For the perception of objects, we must find something in the visual image that corresponds to the boundaries between surfaces. Many such boundaries are marked by differences in luminance. Finding even these is not a straightforward task in real images, as has been found by those who have attempted to devise computer programs to simulate vision. Furthermore, luminance differences are not the only differences that can cause areas to be perceptually segregated. Fig. 12.2*a*, *b* illustrates how striking are the boundaries that can be perceived at the transitions between areas that differ in their orientation or distribution of texture elements, but not in average luminance.

The fact that some differences do *not* lead to perceptual segregation (e.g. in Fig. 12.2*c*) is also interesting. Although you can readily distinguish between individual elements of the two types (Λ and V), this distinction apparently cannot be used effectively by the processes that are responsible for the perception of boundaries. The ability to control perceptual segregation may characterise the most basic analyses, carried out by the perceptual system at an early stage so that structural description of the scene as distinct objects and surfaces can proceed. So for instance, the occurrence of segregation in Fig. 12.2*e*, but not in Fig. 12.2*d* which has similar general statistical properties, has been taken to indicate that some very basic perceptual process can distinguish the closed loop from the open figure.

Two important types of boundary that cannot be illustrated here are those between regions of different stereoscopic depth (which appear in the random dot stereograms discussed in Chapter 10) and those between regions that are in relative movement. Discontinuities

---

be discovered by scrutinising individual elements; no boundary is perceived; (*d*), the texture elements are arbitrary dot-clusters of two types, chosen to have the same second-order statistics (i.e. same distribution of vectors between any pair of dots). It is difficult to perceive the boundary (the upper quadrant contains the different elements); (*e*), as in (*d*) two types of texture elements have the same second-order statistics. This is one of the few cases where such elements give immediate perceptual segregation: one type of element is a closed loop while the other type is open. (*a*, from Julesz, B. (1975) *Scientific American 234*, no. 4, 34; *b*, *c*, from Olson, R. K. & Attneave, F. (1970) *American Journal of Psychology 83*, 1–2; *d*, *e*, from Caelli, T., Julesz, B. & Gilbert, E. (1978) *Biological Cybernetics 29*, 201–4.)

of depth and movement are likely to be very reliable indicators that parts of a visual scene belong to separate objects. (Because of motion parallax, discussed later, relative motion may often in fact be a consequence of relative depth.)

Scene analysis

Determining how the visual field may be divided into areas or surfaces is still some way from determining how it may be divided into objects. To determine which surfaces belong together in Fig. 12.3 requires an analysis of what arrangement in space might be depicted by each junction of lines in the picture. A number of computer programs have been written to produce such analyses of pictures of a 'blocks world'. In looking at the problems encountered in making such *scene-analysis* programs work, we see an indication of the problems that are effortlessly solved by a human perceiver. For example, individual junctions cannot usually be interpreted unambiguously: the two similar 'T' junctions marked in Fig. 12.3 arise from quite different relationships between surfaces, and a single 'T' junction considered alone could be interpreted either way (and several other ways too). However, if a particular interpretation were chosen, it would mean that a particular line ought to be labelled as a convex edge, and this might be incompatible with *any* of the possible interpretations of another junction that included the same line. The programs, therefore, have to work through the junctions in the picture in some systematic way, determining how the interpretation at each point is constrained by previous interpretations, and how each new interpretation may help to decide among those that have already been considered but remain ambiguous.

The end product of such a program is a labelling of each line as convex edge, concave edge, or an outer edge that conceals the 'back' surface of an object, and a set of links between the surfaces that are part of a single object. Such an output is a (partial) example of a structural description. It provides much more detailed information than the classification of the input into one of a set of categories, information of the type we have argued is available to human observers. Of course, it has not solved the pattern classification or recognition problem; indeed, elements in such a description may provide a more appropriate *input* to a pattern classification system than elements of the sensory input, since they are features of objects rather than of images.

A particularly important job in scene analysis is the identification

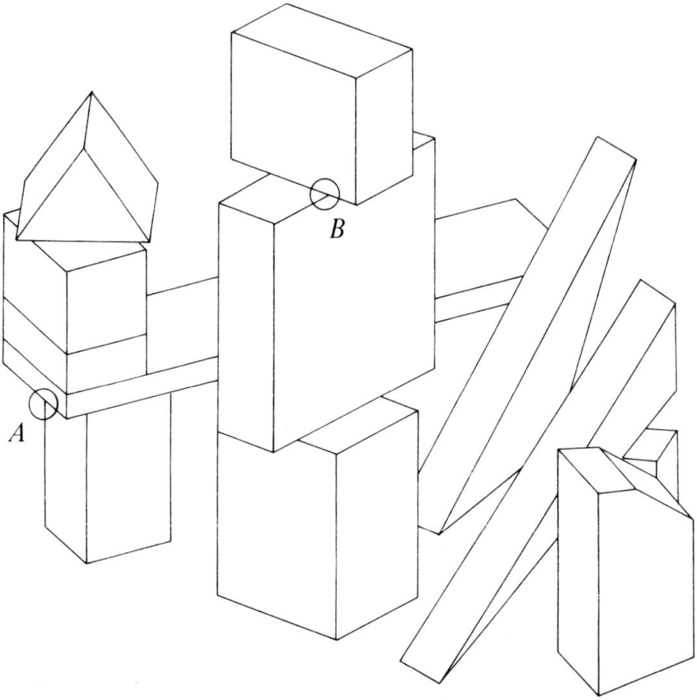

Fig. 12.3. An example of the kind of picture that can be analysed by scene-analysis programs. Such a program could determine how many distinct objects are in the scene, and which areas belong to any single object. It would recognise that the junction at *A* results from an edge which in part occludes a surface with which it is not in contact, and in part occludes the background; while the very similar junction at *B* results from an edge which in part lies directly on a surface and in part occludes a surface with which it is not in contact.

of one region of the image as a background that is not part of any of the discrete objects in the scene. Efficient programs begin the analysis by identifying the outer contour or contours that separate objects from background, since junctions that lie on this contour have many fewer possible interpretations than junctions elsewhere in the picture. Perceptual psychologists had much earlier suggested that segregation of 'figure' from 'ground' is one of the most basic processes in the subjective organisation of the visual field, resulting in an object-like quality in the figure that does not appear in the parts of the field assigned to the background (see Fig. 7.2).

12.3. SPACE PERCEPTION

A scene analysis of the sort described organises an image into objects, but yields only the most qualitative information about their shape ('concave' versus 'convex' but not 'rectangular' versus 'trapezoidal', for instance) or their positions in space ('in front' versus 'behind' only when one object occludes another). Yet we can make quite precise visual judgements of size, shape and distance. The next section considers how we are able to do this.

Sources of depth information

Our awareness of three-dimensional form and distance has been shown to depend on many different sources of information in the visual image, sometimes called 'cues' to depth or distance. Among these the following may be listed.

Binocular disparity in conjunction with convergence, which have already been discussed in Chapter 10. While binocular stereopsis is reliable and very sensitive, it is by no means the only or even necessarily the dominant indicator of depth differences. One-eyed people, amblyopes, and squinters gain an excellent sense of depth from the other, monocular, cues.

Motion parallax: movements of the head or body cause relative displacements of objects at different distances in the visual field, which lead to a strong sense of depth. (The use of successive views to derive the third dimension is geometrically analogous to the simultaneous use of the two eyes' views in binocular stereopsis.) If, instead of the observer, it is the object that moves, the succession of views can still give information about its solid form (the 'kinetic depth effect').

Interposition, i.e. information that one object occludes the view of another, which must be further away. As our discussion of scene analysis should make clear, quite extensive analysis of the overall pattern may be required to recognise that occlusion is occurring.

Effects of perspective: objects of the same size have retinal images that get progressively smaller with increasing distance from the observer. Most obviously, this can allow judgements of distance for familiar objects, such as human beings, whose size is at least roughly known. Even when the true size is not known, however, differences or gradients of image-size carry information about relative distance. One example is the perspective convergence of parallel lines, leading away from the observer, e.g. railway tracks or the edges of a building.

Another is provided by many natural and artificial surfaces that consist of a texture of roughly equally-sized small elements (Fig. 12.4). In this case a gradient of element sizes in the retinal image is a good indicator of the slant of the surface relative to the observer.

Most natural scenes do not consist of isolated objects, but are richly packed with juxtaposed object and surface detail. As a result, even objects that do not themselves provide much information about their distance can usually be located in a framework of perceived space provided by adjacent parts of the field of view, for example, a textured surface with which they are seen to be in contact, as in Fig. 12.4.

The constancies

The same perspective effects that allow us to judge distances mean that image sizes are not a reliable guide to the sizes of real objects. Shapes will also be distorted if they are not viewed head-on; for example, a square viewed at an angle has a trapezoidal image. How, then, can we accurately judge the sizes and shapes of unfamiliar objects?

To make such judgements our perceptual system takes into account information about distance and slant, derived either from cues provided by the image of the object itself, or from the surfaces and objects around it. The distance information is used to set a scale for size information, so that an object is generally judged to be constant in size and shape no matter how it is viewed – a process known as 'constancy' or 'constancy scaling'.

As for many perceptual processes, the existence of constancy scaling is most evident to us when it produces anomalous effects. One demonstration is to form an after-image by looking at an extended bright source, such as a nearby light-bulb. If you now look at surfaces at various distances (e.g. the page of this book and the wall of the room) the after-image appears larger when it is viewed against a more distant surface. Since the after-image originates in a fixed patch of the retina, the changes of apparent size must be the result of a central process acting on this fixed retinal signal: constancy scaling appropriate to the distance of the background surface.

Another constancy of perception despite radical changes in the retinal input is the constancy of lightness (the subjective counterpart of reflectance) and colour. White paper appears white whether it is seen in bright sunlight, in the yellower and weaker illumination of a light bulb, or in moonlight that is dimmer still. In moonlight the

Fig. 12.4. The gradient of texture produced when a surface is viewed obliquely gives a sense of how the surface recedes in depth. The black rectangles, which do not themselves give depth information, can be located by their apparent points of contact with the surface. Note that it is only an assumption that a point of visual contact corresponds to physical contact (see Fig. 12.6). Note also that the difference in apparent distance of the rectangles may lead to a difference in apparent size, because of the operation of size constancy scaling.

actual amount of light reflected is many times less than that from a black object in sunlight (see also Chapter 9).

Chapter 1 has described how the signals in the optic nerve are dominated by local differences in light intensity and carry rather little information about absolute light levels. Since nearby objects will generally be receiving similar illumination, such differences will usually reflect differences in objects' reflectances. Thus lightness constancy is to a large degree implicit in the message received by the brain, unlike size and shape constancy. Even so, the illumination of a surface depends on its slant and its relationship to other objects (consider shadows). It is unlikely that retinal processes alone can give invariant perceived lightness when intensity can vary for these reasons.

Logically, the retinal image of variable size, shape and intensity must be the starting point and the constant object that we perceive must be derived from it. For conscious perception and judgement, however, it is the derived constancy that is primary. It is very difficult to see just how foreshortened an obliquely viewed surface is, and relatively easy to judge its true shape accurately. (Hence the difficulty of perspective drawing for untrained people.) Similarly, it is very difficult to appreciate that the dark lines in a projected lecture slide are returning at least as much light to the eye as the parts of the white screen where no light is falling from the projector.

The constancy of visual direction

The visual location of objects in space requires the perception not only of distance, but also of the direction, left, right, up or down, in which they lie. This too cannot be directly obtained from the retinal image, since eye movements will shift the image of a fixed object around the retina. Despite these movements, perceived visual directions do not change.

Three simple observations help us to understand this *stability of the visual world.* (i) If the eyeball is moved passively (e.g. by the pressure of a finger) the visual world does appear to move; (ii) an after-image, viewed with closed eyes, appears to change its direction as you move your eyes; (iii) the after-image appears static when the eyeball is moved passively.

Observation (ii) indicates that there is indeed some active adjustment of perceived direction that takes into account the changing position of the eyes. Since the after-image is not displaced on the retina, this adjustment leads to an apparent change in direction.

When there is no voluntary eye movement this adjustment does not occur, so image movement across the retina leads to perceived movement in (i), and a stationary image on the retina leads to a static perception in (iii). The passive stretching of the eye muscles by the imposed movement does not affect visual direction, implying that the information about eye position is not derived from proprioceptive input from the eye muscles (the 'inflow theory') but from monitoring the central signals that command the eyes to move (the 'outflow theory').

The outflow theory suggests that, if the eyes were commanded to move but did not do so, the visual world should appear to shift. Recent experiments have shown such an effect when the eye muscles are paralysed by curarisation or local anaesthesia. However, more complete paralysis has been reported to abolish the perceived shift. The control system for eye movements is complex, and just what signals within it are used to provide visual stability is still uncertain.

The idea that information about motor commands (sometimes called a 'corollary discharge' or 'efference copy') is used in perception may be of more general importance: the perception of the spatial framework in which we act requires knowledge of head and body posture as well as of the position of the eyes in the head.

However, all the visual neurons that have been studied to date have receptive fields with fixed *retinal* coordinates. It is quite unknown where or how in the brain visual directions may be represented in a way that takes the efference copy of eye movements into account, though neural signals in the superior colliculus and the frontal eye fields of the cortex have been suggested as candidates for the efference copy.

12.4. SELECTIVE PROCESSES IN PERCEPTION

So far we have spoken as if visual perception was rigidly determined by the retinal input, albeit via processes that may be very complex. But many phenomena of perception can only be understood if we suppose that the observer actively selects the way in which he processes that input.

'Hypotheses' in space perception

One kind of selection is required in dealing with the perspective image of a three-dimensional form. Such an image can be constructed geometrically by taking the points where lines of sight from each

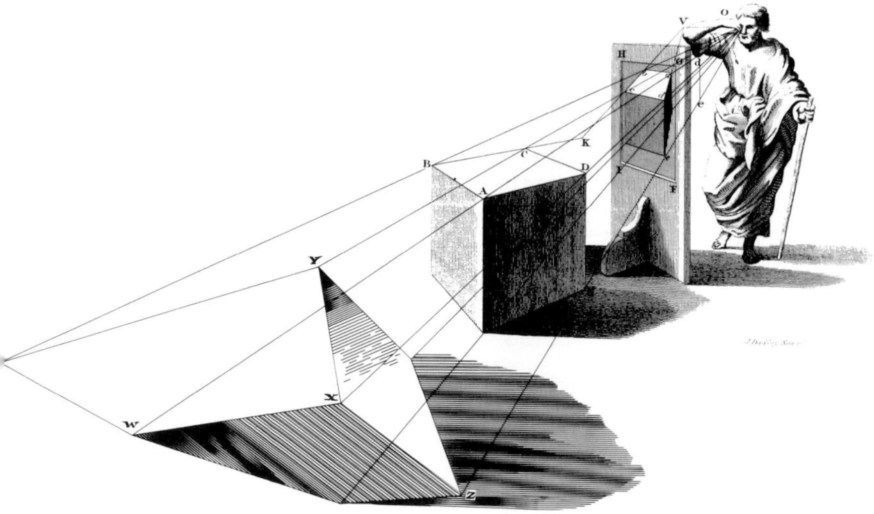

Fig. 12.5. A perspective image of a cube is constructed by taking lines from the viewing point to each point on the cube, and marking the point at which these lines of sight pass through the picture plane. This image exactly reproduces the pattern of light that the cube presents to a (static, one-eyed) observer. It also reproduces the pattern presented by the irregularly shaped solid shown, or any other shape that connects a set of points lying on the same lines of sight. Such a pattern, whether generated by a flat image, a cube, or an irregular solid, is always seen as a cube.

point on the object intersect a frontoparallel 'picture plane' (Fig. 12.5). But, as the figure shows, such a perspective image does not define the form of the object uniquely; in fact, any one of an infinite variety of objects, whose points lie on the same lines of sight, would give the same image. Yet, given a single perspective image of a cube, we always see it as a cube, and it is difficult if not impossible to see any of the distorted shapes which are geometrically equally feasible. This can be described by saying that we form a 'perceptual hypothesis' about the external object which is consistent with the sensory evidence, but is by no means proved by that evidence.

What determines the hypothesis that we adopt? It appears that the infinite range of possibilities is constrained by some implicit knowledge of what properties the object or scene is *likely* to have. In particular, a configuration of edges that could arise from surfaces at right angles is usually assumed to do so. This has been called the 'carpentered world' hypothesis. Its most famous demonstration is the *Ames room*, in which no two surfaces are at right angles but which is constructed in such a way as to give the perspective image of a normal room when viewed from a specified point, and which then appears convincingly rectangular.*

In Fig. 12.3, although the two marked 'T' junctions are interpreted differently, in both cases the edge that continues across the junction is seen as interposed in front of the edge that terminates at the junction. This too is an assumption, of both human and computer scene-analysis, that selects from a wider range of possibilities: in Fig. 12.6*a* it is one of the reasons that we perceive an 'impossible object'. Fig. 12.6*b*, *c* show that the junction in question can be produced by an unlikely alignment in which the apparently interposed edge is in fact further away.

Fig. 12.6 illustrates that, while we may speak of a perceptual hypothesis being selected, it is not freely and consciously selected. The experience of views *b* and *c*, even during a continuous rotation of the real object, does not alter our perception of Fig. 12.6*a*. Perceptual hypotheses rest on their own, restricted, class of perceptual evidence. Fig. 12.6 also implies that the interpretation may be determined quite locally within the scene – by particular junctions of edges and juxtapositions of regions – even though this leads to an overall interpretation that is nonsense.

If the overall perception resulted from the purely local application of fixed selection rules to each part of the scene, 'perceptual hypotheses' would be a metaphor of dubious usefulness. It is more than that for two reasons. First, the hypothesis we arrive at may be based on the interpretation of certain local cues but in turn it may affect how we interpret any other aspect of the scene. So, for instance, the perception of the Ames room as rectangular leads to the far corners of the room being seen as equidistant when in fact one is considerably further than the other. When objects (even people) are placed in these corners constancy scaling is applied to them according to their apparent distance on the hypothesis of a rectangular room, and their perceived relative sizes are therefore distorted. A perceptual hypothesis, then, is based on sensory evidence but it determines how we interpret other sensory evidence. Secondly, in special cases such as Fig. 12.7, more than one hypothesis may be selected by using the same sensory evidence. In these cases, called *multistable percepts*, we switch spontaneously between the alternative perceptions. Note that,

* Full-scale versions of the Ames room can be experienced in the Exploratorium, San Francisco, and in the Science Museum, London.

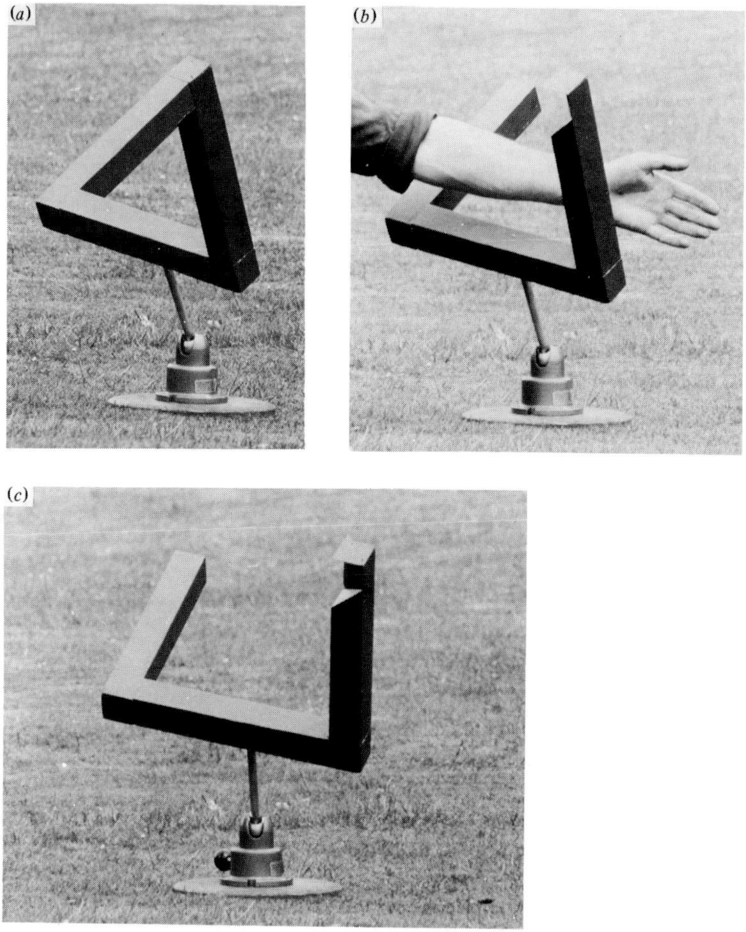

Fig. 12.6. (*a*) The 'impossible triangle'. Each corner if inspected on its own appears to form a sensible three-dimensional configuration but together they appear to make an object that cannot exist. Nonetheless, an object can be made and photographed to produce this image; (*b*) (*c*), two other views of the object, which make clear its three-dimensional form. (*b*) shows how one arm contains a notch shaped so that, when it is seen from the viewing angle used in (*a*), the edges of the notch exactly align with the outer contour of the more distant arm. The resulting junction of regions of the image is perceived as a physical contact and occlusion between surfaces of the object, despite the overall absurdity which results. (By courtesy of R. L. Gregory.)

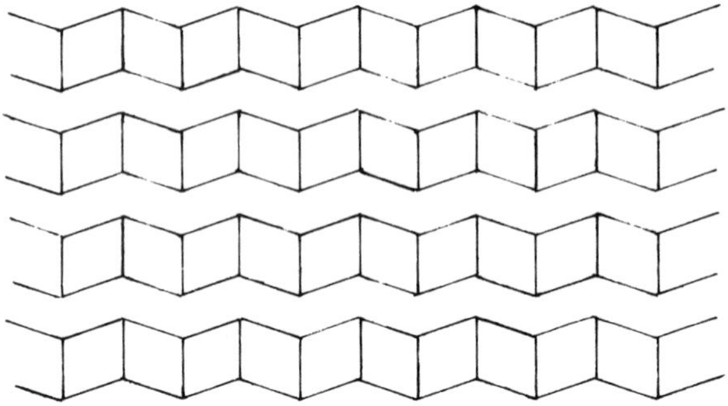

Fig. 12.7. A 'multistable' reversible perspective figure: it can be seen either as a set of zig-zag terraces, with the clear bands as horizontal ledges, or as a stack of thick saw blades, with the clear bands as vertical surfaces. (You may also perceive the figure as a set of aligned, folded strips with the clear bands simply as gaps: two perspective versions of this are possible.) These percepts are alternatives: no part of the figure can be seen in different ways simultaneously, and it is difficult to see different parts in different ways at the same time. Fig. 7.2 illustrates the same principle.

although each junction in Fig. 12.7 changes its perspective interpretation, they almost never do so independently: a hypothesis is selected for the figure as a whole. Multistable percepts provide the most straightforward evidence that perception is not derived from the sensory input by fixed rules; processes within the observer must make an independent contribution (see also Fig. 7.2).

How far visual analysis proceeds in strict sequence from the sensory input ('bottom-up processing') and how far the course of that analysis is guided by its current or expected outcome ('top-down processing') is a fundamental question that arises for almost every aspect of perception. For example, it seems likely that correct segregation of the visual field does not result simply from the local detection of differences and similarities, but that an initial segregation which leads to the recognition of familiar objects will be confirmed and extended, while one that does not may be revised. A tail may be one of the characteristics helping us to recognise a dog; but having recognised something as a dog may also help us decide that an adjacent area is a tail that should be associated with it.

Selection and bias in recognition

Our perception is guided not only by a knowledge of what are likely properties of objects in general – such as the unlikelihood of a nearby surface having a slot that exactly lines up with the outline of a further object – but also by much more specific knowledge about what sorts of objects we are likely to meet in particular situations. Of course, if the stimulus information available to us is sufficiently complete, we will recognise even highly improbable objects without error; but in many real-life situations, our identifications are based on remarkably slight sensory evidence and can be affected a good deal by expectations and biases. For instance, try looking at some unfamiliar handwriting through a mask that shows just a single letter; you will find that in many instances you are quite unable to identify it without moving or enlarging the mask to show the context of the surrounding text. Such context is useful because language is to a large extent predictable.

Perception can use incomplete evidence efficiently by taking account of the fact that some objects are likely: but it might be disastrous if it did not also incorporate the knowledge that some unlikely objects or events are important. Every car driver must at some time have braked for the shadow that was perceived as a person or animal on a poorly lit road. People do not often walk on motorways at night, but it is sufficiently important to detect them that we must be very sensitive to relevant sensory information.

The readiness to recognise probable or significant events might arise in either of two ways. The stimulus analysis we perform might be specially well adapted to extract information needed for detection of the favoured categories; for instance, by using feature detectors that were informative about those categories but less so about others. Alternatively, stimulus information might not be extracted better for one category than for another, but instead we might adjust the criteria of evidence required for different categories, so that we will identify an expected or important event on minimal evidence, but require much more positive evidence to recognise something that is unlikely or insignificant.

Both these kinds of selection occur in human perception. Sometimes they can be distinguished experimentally, since although both methods will lead to easier correct detection of favoured events, they will produce different patterns of errors. If some category has a low criterion of evidence, that category will often be reported as a 'false alarm', since other stimuli will meet the low criterion. If more information is extracted from one category of stimuli, no increase in false alarms need accompany the improved detection.* An analysis of this kind has been performed, for instance, on the 'word frequency effect', that is, the finding that under conditions where correct identification is difficult, common words are more often read (or heard) accurately than are uncommon words. In this case the evidence points to a difference in the criteria of evidence required for the two classes of word, rather than any difference in the amount of information actually extracted from the stimuli.

So far, by 'selective perception' we have meant a greater readiness to recognise a stimulus as belonging to one category rather than to another. A different kind of selection, but at least as important, is that we may vary the type of information that we extract from stimuli of any category. For example, in learning a visual skill such as bird-watching, we are learning to extract those features and stimulus dimensions that differentiate one bird from another. This selective process is applied in the perception of all categories of birds, and does not bias the observer to one category as against another. However, it is not a fixed process; when faced with a different perceptual task (e.g. recognising the faces of friends) different dimensions will be analysed.

One way in which the extracted visual information may be varied is in terms of spatial frequency: we know that different spatial frequency bands are transmitted separately in the visual system (see Chapter 8). Fig. 12.8 shows the same scene spatially filtered into different frequency bands. It is clear that these bands contain different types of information about the scene (gross outline shapes; structure within these shapes; fine surface texture and detail) and it is plausible that for various perceptual tasks, different frequency bands should be selected for processing.

Perception as a process in time

Perception is not an instantaneous process. Many experiments have measured the times required for perceptual tasks, with the aim of analysing how long component processes take. The stimuli have most often been letters and numbers, partly because reading is an

* *Signal detection theory* provides the means of quantitatively analysing detection and false alarm rates. It leads to estimation of two parameters: one (d') is a measure of the signal/noise ratio (Chapter 1, Fig. 1.2) of the information extracted from a stimulus, the other (*beta*) measures the criterion of evidence required for a particular category.

important task that requires a high-speed sequence of visually based decisions.

At any stage of the visual process, activity lasts longer than the original stimulus that caused it. One way of demonstrating this is the 'partial report' experiment. If an array of, say, twelve letters in three rows of four is very briefly exposed, an observer can report only about four letters correctly. However, if a signal is delivered a quarter second after the display telling the observer which row to report, he can report any requested row correctly, implying that the whole array was available to him at that time. This very elementary and short-lived form of memory has been called 'iconic storage' (with 'echoic' storage as an auditory analogue). Persisting activity in the photoreceptors can undoubtedly contribute to partial report performance, but this is unlikely to be the only level at which the visual system retains information. For iconic storage, as evidenced by partial report, not only retains information about static pattern, but also about direction of movement, which must be derived from a sequence over time, and which cannot be retained by the photoreceptors.

In the time that a stimulus is available, either directly or in some form of persistence, the information for identification must be extracted from it. The reason why only four letters can be identified from a brief display is that the stored information decays before any more can be extracted.

The time required to extract visual information can also be assessed in a *visual search* task, where an observer has to find a particular 'target' item in an array of characters. If the observer is asked to search for more than one target (e.g. any *K*, *F* or *U* counts as a target), he will then search more slowly than if a single target (e.g. *F*) is specified. This shows that some part of the process of testing each item in the array is conducted successively for the multiple targets. However, practice with a particular set of multiple targets can lead to performance being as fast as for a single target, implying parallel testing. Thus the timing data can reveal the formation of a new perceptual category, the set of letters that were treated as distinct before practice. It is interesting that in parallel search the observer sometimes finds the target without identifying which particular character was displayed.

With visual tasks at this level, it can be misleading to think of perception as using fixed processing machinery. Rather, individuals can deploy their information-processing systems flexibly to meet the

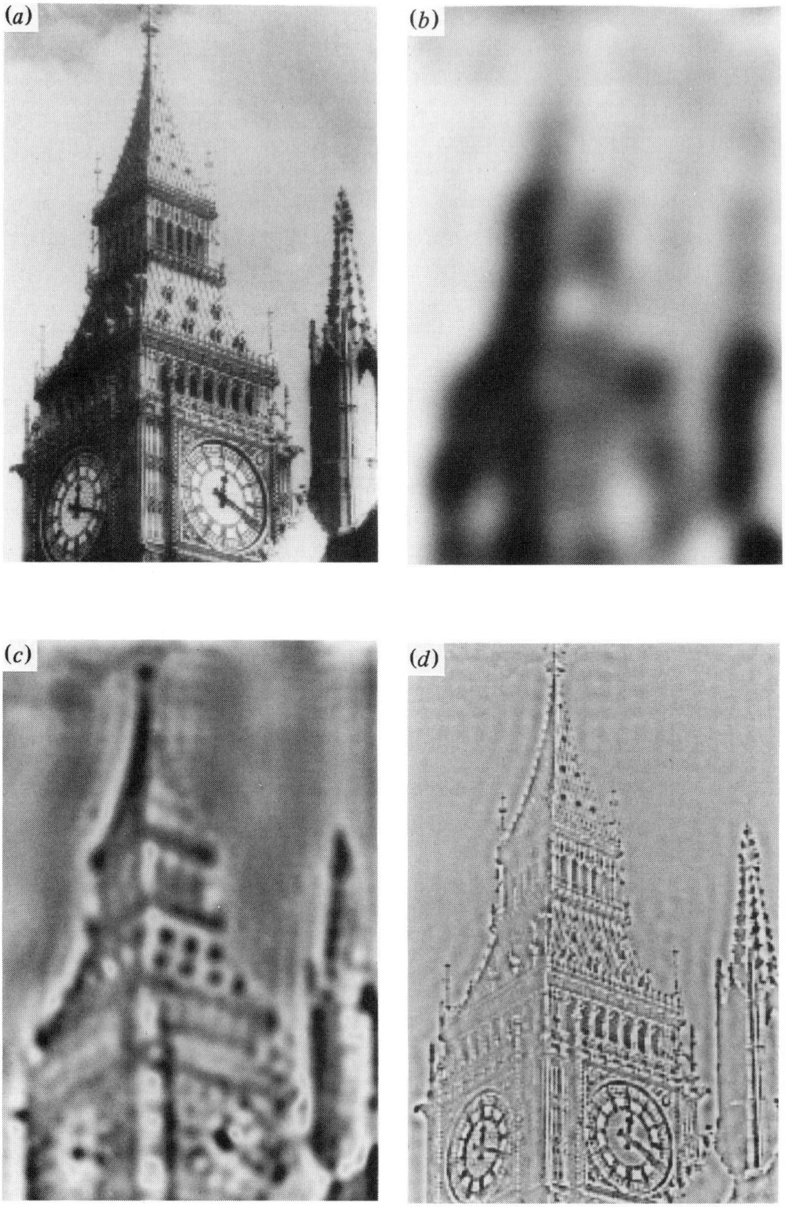
(a)
(b)
(c)
(d)

demands of specific tasks. A striking example of the flexibility of the higher levels of visual function is shown by observers asked to search for a 'zero' among letters. This search is more rapid than that for a letter 'O', even though the physical stimulus is the same. Once again, our understanding of perception must include the idea that the perceiver can actively select a mode of processing.

Perception is an extremely versatile process, and learning to pick out the appropriate sensory cues is an important part of acquiring simple skills, such as driving a motor car. It is even more important for highly developed skills such as those of the police detective or medical specialist, though the owner of such skills is often quite unaware of the processes that guide his perceptions. Among the important lessons from the experiments outlined in this chapter are, first that it is an active process, involving the formation of hypotheses and the selection of cues, and second, that it is fallible: the hypotheses may be wrong.

Fig. 12.8. The photograph at the top left (*a*) has been filtered by computer to give images that contain only restricted bands of spatial frequencies: thus (*b*) contains only frequencies of 0–8 cycles per frame height, (*c*) 8–32 cycles, (*d*) 32–128 cycles. At a viewing distance of 0.5 m (which we suggest), (*b*) corresponds to 0–0.9 cycles per degree of visual angle, (*c*) to 0.9–3.7 cycles per degree and (*d*) to 3.7–14.9 cycles per degree. (For an introduction to the concept of spatial frequency, see Chapter 1, pp. 15–21.)

Even the lowest frequencies (*b*) allow one to identify a building with a spire, and from (*c*) it is clear that the image is of a clock tower; but the highest frequencies (*d*) are needed if one is to tell the time or examine architectural details and we ourselves were surprised to find that the building is identifiable clearly as the clock tower of the Houses of Parliament only in the high-frequency image (*d*). The senior architect of the Houses of Parliament was Charles Barry, but his collaborator was Augustus Pugin, the ardent champion of Gothicism. It is instructive to discover that it is not Barry's basic structure that makes the clock tower so distinctive but rather the imposed detail, the perpendicular tracery, the crockets and finials of the twenty-three-year-old Pugin.

Try the effect of viewing this figure from a greater distance, so displacing all spatial frequencies to higher bands. Notice that (*d*) quickly becomes blank and that (*a*) and (*b*) become very similar.

We are very grateful to Leon Piotrowski for preparing this figure.

12.5. SUGGESTIONS FOR FURTHER READING

General references

J. Frisby (1979) *Seeing*. Oxford University Press. (Provides a well-illustrated introduction to many of the issues discussed in this chapter.)

R. Held, H. W. Leibowitz, H.-L. Teuber (eds.) (1978) *Handbook of Sensory Physiology*. Vol. VIII, *Perception*. Heidelberg: Springer. (A fairly up-to-date compendium on a wide range of topics.)

Specific references

Psychophysical investigation of channels. O. J. Braddick, F. W. Campbell & J. Atkinson (1978) in Held, Leibowitz & Teuber.

Visual function outside cortical area 17. D. C. Van Essen & J. H. R. Maunsell (1983) *Trends in Neuroscience*, **6**, 370.

Texture discrimination and segregation. B. Julesz (1981) *Nature*, **290**, 91.

Pattern recognition. H. B. Barlow, R. Narasimhan & A. Rosenfeld (1972) *Science*, **177**, 569. (A review of some general problems and approaches.)

Scene analysis. P. H. Winston (1977) *Artificial intelligence* New York: Addison-Wesley. (A good view of the programming issues); M. Boden (1977) *Artificial Intelligence and Natural Man* Hassocks; Harvester (Artificial Intelligence in a broader context.)

Constancy of visual direction. Ch. 7 in Howard, I. P. (1982) *Human Visual Orientation*. Chichester: Wiley.

'*Perceptual hypotheses*'. R. L. Gregory (1970) *The Intelligent Eye* London: Weidenfeld & Nicholson.

Selection in perception. D. E. Broadbent (1971) *Decision and Stress* London: Academic Press.

Visual search. P. M. A. Rabbitt (1978) in *Psychology Survey No. 1*, ed. B. M. Foss. London: Allen & Unwin.

CHAPTER THIRTEEN

Basic physics and psychophysics of sound

E. F. EVANS

Though people generally value their sight most highly, hearing is arguably the most important sense for man. For what marks out *Homo sapiens* from other species is his ability to express ideas and concepts, and these he communicates to his fellows chiefly by means of language. This communication occurs first by means of sound, and oral communication remains foremost throughout life. It has been said that a blind person is cut off from the world of *things*, whereas one who is deaf is cut off from the world of *people*.

Yet our understanding of the mechanisms of hearing is less well developed compared with that of the other senses. This has resulted partly because the organ of hearing is so inaccessible – buried, in man, in the *petrous temporal* bone; partly because the important parameters of acoustic signals are less obvious, than, say, those of visual signals, and are measured with considerable technical difficulty and expense; and partly because anaesthetics apparently interfere more with the behaviour of the auditory system than with that of other sensory systems. On the other hand, the application of such basic knowledge as we have on the functioning of the peripheral auditory system has already led to valuable diagnostic tools, and to specific aids designed to compensate for disorders of hearing. There is much more to be known and done.

The multiplicity of terms, measures and scales used in acoustics and psychoacoustics may appear confusing and, at first sight, unnecessary. However, the distinctions are real and important! This first chapter is therefore intended to assist the reader, unfamiliar with the jargon and concepts of acoustics and psychoacoustics, to obtain the necessary background and an understanding of the basic properties of the hearing process.

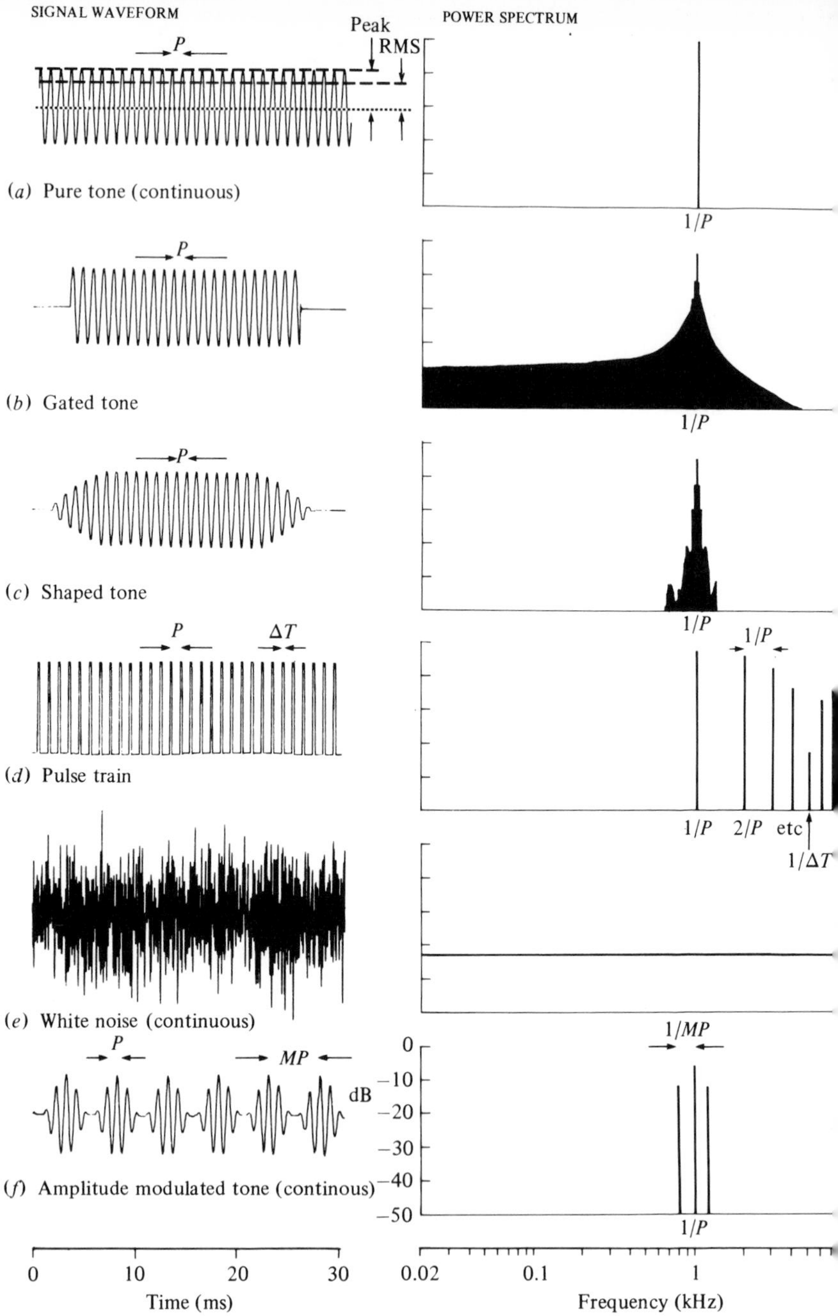

Fig. 13.1. Waveforms and frequency spectra of different signals. The waveforms (on left) describe the excursions of signal (voltage, pressure) with time. Peak indicates the peak amplitude; RMS the root mean square value. *P* indicates the period of the waveform; *MP* the period of modulation. The

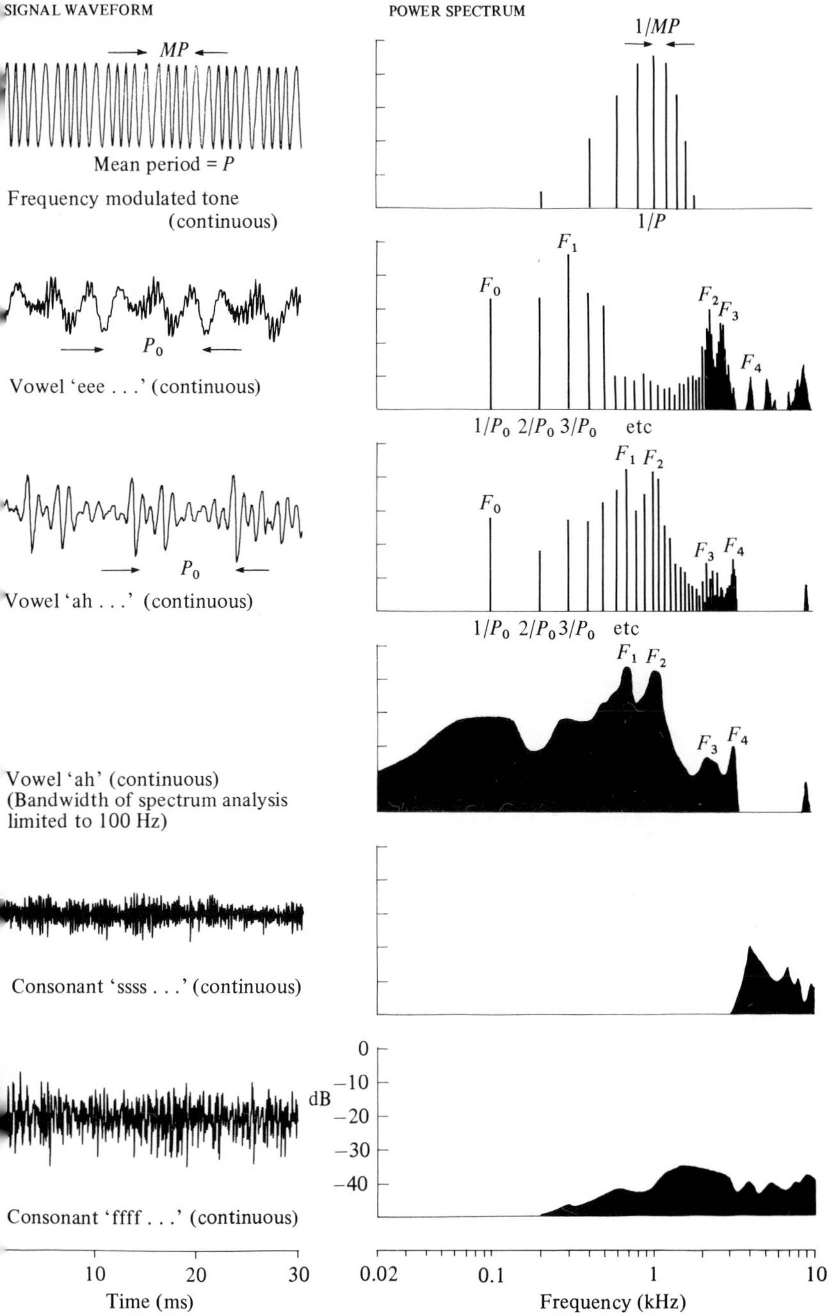

spectra (on right) describe the results of frequency (Fourier) analysis of each signal, i.e.: the relative level of the individual frequency components contained in the signal, plotted on logarithmic power (dB) and frequency scales.

13.1. BASIC PHYSICAL PARAMETERS OF SOUNDS AND THEIR MEASUREMENT

Any vibration, transmitted to the ear via the air or directly via the bones of the head, is in principle capable of generating auditory sensations. In practice, because of the limitations of the hearing mechanism, and the complex transformations that such vibrations undergo in the ear, only a restricted range of sounds is audible. It is therefore important for us to consider what are the relevant parameters of sound vibrations for our ears, and how they are expressed.

Air-borne 'sounds' are relatively small fluctuations in air pressure. At the lowest intensities we can hear, these fluctuations are incredibly small: about 1×10^{-13} of an atmosphere. The fluctuations may be *periodic* (e.g. Fig. 13.1*a–d*, *f–i*), i.e. be regular in time, or *aperiodic*, either random (e.g. Fig. 13.1*e*) or impulsive. Periodic sounds generate musical, tone-like percepts; random sounds, 'noise'; and impulsive sounds, click sensations.

For various reasons, it is convenient to measure the magnitude of acoustic signals by their *power*. Furthermore, because of the immense dynamic range of the ear (greater than $10^{12}:1$), it is convenient to express these powers on a logarithmic scale. This has the added advantage that some important psychophysical and physiological functions are very roughly proportional to the logarithm of the stimulus power. The log unit of power is named the Bel (after Alexander Graham Bell) and it is conveniently divided into ten *decibels* (abbreviated dB). It is important to remember that decibels are not *units* of measurement in the usual sense, but represent *ratios* of powers. Therefore the decibel difference between the magnitude of two signals is given by $10 \times \log_{10} (I_1/I_2)$, where I_1 and I_2 are their respective powers (intensities), or in terms of their fluctuations in pressure (P_1 and P_2 respectively), $20 \times \log_{10} (P_1/P_2)$.*

Thus, a doubling of pressure equals an increase of level of 6 dB; doubling of power: 3 dB. Usually a signal strength is indicated in decibels *relative to some standard*. This is its *level*. In acoustics, the standard adopted internationally is the sound pressure of 20 μPa (1 Pascal = 1 N m^{-2}), roughly the threshold of hearing for a 3 kHz tone (see Figs. 13.2 and 13.3). Acoustic signal levels are conventionally expressed relative to that standard as *dB SPL* (Sound Pressure

* Acoustic power varies as the *square* of pressure. $20 \times \log_{10}(P_1/P_2) = 10 \times \log_{10}(P_1^2/P_2^2)$.

Table 1. *Relations between sound pressure, power, and level, with typical examples drawn from common situations*

Sound pressure ($N\ m^{-2}$ or Pa)	Power (intensity) ($W\ m^{-2}$)	Sound pressure level (dB SPL, i.e. referred to 20 μPa)	Examples and some effects (approximate only)
200	100	140	Jet engine; over-amplified rock group; threshold for pain
20	1	120	Damage to cochlear hair cells
6.32	10^{-1}	110	Threshold for discomfort
2	10^{-2}	100	Motor cycle engine
			Orchestra
6.32×10^{-1}	10^{-3}	90	*fff*
2×10^{-1}	10^{-4}	80	*ff*; busy traffic; shouting
2×10^{-2}	10^{-6}	60	*mf*; normal conversation
2×10^{-3}	10^{-8}	40	*pp*; quiet office
6.32×10^{-4}	10^{-9}	30	*ppp*; soft whisper
2×10^{-4}	10^{-10}	20	Country area at night
2×10^{-5}	10^{-12}	0	Threshold of hearing of young person at 1–5 kHz
6.32×10^{-6}	10^{-13}	−10	Threshold of cat's hearing (1–10 kHz)

Level). Table 13.1 shows, for reference, the relationships between the magnitudes of sound pressure, power (intensity), and SPL, relevant to audition.

All temporal waveforms can be analysed mathematically or physically into their component *frequencies*. The breakdown of complex sounds into their constituent frequency parts is known as *frequency analysis*. It can be accomplished mathematically by *Fourier analysis* (or by digital filtering carried out electronically) or physically by means of *filters* (having resonant elements or circuits). The results of such an analysis on some typical waveforms is shown in Fig. 13.1. The concept is important because the ear itself is capable of a limited but vital frequency analysis (see Chapter 1, p. 15).

The limitations of any frequency analysis are the duration of the signal requiring to be analysed and the bandwidth* of the filters used.

* Bandwidth denotes the range of frequencies accepted by a filter without substantial attenuation; see Fig. 15.1*b*.

Any practical frequency analysis is a compromise between frequency and time resolution. At one extreme is Fourier analysis (equivalent to infinitely narrow filtering of infinitely long signals) as in Fig. 13.1*a*, *d*, *e*, *f*, *g*, *h*, *i*). This yields a *long-term spectrum*, where the frequency components are infinitesimally narrow, hence are said to be *line spectra* as shown on the right-hand side of the figure. Towards the other extreme is the *short-term spectrum*, where the time of occurrence of frequency components is more or less preserved at the expense of lack of precision in the determination of frequency. This is achieved by filtering with finite filter bandwidth: the wider the bandwidth, the poorer the frequency resolution, but the better the resolution of signal components in time (and vice versa). The compromise adopted by the ear seems well-adapted to our needs, for the amplitudes of resolvable frequency components change slowly enough to be followed by the neurons of the auditory system. An analogous comparison, in space rather than time, may perhaps explain the advantage of the local spatial frequency analysis thought by some to be performed in the visual cortex (see Chapter 8, p. 150, and also Chapter 1, p. 17).

Sufficiently rapid changes in a tonal stimulus either in intensity or in frequency produce 'splatter' of energy to other frequencies. In the case of switching a pure tone (Fig. 13.1*b*) on and off, the energy is spread over a wide frequency range. This spread can be minimised by appropriate shaping of the onset and offset of the signal (Fig. 13.1*c*).

Brief pulses produce 'clicks' which, like broad-band (white) noise (Fig. 13.1*e*), contain energy across a wide range of frequencies. In the case of repeated clicks (Fig. 13.1*d*), however, the analysis shows the energy as confined to frequencies (*harmonics*) that are multiples of the click repetition frequency (known as the *fundamental frequency*), as in Fig. 13.1*d*. (Note that because the clicks have finite duration, the energy in the harmonics is not uniform, but there exist minima at frequencies corresponding to the reciprocal of the pulse duration, $1/\Delta T$, $2/\Delta T$...etc.)

Periodic changes in a tonal stimulus, either in frequency or intensity, produce 'splatter' of energy to *discrete* neighbouring frequencies as shown in Figs. 13.1*f* and *g* respectively. The spacing of these frequencies is equal to the modulation rate (1/MP), and in the case of frequency modulated tones (Fig. 13.1g), the number of the adjoining frequencies increases with the depth of modulation.

Speech sounds (Figs. 13.1*h*–*l*) are produced by the action of the many resonances of the vocal tract (mouth, pharynx, nasal passages) on sounds produced by pulsatile excitation of the larynx ('voiced' speech sounds, such as vowels), or by the tongue or lips ('unvoiced' sounds such as most consonants). For continuous steady sounds such as long *vowels* this produces virtually line spectra representing the first 50 or so harmonics of the voice *fundamental frequency* (F_0, Figs. 13.1*h*, *i*). Depending upon the shape of the vocal tract and consequently on the speech sounds being uttered, the amplitudes of the harmonics are emphasised at certain frequencies known as *formants* (F_1, F_2, F_3, F_4 in Figs. 13.1*h*–*j*). The relative spacing of these formants helps to characterise the particular vowel, and their movements help to distinguish certain consonants. The frequencies spanned by the voice fundamental and formants are indicated in Fig. 13.3.

In the production of (unvoiced) consonants, the energy is relatively wide-band, although again, the distribution of energy is an important cue for the identification of the consonant, as demonstrated in the differences between the spectra of 'sss' and 'fff' in Figs. 13.1*k* and *l* respectively.

As already mentioned, the fineness of the frequency analysis depends on the bandwidth of the filters involved. Thus, Fig. 13.1*j* shows the result of analysing the same signal analysed in Fig. 13.1*i* but with filters accepting a band of frequencies 100 Hz wide at each frequency. This means that the individual harmonics (having a spacing of the voice fundamental of 100 Hz) are smoothed out, but the concentrations of energy at the formant frequencies are still clearly visible. The ear's analysis for frequencies above about 1 kHz is similarly limited (see §15.1). Below 1 kHz, however, the ear's filters are sufficiently narrow that some of the harmonic structure can be resolved.

13.2. BASIC PSYCHOPHYSICAL PARAMETERS OF SOUND

Thresholds

Within the frequency range to which the ear is sensitive, a single sound will always be heard if its intensity exceeds a certain level. Below a somewhat lower level, it cannot be discriminated from the background (or internally generated) noise. Between these levels lies an intensity arbitrarily designated '*threshold*': for example, it is commonly taken as the level at which 50% of the sound presentations are correctly identified. In *Clinical audiometry*, the measurements of

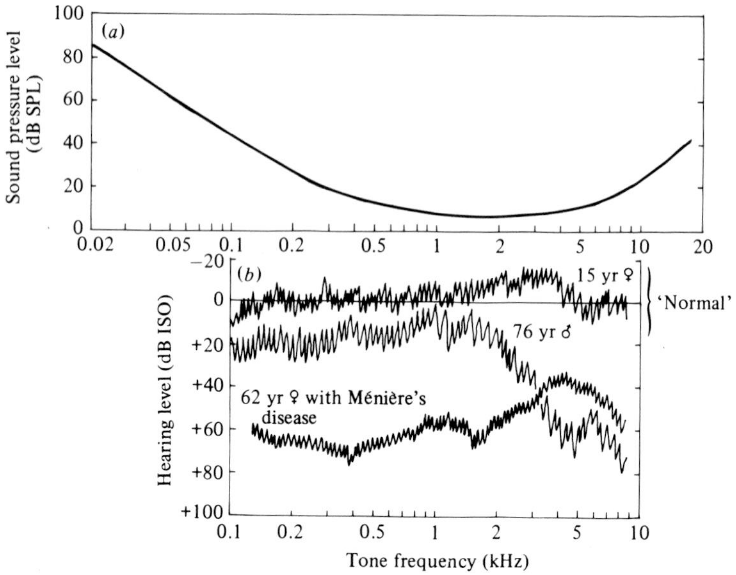

Fig. 13.2. The audiogram, or pure-tone threshold as a function of frequency. (*a*). Pure tone threshold expressed in terms of sound pressure level (dB SPL) at the ear canal under earphone stimulation of one ear. This *minimum audible pressure* (MAP) curve demonstrates the absolute sensitivity of the ear, optimum at the speech frequencies, 0.5–5 kHz. (*b*). Clinical audiograms of 3 subjects by Békésy tracking technique. Each zigzag curve is a tracing of the excursions of tone level resulting from the subject 'tracking' his threshold as the tone frequency is continuously changed, i.e.: he repeatedly increases and decreases the signal level to keep it within the limits of 'just heard' to 'just not heard'. Conventionally, the pure-tone threshold is plotted relative to an internationally accepted standard representing the average hearing of a healthy young adult population (0 dB on scale). Hearing loss appears as a downwards displacement away from zero, as in the case of the patient with Ménières disease. Even in the absence of overt pathology, increasing hearing loss occurs at the high frequencies, with increasing age, as in the case of the 76 year old.

threshold at different frequencies are carried out either by a skilled tester noting the proportion of correct responses at different levels, or automatically by the subject 'tracking' his own threshold, a technique known as Békésy threshold tracking (Fig. 13.2*b*). The threshold so determined, in the quiet, is termed the *absolute threshold.* It can be measured with earphone stimulation of one ear as in Fig. 13.2*a* in direct terms of SPL at the ear canal (i.e. dB with reference to 20 μPa) or, as is conventional for clinical purposes, relative to an internationally agreed average threshold for young persons free from ear disease (Fig. 13.2*b*). In the latter case, the audiometer compensates for the variation with frequency of the average threshold, and for the characteristics of the earphone, to produce for a person with 'normal hearing' threshold values close to zero (i.e. grouped about the abscissa of the plot). Threshold elevations greater than 10–20 dB are considered abnormal. In audiometry, these elevations in threshold are conventionally plotted *downwards* in dB HL (called 'hearing level' or 'hearing loss') as in Fig. 13.2*b*.

It is clear from Fig. 13.2*a* that the absolute threshold is not constant across frequency. In fact, each species (and indeed each individual) has its own threshold sensitivity curve: the sensitivity is least at the extremes of audible frequency, and maximal at some intermediate frequency. In man, the latter occurs at about 0.5–5 kHz, corresponding to the important frequencies for speech perception. (The most sensitive frequency depends on the dimensions of the ear canal and therefore head; thus, in smaller animals such as cats and guinea-pigs, the most sensitive frequencies are higher: 8–10 kHz (see Fig. 14.5.)

The upper frequency limit of hearing in normal persons is highly variable, approaching 20 kHz in young persons, and diminishing with age as indicated by the audiogram of the 76-year-old man in Fig. 13.2*b* (the latter loss is called presbyacusis; cf. presbyopia, Chapter 4). It has been said that above 20 years, the limit diminishes by about 1 Hz per day!

Loudness

The chief physical correlate of loudness is the level of the sound above threshold. (It is not, however, the only factor, as will be indicated in Chapter 15.)

The subjective magnitude of a sound can be quantified in at least two ways. The first is by *matching* the loudness of the test sound against that of a reference sound, for example a pure tone at 1 kHz. Fig. 13.3 shows the result, averaged across many subjects, of carrying out these matches for tones of different frequencies in intervals of 10 dB SPL. The *equal loudness contours* thus obtained are arbitrarily designated to have the same number of *phons* as the SPL of the matching 1 kHz reference tone. Thus a 60 Hz tone at 60 dB SPL is judged to have the same loudness as a 1 kHz tone at 40 dB SPL and is therefore said to have a loudness level of 40 phons. Loudness

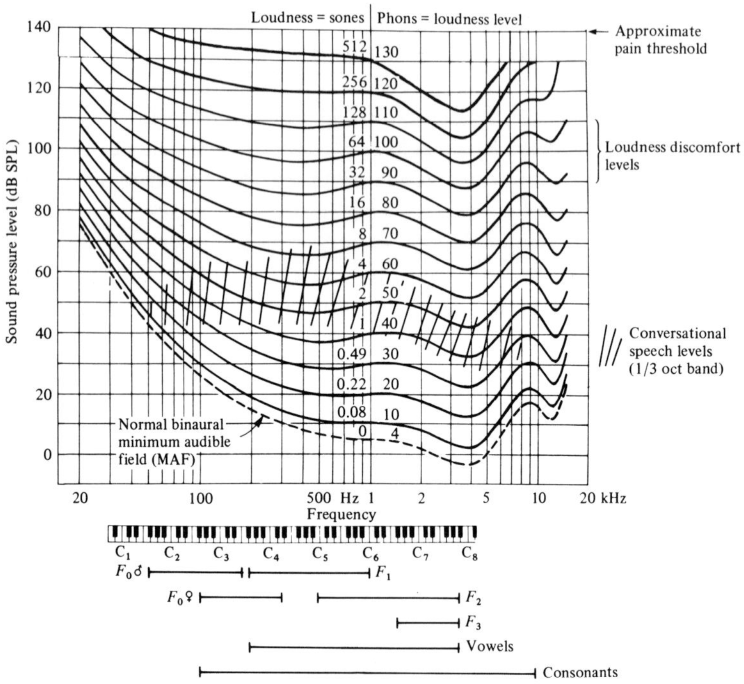

Fig. 13.3. Important subjective functions versus frequency and level. The piano keyboard relates the frequency scale to the musical scale (C_4 is 'middle C'; A_4 = 440 Hz). Curve MAF is the internationally agreed average *minimum audible field*, i.e, the threshold curve measured with the subject facing a loudspeaker in the 'free-field' and averaged across many normal young listeners. Because in this case the sound pressure is measured with a microphone *in place of* the listener's head, the MAF differs from the MAP of Fig. 13.2*a* by virtue of the use of two ears, the diffraction of the sound field by the head, and the resonance of the ear canal. The last two effects combine to enhance further the sensitivity to the speech frequencies, particularly near 4 kHz. The curves above the MAF are the internationally agreed contours of *equal loudness*. Each contour is designated the number of phons equal to the sound pressure level (dB SPL) at 1 kHz. The 40 dB phon contour is arbitrarily assigned to have a loudness (= subjective intensity) of 1 sone. Increase in level by 10 dB doubles the loudness. (This simple relationship does not hold below 40 phon.)

The sound pressure levels of the frequencies in conversational speech measured in 1/3 octave bands at about 0.3–3 m from the speaker are indicated approximately by the shaded area. The frequency ranges covered by the voice laryngeal vibrations (fundamental, *F*o), and the vocal tract resonances (formants: F_1, F_2, F_3) are shown by the lowest horizontal bars. (Equal loudness contours after ISO R 226).

matching ('balancing') techniques are routinely employed in the differential diagnosis of sensorineural hearing loss (Chapter 15, §15.2).

At sufficiently high sound levels, the auditory sensation becomes uncomfortable. This 'discomfort threshold', called the *loudness discomfort level* (LDL), is indicated in Fig. 13.3 at about 100 dB SPL. Above that is the threshold for pain, at 130–140 dB SPL.

The phon, however, whilst used to represent loudness 'level', does not indicate the subjective intensity of a sound. Measurement of this attribute can be carried out by *loudness scaling* techniques in which the subject assigns a number to the perceived magnitude of the sound. These methods give a *power law* relation between the loudness of a sound and its physical intensity, except near threshold. As the exponent of the power law is close to 0.3 for sound levels more than 30 dB above threshold, the loudness function can be simplified to $S = kI^{0.3}$, where S is the loudness estimate, I is the intensity and k is a constant. This means that increasing the sound pressure level by 10 dB leads on average to a doubling of a signal's loudness. It has been internationally agreed to adopt the unit of 1 *sone* as the loudness corresponding to 40 phon. A 50-phon tone sounds on average twice as loud as a 40-phon sound and therefore has a loudness value of 2 sones. The equal loudness contours above 40 phon in Fig. 13.3 have also been labelled in sones.

With headphone listening, the loudness of a signal presented binaurally is almost exactly the sum of the loudness perceived by each ear alone, i.e. it is equivalent to an increase in monaural signal level of about 10 dB.

These scales of loudness and loudness level, determined for the *average subject*, must not be confused with the scale most commonly used in psychoacoustics – the *sensation level* (SL). This simply denotes the physical magnitude (in dB) of a signal above its absolute threshold *for a given individual subject*.

Pitch and timbre

Pitch is the chief psychological correlate of frequency. For pure tones, the ear can locate a signal on a monotonic scale according to its frequency from 'low' (bass) to 'high' (treble) as in Fig. 13.3. The situation is vastly more complicated than this, however, as will be outlined later (§15.1).

'Timbre' is the name given to that quality which distinguishes two steady signals having the same pitch and loudness. It refers to the spectral complexity of the signal, and hence its perceived 'richness', 'mellowness', 'brightness' and so on. Together with the 'attack' (or onset transients of the sound), it allows one to distinguish different instruments of the orchestra playing the same note.

13.3. SUGGESTIONS FOR FURTHER READING

General introductions and reviews

Yost, W. A. & Nielsen, D. W. (1977). *Fundamentals of Hearing*. New York: Holt, Rinehart & Winston.

Durrant, J. D. & Lovrinic, J. H. (1977). *Bases of Hearing Science.* Baltimore: Williams & Wilkins Co. (Physiology and psychophysics, with emphasis on the physics of sound.)

Special references

Sound measurement. Hewlett Packard Acoustics Handbook. Application Note 100. Burns, W. (1973) *Noise and Man.* London: John Murray.

Fourier and signal analysis. Licklider, J. C. R. (1951) Basic correlates of the auditory stimulus, in *Handbook of Experimental Psychology*, ed. S. S. Stephens. New York: John Wiley. Bogert, B. P. (1972) Practising digital spectrum analysis, in *Human Communication: a unified view*, ed. E. E. David & P. B. Denes. New York: McGraw-Hill.

CHAPTER FOURTEEN

Functional anatomy of the auditory system

E. F. EVANS

In this chapter, we consider each of the major elements of the auditory pathway in turn, and summarise its known response properties in terms of its structure. This survey will enable us in the next chapter to account, as far as possible, for the psychoacoustic properties of the auditory system.

Arbitrarily, but conveniently, the auditory pathway is subdivided into the peripheral auditory system, comprising the ear and the primary neurons (auditory or cochlear nerve), and the central auditory system, comprising the nervous pathways and nuclei from the cochlear nuclei onwards (Figs. 14.1, 14.2). The ear itself is conventionally subdivided into the outer ear (pinna and external auditory meatus), middle ear (tympanic membrane, ossicles and associated structures) and inner ear (cochlea), as in Fig. 14.3.

In broad outline, the peripheral auditory system can be considered as a signal 'conditioning' system and spectral analyser (Fig. 14.1). The mechanics of the outer, middle and inner ears transform incident air vibrations into pressure variations in fluids of the inner ear, favouring those frequencies that are important for the organism. In the inner ear, the pressure variations form the input to a bank of filters represented by the mechanical, receptor, and neuronal elements of the organ of Corti.

The central auditory system can in part be considered as a set of parallel systems, parallel in at least two senses. First, signals of differing frequency are represented, in principle, in more or less independent neural 'planes', stacked in the vertical dimension in Fig. 14.1. Secondly, these in turn can be divided into parallel pathways, probably representing different processing subsystems, only two of which, DCN and VCN, are indicated in the plane normal to Fig. 14.1. The former is probably concerned with analysis of the *nature* of a stimulus; the latter its *location* in space.

The diagrammatic layout of these systems seen in Fig. 14.2 is therefore a gross oversimplification of the 'wiring diagram' of the auditory system, expressing ignorance of the detailed anatomy and

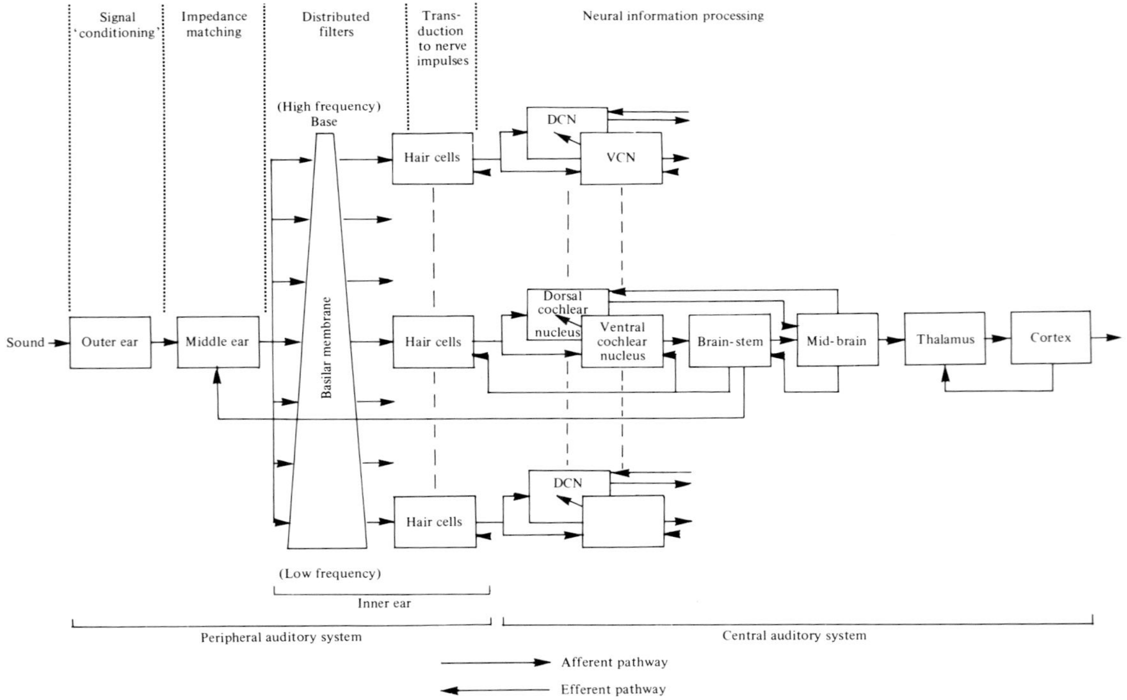

Fig. 14.1. Schematic anatomical and functional map of the auditory system. At the level of the basilar membrane, the system becomes distributed in a spatial dimension (vertical in the figure) that represents frequency. This tonotopic or cochleotopic (see text) organisation is projected to the cochlear nuclei, where further divergence occurs into more than the two subdivisions shown. The cochleotopic dimension is maintained in the neural organisation of the subsequent neural centres. Running in the opposite direction to the afferent pathway (→) is the descending or efferent pathway (←).

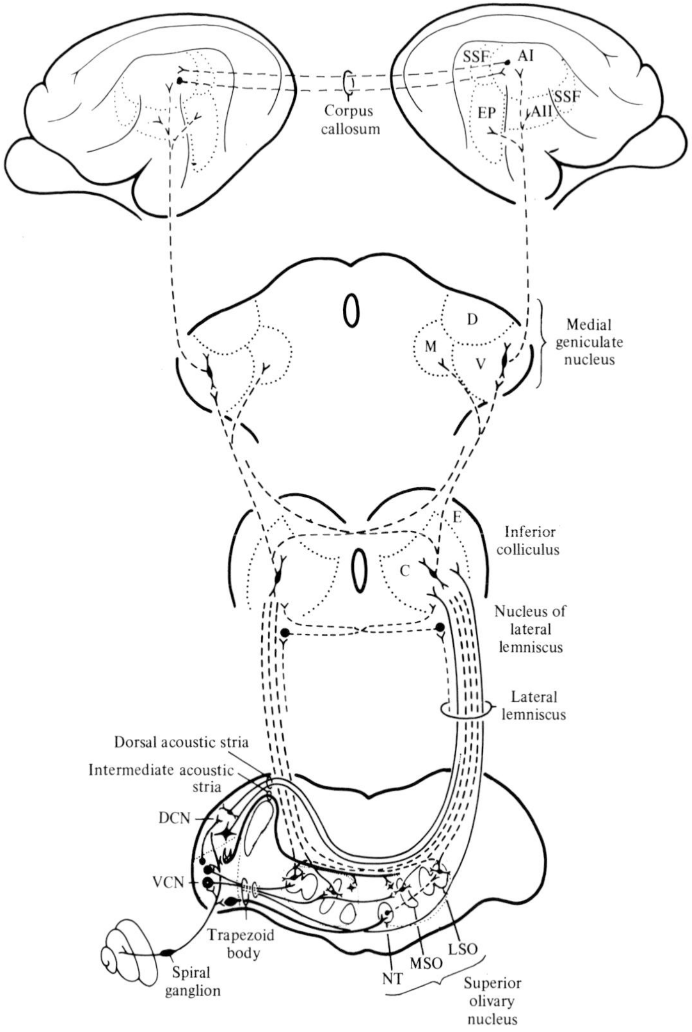

Fig. 14.2. Main ascending anatomical pathways of auditory system. A simplified diagram of the ipsilateral and contralateral projections of one cochlea (bottom left) to the left and right auditory cortex (top) in the cat. DCN: dorsal cochlear nucleus; VCN: ventral cochlear nucleus; NT: nucleus of trapezoid body; MSO/LSO: medial/lateral superior olive; C/E: central and external nuclei of inferior colliculus; M/D/V: medial, dorsal and ventral nuclei of medial geniculate nucleus; AI/AII/SSF/EP: primary, secondary, suprasylvian fringe and posterior ectosylvian areas of auditory cortex. (Modified, after Konigsmark, *Archives of Otolaryngology*, *98*, 403.)

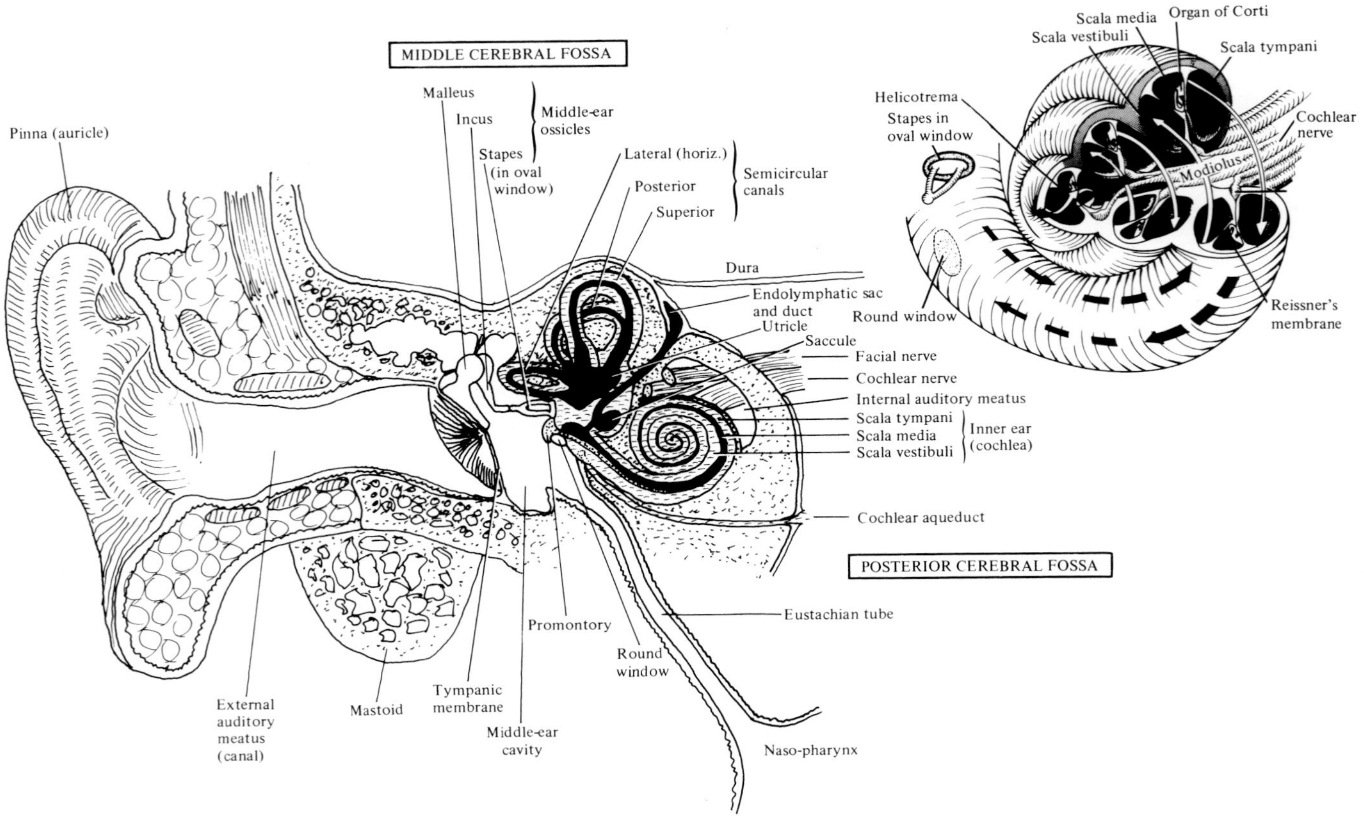
MIDDLE CEREBRAL FOSSA
Malleus
Incus
Stapes
(in oval
window)
Middle-ear
ossicles
Pinna (auricle)
Lateral (horiz.)
Posterior
Superior
Semicircular
canals
Dura
Endolymphatic sac
and duct
Utricle
Saccule
Facial nerve
Cochlear nerve
Internal auditory meatus
Scala tympani
Scala media
Scala vestibuli
Inner ear
(cochlea)
Cochlear aqueduct
POSTERIOR CEREBRAL FOSSA
Eustachian tube
Naso-pharynx
Round
window
Promontory
Middle-ear
cavity
Tympanic
membrane
Mastoid
External
auditory
meatus
(canal)
Scala media
Scala vestibuli
Organ of Corti
Scala tympani
Helicotrema
Stapes in
oval window
Cochlear
nerve
Modiolus
Round window
Reissner's
membrane

function but serving usefully as a basis for description. Furthermore, running as a 'counter-current' to the afferent auditory system, is the *efferent* system (Fig. 14.1). Regrettably, we have little information on the functional interrelationships of these two systems.

Most of our detailed knowledge of the anatomy and physiology of the auditory system comes from studies on animals, particularly the cat. We have far more data available on the peripheral than on the central auditory system, because of the former's relative accessibility and simplicity of structure and function. This is not too unfortunate, for lesions of the peripheral auditory system are the most common cause of hearing loss.

14.1 OUTER EAR

The *auricle* or *pinna*, and the *external auditory meatus* or *canal* together constitute the outer ear (Fig. 14.3). The canal is relatively straight in man, having a diameter of about 0.7 cm and a length of 2–3 cm.

The convolutions of the pinna are of considerable importance for the perception of sound direction both horizontally and vertically and as coming from outside the head.

The shape of the head, pinna and ear canal together modify a plane sound field in such a way that the pressure at the tympanic membrane shows a broad but substantial gain of 10–20 dB at the important frequencies for speech signals: 2–5 kHz. This frequency-dependent gain is much reduced when headphones are used, and is almost lost in the case of the 'insert' ear-pieces of hearing aids.

14.2. MIDDLE EAR

The middle-ear structures are located in an air-filled cavity of the temporal bone, vented to atmosphere via the *eustachian tube* (Fig. 14.3). The cavity is bounded laterally by the cone-shaped *tympanic*

Fig. 14.3. Semi-diagrammatic cross-section of right outer, middle and inner ear of man as viewed from in front. Bone showed as stippled areas; perilymph in osseous labyrinth, dashed areas; endolymph in membranous labyrinth, black areas. Inset shows more precisely the orientation of the cochlea, with its apex laterally and slightly anteriorly. The arrows indicate the continuity of the scalae vestibuli from the stapes-filled oval window via the helicotrema to the scala tympani and the round window. (Inset from Curtis, Jacobson & Markus (1972) *An Introduction to the Neurosciences*. W. B. Saunders Co.)

membrane or *ear drum*, to the upper vertical radius of which is attached the elongated *manubrium* of the first of the *ossicles*, the *malleus* ('hammer'; see Fig. 14.4). The malleus pivots about a horizontal axis through the firm junction between its almost spherical head and the *incus* ('anvil'). Rotations of the malleus—incus combination are transmitted, via the long process of the incus and the *incudostapedial joint*, to the head of the *stapes* ('stirrup'). The two *crura* of the stapes transmit movements to the stapes *footplate*, which fits into the membrane-covered *oval window* of the cochlea and acts like a piston therein. The ossicles of the middle ear are suspended by ligaments and are acted upon by two muscles (Fig. 14.4) the *m. tensor tympani*, attached to the manubrium of the malleus and the *m. stapedius* inserted into the neck of the stapes.

The function of the middle-ear system is to transform efficiently variations in air pressure in the ear canal into pressure variations in the fluids of the inner ear. Since the impedance* of air is very low compared with that of the cochlea (about 1:135), if the energy were presented directly via an air–fluid interface, 97% would be reflected, and only 3% transmitted. Because of the impedance-matching function of the middle ear, an estimated 60% of incident energy is in fact transmitted into the cochlea. This *impedance transformer* operates predominantly by virtue of the substantial difference between the effective area of the tympanic membrane and that of the footplate of the stapes (areal ratio of 17:1). In addition, because the length of the long process of the incus is somewhat shorter than that of the malleus, there is a small mechanical advantage of 1.3:1 in terms of lever ratio. The overall pressure ratio is therefore, in principle, about 1:22 in man. (This represents an impedance ratio of about 1:29.) In practice, this pressure gain is approached only at frequencies intermediate between about 1 kHz and a few kHz in man.† The

* Acoustic impedance at any frequency is the vectorial sum of *reactance* (determined by the opposing effects of mass and stiffness (elasticity) of the system) and *resistance* (produced by friction). It is measured in acoustic ohms. The inverse of acoustic impedance is *admittance*, measured in mhos. The inverse of the stiffness component is *compliance*, i.e.: compressibility. Thus air, being very substantially more compressible than a fluid, has a much lower impedance.

† About 20 kHz in cat and guinea-pig. In this range, the impedance is minimal, and chiefly resistive. For frequencies below this range, the stiffness of the middle-ear and inner-ear membranes (including the tympanic membrane), the middle-ear ligaments and the air in the middle-ear cavities, dominate the impedance, together with a shunting effect of the helicotrema at very low frequencies. For frequencies above the range, the mass component dominates the impedance; this together with reduction of the effective area of the tympanic membrane increasingly limits the transfer.

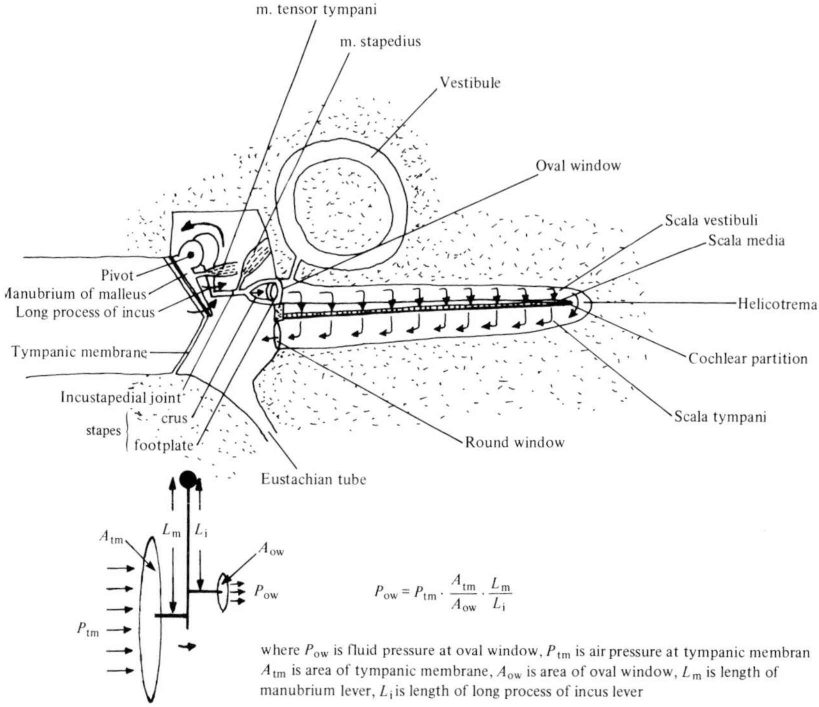

Fig. 14.4. Schematic functional diagram of middle ear and uncoiled inner ear, and middle ear transformer. Vibrations of tympanic membrane are transmitted as rotations of malleus and incus about common axis (normal to page, marked with dot). This produces piston-like movements of the stapes footplate in the oval window, with transmission of the pressure changes across the cochlear partition virtually instantaneously throughout the cochlear length. Arrows indicate direction of movements in response to a compression wave. The lower diagram illustrates transformer action of the middle ear by virtue of the large difference in area of tympanic membrane and oval window, and the (smaller) lever ratio (L_m/L_i). The middle ear muscles are contained in bony canals. The m. stapedius acts sideways on the incustapedial joint to stiffen the ossicular chain.

middle ear therefore, like the outer ear, emphasises intermediate frequencies. The combination of the two almost entirely accounts for the form of the threshold audiogram (Fig. 14.5).

The input impedance of the middle ear can be measured for clinical purposes by determining the attenuation of a low-frequency probe tone, introduced into the sealed external auditory canal from a source

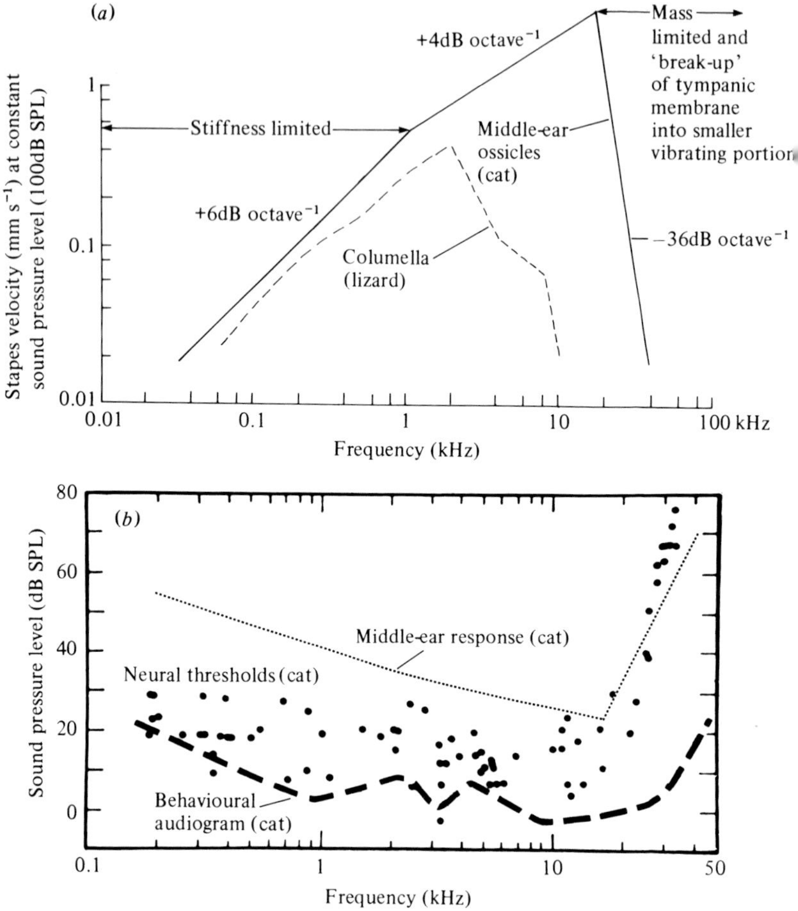

Fig. 14.5. Contribution of middle ear to ear's sensitivity. (*a*). Frequency response of cat's middle ear (continuous line) measured in terms of velocity of stapes motion at constant SPL at the tympanic membrane. (Velocity is used rather than displacement because it represents pressure in the inner ear.) Analogous response in lizard (dashed lines) has much less extended frequency response, probably because the latter possesses only a *columella*, a single peg-like ossicle linking tympanic membrane and inner ear, and not the mammalian ossicular system. (Lizard response from Johnstone & Sellick (1972) *Quarterly Review of Biophysics*, 5, 1; cat response from Wilson & Evans, unpublished data.) (*b*) Comparison between behavioural (audiogram) and neural thresholds of cat and its middle-ear response. (The vertical position of the middle-ear response is arbitrary.) Each point represents the

of known impedance. It is commonly expressed as the reciprocal of the stiffness component, i.e. as compliance, usually in terms of the volume of air having the same compliance. By subtracting out the compliance of the external canal, the compliance of the middle-ear–inner-ear system itself can be estimated. This provides information on the mobility of the ossicular chain, and is a valuable measure clinically, particularly when combined with information on the changes in ear-canal compliance with canal pressure, measurements known as *tympanometry*. Deviations in compliance much above the normal range (of about 0.3–2.5 ml) indicate abnormal mobility of the drum, as would be produced by traumatic interruption of the ossicular chain. Conversely, abnormally low compliance indicates a reduction in the mobility of the middle ear, either from fixation of the stapes footplate as in otosclerosis, or from abnormally high or low pressure in the middle ear (as with a blocked eustachian tube), or from fluid in contact with the tympanic membrane.

These deficiencies in middle-ear transmission produce what is known as *conductive hearing loss*. These losses rarely exceed 60–70 dB because of direct conduction of sound incident upon the head through the skull to the cochlea. This *bone conduction* affords another simple test for lesions of the middle ear. In addition to the audiogram for airborne sound (Fig. 13.2), a bone-conduction audiogram is determined by using a calibrated vibrator pressed to the mastoid bone. The calibration is such that, in the absence of middle-ear disease, the 'bone' and 'air' audiograms coincide. In the presence of middle-ear disease, a 'bone–air gap' appears between the two audiograms, the magnitude of the gap indicating the magnitude of the attenuation in middle-ear conduction.

Transmission of energy through the middle ear can be modified by the middle-ear muscles (Fig. 14.4). The tensor tympani (innervated by the Vth cranial nerve) pulls on the manubrium to increase (as its name indicates) the tension of the tympanic membrane; the stapedius

threshold of an individual cochlear fibre at its characteristic frequency. Much of the ear's sensitivity is therefore accounted for by the middle-ear response. The resonance of the ear canal is responsible for most of the remaining differences, particularly between 1 and 10 kHz. (Audiogram after Neff & Hind (1955) *Journal of the Acoustical Society of America*, *27*, 480, corrected for outer-ear and bulla response; neural thresholds from Kiang (1968) *Annals of Otology*, *77*, 656. After Evans (1975) Cochlear nerve and cochlear nucleus in *Handbook of Sensory Physiology*, vol. V/2, Chapter 1, Springer-Verlag.)

(innervated by the VIIth nerve) pulls on the neck of the stapes at right angles to its 'piston' axis, thereby tending to immobilise the footplate. Both actions increase the stiffness of the middle-ear system, and so attenuate the transmission of energy at the lower frequencies particularly below about 2 kHz, by a factor approaching 30 dB at 100 Hz. The muscles of the middle ear (mainly or only the stapedius) contract *reflexly* in response to high level (> 80 dB SL) sounds received *by either ear* with a latency of effect on middle-ear transmission of 15–150 ms and about 100–500 ms for maximum action, depending upon stimulus level. Note that the reflex can be evoked in one ear by acoustic stimulation of the other. This property is useful for diagnostic purposes.

The reflex can protect the inner ear against continuous intense sounds. Attenuations of 0.6–1 dB for every dB increase in acoustic input have been measured, directly in animals and indirectly in man, for SLs above 100 dB. More significantly, the middle-ear muscles are contracted just prior to and during vocalisation in order to reduce self-stimulation. In man, the preferential attenuation of lower frequencies means that the acoustic reflex would be effective in reducing the masking of high-frequency signals by low-frequency sounds. This is of potential importance for reducing the masking caused by the high-energy low-frequency components of one's own voice on the less intense higher-frequency components (e.g. in consonants) of simultaneously received speech from another speaker. The threshold and time course of the middle-ear muscle reflex, elicited by tone pips, can be determined by dynamic impedance measurements as above, and is clinically useful in evaluating ossicular function and the integrity of the cochlear nerve and brainstem nuclei.

14.3. INNER EAR

Morphology and biochemistry

The inner ear is an extremely delicate organ buried, in man, in the hardest bone in the body, the *petrous temporal bone* (Fig. 14.3). It consists of a series of passages, the *osseous labyrinth*, containing a complex system of sacs and tubes, the *membranous labyrinth* (which appears black in Fig. 14.3) surrounded by a fluid known as *perilymph*. The membranous labyrinth houses the receptor cells of the vestibular and cochlear systems, and contains a fluid of different composition from the perilymph, the *endolymph*.

The cochlear component of the labyrinth is coiled like a snail shell (hence the name *cochlea*), around a central axis, the *modiolus*, occupied by the nerve trunks. It has nearly three turns in man. Its base lies at the oval and round windows, where it connects with the vestibular labyrinth, and its apex lies more anteriorly and laterally than appears in Fig. 14.3 (see inset). The membranous labyrinth partitions the coiled labyrinth into three channels, shown diagrammatically in Figs. 14.3 and 14.4, and in cross-section in Fig. 14.6. The more anterior channel, the *scala vestibuli*, communicates with the vestibular labyrinth (Fig. 14.3) and hence the oval window, and therefore receives the acoustic input to the cochlea (Fig. 14.4). The more posterior channel, *scala tympani*, terminates in the membrane-covered *round window*, which acts to release intracochlear pressure.

The middle channel, the *scala media*, or *cochlear duct*, contains the acoustic sensory epithelium, the *organ of Corti* supported by the *basilar membrane* which, with the spiral lamina, forms the *cochlear partition*. The length of the basilar membrane is about 34 mm in man (compared with about 20 mm in cat and guinea-pig). It tapers in width from about 500 μm at the apical (helicotrema) end to about 100 μm at the basal end where, conversely, the scalae are largest. Here, the osseous spiral lamina spans most of the scalar width.

The organ of Corti in mammals is about 100 μm in thickness and consists of two types of sensory *hair cells* rigidly attached to the basilar membrane by supporting cells (*Deiter's cells*) and the *pillars* (Fig. 14.6). The hair-bearing ends of the hair cells are held firmly together in the rigid *reticular lamina*, continuous with the heads of the pillars. There is a single row of *inner hair cells* and 3–5 rows of *outer hair cells* (Figs. 14.6, 14.7) each row containing about 100–130 hair cells mm^{-1}. The hair cells bear stiff sensory hairs or *stereocilia* (3–6 μm in length), composed of filaments of actin. The stereocilia are arranged in characteristically shaped rows: in a shallow U formation, for the inner, and in a V or W formation for the outer hair cells (Fig. 14.7). The cilia of the outer hair cells are embedded in the *tectorial membrane*, a gelatinous structure overlying the hair cells and anchored to the outermost supporting cells by a marginal net (Fig. 14.6). Those of the inner hair cells appear not to be embedded at all in some species (guinea-pig, primates) or not firmly (cat).

As indicated in Fig. 14.3, the perilymph-filled spaces of the labyrinth communicate with the posterior cranial fossa via the *cochlear aqueduct* leading to the scala tympani near the round

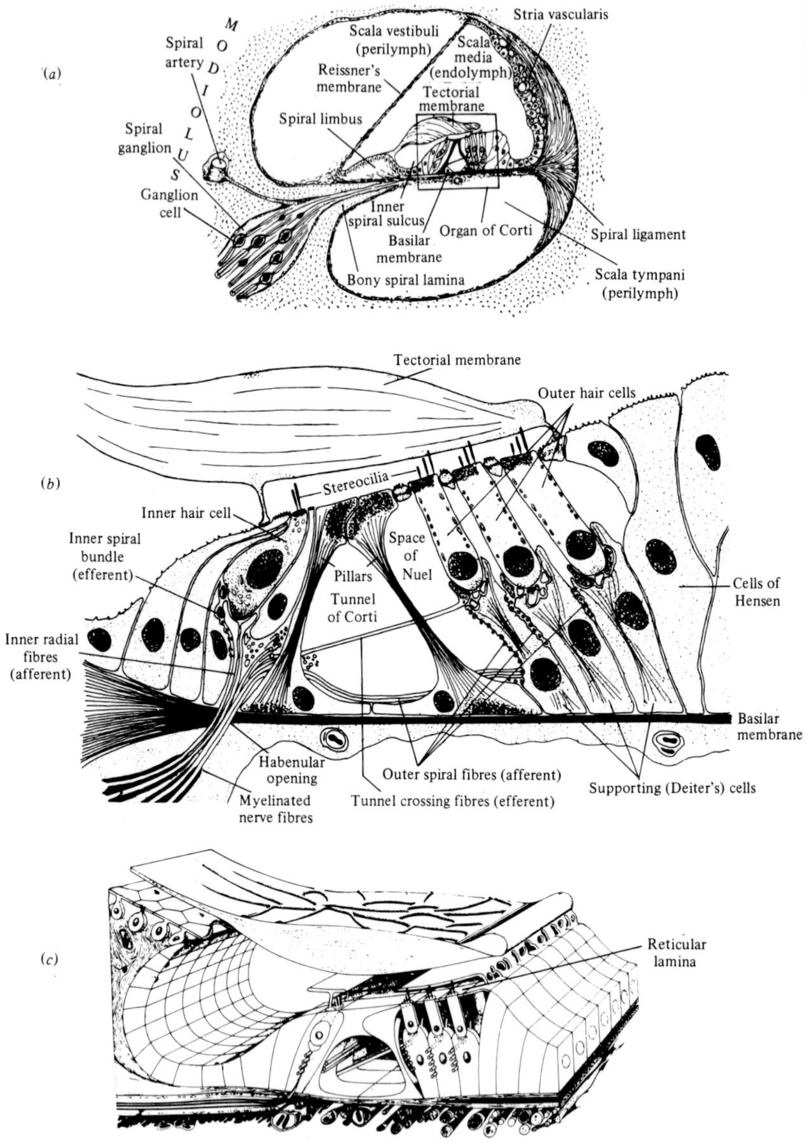

Fig. 14.6. Detail of cochlear partition and organ of Corti. (*a*) Cross-section through cochlear duct, modiolus of cochlear spiral being to the left of the diagrams. (For orientation see inset of Fig. 14.3.) (After Rasmussen (1933) *Outlines of neuro-anatomy*. William Brown Co.) (*b*) Enlarged view of organ of Corti. (After Durrant & Lovrinic (1977). *Bases of Hearing Science*. Williams & Wilkins Co.) (*c*) Three-dimensional view of hair cells and tectorial membrane. (After Lim (1972) *Archives of Otolaryngology*, *96*, 199.)

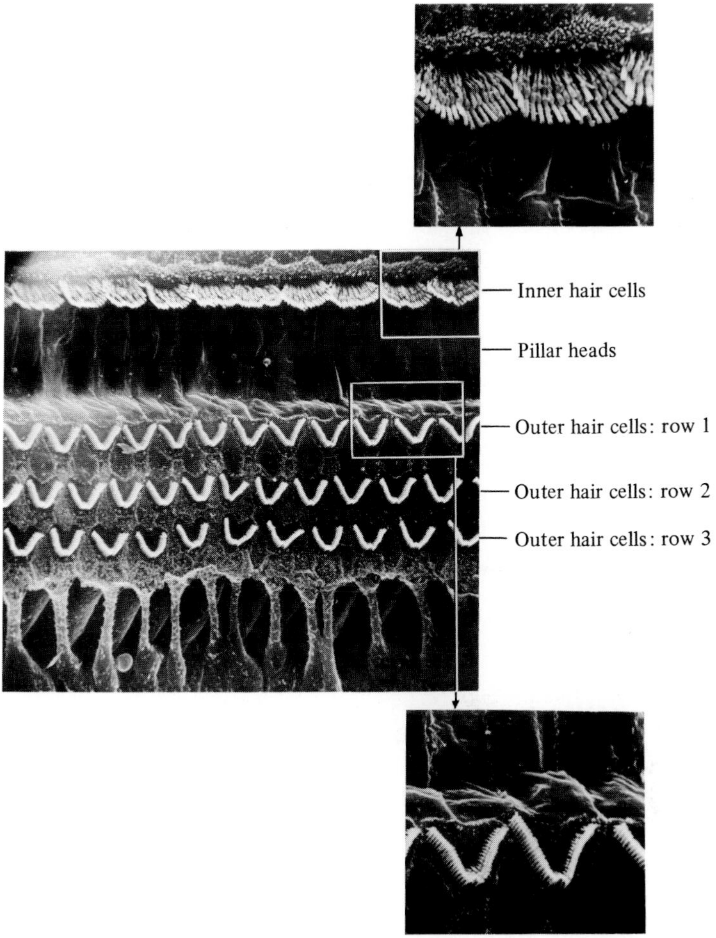

Fig. 14.7. Scanning electron microscope views of the surface of the cochlear hair cells after removal of tectorial membrane. Insets show enlarged views of the cilia (hairs) of the inner and outer hair cells. (From Bredberg *et al.* (1972) *Acta Otolaryngology, Supplement*, 301.)

window. The composition of the perilymph is therefore virtually identical to cerebrospinal fluid, i.e. high in sodium ions (about 150 mM l^{-1}) and low in potassium ions (3–4 mM l^{-1}). In contrast, endolymph contains a high concentration of potassium ions (about 150 mM l^{-1}) and a very low concentration of sodium ions (1–2 mM l^{-1}). It is likely that the cells of the *stria vascularis* (Fig. 14.6) contain an electrogenic ion pump, which actively transports potassium ions into, and sodium ions out of the endolymph, and is responsible for the standing intracochlear potential (see *Cochlear potentials*).

Surrounding the outer hair cells and the outer pillars are a series of spaces which interconnect with that between the pillars, the *tunnel of Corti* (Fig. 14.6). They are filled with *cortilymph*. This probably has a similar composition to the perilymph. Thus, with the exception of their hair-bearing ends, the hair cells have a similar extracellular milieu to that of cells elsewhere in the body. On the other hand, the high concentration of potassium ions in the region of the apical (hair bearing) ends of the hair cells may be essential to mechanoreception. It is also found in the lateral-line systems of fishes and amphibia.

The blood supply to the labyrinth (from the basilar artery directly, or via the anterior inferior cerebellar artery) enters the cochlea by way of the internal auditory meatus. It is therefore easily occluded, for example by tumours of the VIIIth nerve. There are no blood vessels in the organ of Corti itself, but a spiral artery shown in Fig. 14.6 runs along the basilar membrane in some species including man. The organ of Corti is extremely sensitive to reduction in its oxygen supply, which comes mainly from capillary networks in the vicinity of the spiral lamina, under the tunnel of Corti, and in the spiral limbus.

Cochlear mechanics

The piston-like movements of the stapes footplate in the oval window create fluctuations in pressure in the perilymph of the scala vestibuli (Fig. 14.4). These are transmitted with virtually no delay throughout the scala vestibuli and, through Reissners membrane, to the scala media. Because the membrane of the round window lacks stiffness, the pressure in the scala tympani is virtually constant at that of the middle ear. Hence, a temporally varying pressure difference is established across the cochlear partition. Because the stiffness of the basilar membrane changes along its length as the width tapers from the apical to the basal end, these pressure differences set up *travelling waves* in the membrane itself (Fig. 14.8).* These waves

* It is important to point out that a travelling wave motion of the basilar membrane does not necessarily imply transfer of energy along the membrane itself (in the manner of a wave travelling along a 'flicked' rope). The degree of coupling between segments of the basilar membrane along its length is slight, and it is the pressure difference across the length of the cochlear partition that is the driving force for the travelling wave. This means that the travelling wave will propagate from base to apex irrespective of where the pressure changes originate (as in bone conduction via the bones of the skull), and accounts for the fact that travelling waves produced by low frequencies are not impeded by damage to or calcification of the basal part of the cochlear partition.

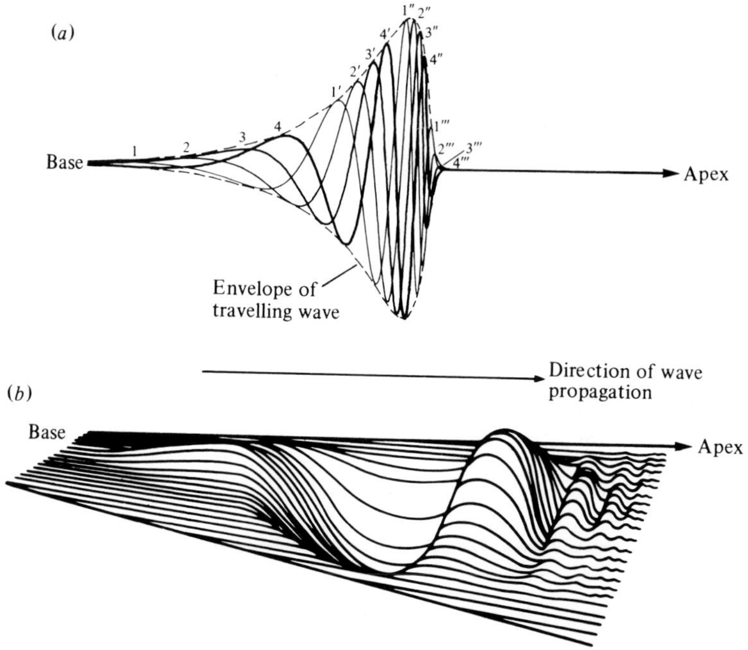

Fig. 14.8. Diagrams of travelling wave motion of the basilar membrane. In both cases, the amplitudes of the motion have been exaggerated for clarity. Even above the threshold of discomfort (120 dB SPL) the amplitude of vibration would be less than 1% of membrane width. (*a*) Two-dimensional representation of travelling wave at instants of time corresponding to 1/3 period (1, 2, 3, 4, 1′, 2′, 3′, 4′, etc.) (Courtesy of G. J. Sutton.) (*b*) Three-dimensional representation of travelling wave in small segment of basilar membrane, at one instant in time. (From Tonndorf (1960) *Journal of Acoustical Society of America*, *32*, 238.) Note wave travelling from base (to left) toward apex (out of view to right), with its *envelope* having a maximum at a given location, and a sharp cut-off towards the apex.

propagate from the basal to apical end, and as they do so, they grow in amplitude gradually to a peak, then rapidly collapse. The position of the peak amplitude depends upon the stimulus frequency, so that a map of peak frequencies can be established for the basilar membrane, the highest peak frequencies being represented at the basal end and the lowest at the helicotrema. The velocity and the wavelength of the travelling wave decrease with distance from the stapes, both decreasing very rapidly beyond the point of peak motion, as does the amplitude of vibration.

At any one point on the basilar membrane, the amplitude of

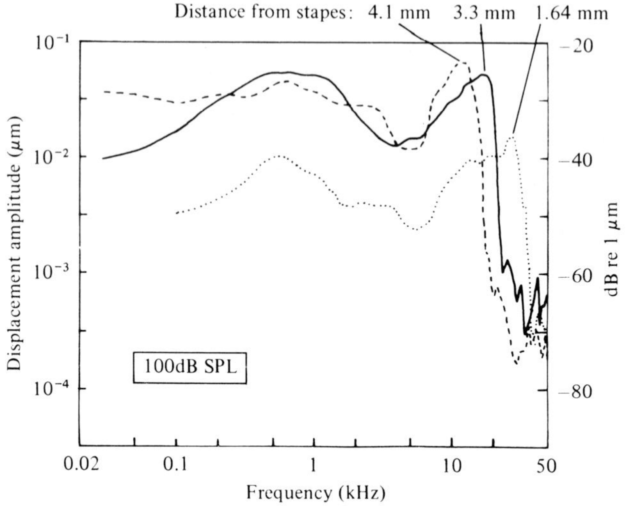

Fig. 14.9. Frequency response of individual basilar membrane locations. Amplitude of vibration *vs.* frequency of three places on guinea-pig basilar membrane, as recorded by capacitive probe *for constant sound pressure at tympanic membrane* (*100 dB SPL*). Location of places: 1.64, 3.3 and 4.1 mm from stapes. Each curve represents the amplitude, at a given location, of the envelope of the travelling waves (as in Fig. 14.8) corresponding to the frequencies examined. The smooth curve of Fig. 14.8 is therefore transformed (from amplitude as a function of position at a given frequency) to response as a function of frequency at a given location, and is 'distorted' by the response of the middle ear (pressure at the tympanic membrane being constant across frequency). This produces a low-pass filter characteristic: little change of the vibration amplitude at low frequencies; rapid cut-off at high frequency side of frequency of maximum vibration. (From Wilson & Johnstone (1972) in *Hearing Theory*, p. 172, IPO Eindhoven.)

motion shows broad tuning with frequency (Fig. 14.9). Unfortunately, not all studies are in agreement on the degree of tuning, and on the linearity of vibration.* The majority, however, indicate that the motion of a point on the basilar membrane is more or less *low-pass* in character (Fig. 14.9). This follows from the shape of the envelope of the travelling wave (dashed line in Fig. 14.8): all locations on the basilar membrane can be more easily activated by frequencies lower than their peak frequency than by higher frequencies.

The peak movements of the basilar membrane are some 30 times greater in amplitude than those of the stapes footplate. Even so, they

* Very recent studies suggest that the tuning shown in Fig. 14.9 represents that of the *passive* basilar membrane.

are extremely small, of the order of 0.01–0.1 μm (10–100 nm) at 100 dB SPL. Thus, at threshold (ca. 0 dB SPL), they must be less than atomic dimensions, on the assumption of linearity of vibration!

It is generally held that motion of the cochlear partition away from and towards the tectorial membrane induces a shear motion of the hairs of the hair cells in a radial direction (i.e. across the cochlear partition). It must be emphasised, however, that because the movements of the cochlear partition are so small compared with its dimensions, and because the physical properties of the tectorial membrane and the nature of contact between it and the cilia of the inner hair cells are not yet clear, it is as yet impossible to measure the exact motion of the cilia. It is assumed that movements of the stiff cilia are the necessary antecedents of the electrical changes in the hair cells to be described below. However, it is at present impossible to decide how this occurs (see Fig. 14.6): whether direct brushing of the cilia against the tectorial membrane is involved, as is likely for the outer hair cells, or whether, in the case of the inner hair cells, the cilia are displaced by the motion of fluid in the subtectorial space, or whether other, complex forms of 'micromechanics' of tectorial membrane, fluid and hair cells are involved. This is an important question, for it appears that the tuning of the responses of inner hair cells and cochlear nerve fibres is sharper than that of the basilar membrane (see p. 277), and that the presence of a tectorial membrane and outer hair cells is necessary for the sharp tuning.*

Cochlear potentials

Standard potentials. Fig. 14.10 summarises the results of exploration of the cochlea by microelectrodes. The d.c. potential in the endolymph of the scala media (endolymphatic potential) is +80 mV relative to that of the perilymph or a remote electrode. This potential is probably maintained (against a potassium ion diffusion potential in the opposite direction) by the electrogenic pump (E_{SV}) in the stria vascularis. The pump may be that responsible for the transport of potassium ions into and sodium ions out of the endolymph. Abolishing the activity of the pump, by anoxia, leads to a rapid reduction and reversal of the endolymphatic potential.

This 'battery' (E_{SV}) appears to act in conjunction with the transmembrane battery (E_M) of the hair cells, to produce a standing current flow through the hair cells. The latter battery (E_M) is the classical ionic diffusion potential (about −60 mV) responsible for the negative intracellular potential of cells in general, based on the ratio

* Recent evidence emphasises the active role of the organ of Corti, including the basilar membrane, in accounting for sharp cochlear tuning.

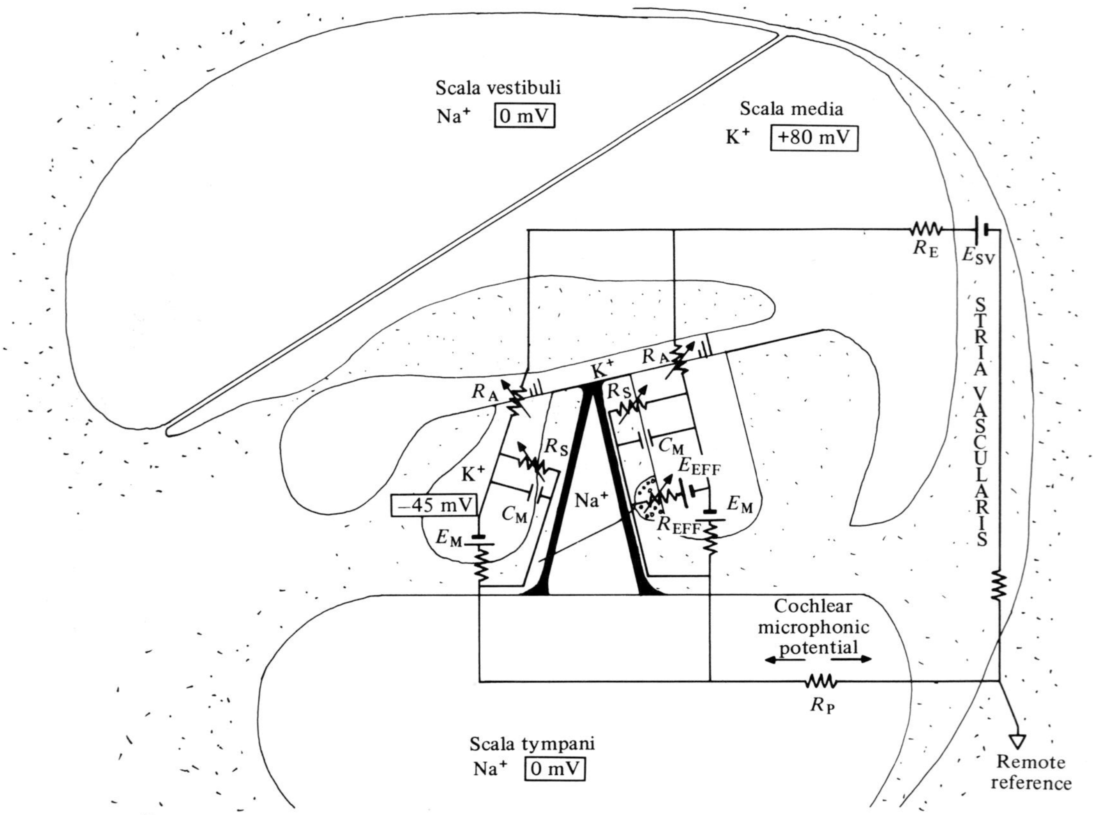
Scala vestibuli
Na+ 0 mV
Scala media
K+ +80 mV
R_E
E_SV
STRIA VASCULARIS
R_A
K+
R_S
C_M
E_EFF
E_M
R_EFF
Na+
-45 mV
Cochlear microphonic potential
R_P
Scala tympani
Na+ 0 mV
Remote reference

of potassium ion (and/or chloride ion) concentrations between the hair cell interior and the cortilymph.

Receptor potentials, transduction, and generation of action potentials. As indicated above, the form of the mechanical stimulus to the hair cells of the mammalian cochlea is not known precisely. However, by analogy with hair cells in vestibular and lateral-line systems, it must involve displacement of the cilia. Intracellular recordings from such hair cells (in the mammalian cochlea so far mainly from the inner hair cells) show that a sufficient stimulus produces a depolarising *receptor potential* in the inner hair cell (Fig. 14.11), accompanied by a decrease in the resistance between the interior and exterior of the cell. The receptor potential is generally a distorted version of the mechanical stimulus waveform (Fig. 14.11*a*). At frequencies higher than about 300 Hz in the mammalian inner hair cell, the low-pass filtering characteristic of the cell membrane (represented by C_M and the resistances of Fig. 14.10) progressively attenuates the a.c. component of the receptor potential, leaving predominantly a depolarising d.c. component for the frequencies above a few kHz (Fig. 14.11v). As will be shown later there is evidence for a cyclical component of excitation of cochlear nerve fibres known as *phase locking*. This is probably attributable to the a.c. component of the receptor potential; it attenuates with frequency above 1 kHz, but is still evident up to 5 kHz.

Fig. 14.10. Simplified diagram of electrical potential generators and current paths in cochlea. E_{SV} electrogenic pump in stria vascularis, responsible for the endolymphatic potential of +80 mV. Displacement of the hair cells in the effective direction probably reduces the series resistance through the apical, hair-bearing end of the hair cells (R_A) or the hair cell transmembrane permeability (shunt resistance, R_S), producing the *depolarising* receptor potential recorded intracellularly (Fig. 14.11) superimposed as the standing resting potential of about −45 mV. This in turn results from the shunting currents through R_A and R_S reducing the effects of the transmembrane diffusion potential battery (E_M, probably about −60 mV). In the case of the outer hair cells, activity of the efferent terminals decreases R_{EFF} hence *hyperpolarises* or stabilises the intracellular hair cell potential.

The receptor currents flow through the low extracellular resistance paths R_E and R_p (endolymph and perilymph) thus generating small extracellular potentials, such as the *cochlear microphonic* potential, recorded in the extracellular spaces and outside the cochlea (Fig. 14.12).

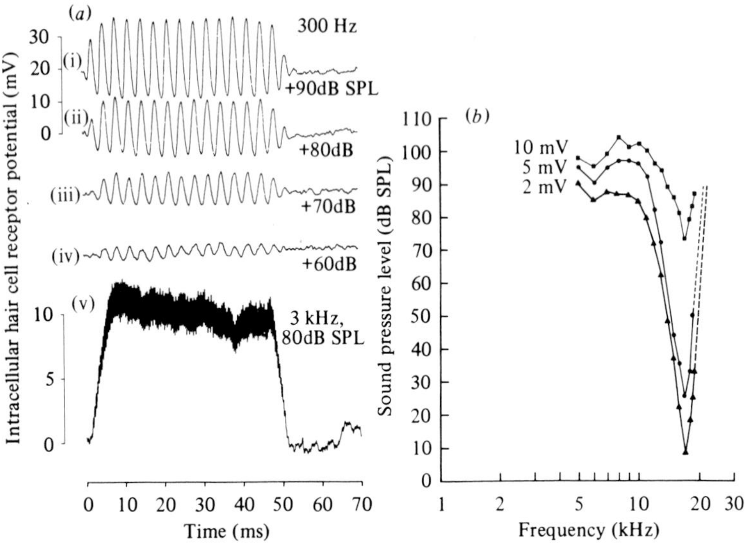

Fig. 14.11. Intracellular potentials from mammalian inner hair cells. (*a*) Intracellular potential waveforms recorded by a microelectrode in an inner hair cell in the basal (high frequency turn) of a guinea-pig: (i)–(iv) to a tone burst of low frequency (300 Hz), and at different sound-pressure levels. The waveform of the stimulus, distorted at high sound levels, can be seen in the response. (v) response to 3 kHz tone. Note small a.c. response superimposed upon a substantial (*c.* 12 mV) depolarising potential. (*b*) Variation in stimulus level required to keep d.c. receptor potential constant across frequency for three voltages (isovoltage frequency tuning curves). ((*a*) courtesy of Merzenich, Russell & Sellick, and (*b*) after Russell & Sellick, (1978) *Journal of Physiology*, *284*, 261.)

The amplitude of the receptor potential in mammalian inner hair cells depends on the intensity and frequency of the sound stimulus. Each hair cell has a most sensitive frequency (17 kHz in Fig. 14.11*b*); at progressively higher intensities, the band of frequencies evoking a response grows progressively wider until at the maximum sound level, frequencies from 1 to 19 kHz evoke a response, and the maximum d.c. receptor potential approaches 20 mV. By choosing a constant response amplitude, an isoresponse intensity–frequency function can be plotted, as shown in Fig. 14.11*b*. This shows how sharply tuned are the receptor potentials of inner hair cells.

In the reptilian cochlea there is evidence that the receptor potential is a component in a resonance mechanism within the hair cells themselves. This mechanism is possibly ionic or electromechanical in nature and is responsible for the sharp tuning. Whether this is the case also in the mammalian inner hair cells is not known.

By contrast, outer hair cells do not appear to have d.c. receptor potentials; so far only a small (few mV) a.c. voltage has been recorded. This appears to be as broadly tuned as the amplitude of vibration of the basilar membrane, and appears likely to be the origin of the extracellularly recorded cochlear microphonic (see below).

It is not certain at present how the hair cell receptor potential is generated. In the most widely accepted explanation (the Davis model), displacement of the hairs alters the resistance across the *apical* end of the hair cell (R_A in Fig. 14.10) and thus alters the standing current, driven through the hair cell by the two batteries referred to above. Deformation of the hair-bearing surface membrane in one direction brings about a reduction in its resistance and a consequent increase in permeability to potassium ions, and hence a depolarisation of the hair cell. In the alternative explanation, an increase in permeability occurs to ions *across the receptor cell membrane lying below the reticular lamina and in contact with cortilymph* (i.e. decrease in R_S in Fig. 14.10). In this case the ion species could be calcium. Both models predict reduction of membrane resistance with depolarising receptor potentials.

As in the case of other receptor cells, the depolarisation of membrane potential releases a chemical transmitter, in the mammalian cochlea probably an amino acid such as aspartic acid or glutamic acid, and this in turn generates characteristically irregular excitatory postsynaptic potentials (EPSPs) in the terminals of cochlear nerve fibres with a synaptic delay of about 0.5 ms. EPSPs larger than a threshold value evoke propagating action potentials, possibly at the point of myelination of the inner radial fibres (beneath the habenular openings, Figs. 14.6, 14.13). The average timing of the EPSPs and action potentials reflects periodicities in the stimulus waveform for frequencies up to about 5 kHz (see p. 284).

The large *efferent* synapses on the base of the outer hair cells liberate acetylcholine as transmitter, which probably opens an ionic channel (R_{EFF} in Fig. 14.10) to chloride ions, thus producing a *hyper*polarisation of the hair cell.

Extracellularly recorded cochlear potentials. It will be evident from Fig. 14.10 that the conductance changes (in either R_A or R_S) responsible for the receptor potentials will generate small time-varying

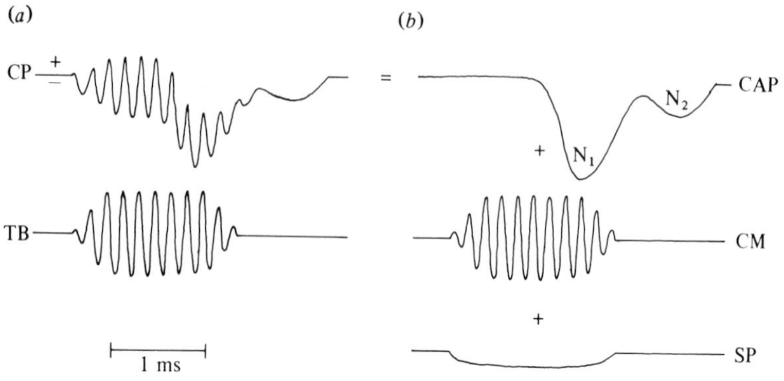

Fig. 14.12. Gross cochlear potentials recorded by an electrode on the cochlea. (*a*) Waveform of cochlear potential (CP) evoked by a short tone burst (TB). (*b*) This waveform can be analysed into three components: gross cochlear action potential (CAP) – a neural response; the a.c. cochlear microphonic (CM) – a receptor response; the d.c. summating potential (SP), probably also a receptor response.

potentials in the scala media, scala tympani and adjacent tissues. These can be easily recorded, together with potentials arising from synchronised action potentials in the cochlear nerve. These cochlear potentials are (Fig. 14.12): the *cochlear microphonic*, *summating potential*, and *gross cochlear action potential*. The cochlear potentials, being more accessible, have been more extensively studied than the intracellular potentials. They can be recorded by an electrode in the cochlear fluids (Fig. 14.10), or on the external wall of the cochlea, and (at very low amplitude) even in the external auditory meatus, and are valuable clinically as the *electrocochleogram*.

The *cochlear microphonic* follows the sound waveform with virtually no latency (the polarity depending upon the electrode location), and is likely to represent, for a single electrode near the round window, the sum of hair cell currents from a large portion of the cochlea, predominantly of the basal turn. The *summating potential*, on the other hand, is likely to reflect the asymmetry of the intracellular receptor potential. There is good evidence to suggest that the cochlear microphonic potential is predominantly generated by the *outer* hair cells.

Because the *gross cochlear action potential* represents the sum of *synchronised* individual action currents of a large number of cochlea fibres, it is observed only at the onset of acoustic stimuli (particularly to transients such as clicks) subject to a delay of 0.5–1 ms, representing synaptic and conduction delay.

In recordings direct from the inside or from the surface of the cochlea in animals, the cochlear potentials are relatively large (maximum of about 1 mV for cochlear microphonic and cochlear action potential respectively). To obtain the electrocochleogram in patients, however, averaging is required to extract the small cochlear potentials (1–30 μV) from background electrical noise. This entails the use of an *averaging computer* to sum the responses to many repetitions of the signal: the cochlear potentials, being time-locked to the stimulus, sum, whereas the background electrical noise tends to cancel.

Innervation of hair cells

Afferent. The great majority of the 30000–50000 afferent fibres in the cochlear nerve arise from the inner hair cells (95% in cat, 85–90% in guinea-pig); see Fig. 14.13. Each of about twenty synapses at the base of each inner hair cell gives rise to an *inner radial fibre*, which immediately leaves the organ of Corti (i.e. radially) via an adjacent or neighbouring opening in the *habenular perforata* into the spiral lamina. Each fibre thus innervates one (in some cases in the guinea-pig two to three) inner hair cell only.

In contrast, the much more numerous outer hair cells are innervated by only 5–10% of the cochlear fibres, which run spirally on their central course (Fig. 14.13). These *outer spiral fibres* innervate about ten outer hair cells in the first 100 μm or so of their spiral course (Fig. 14.13). They then run in the same direction centrally (always apically), without synapsing for considerable distances (0.6 mm in the basal turn of the cat) down the supporting (Deiter's) cells before they cross the floor of the Tunnel of Corti to penetrate a habenular opening, as indicated by the thick lines in Fig. 14.13. Each outer hair cell makes contact with the terminals of several outer spiral fibres.

The afferent fibres from the inner hair cells become myelinated below the habenular openings and pass to their bipolar ganglion cells situated in the spiral *Rosenthal's canal* in the modiolus (see Fig. 14.6). Their central (myelinated) axons leave the cochlea via the modiolus and internal auditory meatus (Fig. 14.3). The fibres from the outer hair cells remain unmyelinated. There is some doubt whether they have functionally significant axons central to their ganglion cell bodies, and what function they serve.

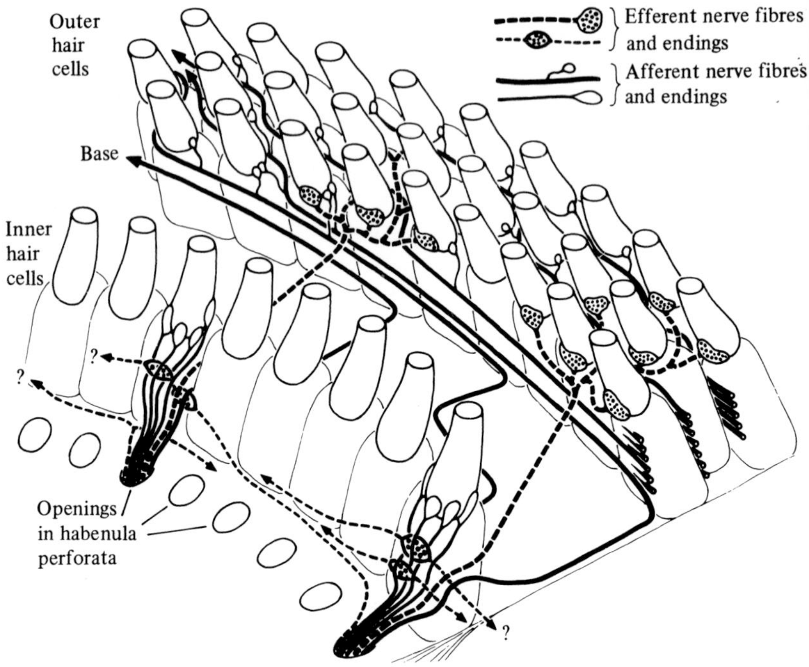

Fig. 14.13. Innervation of inner and outer hair cells. Diagram of surface view of hair cells. For clarity, only a few inner radial fibres ending on the inner hair cells are shown, and a few efferent endings on the outer hair cells. The full spiral extent of the outer spiral fibres cannot be shown on this scale. Note afferent inner radial fibres (shown as thin lines) ending on inner hair cells, with efferent synapses terminating on the afferents themselves; afferent outer spiral fibres (thick lines) crossing the floor of the tunnel of Corti and ascending the supporting cells, with the efferent synapses ending on the outer hair cell. (From Spoendlin in *Frequency Analysis and Periodicity Detection in Hearing*. p. 2. Sijthoff, 1970.)

Efferent. A very substantial proportion of the synaptic terminals in the cochlea, particularly on the outer hair cells, are *efferent*, that is, are part of the descending, centrifugal, auditory pathways (Fig. 14.1). The *olivocochlear bundle*, as its name implies, arises in the superior olive region of the brainstem as *uncrossed* (ipsilateral) and *crossed* (contralateral) bundles and enters the cochlea via the vestibular nerve. The crossed bundle runs predominantly but not exclusively to the outer hair cells (Fig. 14.6) ending in huge terminals (Fig. 14.13), dwarfing and outnumbering those of the afferents. The remainder of

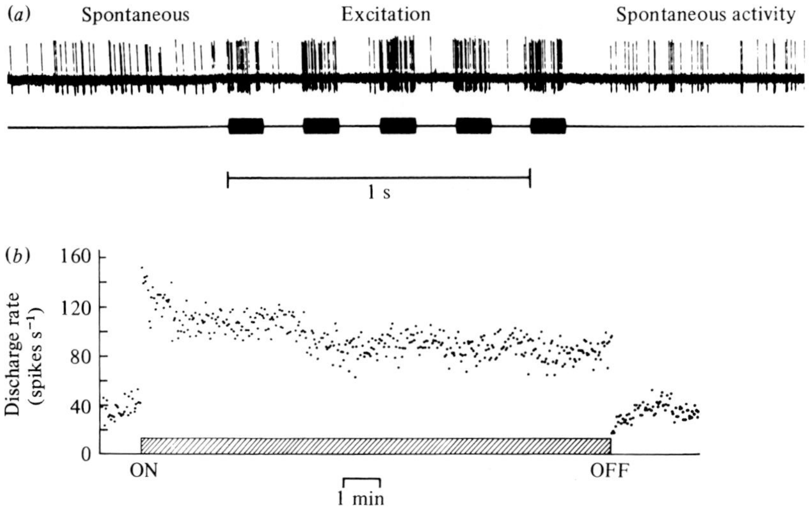

Fig. 14.14. Response of single fibre of cochlear nerve to sound. (*a*) Spontaneously active fibre responds to 5 tone bursts by producing bursts of action potential spikes. Note reduction in activity (off-suppression) following each burst of activity. (Guinea-pig: From Evans (1975). Cochlear nerve and cochlear nucleus. In *Handbook of Sensory Physiology*. Vol. V/2. Chapter 1. Springer-Verlag.) (*b*) Time course of response to a continuous tone of 13-min duration. Each point indicates the number of spikes counted in 1 s. Note discharge rate is maximum at onset of stimulus, and progressively reduces thereafter (adaptation); transient reduction in probability of discharge on terminating stimulus (off suppression). (Cat: After Kiang *et al.* (1965). *Discharge patterns of single fibres in the cat's auditory nerve*. M.I.T. Press.)

the efferent fibres form the *inner spiral bundle* ending on the inner hair cell afferent fibres themselves (Fig. 14.13).

14.4. COCHLEAR NERVE

There is now available a very large body of quantitative data on the responses of individual cochlear nerve fibres (mainly in cat, guinea-pig and squirrel monkey) to a wide variety of stimuli. In the absence of sound stimulation, the majority of cochlear fibres are spontaneously active (*spontaneous activity*, Fig. 14.14). To tones of appropriate frequencies and intensity, all fibres give a characteristic response: *excitation* lasting the duration of the stimulus, after a latency of 1–10 ms (Fig. 14.14*a*). Both the spontaneous and the evoked discharges are characteristically irregular.

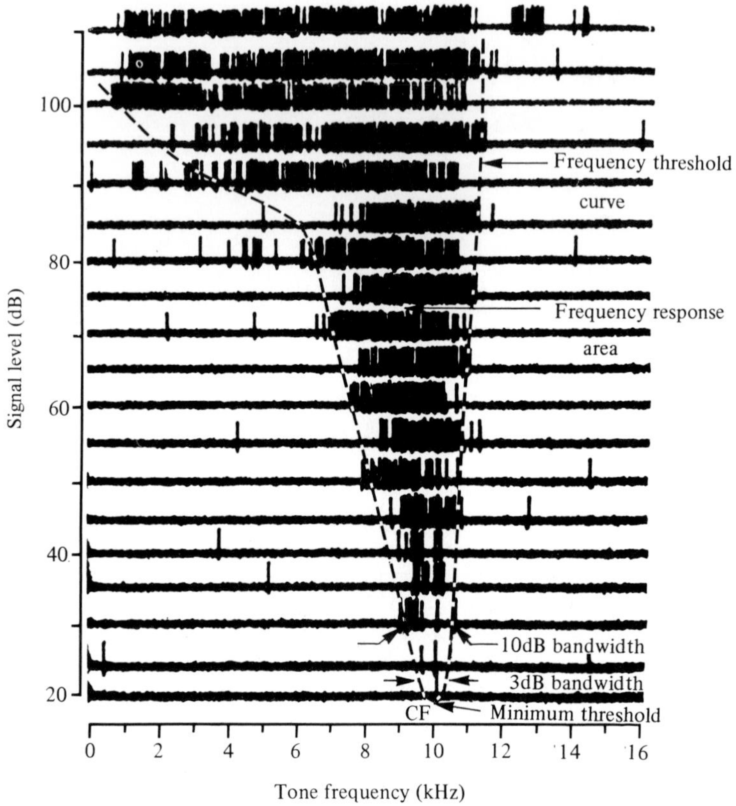

Fig. 14.15. Frequency response area of single fibre in cochlear nerve. Frequency response area is mapped out by sweeping a continuous tone successively upward and downward in frequency, increasing the tone level by 5 dB at the end of each sweep. At the lowest intensities, the fibre responds to a narrow band of frequencies at 10 kHz. This is its *characteristic frequency* (CF). At higher levels, the excitatory frequency band becomes progressively wider. For fibres of this high characteristic frequency, the response area is asymmetrical, extending on the low frequency side to form a low frequency 'tail' above about 80 dB SPL. The boundary of the frequency response area is called the frequency threshold curve (FTC) or 'tuning curve'. (Minor deviations of the responses from the curves are caused by 'off-suppression' effects working in opposite directions in each sweep.) Note linear frequency scale. (Guinea-pig: from Evans (1972) *Journal of Physiology*, *226*, 263.)

In common with all receptor neurons, the rate of discharge evoked by a continuous stimulus *adapts* in time (Fig. 14.14*b*). The decrement in rate is approximately related to the log of time. On termination of the stimulus, a brief period of depression of spontaneous activity (and excitability) ensues.

Frequency selectivity

The range of frequencies that will evoke responses from any given cochlear fibre is relatively limited. Fig. 14.15 shows the responses of a single fibre innervating the 10 kHz place in the organ of Corti, to sweeps in continuous tone frequency, in each direction. At the lowest signal level illustrated (+20 dB) the fibre responds, at its *minimum threshold*, to 10 kHz. This frequency of maximum sensitivity is termed the fibre's *characteristic frequency* (CF). At progressively higher stimulus levels, the effective frequency band grows wider, so that at the highest level tested, 90 dB above minimum threshold, frequencies from 1 to 11 kHz evoke responses. The range of effective frequencies and intensities map the *frequency response area* of the fibre (equivalent to the *receptive field*) and the threshold boundary, the *frequency threshold* ('tuning') *curve* (FTC). Fig. 14.16 depicts families of FTCs from cochlear fibres originating from different points along the cochlea, from guinea-pig, cat and monkey. For those fibres with characteristic frequencies above 2 kHz, the FTCs are asymmetrical, and represent quite remarkable filters. They have very steep cut-off slopes on the high-frequency side, and low-frequency cut-offs that are somewhat less steep at lower sound levels, but which suddenly decrease in slope, above 70–100 dB SPL, thus forming a low-frequency 'tail' to the response area. At least for 40–60 dB or so above minimum threshold, the cochlear fibres act as narrow band filters, having bandwidths ranging from about $\frac{1}{2}$ octave at 0.2 kHz to $\frac{1}{10}$ octave at 20 kHz. Their shapes match those of the isoresponse curves of the receptor potentials of inner hair cells (Fig. 14.11), and as we shall see, the analogous psychoacoustic 'tuning curves' (Fig. 15.1).

In contrast, the tuning of the cochlear fibres is sharper than existing measurements of the amplitude response of the basilar membrane (dotted lines in the lower part of each panel of Fig. 14.16).* At least part of this difference in tuning may be due to the action of a metabolically active, so-called 'second filter' mechanism

* Recent studies suggest that the basilar membrane tuning is much sharper than that indicated in Figs. 14.9 and 14.16, and that active mechanisms are involved. For references, see p. 306.

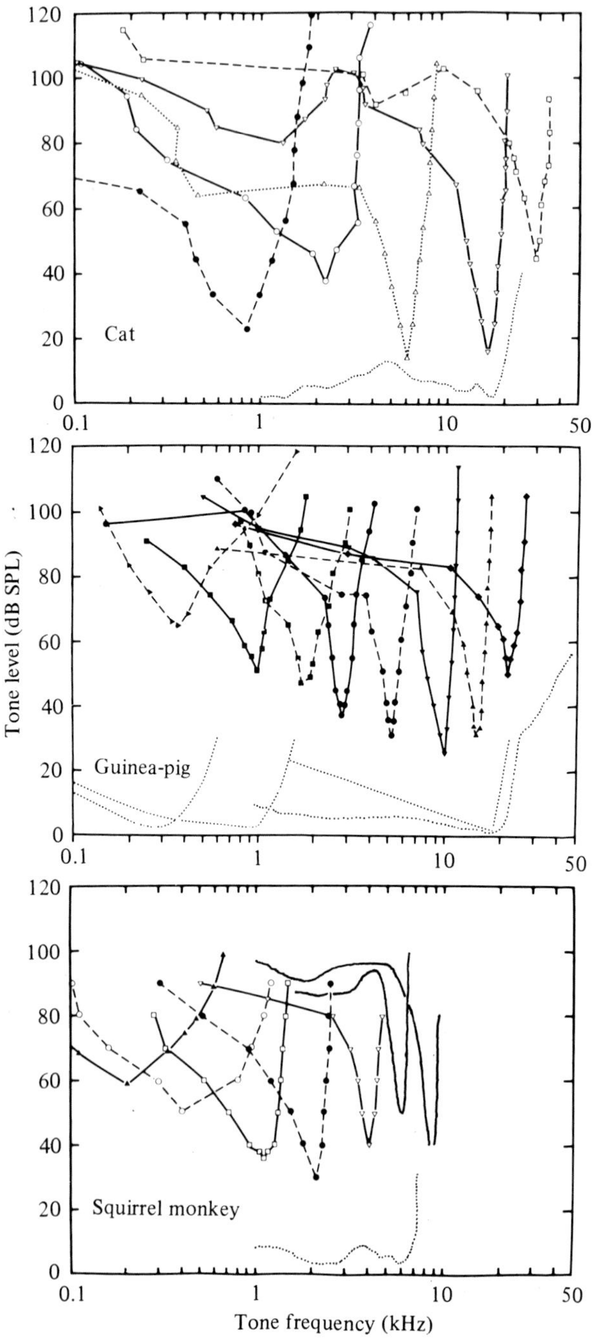
Cat
Guinea-pig
Squirrel monkey
Tone level (dB SPL)
Tone frequency (kHz)
0
20
40
60
80
100
120
0.1
1
10
50

in the cochlea (the first filter being that of the basilar membrane). Some candidates proposed for this frequency sharpening mechanism are the 'micromechanics' of the hair-cell: tectorial membrane region (p. 267), electrical tuning of the hair cells (p. 270), and even an active motion of the hair cells or the cilia themselves.

The minimum threshold (i.e. at the FTC tip) of cochlear fibres follow closely the behavioural audiogram of the species (Fig. 14.5*b*). As indicated earlier (p. 257), this overall frequency sensitivity is determined largely by the characteristics of the outer and middle ears. Within an individual ear, the range of minimum thresholds at any frequency is relatively restricted, to 20 dB or so. Interestingly, the most sensitive fibres tend to have higher rates of spontaneous activity than the less sensitive.

The representation of frequency along the cochlear partition and the discrete innervation patterns of the inner radial fibres imply that the characteristic frequency of a cochlear fibre depends upon where it originates. Fibres innervating the apical end have the lowest characteristic frequencies (in the cat about 0.1 kHz) and those from the basal end, the highest (in the cat about 50 kHz). Because this mutual relationship of cochlear fibres is preserved in the cochlear nerve trunk (in the manner of a spiral) and in its termination in the various subdivisions of the cochlear nucleus (§14.5), and because cochlear fibres are relatively narrowly tuned, the peripheral auditory system is *cochleotopically* and *tonotopically* organised, i.e. stimuli of different frequencies (*tonos*) evoke activity at systematically ordered positions (*topos*) along the neural array. Thus, in principle, the frequency of stimulus is coded by the *place* of neural activity (p. 315).

Fig. 14.16. Families of frequency threshold curves for single cochlear nerve fibres in cat, guinea-pig and squirrel monkey. Each curve represents the frequency threshold curve of a different fibre, the fibres chosen to cover a wide range of characteristic frequencies, on a logarithmic frequency scale. Below the neural curves are arbitrarily positioned analogous response curves (dotted lines) for the basilar membrane, to show the difference in tuning between basilar membrane (low-pass) and the cochlear nerve fibres (narrow-band). Neural data: cat, Kiang *et al.* (1967) *Journal of the Acoustical Society of America*, *42*, 1341; guinea-pig, Evans (1972) *Journal of Physiology*, *226*, 263; squirrel monkey, Rose *et al.* (1971) *Journal of Neurophysiology*, *34*, 685; Geisler *et al.* (1974) *Journal of Neurophysiology*, *37*, 1156. Basilar membrane data: cat, Evans & Wilson (1975) *Science*, *190*, 1218; guinea-pig, Wilson & Johnstone (1972). *Hearing Theory*. IPO Eindhoven; squirrel monkey – Geisler *et al.* (1974) *Journal of Neurophysiology*, *37*, 1156.

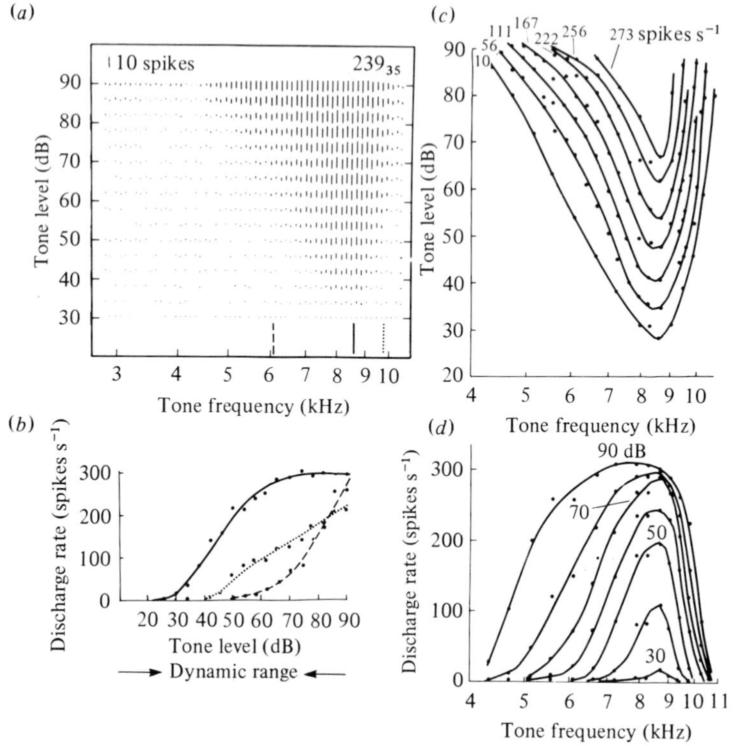

Fig. 14.17. Response of single cochlear nerve fibre as function of frequency and intensity. (*a*) Frequency response 'map' of fibre. The length of each vertical line indicates the average number of spikes evoked by a 50-ms duration stimulus at the frequency and intensity indicated by the centre of the line. (*b*) Vertical 'sections' through *a* at frequencies indicated by dashed, continuous and dotted lines respectively. These are known as rate–intensity or rate–level functions. Note restricted *dynamic range* over which the fibre can signal intensity at its characteristic frequency, and *saturation* of discharge rate at high levels. (*c*) Isorate contours. Each contour indicates the tone frequencies and intensities evoking a given discharge rate. These are obtained by taking horizontal 'sections' through a family of rate-level functions as in (*b*). (*d*) Iso-intensity contours, i.e. horizontal 'sections' through (*a*). Note flattening of the contours at the highest levels, because of saturation of the response, and a shift in frequency of maximum response. (Cat: from Evans (1978) *Audiology*, *17*, 369.)

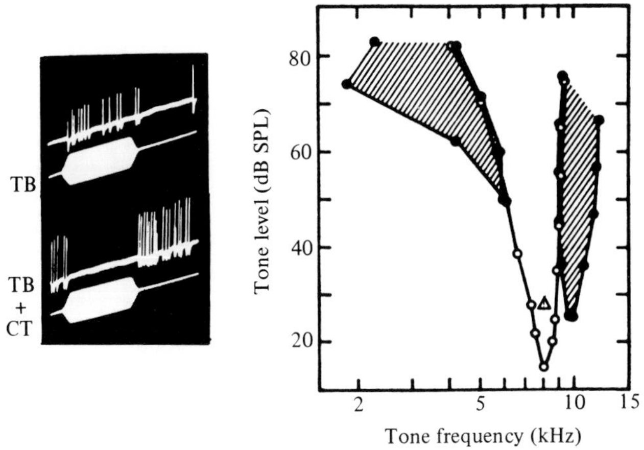

Fig. 14.18. Lateral suppression of cochlear fibre response. The response to stimulus tone (indicated by the triangle at 8 kHz, the fibre's characteristic frequency) can be suppressed by a second stimulus falling within the shaded areas. The suppression areas slightly overlap the frequency threshold curve (unshaded area) so that a tone burst that excites the fibre when presented alone (TB) suppresses the response to a continuous exciting tone (CT). (From Evans (1975). *Cochlear nerve and cochlear nucleus*, in *Handbook of Sensory Physiology*. Vol. V/2 Chapter 1, Springer-Verlag; After Nomoto *et al.* (1964). *Journal of Neurophysiology*, *27*, 768; Arthur *et al.* (1971) *Journal of Physiology*, *212*, 593.)

The above description of cochlear fibre tuning refers to *threshold* characteristics. The suprathreshold response is also tuned, up to a point (Fig. 14.17*c*, *d*). Most fibres have a limited dynamic range over which they can signal the level of a tone stimulus of given frequency. Thus, at the characteristic frequency of the fibre illustrated in Fig. 14.17, stimulus levels between about 25 and 70 dB SPL evoke a monotonically increasing discharge rate (Fig. 14.17*b*). Below this range, the stimuli are subthreshold, and above, the unit's discharge rate *saturates*, owing to some inherent non-linearity in the transducer and/or synaptic mechanisms. This limited dynamic range makes it difficult to understand how the intensity of signal components is peripherally coded; this problem is discussed in Chapter 15.

Another instance of cochlear non-linearity is the phenomenon of *two-tone or lateral suppression* (Fig. 14.18). Here, the response of a cochlear fibre to a tone (at the triangle in Fig. 14.18) can be suppressed by a tone falling within frequency regions flanking the

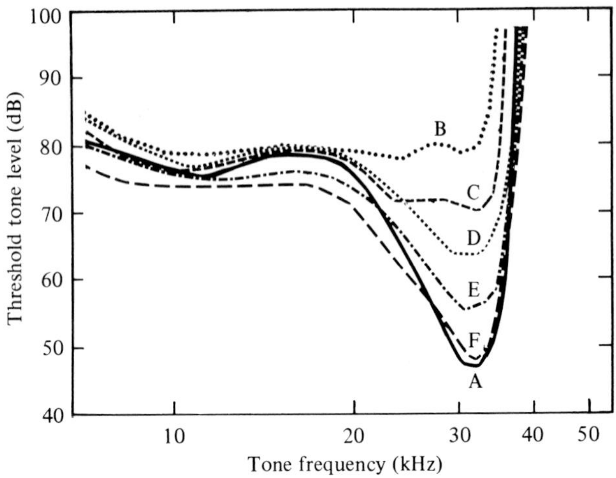

Fig. 14.19. Physiological vulnerability of cochlear tuning. Effects obtained on the tuning of a single cochlear nerve fibre in cat, from an intra-arterial injection of furosemide, an ototoxic diuretic known to cause reversible hearing loss in man. A (continuous line) is the fibre's frequency–threshold curve before injection, B and C are the curves obtained 1 and 2 min respectively after injection. Note the progressive loss of the low threshold, sharply tuned 'tip' segment of the frequency threshold curve until, at C, only a high threshold, broad curve remains. Curves D, E and F indicate that the effects on the neural tuning are reversible at this dosage. They were obtained 5, 7 and 20 min after the injection. (From Evans & Klinke (1974) *Journal of Physiology*, *242*, 129P.)

response area, as indicated by the hatched areas in Fig. 14.18. Because these suppressive areas slightly overlap the frequency threshold curve, the suppressive stimulus itself can actually excite the fibre in the absence of the CF tone (Fig. 14.18). The mechanism underlying this phenomenon is not clearly understood.

Physiological vulnerability of cochlear frequency selectivity

The cochlear fibre responses (and hair-cell receptor potentials) are sharply tuned and highly sensitive only if the cochlea is in normal physiological condition. Deterioration in the condition of the cochlea can produce selective loss of the low-threshold, sharply tuned segment of the FTC (Figs. 14.19, 14.20). This loss may be acute, i.e. short-lived and reversible, such as when it results from brief hypoxia, or from systemic administration of doses of ototoxic agents such as

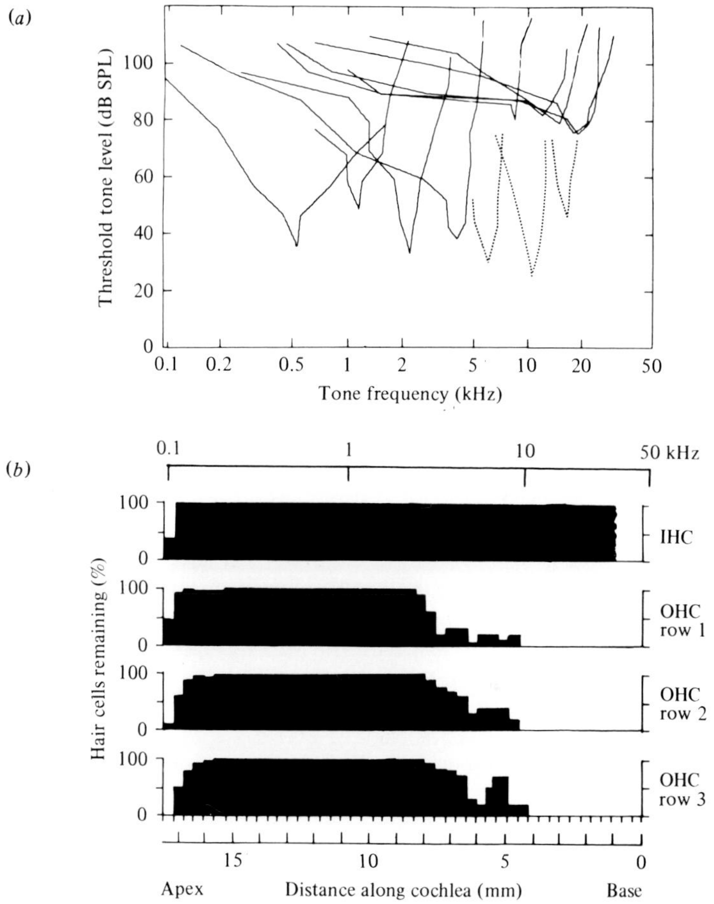

Fig. 14.20. Effects of hair-cell damage on cochlear fibre tuning. (*a*) Continuous curves are the frequency–threshold curves obtained from cochlear nerve fibres in a guinea-pig ear damaged by long-term injections of kanamycin, an antibiotic of the ototoxic streptomycin family. (*b*) The black areas below indicate the proportion of inner and outer hair cells remaining in the cochlea at the time of the physiological recording. All of the outer hair cells are missing from the basal half turn (first 4–5 mm) of the cochlea. The frequency–threshold curves of fibres with characteristic frequencies corresponding to the damaged regions have high thresholds and are broad; the dotted outlines indicate the appearance of normal curves. (From Evans & Harrison (1976) *Journal of Physiology*, *256*, 43P.)

furosemide (a powerful diuretic used clinically in renal failure, Fig. 14.19). On the other hand, permanent loss of tuning can be caused by local damage (e.g. haemorrhage), or by the systemic administration of the aminoglycoside antibiotics (streptomycin, neomycin, gentamycin, kanamycin) as shown in Fig. 14.20*a*. Morphological evidence of damage to the organ of Corti often occurs in these cases, particularly after long-term administration of agents like kanamycin (Fig. 14.20*b*). The basal turn and especially the outer hair cells are generally affected first as shown by the *cochleogram* of Fig. 14.20*b* indicating the percentage of hair cells remaining. Curiously, under these conditions, the threshold and tuning properties of the cochlear fibres recorded from the same cochlea correlate more with the presence and absence of the *outer* hair cells than with the *inner* hair cells that the fibres innervate (assuming that the fibres recorded from originate from the inner hair cells, and that the latter are *functionally* normal). This suggests that the normal sharply tuned properties of the inner hair cells depend in some way upon the integrity of the outers. This could afford the outer hair cells with an important modulating function, in view of their apparently insignificant afferent, but powerful efferent innervation.

This *physiological vulnerability* of the cochlear filter is of considerable significance for our understanding of the nature of sensorineural hearing loss of cochlear origin (p. 311).

Temporal patterning of discharges: coding of stimulus period

So far, we have limited discussion of cochlear-fibre responses to the *mean rate* of the discharges. For suprathreshold stimuli of frequency below about 5 kHz, impulses in cochlear fibres show a temporal pattern (Fig. 14.21); there is a preference for the fibre to discharge in a given half cycle of the stimulus period (Fig. 14.21*a*). This is shown more clearly in the *period histogram* of Fig. 14.21*b*, where the number of discharges per unit of time across the cycle of the stimulus is accumulated, and plotted as a function of the stimulus phase. The phenomenon is known as *phase locking* of the discharges. Thus, the probability of discharge is roughly proportional to the half-wave rectified stimulus waveform (Fig. 14.21*b*). This 'phase-locking' can be demonstrated by another type of histogram (Fig. 14.21*c*) where the number of *intervals* of different durations between action potentials are accumulated – the *interspike interval histogram*. This shows an exponential relationship between the number of intervals and their

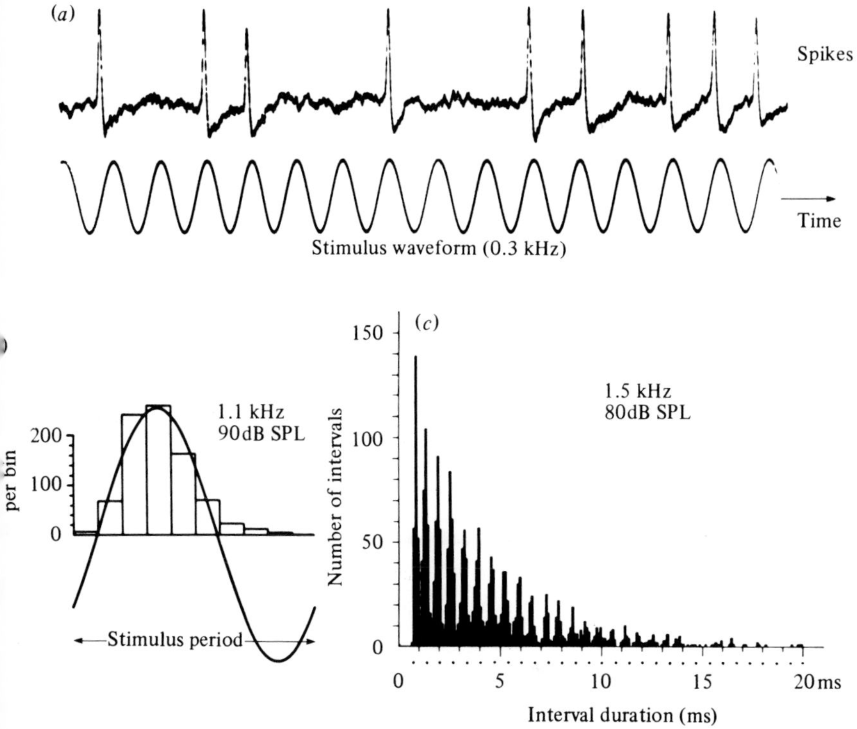

Fig. 14.21. Temporal patterning of discharges of single cochlear nerve fibres. (*a*) Response in time of a single cochlear fibre to a low-frequency continuous tone. Note that the spikes are 'phase-locked' to a given half-cycle of the stimulus waveform. (Guinea-pig: from Evans (1975). *Cochlear nerve and cochlear nucleus* in *Handbook of Sensory Physiology*. Vol. V/2 Chapter 1, Springer-Verlag.) (*b*) 'Period histogram' of responses similar to those in (*a*). The height of each bin indicates the number of spikes falling in that portion of the period of the waveform. Note that the histogram approximately follows a half-wave rectified version of the stimulus waveform (which has been phase-shifted to match the histogram). (Squirrel monkey: Rose *et al.* (1971). *Journal of Neurophysiology*, *34*, 685.) (*c*) Interspike interval histogram indicating number of intervals of given duration obtained from response of a cochlear fibre to continuous tone of 1.5 kHz. The separation between the histogram peaks reflects the period of the stimulus (indicated by the dots under the abscissa). (Squirrel monkey: from Rose *et al.* (1968) in *Hearing Mechanisms in Vertebrates* p. 144. Churchill.)

duration,* modulated so that the intervals chiefly represented are those corresponding to the *period* of the stimulus and its multiples.

Interestingly, some 'phase-locking' occurs in fibres at stimulus levels as much as 20 dB *below* the threshold for a change in the mean discharge rate. Phase-locking also persists at sound levels *above those causing saturation of the mean discharge rate* (although the *degree* of phase-locking saturates at lower levels).

Complex waveforms comprising several frequencies below 5 kHz, produce responses where the waveforms, after modification by the cochlear filter, are represented in the period histogram and in the distribution of interspike intervals. This representation persists relatively unaffected at sound levels above saturation of the mean discharge rate, and is therefore remarkably robust. This temporal information is certainly important for the localisation of the sources of a sound (§15.4) but whether and how it is used in analysing frequency is not known.

Response to click stimuli

Click stimuli, generated by brief pulses, are in principle wide-band. They can therefore excite every cochlear fibre, providing the energy filtered by the cochlear filter exceeds the threshold of that fibre. As expected, for fibres of low CF, the 'ringing' of the cochlear filter is manifested in the temporal response patterns of the discharges as a periodic enhancement of discharge probability (Fig. 14.22*a*, *b*). For fibres with CF higher than about 4–5 kHz (the approximate limit of 'phase-locking'), this periodicity is blurred out.

Summary

This brief account of the response properties of cochlear fibres has emphasised a number of features that make the task of predicting their responses to complex sound more straightforward. *To a first approximation*, these responses can be predicted on the following basis: (i) *linear filtering* with a filter shape corresponding to the frequency threshold curve; (ii) half-wave *rectification* of the waveform; (iii) a *logarithmic transform* to represent the approximately linear

* This exponential relation is expected for a Poisson process in which the occurrence of an impulse is equally likely at any time, the *probability* of occurrence being the only quantity that can be specified. For cochlear nerve fibres this probability is primarily determined by the frequency and intensity of the sound, but is also modulated by its phase as shown in Fig. 14.21. Note that there is a short period following an impulse within which the probability of a second impulse is reduced by refractoriness.

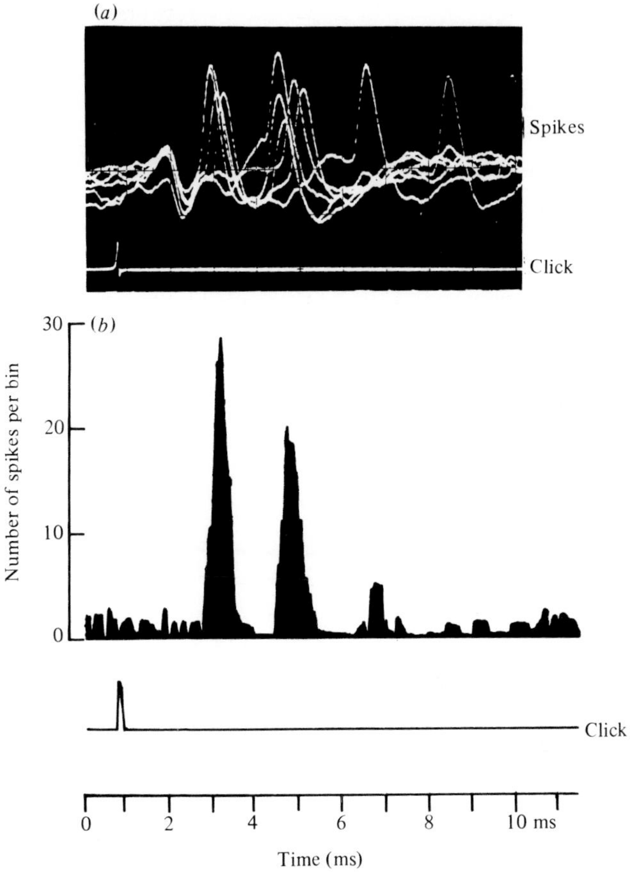

Fig. 14.22. Response of single fibres in cochlear nerve to click stimuli. (*a*) Superimposed records of response of low-frequency fibre (characteristic frequency, 0.5 kHz) to 6 click stimuli. Note grouping of discharges at preferred times following the click. This indicates the 'ringing' properties of the cochlear filter. (*b*) Post-stimulus time histogram of many responses (as in *a*). Height of each bin indicates the number of spikes elicited at the given time interval following the click stimulus (at 1 ms). Note periodicity of discharge probability having a period (2 ms) that corresponds to the reciprocal of the fibre's characteristic frequency (0.5 kHz). Thus the 'ringing' frequency and characteristic frequency both reflect the same cochlear filter.

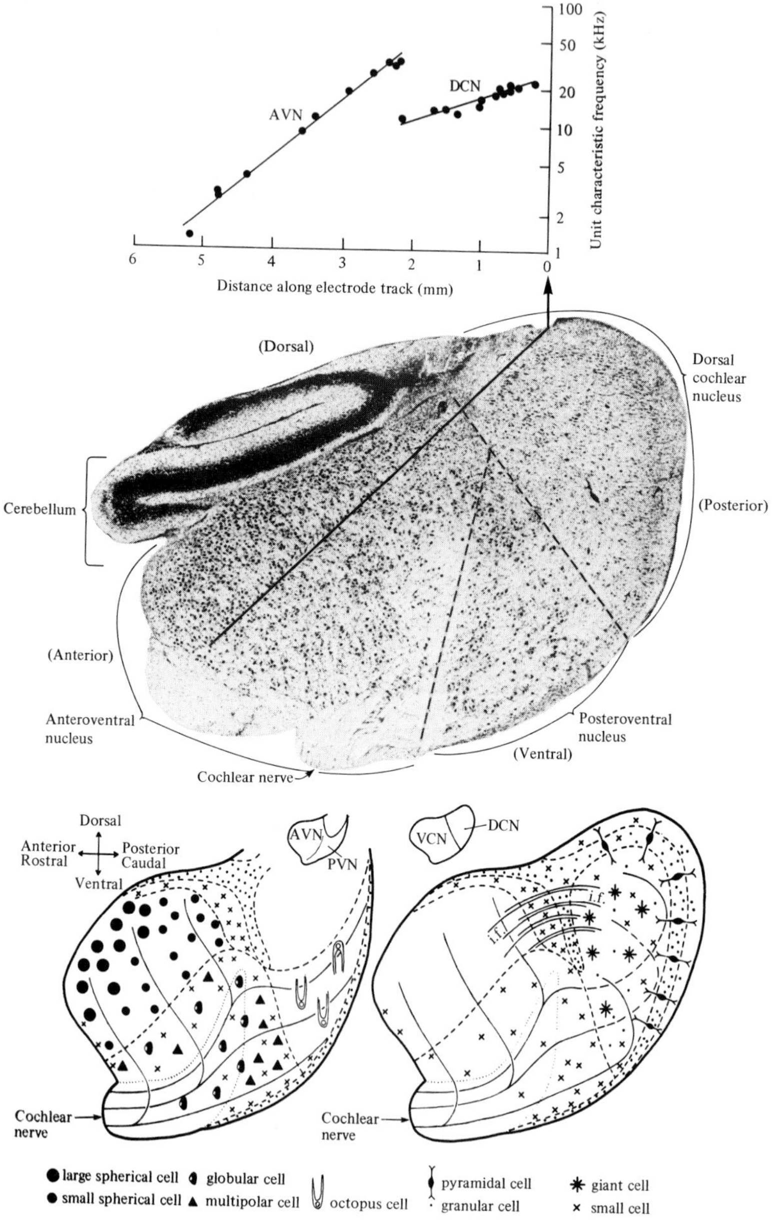
100
50
20
10
5
2
1
Unit characteristic frequency (kHz)
AVN
DCN
6 5 4 3 2 1 0
Distance along electrode track (mm)
(Dorsal)
Dorsal cochlear nucleus
Cerebellum
(Posterior)
(Anterior)
Anteroventral nucleus
Posteroventral nucleus
(Ventral)
Cochlear nerve
Dorsal
Anterior Rostral
Posterior Caudal
Ventral
AVN
PVN
VCN
DCN
i.f.
i.f.
Cochlear nerve
Cochlear nerve
large spherical cell
small spherical cell
globular cell
multipolar cell
octopus cell
pyramidal cell
granular cell
giant cell
small cell

relation between discharge rate and decibel stimulus level within the limits of threshold and saturation; (iv) a low-pass (smoothing) filter which limits the frequency range over which 'phase-locking' occurs; (v) a *probabilistic spike generator*, in which the probability of discharge is a function of the filtered, rectified and transformed waveform.

The cochlear nerve can therefore be considered to represent a *filter bank*, each fibre representing a narrow band-pass filter covering overlapping but slightly different frequencies from its neighbour. Both the mean discharge rate (below saturation) and, for stimuli with frequency components below about 5 kHz, the temporal discharge patterns of each fibre, are determined by the relative magnitude of the frequency components passed by its filter.

14.5 COCHLEAR NUCLEUS

Organisation

In the cochlear nucleus we see the first signs of divergence and parallel specialisation of functions in the auditory system (Figs. 14.1, 14.2). The cochlear nerve bifurcates on entry to the nucleus, sending terminals in *parallel* to the three main subdivisions of the nucleus (Fig. 14.23): the antero- and postero-ventral divisions, and the dorsal division. Because of the orderly arrangement of the cochlear fibres and of their branching, each subdivision is tightly cochleotopically organised (Fig. 14.23): i.e. the cells of each subdivision have characteristic frequencies that are logarithmically related to linear distance

Fig. 14.23. Anatomical organisation of cochlear nucleus. The cochlear nerve enters the nucleus anteriorly and ventrally. Centre of figure: sagittal section though cat cochlear nucleus stained to show cell bodies (Nissl stain). Dashed lines indicate major subdivisions: ventral cochlear nucleus (VCN), subdivided into anteroventral (AVN) and posteroventral (PVN) nuclei, and dorsal cochlear nucleus (DCN). Note the lamination of the dorsal nucleus below the surface, and the separation of the more homogeneous anterior and posterior divisions of ventral nucleus (by root of cochlear nerve). Lower third: major classes of cell types and their locations, according to Osen (1970) *Italian Archives of Biology*, *108*, 21). i.f., intrinsic fibres interconnecting anteroventral and dorsal nuclei. Upper third: plot of characteristic frequencies of cochlear nucleus cells at the position indicated along the electrode track (from dorsal to ventral) shown as continuous straight line in the centre figure. Note tight cochleotopic organisation of each major subdivision.

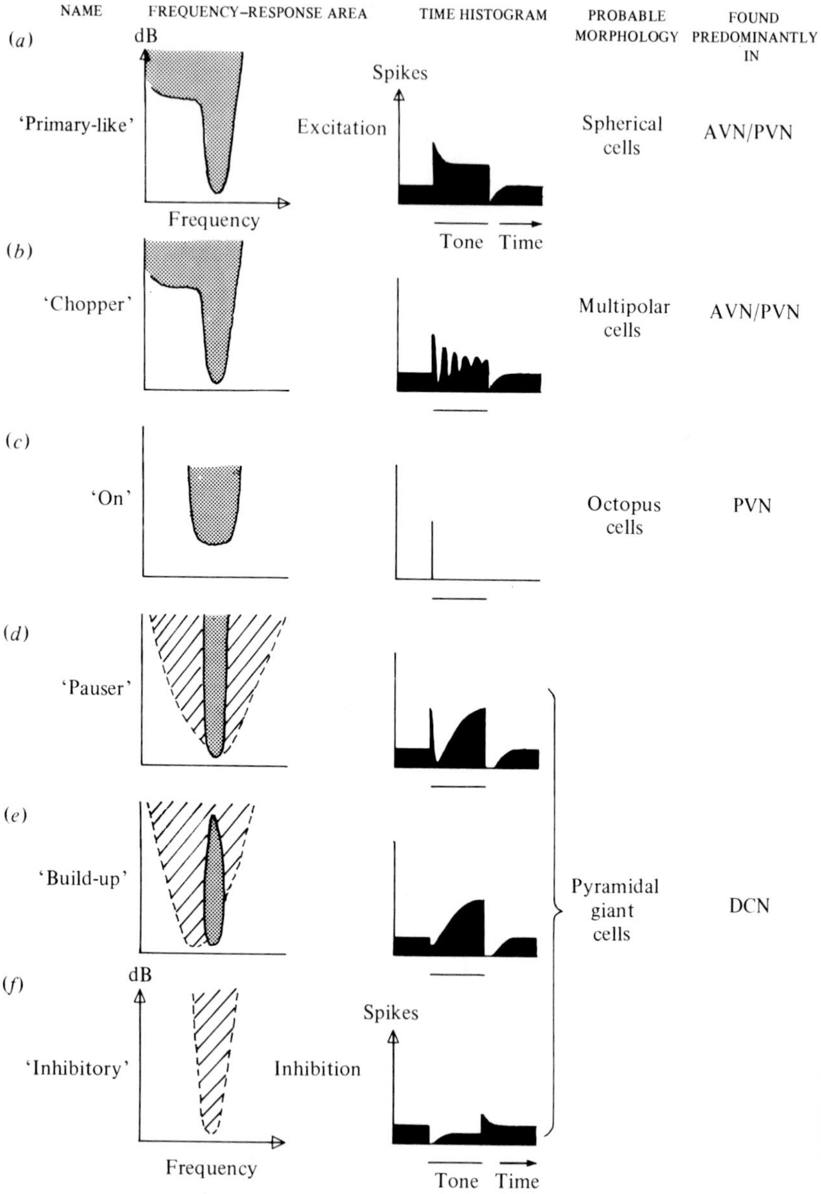

Fig. 14.24. Diagram of frequency and time responses of major types of cells in cochlear nucleus. Left-hand column: diagrams of frequency response areas (axes as in *f*) indicating excitatory (stippled) and inhibitory (shaded)

in the nucleus. (In Fig. 14.23 the electrode track passes only through the dorsal and anteroventral divisions.)

Each subdivision is characterised by a particular distribution of the different cell types (Fig. 14.23). Thus the ventral divisions are the more homogeneous, and the dorsal is the more complex, being laminated. Fig. 14.23 also illustrates the strong bundle of intrinsic fibres (i.f.) connecting the dorsal and ventral divisions.

Subdivision of function in the cochlear nucleus

As might be expected, the different cell morphologies reflect different synaptic dispositions and therefore response properties. These in turn presumably imply specialisations of function (see below). The main correlations can be summarised as follows (Fig. 14.24). The majority of the cells in the ventral nucleus have discharge properties not very different from those of the cochlear fibres. They are hence termed 'primary-like'. Exceptions are the 'chopper' and 'on' cells. The latter are the only type of response found in the 'octopus cell' region. These cells, because of their large dendritic extensions cutting across a relatively large extent of the tonotopically organised afferent array, are also unusual in that they have very broad frequency response areas, and can follow the individual pulses or cycles of click or tone stimuli at repetition rate of up to 800 s^{-1}.

In the dorsal nucleus, more complex responses of longer latency are found and inhibitory inputs are particularly apparent. These inhibitory inputs are such that a *single* tone can reduce – often completely – the cell's activity, as in Fig. 14.25. In some cases, the termination of the inhibition is accompanied by an 'off' response as in Fig. 14.25*a*. It is thus different from the 'two-tone' suppression at the cochlear nerve level, and is likely to result from lateral inhibitory synaptic mechanisms, analogous to those in the retina (Chapter 6). The intrinsic fibre projection (i.f. in Fig. 14.23) from the ventral nucleus represents a delayed inhibitory input to the dorsal nucleus cells.

---

frequency–response areas of each type of cell. Inhibition alone (*f*), or combined with excitation (*d*, *e*), is found predominantly in the morphologically more complex dorsal nucleus. Middle column: diagrams of peristimulus-time histograms of response to characteristic frequency tone of each type of cell (axes as in *f*). The shape of the PST histogram gives rise to the 'nick-name' of each cell type. Right-hand column: probably cell types and location.

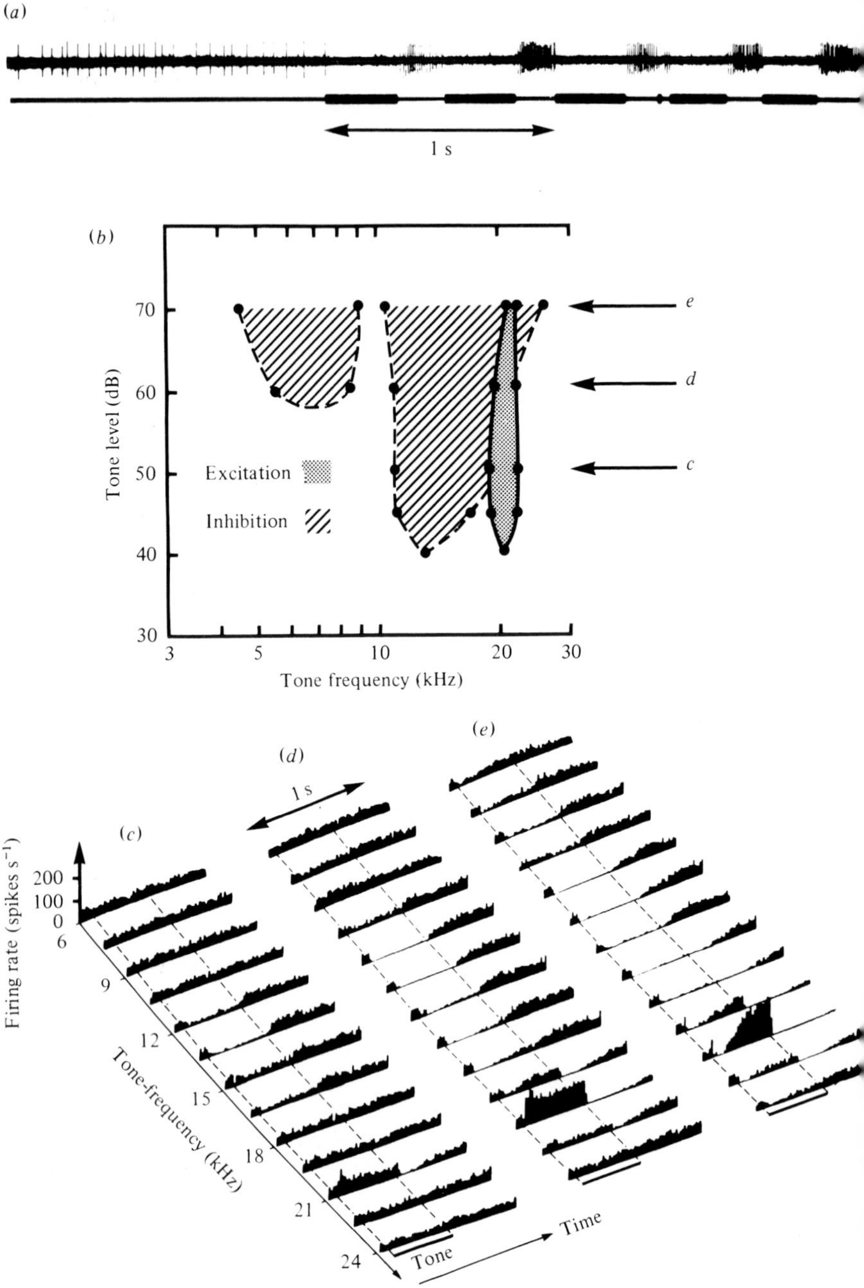

(a)
1 s
(b)
Tone level (dB)
70
60
50
40
30
3
5
10
20
30
Tone frequency (kHz)
Excitation
Inhibition
e
d
c
(c)
(d)
(e)
1 s
Firing rate (spikes s⁻¹)
200
100
0
6
9
12
15
18
21
24
Tone-frequency (kHz)
Tone
Time

In the frequency response areas, the extent of inhibition ranges from one or two relatively restricted 'sidebands' contiguous with an excitatory response area ('pauser', Fig. 14.24), through extensive sidebands restricting the width of the excitatory area at high levels (Fig. 14.25), to a solely inhibitory response area ('inhibition', Fig. 14.24). This separate mapping of excitatory and inhibitory response areas does not imply that they are independent; on the contrary, the 'excitatory' responses are commonly a complex interaction of excitation and inhibition (as in the time histograms of Fig. 14.25). In addition, the inhibition commonly produces a non-monotonic relation between discharge rate and intensity, in contrast to the situation at the cochlear nerve: in other words, the discharge rate at high intensities may be lower than at intermediate intensities.

This lateral inhibition may enhance temporal contrast, and contrast across the frequency spectrum; it may increase the dynamic range at this level (p. 323), and increase sensitivity to frequency *change* in one direction, depending upon the asymmetry of inhibition. (This is presumably analogous to the enhancement, by asymmetrical lateral inhibition, of response to one direction of movement in the rabbit retina; see Chapters 1 and 8.)

There are at least three parallel subsystems at work in the cochlear nuclei, each having its own specific output pathway (Figs. 14.1, 14.2). The trapezoid body, forming a *ventral acoustic pathway*, carries 'primary-like' information from the ventral cochlear nucleus to the superior olivary nuclei. Likewise, the *intermediate acoustic stria* conveys the 'on' type response of the octopus region to the superior olivary nucleus. These two pathways specialise in transmitting, with the minimum of delay, information necessary for the correlation of

Fig. 14.25. Complexity of neural inhibition in two neurons from dorsal cochlear nucleus. (*a*) 'Off' response following inhibition of first unit's spontaneous discharge by tone at characteristic frequency. (*b*) Map of frequency–response areas of excitation (stipple) and inhibition (cross-hatch shading) of second neuron. (*c*, *d*, *e*) Array of peristimulus-time histograms showing time course of responses mapped out in *b*. Note complexity of time course of inhibition during and after tones. Note also that 'excitatory' response areas e.g. at 21 kHz, 70 dB SPL (*e*) represent a complex mixture of excitation and inhibition in time. This figure also indicates that a cell's response type depends on frequency and intensity (e.g. 'pauser', 'inhibition', 'build-up'). (Cat: after Evans & Nelson (1973). *Experimental Brain Research*, *17*, 402.)

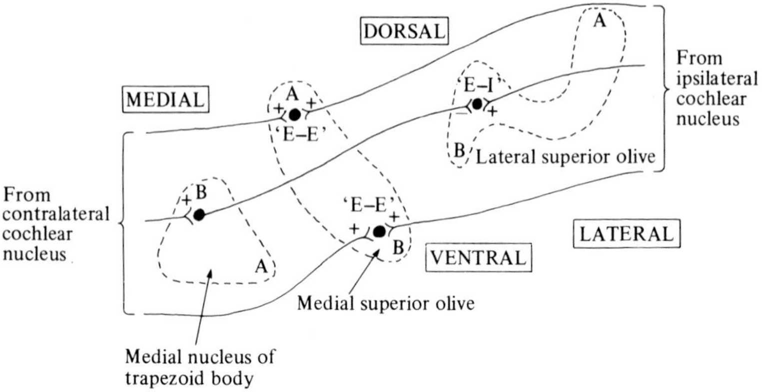

Fig. 14.26. Organisation of Superior Olivary Complex. Right superior olive of cat. Each subdivision is organised cochleotopically, i.e. the characteristic frequencies of the constituent cells are arranged so that in the medial nucleus of the trapezoid body and lateral superior olive, low frequencies (the cochlear apex, A) are represented laterally and the cochlear base (B) medially. The reverse is the case in the medial superior olive. The inputs to the E–E cells and E–I cells are shown.

inputs from the two ears (see next section). They are, presumably, chiefly concerned with the neural analysis of the *position* of stimuli in acoustic space. In contrast, the *dorsal acoustic pathway*, formed by the *dorsal acoustic stria* from the dorsal cochlear nucleus, is presumably chiefly concerned with the analysis of the *pattern* of stimuli. Whether the separate functional identities of these pathways are retained at higher levels of the auditory system is unknown.

14.6. SUPERIOR OLIVARY NUCLEI

This collection of nuclei (Fig. 14.26), lying in the ventral pons, receives input from both ears, and therefore represents the lowest neural level at which correlations are made between signals arriving at the two ears. This appears to be its chief function.

Three modes of binaural interaction are found in this nucleus. The first two are mutually exclusive, involve cells having characteristic frequencies above 1 kHz, and are concerned with correlating *intensity* between the two ears. In the first mode, characteristic of cells having excitatory inputs from both ears (called *E–E cells*) the cells are more sensitive to variations in binaural intensity (i.e. variations that are

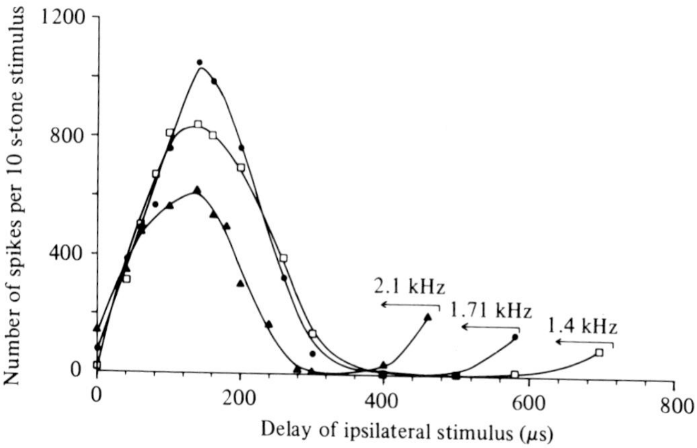

Fig. 14.27. 'Critical delay' cell. Response of a cell in inferior colliculus as a function of the time delay between signals arriving at the two ears. A response maximum occurs at the so-called critical delay of 160 μs largely irrespective of the frequency of the tone (or its level). Similar responses are found in the superior olive. (Cat: After Rose *et al.* (1966). *Journal of Neurophysiology*, *29*, 288.)

correlated at the two ears) than to differences in intensity between the two ears. In the second mode, the converse is the case: the cells are more sensitive to changes in the intensity *difference* between the two ears than to changes in the *average* binaural stimulus level. These are cells receiving excitation from one ear and inhibition from the other (*E–I cells*). This second mode is particularly relevant for the localisation of high-frequency sounds, where head-'shadowing' and the effects of the pinna cause significant differences in interaural intensity, depending upon the location of the source (§15.4).

The third mode of interaction is found in both E–E and E–I type cells having characteristic frequencies below about 1 kHz: sensitivity to the interaural *time delay* between sounds reaching the ears. In the case of the E–I cells, whether the cell is predominantly excited or inhibited depends upon the interaural delay. For some cells, the maximum response occurs at a given interaural delay largely irrespective of the intensities at the ears, and the frequency of the signal. These have been termed 'critical delay' cells (see Fig. 14.27). For others, the interaural time and intensity differences interact in such a way that the effects of one can be traded against the other. Thus, the effects of increasing the intensity at one ear can be offset by delaying the sound to that ear. This '*time-intensity trading*' is of the same order of magnitude as that required in psychophysical experiments to maintain a sound image in the same subjective location.

Both medial and lateral divisions of the superior olive are tonotopically organised, indicating a precise convergence of input fibres from the cochlear nuclei of the two sides.

14.7. INFERIOR COLLICULUS AND MEDIAL GENICULATE NUCLEUS

The auditory system differs from the visual system particularly at this level, in that the *inferior colliculus* is an obligatory station on the pathway to the cortex (Figs. 14.1, 14.2). The fibres of the lateral lemniscus end predominantly in its central nucleus, sending collaterals to the thin surrounding external (pericentral) nucleus which also receives input from the central nucleus. Both nuclei are cochleotopically organised.

As expected from the convergence of dorsal and ventral pathways and pathways from the superior olives, a great variety of types of response are encountered at this level. Of particular note are neurons that do not respond to steady tones but can be activated by frequency or amplitude modulation. In the bat, neurons have been described that are sensitive even to the *form* of the amplitude modulation. These neurons may subserve the analysis of the rate of rise of stimulus transients. As in the superior olive, the pattern of excitation and/or inhibition in many cells depends upon binaural influences (Fig. 14.27). However collicular neurons, by comparison, demonstrate a higher degree of sensitivity to interaural time or intensity differences. Interestingly, relatively few of the neurons sensitive to interaural time differences show phase-locked responses at low-frequency tones.

Another type of binaural processing has been reported at this level. This is the presence of cells sensitive to the *direction of virtual movement* of a train of click stimuli. Dichotically presented trains of clicks, having changing interaural time delays (simulating movement of a sound source), can evoke a response *only* if the direction of virtual movement is in a given direction, i.e. toward *or* away from the midline.

While, surprisingly, the role played by the colliculus in the localisation of sources of sound has not been established very clearly in experiments involving lesions therein, it may be responsible for the acoustic control of head and eye movements, mediated by the superior colliculus, in orientation towards sounds.

The *medial geniculate nucleus* appears to be organised in a similar manner to other thalamic nuclei, particularly the lateral geniculate and ventroposterior nuclei. The afferents, ascending from the inferior colliculus (Fig. 14.2), end in a small number of large terminals in aggregations on the dendrites of the geniculate cells of the ventral nucleus, the main 'through-pathway' to the cortex. The other divisions (medial and dorsal nuclei) receive input also from elsewhere than the inferior colliculus and are probably multimodal. They project predominantly to the secondary auditory cortex and other cortical areas.

Like the central nucleus of the inferior colliculus, the ventral nucleus is laminated, with a cochleotopic organisation. Low characteristic frequencies are represented in the outermost laminae and high in the innermost. Response areas are similar to those found at the cochlear nucleus level, with a high proportion of neurons being inhibited by tones. Intranuclear inhibition tends to emphasise the onset of stimuli and to curtail the sustained excitation seen at lower levels.

14.8. AUDITORY CORTEX

Organisation

The auditory cortex in primates is largely buried in the sylvian (lateral) fissure and located mainly on the inner surface of the superior temporal gyrus (Fig. 14.28*b*). In lower mammals such as the cat, it is conveniently spread out over a large part of the lateral surface of each hemisphere (Fig. 14.28*a*).

It is divided into a central core area, called primary cortex or AI, and various, surrounding, areas forming an auditory 'belt' (CM, L, RL in Fig. 14.28*b*; AII, Ep, SSF in Fig. 14.28*a*). AI receives the most direct, short-latency input from the medial geniculate nucleus. The surrounding belt areas, while also receiving input from the ventral division of the geniculate, receive projections from other thalamic nuclei. Consequently, some of the belt areas are polysensory areas.

The thalamic projections to most if not all these areas are each topographically organised. Each area contains a complete representation of the cochlea, although the scale and direction of the representation differs from area to area. In the primary area (AI) of the cat, the cochlear base is represented anteriorly and the apex posteriorly (Fig. 14.28*a*); in the primate, the representation is reversed (Fig. 14.28*b*). The basal region of the cochlea is relatively over-represented in AI, so that in cat and monkey over half of the

(*a*) Cat

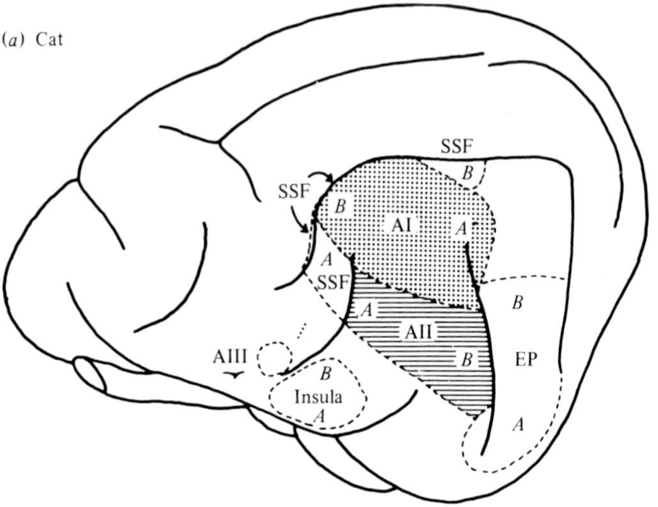

(*b*) Macaque monkey

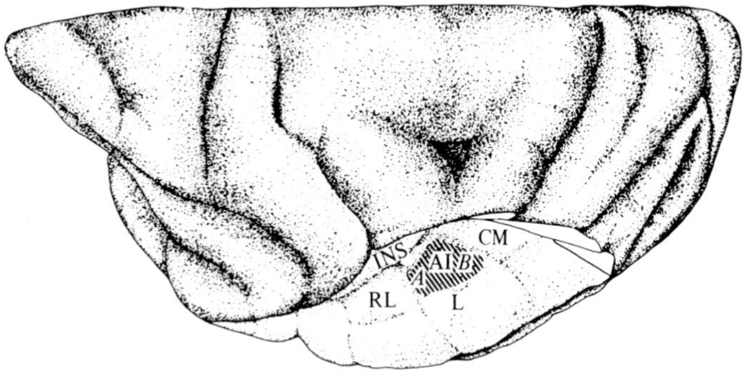

Fig. 14.28. Functional organisation of auditory cortex of monkey and cat. (*a*) Cat: view of left hemisphere from side. (*b*) Macaque monkey: view of left hemisphere from above. The parietal cortex has been removed to show auditory cortex spread out on the inner surface of the superior temporal plane. In both cases, the central primary cortex, AI, which receives the most direct input from the ventral nucleus of the medial geniculate nucleus, is surrounded by a 'belt' of cortex (AII, Ep, SSF, in (*a*); CM, L, RL, in (*b*). *A* and *B* indicates order of projection of fibres related to apical and basal ends of cochlea. (After Woolsey (1960). In *Neural Mechanisms of the Auditory and Vestibular Systems*, p. 165. Thomas; Brugge & Merzenich (1973). In *Basic Mechanisms in Hearing*, p. 745. Academic Press.)

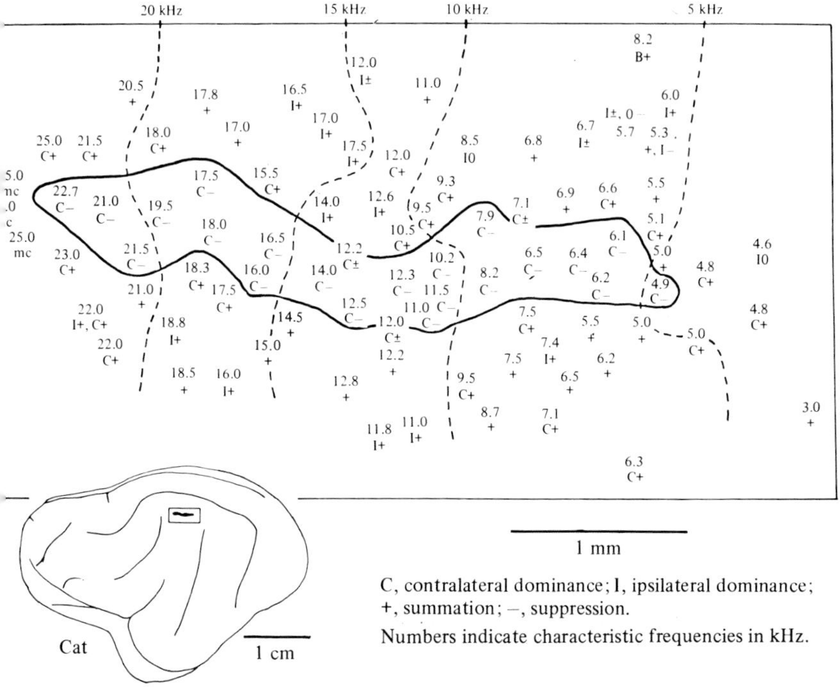

Fig. 14.29. Ear dominance 'slabs' in cat auditory cortex. Area (slab) enclosed in solid outline contains cells whose excitatory response is dominated by the contralateral ear, and suppressed by the ipsilateral ear. Areas above and below are summation cortex where the responses to the two ears are excitatory and summate, the contralateral (C) or ipsilateral (I) ears dominating the response. (From Imig & Adrian (1977). *Brain Research*, *138*, 241.)

area corresponds to cochlear regions of 10 kHz and above. In some species, substantial over-representation of frequencies of particular interest occurs, e.g. in the bat, for those concerned with echolocation. Intrinsic connections of the cortex, from one hemisphere to the other via the corpus callosum, are ordered so that homologous areas are interconnected. These connections end in cortical layer III, the afferent projections from the geniculate ending in layer IV. The primary area (AI) is reciprocally connected to some of the belt areas (AII, CM).

The primary area is also vertically organised into at least two overlapping systems of planes and slabs (Fig. 14.29), though this organisation is not so clear-cut as in the visual and somatic cortical receiving areas. Lying at right angles to the cochleotopic axis of the primary area are planes, which extend the width of the area, each plane corresponding to a different locus of origin in the cochlea. Cortical cells in one of these planes tend to have similar characteristic frequencies. However, there is little sense in which the planes can be said to be frequency-specific: in unanaesthetised animals, a significant degree of variance of characteristic frequencies is encountered; many of the cells have broad response areas covering several octaves; and many of the cells are not responsive to pure tones at all. Orthogonal to these 'isofrequency' projection planes are slabs (e.g. solid outline in Fig. 14.29), which have a common binaural interaction pattern. There is anatomical evidence in the cat for the existence of two slabs in each hemisphere, where callosal fibres couple homologous areas of primary cortex. These are likely to correspond to the areas in Fig. 14.29 marked with a plus sign, where binaural *summation* occurs, i.e. the response to the binaural inputs are greater than those to the monaural inputs, although generally the contralateral or the ipsilateral ear dominates the response (C+ or I+). In the intervening areas where the callosal input is sparse, are situated the *suppression slabs*, one of which is shown in Fig. 14.29. Here, the contralateral input is dominant, the ipsilateral input exerting a suppressive effect on the cell's response to the contralateral input.

Selective response properties

A great variety of response types are encountered in the primary cortex, some of which are illustrated in Fig. 14.30 (A–F). In contrast to the situation at the auditory periphery, however, where all neurons respond to tones and wide-band signals (noise, clicks), not all cortical units respond to these unstructured stimuli. Half do not respond to wide-band noise in spite of having wide frequency response areas. On the other hand, the great majority can be activated by sound complexes structured in certain ways.

About 10% of cells will respond to tones only if their frequency is changing (Fig. 14.30*f*, *g*). Their response usually is to a preferred direction of frequency change: either up or down (Fig. 14.30*f*, *g*). In some cases it is to both directions of change. The rate of change of frequency and/or the repetition rate, if the modulation is periodic, are also important.

In addition to this selectivity for frequency change, direction and rate, certain cortical cells are selective for the temporal 'shape' of

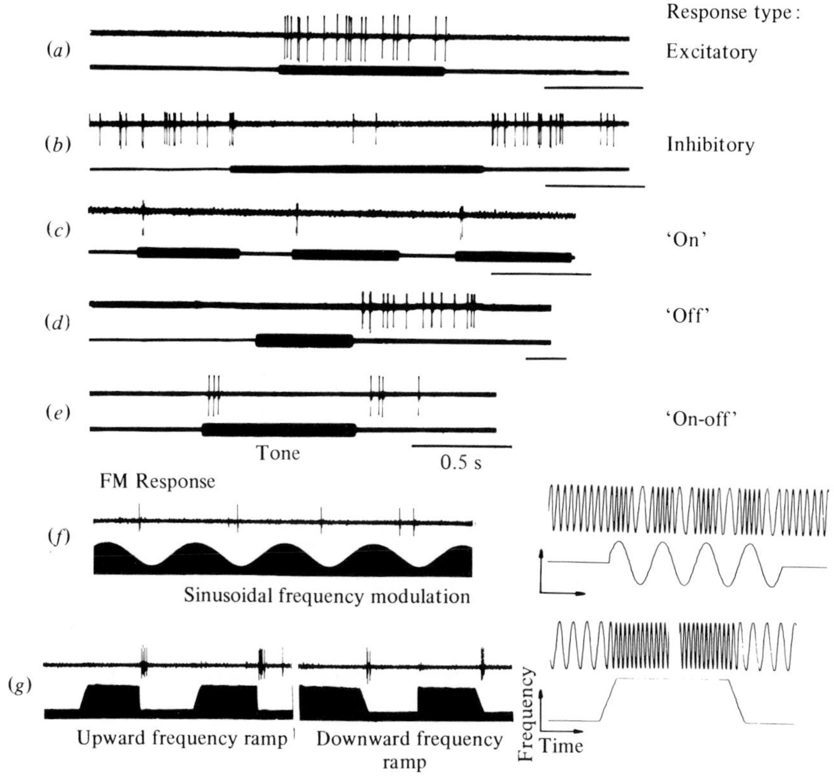

Fig. 14.30. Variety of response types of cells in primary auditory cortex. (*a*)–(*e*) Types of response to steady tones. (*f*, *g*) Cell responding to frequency-modulated, not steady tones. Note response selectivity to direction of frequency sweep: in the downward direction, not to upward sweeps. Black envelope indicates excursions of frequency as illustrated by the waveforms to the right. (Cat, after Evans & Whitfield (1964). *Journal of Physiology*, *171*, 476; Evans (1968) in *Hearing in Vertebrates*, p. 272. Churchill.)

the amplitude envelope of sounds. Others appear to be selective for noise or click stimuli and others for stimuli occurring within a narrow range of stimulus intensities.

In primates that have developed an extensive repertoire of socially meaningful vocalisations, such as squirrel monkeys, the cortical units are especially responsive to these calls. In fact, a small percentage of cells are selective for *individual* calls and remain unresponsive to other calls, tonal stimuli, noise, etc.

Some auditory cells, therefore, like their counterparts of other modalities (Chapter 1), can be said to be *abstractive* for certain features of complex stimuli which may have particular biological significance for the animal. Thus cortical cells are capable of providing 'answers' to the following questions: is the stimulus a noise, a click, a tone, or a species-specific call? Is the stimulus on? Has it just commenced? Has it just been terminated? What is its duration and repetition rate? Is the frequency changing? If so, in which direction; at what rate? How rapidly is the amplitude rising? Where is the stimulus located in space? Is it moving? And so on.

In certain bats, neurons selective for a particular feature are grouped in a restricted cortical area. These are units selective for changes in frequency that are characteristic of their calls used for echolocation. This is a highly specialised function, and may, like other such specialisations in the bat, not be generalisable to other mammals, where there is no discernible order to the location of these selectivities.

In man, there is psychophysical evidence for 'channels' specific for modulation at different frequencies. Exposure to repeated stimulation at a given modulation frequency or in one direction of frequency change, can raise the threshold of detection for those particular parameters. From similar experiments with speech sounds, there is evidence for feature selectivities in the human auditory system for certain of the acoustic cues important for the recognition of the parts of speech (§15.3).

Functional effects of cortical damage

It is perhaps not surprising, in view of the relative lack of specificity of cortical cells for frequency and intensity *per se*, that bilateral ablation of auditory cortex in cat and monkey leaves minimal deficits (upon relearning) in the behavioural audiogram, and in the discrimination and generalisation of frequency and intensity. The deficits are in the ability to localise or lateralise sounds, or distinguish changes in their duration, temporal pattern (order or sequence), or spectral composition. Animals relearn with difficulty to distinguish the direction of change of frequency. Similarly, in man, damage to the temporal lobe has been found to produce deficiencies in localisation in the contralateral sound field. However, in man, there is evidence of some hemispheric asymmetry: damage to the temporal lobe of the dominant hemisphere (the left in the case of a right-handed person) produces deficits predominantly in the recognition of speech; whereas damage to the non-dominant lobe affects recognition of the timbre of sounds, tonal sequence and pattern, i.e. some of the attributes of music.

14.9. DESCENDING, EFFERENT, PATHWAYS

Organisation

The ascending auditory pathway is accompanied, throughout its length, by a descending, efferent set of pathways (Fig. 14.1). One short pathway establishes reciprocal connections between the primary auditory cortex and the ventral nucleus of the medial geniculate nucleus. The second pathway runs from primary and surrounding cortical areas to the central and external nuclei of the inferior colliculus. From there it diverges to supply the dorsal cochlear nucleus, and the periolivary nuclei. The latter in turn give rise to the *olivocochlear bundle*. The ventral cochlear nucleus receives efferent connections from the lateral region of the superior olive.

Function

While the effects of electrically stimulating the efferent system are well known, paradoxically its action in the behaving animal is less clear. Stimulation of the efferent pathways can produce excitatory and inhibitory effects in cells of the cochlear nucleus, predominantly inhibition in the dorsal, and excitation in the ventral nuclei. However, stimulation of the olivocochlear bundle results only in *inhibition* of the evoked activity in cochlear fibres; this effect is mediated by acetylcholine release at the efferent endings in the cochlea. These inhibitory actions are frequency specific: they increase the fibre's threshold (by up to 25 dB in some fibres) predominantly at the characteristic frequency, thus reducing the sharpness of tuning. This efferent inhibition does not, however, affect significantly the spontaneous or the saturated discharge rates of fibres.

The fibres of the olivocochlear bundle itself are inactive in anaesthetised animals, and most exhibit no spontaneous activity in the absence of anaesthesia. However, in the latter case, tones evoke a slow regular discharge beginning after a relatively long latency of 5–40 ms.

In contrast to these clear effects of the olivocochlear bundle on cochlear activity, surgically interrupting the bundle appears to be without effect on auditory *behaviour*. Thus, no conclusive effects have been found on absolute behavioural thresholds, thresholds in noise, frequency discrimination or selective attention. There is some evidence, however, that modification of efferent activity to the cochlear nucleus may affect an animal's ability to detect signals in noise. Inhibition of responses in the lateral lemniscus has been observed in behaving animals immediately before vocalisation, suggesting that one important – possibly the only – role of the efferent system (like that of the middle-ear muscle system) is to reduce activation of the auditory system in response to sounds generated by the animal itself.

14.10. EVOKED POTENTIALS FROM THE HUMAN AUDITORY SYSTEM

In clinical practice, useful information can be provided by recording electrical activity from the human auditory system without the invasive means required to investigate the response of single neurons in animals. The *electrocochleogram* has already been referred to (§ 14.3). For information on the electrical activity of more central structures sound-evoked potentials can be obtained (with an averaging computer) between an electrode simply placed on the scalp over the vertex of the head and one on the skin over the mastoid or on the ear lobe. At least 15 waves have been identified in response to click stimuli, and are conventionally labelled as in the diagrammatic representation (on a logarithmic time scale) of Fig. 14.31. Waves I to VI are the so-called *brain-stem evoked responses*, and some of them have been tentatively identified as indicating activity in a specific location, e.g. I: cochlear nerve; II: cochlear nucleus; III: superior olive; IV/V: inferior colliculus. The sites of origin of the so-called middle-latency waves, (N_o–N_b) and the long latency potentials (P_1–N_2) are very uncertain. The medial geniculate nucleus and the auditory cortex have been implicated in the generation of waves N_o and N_a respectively. However, with the exception of the cochlear nerve, it is unlikely that the various waves can represent components *specific* to the nuclei, in view of the wide range of latencies encountered in the responses of individual neurons in each nucleus in animal experiments, from the cochlear nucleus onwards. It is possible that the waves represent the synchronous activity of those neurons with the shortest latency (predominantly found in the ventral acoustic pathway, p. 293) in the *nerve fibre tracts connecting the nuclei.*

The recording of these responses, particularly the early waves I to V, is proving to be a useful clinical tool. First, it provides information

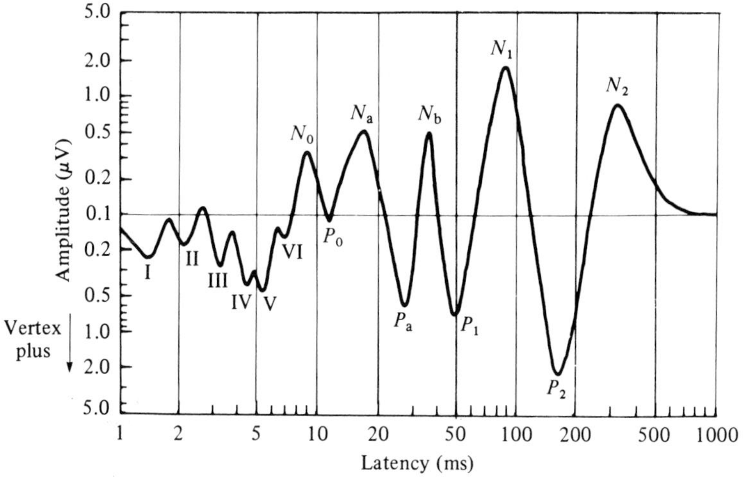

Fig. 14.31. Human auditory evoked potentials. Diagrammatic representation of the potentials obtained with the aid of an averaging computer), between an electrode on the vertex of the head and over the mastoid or on the ear lobe. Waves I to VI are called the short-latency or brain stem evoked responses; waves N_0 to N_b the middle-latency responses, and waves P_1 to N_2 the late-latency responses. (From Picton *et al.* (1974). *Electroencephalography and clinical Neurophysiology*, *36*, 179.)

of value for localising lesions within the central auditory pathways. Second, it allows objective measurements of hearing threshold to be made without invasive procedures in non-communicating children and adults. The long-latency waves are markedly affected by the state of arousal or sedation, and are therefore less useful clinically. However, the longest latency waves (e.g. N_2 and later) are systematically affected by cognitive processes and may turn out to be of value in the investigation of language disorders.

14.11. SUGGESTIONS FOR FURTHER READING

General introductions and reviews

Yost, W. A. & Nielsen, D. W. (1977) *Fundamentals of Hearing*. New York: Holt, Rinehart & Winston. (Well illustrated, clear introduction to the peripheral auditory system.)

Durrant, J. D. & Lovrinic, J. H. (1977) *Bases of Hearing Science*. Baltimore: Williams & Wilkins Co.

Special references

Outer, middle and inner ear. Dallos, P. (1973) *The Auditory Periphery.* New York: Academic Press. Relevant chapters in *Handbook of Sensory Physiology*, vol. 5, pt 1 (1974), ed. W. D. Keidel & W. D. Neff. Berlin: Springer-Verlag.

Sharp tuning of basilar membrane. Sellick, P. M., Patuzzi, R. & Johnstone, B. M. (1982) *Journal of the Acoustical Society of America*, **72**, 131–41; Khanna, S. M. & Leonard, D. G. B. (1982) *Science*, **215**, 305–6.

Cochlear hair cells. Russell, I. J. & Sellick, P. M. (1978) *Journal of Physiology*, **284**, 261.

Electrical resonance of hair cells. Crawford, A. C. & Fettiplace, R. (1981) *Journal of Physiology*, **312**, 377.

Cochlear nerve and nucleus. Kiang, N. Y-S., Watanabe, T., Thomas, E. C. & Clarke, L. F. (1965) *Discharge patterns of single fibers in the cat's auditory nerve.* Cambridge, Mass: MIT Press. Evans, E. F. (1975) Cochlear nerve and cochlear nucleus. In *Handbook of Sensory Physiology*, vol. 5, pt 2, ed. W. D. Keidel & W. D. Neff. Berlin: Springer-Verlag.

Superior olive, medial geniculate nucleus and cortex. Relevant chapters in *Handbook of Sensory Physiology*, vol. 5, pt 2, ed. W. D. Keidel & W. D. Neff (1975) Berlin: Springer-Verlag.

Feature- and call-specific neurons in auditory pathway. Evans, E. F. (1974) In *The Neurosciences: Third Study Program*, ed. F. C. Schmitt & F. G. Worden. Cambridge, Mass: MIT Press.

Binaural columns in primary auditory cortex. Imig, T. J. & Adrian, H. O. (1977) *Brain Research*, **138**, 241–57.

CHAPTER FIFTEEN

Functions of the auditory system

E. F. EVANS

15.1. PROCESSING OF FREQUENCY

Frequency analysis

If the individual components of a complex sound are sufficiently far apart in frequency, they can be heard as separate, as in the notes of a musical chord. This is *Ohm's acoustic law*. Closer together they are heard as one sound. The ability of the ear to resolve sounds into their constituent (*simultaneously* present) frequency components, is termed *frequency analysis or frequency selectivity*. It is essential if the ear is to be able to follow simultaneous changes in the frequency and intensity of the individual components of a complex sound, as, for example, in speech sounds (Fig. 13.1*h*–*l*).

In many ways, the ear acts as if it contains a bank of narrowly tuned filters. Psychophysical estimates of the shapes of these filters (as in Fig. 15.1*a*) resemble those of the filters represented by the frequency threshold curves of cochlear nerve fibres (as in Fig. 14.16), and many of the frequency-selective properties of the ear can be accounted for (to a first approximation) on this basis. To understand some of the ear's frequency selective properties, it is helpful to consider these neural and psychophysical frequency threshold 'tuning' curves as representing linear filter functions, as in Fig. 15.1*b*. Here the curve for a single cochlear fibre is drawn, inverted, on *linear* power and frequency scales. It has a central '*pass band*' about the characteristic frequency (6.4 kHz) and (on this scale), nearly symmetrical '*skirts*' or filter cut-offs (for the first 10 dB or so). In a linear filter, the width of the pass band can be usefully expressed as the *effective bandwidth* (Fig. 15.1*b*). This is the bandwidth of the rectangular filter having the same area; it is approximately the half-power (3 dB down) bandwidth. The concept of effective bandwidth is important because with multicomponent or broad-band signals, it is the frequencies falling within the effective bandwidth that dominate the filter's response, compared with those falling outside.

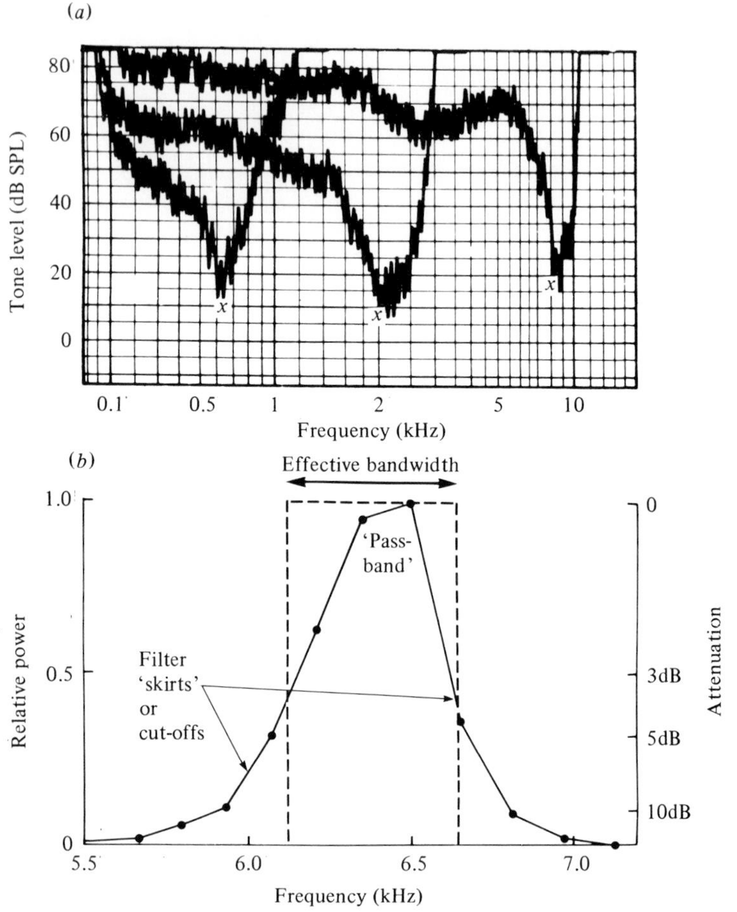

Fig. 15.1. Psychophysical and physiological tuning. (*a*) Psychoacoustic 'tuning curves' obtained by human subject. Each outline encloses the frequencies and levels of a tone required to mask a second, low-level tone at constant frequency and level indicated by *X*. (The masked threshold outlines were determined by a Békésy tracking technique, as in Fig. 13.2*b*). Note similarity with neural frequency threshold curves of Fig. 14.16, (in spite of the possibility that complicating factors are likely to be involved in the psychophysical determination). (From Zwicker (1974). In *Facts and Models in Hearing*, p. 132. Springer-Verlag.) (*b*) Physiological filter curve. Frequency threshold curve of a cochlear nerve fibre expressed as a filter function, on linear power and linear frequency coordinates. The 'effective bandwidth' is the bandwidth of the rectangular filter having the same area; it is approximately the half power (3 dB down) bandwidth. (Cat: from Evans & Wilson (1973), in *Basic Mechanisms in Hearing*, p. 519. Academic Press.)

The narrower the effective bandwidth, and the steeper the skirts ('cut-offs'), the more frequency selective is the filter, i.e. it accepts a narrower band of frequencies and better rejects others further away. Frequencies falling within the effective bandwidth will be least attenuated and will therefore be able to sum, to determine the threshold and suprathreshold response and interact in various ways such as interfering to form beats. (This must not be taken to mean that, say, two frequencies wide apart cannot interact to produce 'beats' reflected in the discharge of an auditory neuron, providing they fall within the skirts of its response area. But their effect will in principle be 'swamped' if other stimulus components fall within the effective bandwidth.)

In spite of the cochlear non-linearities outlined in §14.4 the filtering properties of cochlear fibres are surprisingly linear, especially for broad-band noise stimuli within the dynamic range (40 dB or so above threshold). This means, for example, that given a fibre's effective bandwidth, one can predict its threshold to noise stimuli. The wider the effective bandwidth, the more energy filtered from a wide-band noise, and the less the difference between tone and noise thresholds. In cochlear hearing loss, the neural effective bandwidths become pathologically wide, by up to an order of magnitude. This may have important consequences for trying to hear signals like speech, especially in noisy backgrounds (see section on Masking, later in this chapter).

The sharpness of filters can be expressed in a number of ways, in terms of the effective bandwidth or its reciprocal the 'Q' factor (characteristic frequency divided by effective bandwidth).* The Q values derived from cat cochlear fibres, from human psychophysical tuning curves and other psychophysical measurements (when corrected for the different methods and assumptions involved), are approximately the same, about 10 for frequencies above about 0.5 kHz. The values of effective bandwidths for cat cochlear fibres are plotted in Fig. 15.2, together with a number of other measures of psychophysical frequency selectivity.

The best known index of psychophysical frequency selectivity is the so-called *critical band*. The values for cat and man are shown in Fig. 15.2. The critical band can be considered analogous in some ways to the effective bandwidth of the auditory filter: signals falling within the critical bandwidth cannot be easily resolved, i.e. heard as

* Physiologists (and psychophysicists) often use the 10 dB bandwidth and the corresponding $Q_{10\,dB}$ factor. The $Q_{10\,dB}$ is approximately half the Q value.

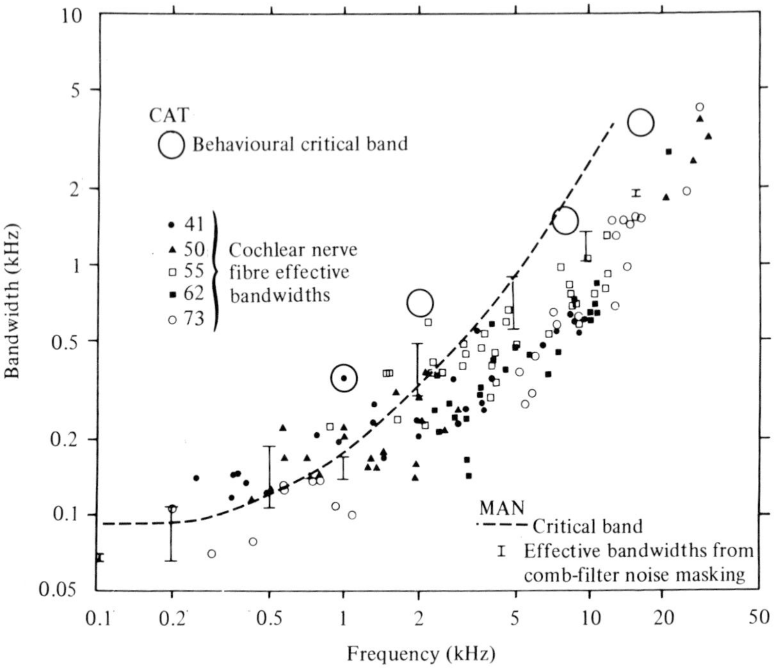

Fig. 15.2. Comparison of neural and psychophysical effective filter bandwidths. Each small symbol represents the effective bandwidth of an individual cochlear nerve fibre filter (as defined in Fig. 15.1) plotted against its characteristic frequency. Individual cats are identified by different symbols. Dashed line represents the 'critical band' of human hearing, one estimate of the bandwidth of human auditory filters determined psychophysically. Human effective bandwidths determined by comb-filtered noise masking (by measurements as in Fig. 15.3) are also included. Open circles: behavioural measurements of 'critical band' in the cat (data from Pickles (1975). *Acta oto-laryngologica*, *80*, 245.). (After Evans & Wilson, in *Basic Mechanisms of Hearing*, p. 519. Academic Press, 1973.)

separate, and therefore tend to sum for threshold and loudness (p. 320) and to interact in various ways.*

Interestingly, there is significant variation in effective bandwidths from subject to subject (animals as in Fig. 15.2; and humans) i.e. some individuals are sharply tuned (e.g. cat number 73 in Fig. 15.2) and others are more blunt!

* The bandwidths of the filters used in analysing the speech sound of Fig. 13.1*j* correspond approximately to the critical band for frequencies up to a few kHz.

Auditory frequency selectivity can be measured directly in a manner analogous to that used for determining spatial frequency resolution in vision (Chapters 1 and 8), using an acoustic grating: *comb-filtered noise.* This is broad-band noise filtered (or generated) so that it has a sinusoidal *spectrum*, i.e. evenly spaced peaks of energy across *frequency*, like the teeth of a comb (Fig. 15.3). The contrast sensitivity functions so determined psychophysically and physiologically (in cochlear nerve and nucleus) agree well (Fig. 15.3*c*). This again suggests that auditory frequency selectivity is already largely determined at the cochlear nerve level, or is at least limited there.

Because auditory frequency selectivity is peripherally limited, and because the cochlear filtering is physiologically vulnerable (p. 282), a patient's frequency selectivity commonly deteriorates when the cochlea is damaged by noise, hypoxia, drugs (Figs. 14.19, 14.20), or infection, etc. This deterioration manifests itself psychoacoustically in changes in a number of complementary aspects of frequency selectivity: the critical bands increase in size; the psychophysical tuning curves become broader; there is increased spread of masking (p. 324); the loudness function becomes abnormal, a phenomenon termed 'recruitment of loudness' (used as a diagnostic test, p. 323); and there is some deterioration in the intelligibility of speech (particularly in background noise) as will be described later.

In a number of ways, therefore, cochlear fibres act *to a first approximation* as linear filters for complex stimuli. Providing that a complex signal (like speech) contains frequency components separated by more than the effective bandwidth of the fibres, then, *in principle*, they can be resolved, each resolvable frequency being signalled by fibres of that (characteristic) frequency. This means that these resolvable components, such as the formants of speech, can be coded by the distribution of responses across the tonotopic array of fibres. However, these statements are true only to a first approximation: because of the relatively restricted dynamic range of the majority of individual cochlear fibres (Figs. 14.25, 15.4), and the relatively restricted distribution of the thresholds of the sharply tuned fibres (lower limits of the dynamic range lines in Fig. 15.4), the great majority are saturated at medium to high intensity levels and are therefore apparently not able to signal, *by their mean discharge rate* at least, the presence of close but separable frequency components in a stimulus (see Fig. 14.17*d*).

This problem is illustrated in another way in Fig. 15.5, in which are plotted the distributions of evoked discharge rates across the

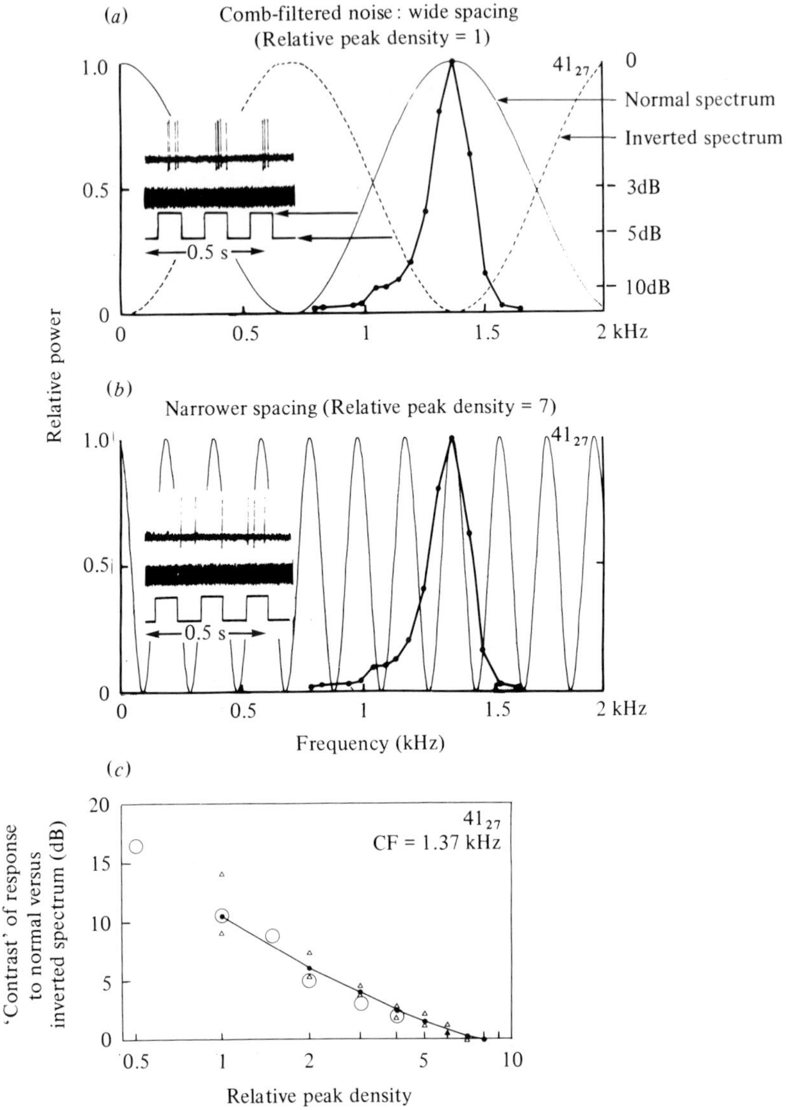

Fig. 15.3. Grating acuity of the auditory system. (*a*, *b*) Power spectra of *comb-filtered noises*, compared with cochlear nerve filter function. The former have a sinusoidal distribution of energy with frequency, i.e. evenly spaced peaks and valleys of energy. Two examples are shown of coarse (*a*) and fine (*b*) spacing, and for the former the *inverted* spectrum. The cochlear fibre filter function (filled circles and thick lines) is obtained as in Fig. 15.1.

tonotopic array of cochlear fibres, for a 1 kHz tone (Fig. 15.5*b*) and for an 8 kHz tone (Fig. 15.5*a*), at different sound levels. Whereas at low sound levels the activity corresponding to each tone is well circumscribed, for even moderate levels in the case of the 1 kHz tone (45 dB SPL, Fig. 15.5*b*) a considerable extent of the cochlear array is activated. At levels of 70 dB SPL, it is difficult to see how, in these discharge rate patterns, any information could be conveyed on whether there was one or several frequency components present in the range from 0.3 to 10 kHz! At higher frequencies, e.g. 8 kHz (which admittedly are more optimal for the cat), the problem still exists, though it is less acute (Fig. 15.5*a*).

This is a serious problem for theories of the coding of frequency on the basis of 'place' of neural activity. It may be overcome in a number of possible ways. The first is that the place coding at high sound levels may be handled by a small number (about 10%) of cochlear fibres whose dynamic ranges are larger than the majority (indicated by the arrows in Fig. 15.4) and which incidentally have low spontaneous discharge rates. The second is that lateral ('two-

(*c*) Measurements of neural and psychophysical ability to determine the difference between normal and inverted comb-filtered noise spectra, i.e. the apparent 'contrast' of the acoustic grating, as a function of its fineness of peak spacing. (The 'relative peak density' indicates the number of spectral peaks between zero and the characteristic frequency of the fibre or frequency of psychophysical measurement.) The neural measurements are on the single cat cochlear nerve fibre, the filter function of which is shown in *a*. Each measurement (each open triangle) indicates the equivalent difference ('contrast') in noise power required to evoke the difference in response actually obtained (see insets) between the normal and inverted comb-filtered noise spectra, for each spectral spacing. The human psychophysical measurements (open circles) indicate the differences in threshold of a tone, at the measurement frequency (1 kHz), masked by the normal and inverted spectra respectively. (Filled circles: calculated values based on linear filtering with filter function determined as in *a*.) At coarse comb-filter spacings (*a*) the response contrast is maximum (note discharge, in inset, to normal spectrum i.e: when peak of noise energy coincides with the cochlear fibre FTC); it decreases as the spacing between the spectral peaks is made finer, the contrast decreasing until the limit of grating acuity is reached (at about peak density = 7 as in (*b*); note absence of contrast in discharge in inset between normal and inverted spectrum.) (From Evans & Wilson (1973), in *Basic Mechanisms of Hearing*, p. 519. Academic Press; Evans (1978) *Audiology*, *17*, 369; Psychophysical data from Pick *et al.* (1977), in *Psychophysics and Physiology of Hearing*, p. 273. Academic Press.)

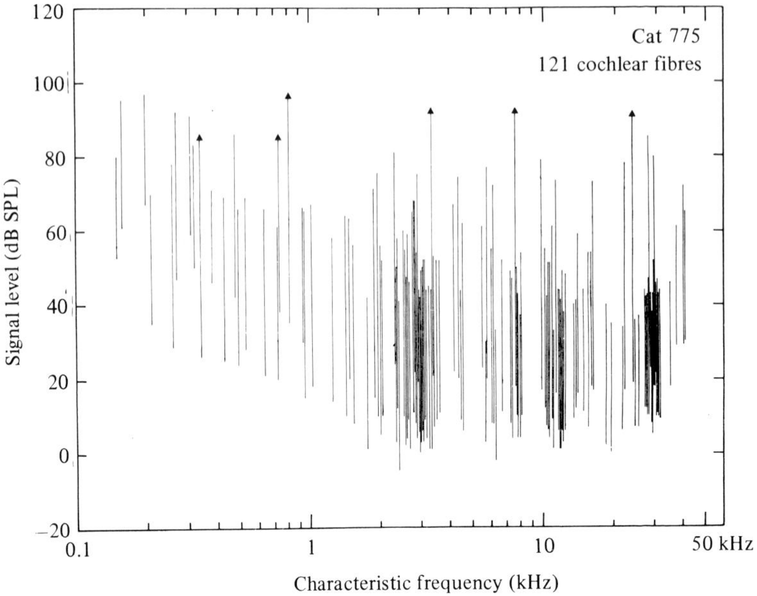

Fig. 15.4. Dynamic ranges of discharge rate of 121 cochlear fibres in one ear. Each vertical line represents the dynamic range (see Fig. 14.17*b*) of a single cochlear nerve fibre. The extent of each line indicates the range of tone intensities at the characteristic frequency of the fibre, between threshold (lower limit of line) and saturation of the discharge rate (upper limit of line). The great majority of fibres are saturated at moderate sound levels, i.e.: their dynamic ranges end at about 40–60 dB SPL. Arrows indicate cochlear fibres not completely saturated at highest sound levels employed in the experiment. (Cat: Palmer & Evans (1979) *Experimental Brain Research*, *Suppl.* II, 19.)

tone') suppression may enhance the separation of activity in the neural array at the level of the cochlear nerve in the way that lateral *inhibition* certainly does at the cochlear nucleus level (§ 14.5). A third possibility is that central mechanisms may utilise information on the relative level of frequency components that is coded in the temporal patterning of the discharges of fibres at the appropriate place (p. 284), though for stimulus components above 5 kHz classical phase-locking mechanisms cannot be involved.

Frequency discrimination; pitch

The ear is exquisitely sensitive to small changes in the frequency of single or multiple component stimuli. The *just noticeable difference*

(JND) or *difference discrimination limen* (DL), for frequency (ΔF), depends both on the frequency and on the sensation level of a signal (Fig. 15.6*a*). In contrast to the critical band, it represents a very small fraction of the stimulus frequency, ranging from less than 0.3% at intermediate frequencies (0.5–2 kHz) to about 1.5% at 10 kHz (Fig. 15.6*b*).

The discrimination of frequency differences of this kind is probably mediated by two types of neural code acting conjointly, when the frequencies concerned fall in the range 0.1–5 kHz (where the frequency discrimination is most acute). The first is the *place* of activity in the peripheral auditory system, i.e. which neurons in the tonotopic array (pp. 279, 289) are active. Because the frequency response areas are asymmetrical for cochlear fibres of characteristic frequencies above about 2 kHz, and their *high-frequency* cut-offs are steep (almost vertical) (see Fig. 14.16), the *low-frequency* boundary of activity in the tonotopic array of cochlear fibres is very sharp (Fig. 15.5*a*), in contrast to the high-frequency boundary. The other second type of neural code likely to be involved is the *periodicity* of the neural discharge patterns (p. 284), although the neural mechanisms required to decode this periodicity are unknown. For frequencies above 5 kHz, the periodicity (phase-locking) has largely dropped out, so only the 'place' mechanism is left, where admittedly it is most accurate. For frequencies much below 1 kHz on the other hand, the frequency response areas do not have sharp cut-offs (Fig. 14.16), and the place mechanism is of little value at moderate to high stimulus levels (Fig. 15.5*b*). Here, and in other circumstances where *only* place or *only* periodicity cues are operative, the frequency difference limen deteriorates to approach a few per cent.

Pitch is not a simple correlate of frequency. As every musician knows, the pitch of even a pure tone depends upon not only its frequency but to a small extent also on its intensity, and the pitch of most natural sounds and those produced by musical instruments is not a simple function of their frequencies. Furthermore, the pitch of a pure tone or tone complex may be perceived as different between the two ears. Normally, this amounts to less than 1%, and is not noticed. In cochlear pathology, however, this *diplacusis* can become substantial.

For complex (e.g. naturally occurring) sounds containing several component frequencies, there are two well recognised modes by which we perceive pitch. The first (termed the '*analytical*' *mode*), is where we hear the pitches of the *individual* frequency components of

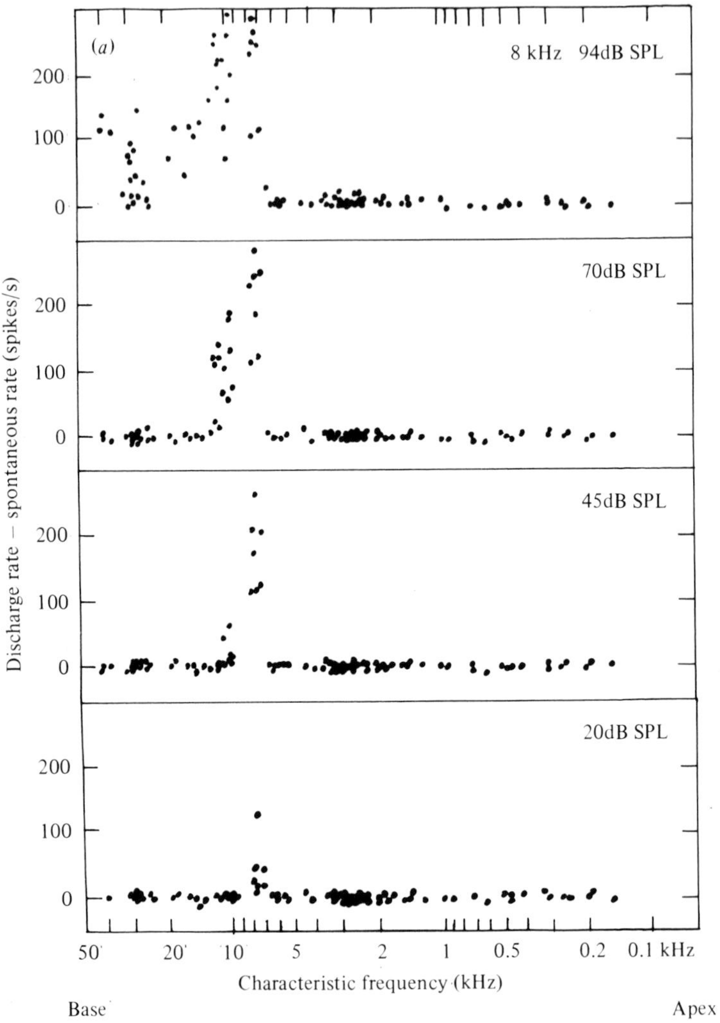

Fig. 15.5. Distribution of evoked activity across cat cochlear fibre array in response to single frequencies. Each point represents the increase in discharge rate above spontaneous rate given by a cochlear fibre plotted at its characteristic frequency (and therefore position along the tonotopic array). Left half: Responses to 8 kHz tone at four sound levels (Evans & Palmer, unpublished data). Right half: Responses to 1 kHz tone at three sound levels (from Kim & Molnar (1979). *Journal of Neurophysiology*, *42*, 16.) In the case of 1 kHz signals in the cat, the frequency threshold curves are broad and symmetrical; hence there is an extensive spread in location of neural activity in both directions with increase in stimulus level. In the case of 8 kHz signals, about the most sensitive frequency in the cat, the frequency threshold curves

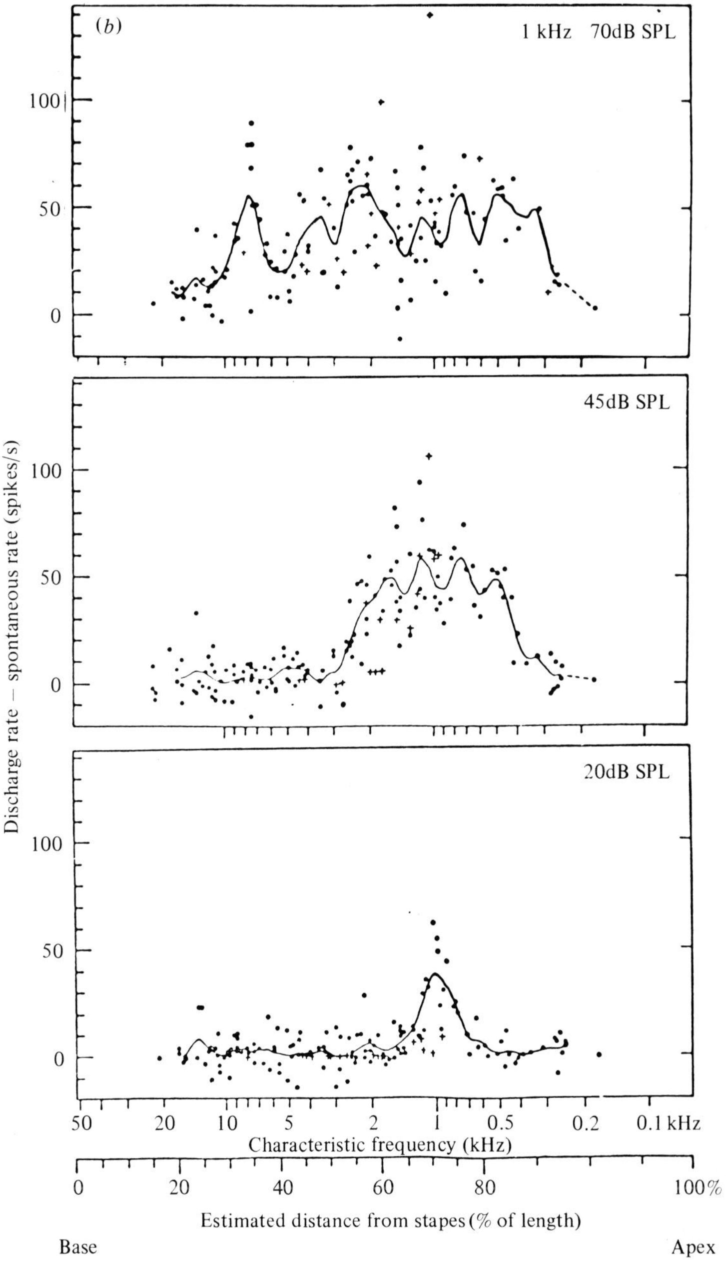

have steep high-frequency cut-offs (see Figs 14.15, 14.16), hence the spread of activity with level is limited towards higher characteristic frequencies only. There is a sharp demarcation between active and non-active regions at all sound levels.

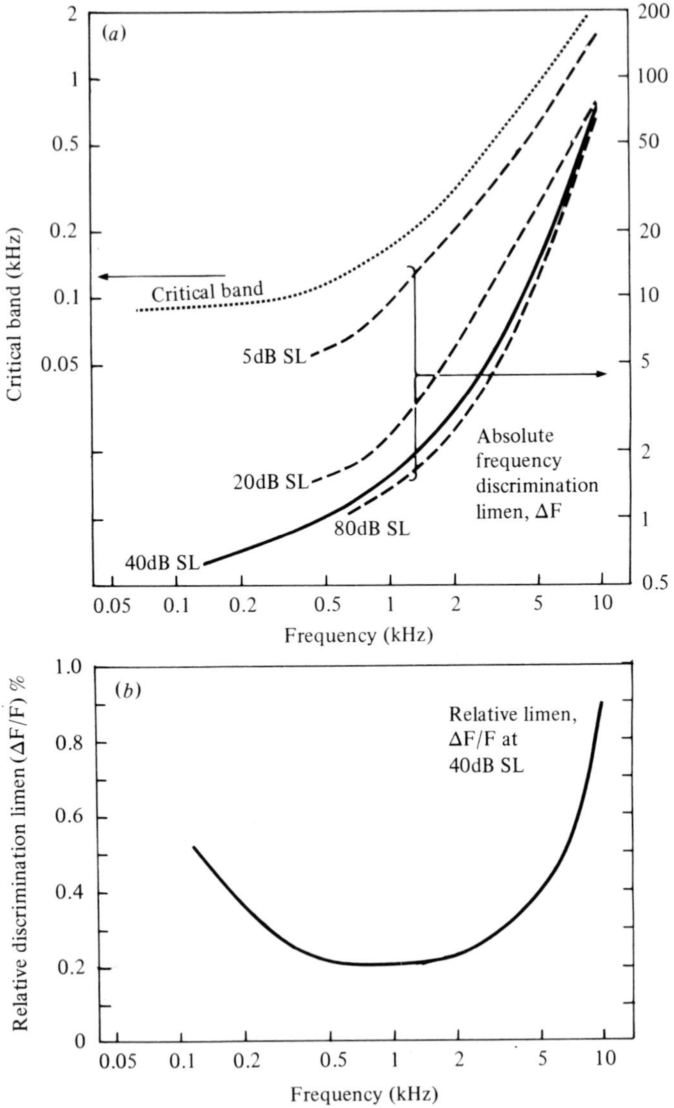

Fig. 15.6. Frequency discrimination in man. (*a*) frequency discrimination limen (just noticeable difference in frequency) as a function of frequency, at different sensation levels (SL). Note the improvement in discrimination with sensation level up to about 40 dB SL (continuous line) and the approximate similarity between the form of the curve at 5 dB sensation level and the width of the critical band (dotted line). (*b*) Relative frequency discrimination as

the signal as those of pure tones (termed by some '*spectral*' *pitch*), whereas the second, termed the '*synthetic*' *mode*, is the more familiar case in music where we hear a unified pitch of the tone complex as a single 'gestalt' (termed '*virtual*' *pitch*).* A strong example of the latter pitch is that of a train of pulses (Fig. 13.1*d*). This produces a series of frequency components which occur at multiples of the fundamental, repetition frequency ($1/P$) and are known as (2nd, 3rd etc) harmonics of the fundamental (F_0). While it is possible, when listening carefully in the analytical mode, to 'hear out' several of the lower harmonic components individually, the most compelling percept (synthetic mode) is of a unified pitch – that of the fundamental. The interesting aspect of this joint perception of the harmonics is that one can remove many of them *including the fundamental* without disturbing the pitch heard. This, the case of the 'missing fundamental', is encountered in musical instruments such as the oboe where there is very little energy present at the frequency corresponding to the actual pitch heard.

Both 'place' and 'periodicity' mechanisms are probably involved in determining the pitch of those sounds having the most salient or clear pitches, such as pure tones up to 5 kHz and those tone complexes that yield a strong virtual pitch. In situations where place or periodicity mechanisms are likely to be operating exclusively, the 'pitchiness' is less salient, as in the case of pure tones above 5 kHz (presumably 'place' alone) and amplitude-modulated white noise or electrical stimulation of the cochlear nerve (presumably 'periodicity' alone).

Whereas the nature of the place and periodicity cues in pure tone pitch is obvious, it will not be so for virtual pitch. Here, the most prominent intervals between spike discharges still correspond to the pitch(es) heard. The utilization of 'place' information is more complex, however. It seems likely that the fundamental (whether present or missing) could be 'inferred' by some sort of pattern recognition process, from the pattern of frequencies represented in the neural array (corresponding to the frequency components

---

a function of frequency at a moderate sensation level (40 dB SL). Note decrease in relative acuity at frequencies below 0.5 kHz and above 2 kHz. (Data from various sources in Wier *et al.* (1977). *Journal of the Acoustical Society of America, 61*, 178.)

* Other terms are commonly used such as 'periodicity pitch', 'residue', but these are not appropriate under certain conditions.

resolved). Such a process must be located centrally, but where, and what its nature would be in neural terms is not established.

Timbre

Timbre depends on the spectral character of the sound, and therefore on the mechanisms already outlined in § 15.1. In addition, the *envelope* of the waveform is involved. The rate of amplitude modulation of a signal can be detected up to several hundred hertz, corresponding to the range over which peripheral auditory neurons (cochlear nerve and cochlear nucleus) can 'follow' the envelope in their discharge patterns.

15.2. INTENSITY PROCESSING

Factors determining loudness

As mentioned in Chapter 13, the loudness of a sound depends not only on the sound level, but also on other factors.

One factor is duration. The intensity of tones or noise bursts has to be increased by about 3 dB for every halving of their duration below about 200 ms, in order to keep the loudness constant. The ear thus appears to behave as an integrator of energy for loudness purposes with a time constant of about 100 ms. The time constant of integration of peripheral neurons is only of the order of a few milliseconds; therefore this effect must be central.

Another factor is signal bandwidth (the width of the band comprising the frequencies in the signal). Frequency components falling within the 'critical band' of p. 309 sum in loudness, so that if the overall level of the noise band is kept constant whilst its bandwidth is increased (by halving its spectrum level – i.e. the energy per hertz – for every doubling of its bandwidth) the loudness is judged to be constant. Outside the critical bandwidth, the loudness level increases with bandwidth. Providing complex signals have a bandwidth of less than the critical band, their loudness is equal to that of a pure tone, centred on the band, of equal power.

Loudness is thus a complex subjective correlate of several physical parameters of sounds, and it is not a simple matter to devise techniques to measure the 'loudness' of sounds in noisy environments such as factories or airport neighbourhoods.

A somewhat analogous problem arises when we attempt to predict the risk that different sounds will damage the structure and function of the inner ear. Different persons have differing susceptibilities to noise damage depending upon their age and the condition of their inner ears (e.g. degree of cerebrovascular disease; exposure to ototoxic drugs such as streptomycin, neomycin, kanamycin). However, it is possible, for most sounds encountered in industry, to predict *statistically* the degree of hearing loss on the basis of the *total sound energy* encountered, after suitable frequency weighting (known as 'A' weighting, approximating to the 40 phon equal-loudness curve of Fig. 13.3). There is thus a 'trade-off' between the *duration* of exposure to a sound and its *power*, for a given probability of damage. To avoid hearing loss at the frequencies to which the ear is most sensitive and vulnerable (4 kHz), a conservatively safe noise exposure level is 73 dBA for an eight-hour working day for 40 years. It has been estimated that this would produce no more than 5 dB of hearing loss in any member of the total population at risk. For various economic and practical reasons, current international standards adopted in Europe consider an eight-hour/day exposure to a continuous noise of 90 dBA to be of no 'material risk'.

Discrimination of loudness

To a first approximation, a change in signal intensity can be discriminated if the difference in intensity exceeds a constant fraction of the signal's intensity. If the just-noticeable difference (JND) be ΔI, then $\Delta I/I$ is constant, to a first approximation. This is *Weber's law* (see also Chapters 1 and 7). Fig. 15.7 indicates that while Weber's law holds for broad-band noise signals at levels of 20 dB to 100 dB above threshold (dotted line) it does not do so near threshold, nor does it apply exactly to pure tones. For the latter, the Weber fraction reduces gradually from about 1.5 dB at 5 dB above threshold to about 0.5 dB at 80 dB above threshold.

The neural cues on which the perception of loudness and its discrimination are based are probably multiple (as with pitch). The number or extent of fibres active, and their discharge rate, are the two most obvious candidates. As indicated above (§15.1 and Fig. 15.4) the discharge rate of the great majority of cochlear fibres saturates at moderate sound levels. Above these, the spread of activity into unsaturated fibres of neighbouring characteristic frequencies (as in Fig. 15.5) could afford the appropriate cues for loudness and loudness discrimination of single component (pure tone) signals. This is the '*population*' *model* for the neural coding of loudness, and could account for its enormous dynamic range of 100

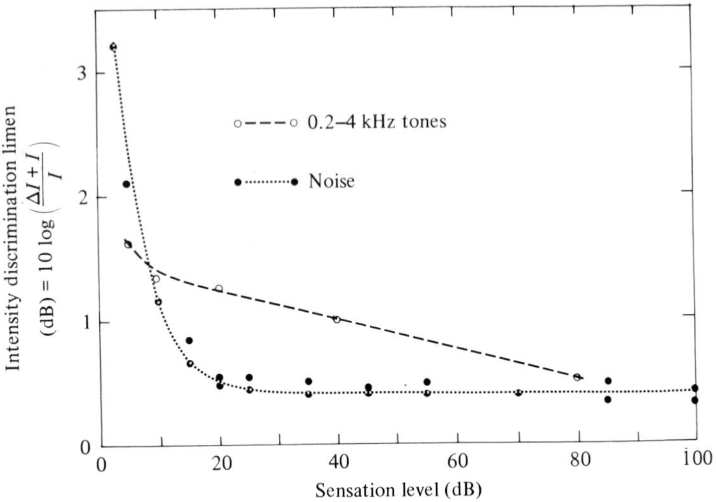

Fig. 15.7. Intensity discrimination in man. The just-noticeable difference of intensity for human subjects as a function of sensation level, for pure tones (dashed line) and white noise (dotted line). Note that above 20 dB SL the JND continues to improve with level for tones but only very slightly for noise. (Noise data from Miller (1947) *Journal of the Acoustical Society of America, 19*, 609; tone data from Jesteadt *et al.* (1977). *Journal of the Acoustical Society of America, 61*, 178.)

dB or more (Fig. 15.7) for *single* components. However, we have the same problem already encountered in considerations of frequency selectivity (p. 311): how the levels of the individual components in sounds (like speech) having *multiple* frequency components, could be coded. Furthermore, the difference limen for intensity remains unaffected at high levels by the addition of simultaneous high and low pass masking noise (known as 'band-stop' noise), which would be expected to limit the spread of activity to neighbouring fibres required by the population model. How the absolute level of components is coded under such conditions is not clear. Again, it could be by the 10% of fibres, illustrated in Fig. 15.4 and referred to on p. 313, remaining unsaturated at high signals levels. However, one would then expect some deterioration in the difference limen, as these fibres 'took over' from the majority, particularly as the rate of their increase in discharge with intensity is low compared with that of fibres at lower stimulus levels. There is no evidence of such an effect – intensity discrimination if anything *improves* as level is increased (Fig. 15.7). Neither can classical 'phase-locking' serve to signal absolute level, although it can represent the level of adjacent components relative to one another. It is conceivable that other aspects of the fine time structure of cochlear fibre discharges convey the relevant information to higher levels of the auditory system. Certainly, at the level of the dorsal cochlear nucleus, many cells are found which, by virtue of extensive lateral inhibition, are able to signal differences in the intensity of components in a multicomponent sound, over a dynamic range as great as that of Fig. 15.7.

Interesting derangements of the coding of the intensity of pure tones occur in hearing loss of cochlear origin. The discrimination of increments in intensity, paradoxically, can be *more* acute than normal. This is an aspect of the so-called '*recruitment of loudness*', characteristic of cochlear pathology, where the loudness of supra-threshold tones grows more rapidly with increase in stimulus level in the affected than in the normal comparison ear. It is likely to be related to the abnormally rapid spread of activity across the neural array with increase in intensity, which would be expected to occur on account of the broader tuning of cochlear fibres in cochlear pathology, and therefore greater overlap of their frequency response areas. Furthermore, the discharge rate versus intensity functions are often steeper than normal in these fibres.

Loudness adaptation

When a signal of constant intensity just above threshold is maintained for a sufficient length of time, its loudness is perceived to decline. This occurs rapidly at first, then progressively more slowly. Measurement of the time course of this loudness adaptation forms the basis of one clinical test (the 'tone decay' test) for lesions affecting the cochlear nerve. In normal ears the loudness adaptation is small, amounting to less than 15 dB for 1-min exposure to a continuous tone. It is presumably related to physiological adaptation of the discharge rate of cochlear fibres in response to continuous tones (Fig. 14.14*b*). On the other hand, for ears where long-term pressure damage has occurred to the cochlear nerve by virtue of an *acoustic neuroma* (a tumour of the nerve-sheath occurring in the internal acoustic meatus), the adaptation can be excessive (for example, more than 30 dB in 1 min). This is an important diagnostic test.

Masking

The audibility of one signal is, under certain circumstances, diminished by the presence of another. If the first signal is not heard at all, it is said to be *masked* by the other.

Generally, the closer the frequency of the *masker* is to that of the *maskee* signal, and the greater the intensity of the masker, the greater will be the degree of masking. Maskers lower in frequency than the maskee signal are more effective than those higher in frequency. Masking is maximum when both maskee and masker occur simultaneously; however significant masking also occurs when the masker and maskee are completely separated in time; for up to about 100 ms when the masker precedes the maskee ('forward masking') and for a smaller duration under reversed conditions ('backward masking').

With the exception of backward masking, all of these effects can be found in neural responses at levels as low as the cochlear nerve. A 'masker' of sufficient energy falling into the response area of a cochlear fibre can dominate the response of the fibre both in terms of discharge rate and (at low frequencies) the temporal discharge patterns ('phase-locking') of the fibre, in the presence of other signals that would evoke a response on their own. The latter are thereby masked. Thus, the relationship between the frequencies of masker and maskee (and particularly the asymmetry of masking) are primarily determined by the shape of the cochlear nerve frequency area (as in tone on tone masking, Fig. 15.1*a*), although other factors (particularly certain non-linearities) are also involved at this level; and at higher levels, *inhibition* also occurs.

As would be expected from the above discussion, the degree of masking of a narrow-band signal (such as a tone) by a wide-band signal (such as white noise) depends *predominantly* upon the energy falling in the effective bandwidth of the cochlear filter. In ears where damage to the cochlea has occurred, the neural effective bandwidths become abnormally wide (p. 311). This means, *inter alia*, that signals are more easily masked by wide-band maskers, and is one reason why patients with such hearing loss have particular difficulty in perceiving speech in noisy surroundings.

Auditory fatigue

Stimulation at sufficiently high sound levels can also reduce the sensitivity of hearing *after* the stimulation ceases. If the stimulation is not excessive, sensitivity fully recovers, given time. The time course of the recovery of this '*temporary threshold shift*' (*TTS*) beyond the first 2 min, is approximately linearly related to log time, the speed of recovery depending upon the total energy (i.e. intensity × time) delivered. Thus, in severe cases, it may take several hours or days for complete recovery of threshold. For fatiguing tones of level above about 90–100 dB SPL, the loss of sensitivity becomes rapidly more severe. As indicated earlier, if the energy of stimulation is excessive, part of the threshold elevation may be irreversible i.e. leave a *permanent threshold shift* (*PTS*).

As expected, the frequency at which the temporary threshold shift is maximal is close to that of the frequency of the fatiguing stimulus, at low levels. However, at high levels, the frequency of maximum threshold shift can be as much as a half octave or more *above* the fatiguing frequency.

All of these effects can be found on the threshold of cochlear nerve fibres.

15.3. PROCESSING OF SPEECH

Speech perception may be regarded as a special case of auditory pattern recognition. Linguists commonly describe speech in terms of units called *phonemes* (corresponding to the percepts of consonant and vowel sounds) which are signalled in speech by multiple acoustic cues (Fig. 15.8). For vowels (e.g. 'I' in Fig. 15.8), the relative spacing of the formants is most important; in the case of diphthongs ('I' in Fig. 15.8), also their *changes* in frequency. Consonants are generated during contact or changes in contact between mouth structures (especially tongue and lips). They are therefore characterised by more or less broad-band ('noisy') emissions (e.g. 'S' in Fig. 15.8) and changes in formant frequencies. The distribution of energy within those bands, and their time relation to the following vowel are important (e.g. silent gap in the 'K', as in the other stop consonants; p, t, b, d, g) as well as whether the vocal tract is being excited by the larynx as in so called 'voiced' sounds (e.g. b, d, g) compared with unvoiced sounds (e.g. p, t, k).

It is important to note that the process of relating acoustic cues to phonemes is extremely complex: consonants, especially, are not signalled by an invariant group of acoustic cues. Which cue is relevant is determined by the context: thus the same acoustic cue (e.g. a noise burst at 1440 Hz) can signal different phonemes depending upon which vowel follows it (e.g. p before i; k before a). Likewise,

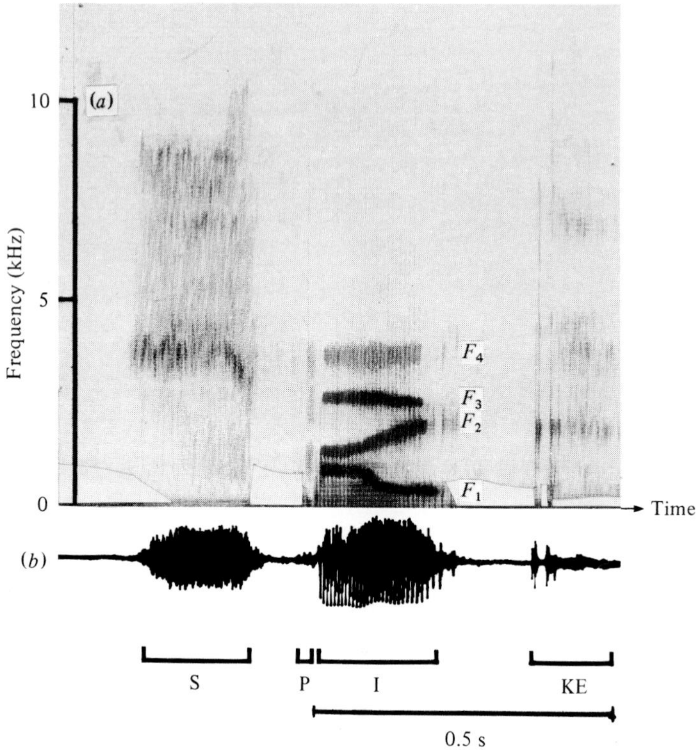

Fig. 15.8. Frequency and time analysis of running speech. (*a*) Spectrogram of the word 'SPIKE'. The spectrogram indicates, by relative blackness, the relative energy levels of the frequency components in the sounds, as a function of time. (*b*) Waveform envelope of the speech sounds as recorded by a microphone. Note the broad-band character of the consonants, S, P, K, with energy predominantly located above 4 kHz in the case of the S, and below in the other cases. In the vowel, I, the concentrations of energy into the 4 formants (F_1–F_4) can be seen. The spacing and movements of F_1 and F_2 with time are the most important cues for the vowel. Note also the laryngeal pulses of the *voiced* vowel in the waveform trace appearing as vertical striations in the spectrogram. Note also the brief silent period within the KE complex. (From Evans (1974). in *The Neurosciences: Third Study Program*, p. 131. M.I.T. Press.)

two contiguous phonemes can be signalled by a single acoustic cue. This allows a very large number of phonemes to be transmitted in a given time in spite of the temporal limits set by the vocal tract in production, and by the auditory system in analysis. Furthermore, the same phoneme is signalled by different acoustic cues according to context. Thus, there is no 1:1 correspondence between the physical nature of speech sounds (called acoustic cues or features) and what is heard (called phonetic features). This is a general problem in acoustic pattern recognition as it is in the perception of the pitch of complex sounds.

The peripheral auditory system performs primarily a spectral analysis of the speech signals. Thus, at low sound levels at least, a 'spectrographic' representation of the sounds, on the lines of that in Fig. 15.8, occurs, where the distance along the ordinate represents distance along the tonotopic neural array, and the intensity of the spectral energy is represented by the evoked rates of discharge of the neurons (as in Fig. 15.5). The effective bandwidths of the cochlear fibre filters (forming the peripheral neural filter bank) are similar to or narrower than those used to produce the spectrum of Fig. 13.1*j*, at least for characteristic frequencies up to a few kilohertz. The bandwidth of these filters is such that a good compromise is made between *narrow bandwidth*, with consequent loss of temporal resolution (the spectral features, though highly resolved in frequency, would be 'smeared out' in time), and *high temporal resolution* (necessitating wide-band filters and consequent smearing of the spectral details). The time resolution of the cochlear nerve filters is less than about 5 ms. Thus, the peripheral analysis represents, in the neural activity patterns, a running *short-term spectrum* of the speech sounds (see also p. 244).

At moderate to high speech levels, however, we encounter the same problem of the limited dynamic range of the majority of cochlear fibres, discussed on p. 311. The possible solutions, discussed there, are likely to apply also in the case of speech.

The central levels of the auditory system are organised in a manner that could serve the purpose of extracting certain salient features from the peripheral running spectrum. For instance, the neurons at the cortical level, having selective sensitivities for the direction and rate of frequency changes, could serve to distinguish the formant transitions important for making distinctions among certain vowels and among certain consonants. Likewise, the neural sensitivity for temporal patterning, for the form of the amplitude envelope of signals, and for other dynamic properties of complex sounds could be valuable; evidence for such a role is given by the results of cortical ablation (p. 302).

An interesting question is whether, in the human brain, there are neurons as selective for certain of the acoustic features of speech sounds as some neurons in the auditory cortex of squirrel monkeys are for the acoustic features of their vocalisations. As mentioned on p. 302 there is evidence that in right-handed individuals, mechanisms exist in the cerebral cortex of the *left* hemisphere that are specialised for the analysis of human speech sounds, in contrast to melodic patterns, for example, which appear to be analysed in the right hemisphere. Furthermore, some of the acoustic features of certain speech sounds having rapidly changing spectra (such as the stop consonants b, d, g), when presented repeatedly, produce a so-called 'adaptation' of the auditory system in such a way that the identification of the phonemes is changed: thus, after /ba/ has been played 40 times in succession to a listener, he is more likely to hear as /da/ an intermediate sound that would be equally often heard as /ba/ and as /da/ by the unadapted subject. This appears to be a central process, probably again in the left hemisphere, and could represent the habituation of 'feature sensitive' neurons.

The analysis of the voice pitch fundamental, on which the intonation patterns of speech depends (e.g. the rise in pitch that signals a question; the fall in pitch that signals a statement) is probably accomplished in two ways. One involves 'periodicity' cues in the neuronal discharges, generated either by the fundamental energy itself, or if the fundamental is filtered out (as on the telephone), carried by the amplitude modulation of the formants (p. 320). The second mode would operate on the 'place' coded (resolved) harmonics of the voice fundamental, interpreted by the sort of pattern recognition process mentioned on p. 319.

Lesions of the auditory system affect speech perception in various ways depending upon the level at which they occur. Disease of the conductive system of the middle ear simply attenuates all sounds and requires straightforward amplification (as by a standard type of hearing aid) in order to raise the level of the sounds above threshold. Hearing loss due to disorders of the inner ear however, as has been indicated above, not only involves elevation of threshold, but deterioration in frequency selectivity (p. 311). The latter has the following effects on the analysis of spech sounds: reduction in the ability to resolve closely spaced formants (interfering particularly with the distinctions between certain vowels); greater masking by the relatively intense first formant on the higher formants; a greater masking effect of background noise on speech sounds (p. 324); disturbance in the coding of the intensity of the speech sounds (recruitment of loudness, p. 323). There is also some evidence of defective temporal discrimination even in cochlear hearing loss. In addition, cochlear hearing loss generally involves the higher frequencies to a greater degree than the lower (Ménière's disease being an important exception); hence the higher-frequency, lower-level sounds – consonants – which bear most of the speech information, tend to be lost. Simple amplification, though raising levels above the elevated threshold, will not compensate for these deficits in auditory processing, with the consequence that the speech remains indistinct. Schemes of electronic speech processing are being investigated in order to circumvent some of these problems for the next generation of hearing aids.

Attempts are now being made to embody some of the principles by which speech is analysed by the peripheral auditory system in the design of arrays of stimulating electrodes for insertion into the inner ear of totally deaf patients, in an attempt to restore the intelligibility of speech. These *cochlear implants* will have to recreate essentially normal distributions of activity in surviving nerve fibres located in at least ten separate frequency 'places' along the cochlear partition, if even rudimentary speech perception is to be achieved. At present, single channel implants, or implants having a few channels, are being evaluated. These can provide only information on the rhythm, amplitude envelope, and fundamental frequency of sounds, but appear to be valuable in helping patients to identify important environmental sounds (like the telephone bell), to control their voice, and to aid in lip-reading.

Lesions at the cortical level (usually from cerebrovascular accidents – 'strokes') affect mechanisms of speech perception, in a variety of ways, at present not well understood. Some lesions impair the ability to distinguish between different complex acoustic features (a condition known as *auditory agnosia*). Others manifest their effects at the *linguistic* level. This means, for example, that there is difficulty in assigning a linguistic *meaning* to a sound, e.g. in recognising the *name* of an object (*receptive aphasia*).

15.4. PERCEPTION OF SOUNDS IN SPACE

The pinnae are necessary for hearing sounds as originating *outside* the head: witness the effect of listening to a conventional stereo broadcast over headphones – to most people, it sounds *inside* the head. However, if the broadcast microphones are inserted in a dummy head having 'pinnae', then with headphone-listening the sounds are dramatically projected into space (so-called 'binaural stereo'). On the pinnae we depend for our ability to locate sound in *elevation*, i.e. height. This probably results from the corrugations of the pinnae, which modify the spectrum of the incoming signal by adding reflections, in different ways, according to the elevation of the sound.

As would be expected from the response properties of neurons in the superior olive (p. 294), two mechanisms are involved in our location of sounds in *azimuth*, i.e. to left or right of the midline. The first, which alone operates for tones having frequencies above about 1.5 kHz, exploits the *intensity differences* between signals arriving at the two ears. The higher the frequency, the greater the effects of 'shadowing' by the head of a sound field approaching from one side. The attenuation afforded by this can be as large as 20 dB. The 'E–I' neurons of the superior olive and higher levels are sensitive to such interaural differences. The second mechanism extracts *temporal or phase differences* between signals arriving at the two ears. Human beings are able to discern a shift in sound source from the midline corresponding to an interaural delay as small as 10 μs. A delay of more than 650 μs localises a sound entirely at one ear. Again, the cells in the superior olivary nucleus and higher levels, having characteristic frequencies below about 1–2 kHz, and having sensitivities to interaural phase, presumably subserve this effect. The 'phase-locking' of impulses, in the cochlear nerve (p. 284) carries the temporal information to the superior olive. Within a limited range of the midline it is possible to 'trade' interaural delay for intensity, to maintain constant the perceived location of a sound. This is also the case in terms of the responses of certain of the neurons in the superior olive (p. 295).

Fortunately, sounds are perceived as coming from a single location even in the presence of strong confusing echoes. Two transient sounds are 'fused', i.e. heard as one, if the delay between them is less than about 5 ms, even if their sources are different. In the latter case, the location of the fused sound corresponds to the location of the sound received *first* unless the time interval between the sounds is less than 1 ms). This phenomenon is called the *precedence effect*. It serves the useful purpose of reducing the unwanted effects of reflections in a reverberant room. The mechanisms underlying the precedence phenomenon are likely to be essentially cortical. Removal of the *cortex of one hemisphere* in animals disturbs this precedence effect on the contralateral side.

'Two ears are better than one', not only for the purpose of the location of a sound source in azimuth, but also for enabling a speaker to be heard more clearly in a crowd, or an instrument to be heard out from an orchestra. This is known as the 'cocktail party' effect, and its underlying nature is complex. One important factor involved is *binaural unmasking*. The degree to which one sound masks another depends not only upon their relative frequency and intensity but also upon their relative locations (actual or virtual): if both coincide, the masking is maximum; if these are separated, the masking is minimum. Masking-level differences of as much as 15 dB can be obtained in this way. Curiously, this means that if a tone and noise are presented to one ear and the noise level adjusted until the tone is masked, then adding the same noise to the other ear will allow the tone to become audible. (In the latter case the tone is lateralised at one ear, the noise in the midline.) The neural mechanisms underlying this important phenomenon are not known, but it behaves as if a sound introduced into one ear can subtract from that in the other ear.

As indicated, the binaurally dependent precedence and unmasking effects are of importance for enabling us to hear an individual in a company of competing speakers, in a reverberant room. Patients with cochlear hearing losses appear to lose the benefit of these mechanisms, and this further adds to their disabilities in hearing speech against competing backgrounds.

15.5. SUGGESTIONS FOR FURTHER READING

General references

Moore, B. C. J. (1982) *Introduction to the Psychology of Hearing*, 2nd edn. London and New York: Academic Press. (A comprehensive introduction to psychoacoustics.)

Green, D. M. (1976) *An introduction to hearing*. Hillsdale, N.J.: Lawrence Erlbaum Associates.

Plomp, R. (1976) *Aspects of tone sensation*. London: Academic Press.

Special references

Speech. T. H. Bullock (ed.) (1977) *Recognition of complex acoustic sounds.* Berlin: Dahlem Konferenzen. Flanagan, J. L. (1972) *Speech Analysis, Synthesis and Perception.* Berlin: Springer. Fletcher, H. (1953) *Speech and Hearing in Communication.* New York: van Nostrand.

Neural analysis of complex sounds; speech; adaptation to speech sounds. Ainsworth, W. A. (1976) *Mechanisms of Speech Recognition.* Oxford: Pergamon Press.

Correlations between physiology and psychoacoustics. Evans, E. F. & Wilson, J. P. (1973) *Basic Mechanisms in Hearing*, pp. 519–51. New York: Academic Press. Evans, E. F. & Wilson, J. P. (1977) (eds) *Psychophysics and Physiology of Hearing.* London: Academic Press. Evans, E. F. (1978) *Audiology*, **17**, 369–420.

Audiology, pathophysiology, rehabilitation. Eagles, E. L. (1975) (ed.) *Human Communication and its Disorders.* New York: Raven Press. Ballantyne, J. (1977) *Deafness* (3rd Edn) Edinburgh: Churchill Livingstone. Henderson, D. *et al.* (eds) (1976) *Effects of Noise on Hearing.* New York: Raven Press. Ludvigsen, C. & Barfod, J. (eds) (1978) *Sensorineural hearing impairment and hearing aids. Scandinavian Audiology*, Suppl. 6. Ballantyne, J. C., Evans, E. F. & Morrison, A. W. (1978) Electrical auditory stimulation in the management of profound hearing loss. *Journal of Laryngology and Otology*, Suppl. 1.

CHAPTER SIXTEEN

The vestibular sensory system

A. J. BENSON

The vestibular sense, unlike the sensory systems described in other chapters, is not one of the classical senses. Indeed, argument persisted well into the twentieth century about the sensations engendered when the vestibular receptors of the inner ear were stimulated. Is the sensation of turning (the vertigo), such as is experienced immediately after a prolonged spin, the direct consequence of a change in the afferent discharge from sensory receptors in the vestibular apparatus? Or is it a result of the stimulation of muscle, joint and cutaneous receptors by the postural adjustment that is the primary response to the vestibular stimulus? The demonstration, within the last decade, that vestibular afferents project to the cerebral cortex gives neurophysiological support to the view that there are characteristic sensations and percepts that depend, primarily, on information provided by the vestibular apparatus. Nevertheless, the main function of this end-organ is to provide information about the movement and orientation of the head that is used, principally, in the regulation of motor activity at a sub-cortical level. Man is not normally aware of the continual stream of information coming from the sensory epithelium of his vestibular apparatus. It is only when these receptors are stimulated in an atypical manner, (e.g. on stopping from a prolonged spin, or by disease of the inner ear) that he may become conscious of sensations of body motion and attitude that do not accord with the veridical percepts obtained through other sensory modalities.

The functional role of vestibular mechanisms, rather than characteristics of vestibular sensations, may be illustrated by the results of complete loss of vestibular functions. In man this is a relatively rare condition, but it is seen as the unfortunate consequence of the administration of ototoxic drugs (like streptomycin) or of disease processes. One of the most significant features of such a patient is that he may be quite unaware that he has any sensory deficit, a finding that reflects an essential difference between the vestibular modality and those 'exteroceptive' systems, like sight and hearing, that provide us with information about events outside our bodies. Observation and more specific questions do, however, reveal certain deficits. The person without vestibular function has difficulty in maintaining his balance – his postural equilibrium – when required to walk on an irregular or compliant surface (such as a mattress), when asked to stand with the feet close together or on one leg, or when deprived of visual cues. Such patients can cope quite well with normal locomotor activities, provided they can see where they are going and have adequate visual cues for orientation to gravity, but once such cues are removed they are unable to maintain their balance. The other salient symptom, commonly described by such patients, is an impairment of visual acuity during cyclical head movements, as when walking or when travelling in a vehicle over rough ground. The visual disturbance is characterised by an apparent bobbing displacement of the visual field, oscillopsia, which is coincident with the head movement, though, typically, there is no loss of visual acuity when the head is moved slowly.

Other sensory deficits associated with a loss of vestibular function are likely to be revealed only by special tests. The patient, when seated on a turntable and blindfolded, is unable to detect the onset or decay of rotation, unless the acceleration is so intense that the motion is detected because cutaneous and other somatosensory receptors are adequately stimulated by inertial forces. However, his ability to detect body attitude with respect to the forces of gravity (the basic reference of verticality) may be only slightly degraded until somatosensory cues are substantially reduced by immersion in water; he is then quite unable to differentiate up from down, unless provided with visual cues. The only benefit enjoyed by those without functional labyrinths is a complete immunity to motion sickness.

The study of man and experimental animals with vestibular lesions has firmly established that the primary function of the vestibular apparatus is to transduce both angular and linear movements of the head and to signal its attitude relative to the specific force of gravity, the gravitational vertical. This information is used to maintain postural equilibrium, in conjunction with information from other widely distributed mechanoreceptors signalling forces acting on or within the body as well as the relative orientation of limb and body segments. Visual cues also play an important role in equilibratory mechanisms, and on occasion can be the overriding input in the hierarchy of control. Visual and vestibular mechanisms are interdependent: information about head movement from the vestibular

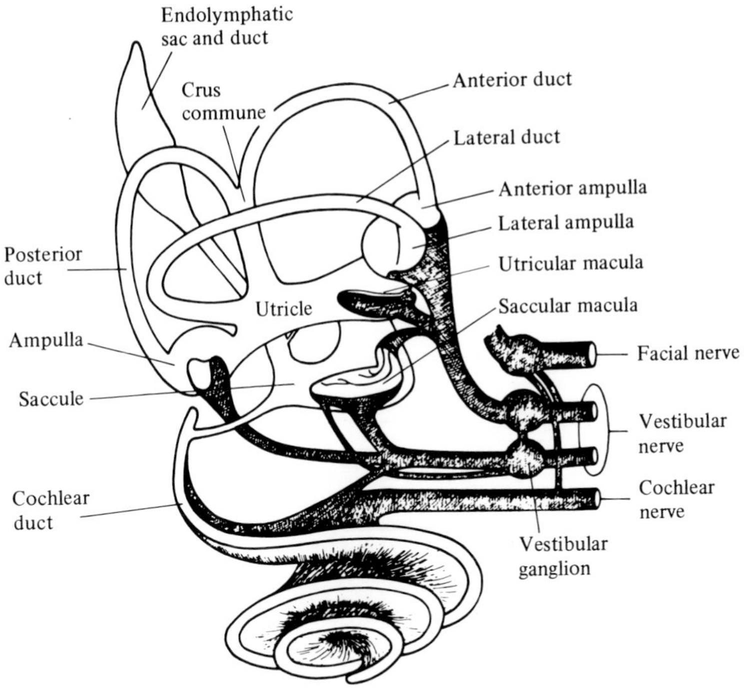

Fig. 16.1. Diagram to show the principal structures of the membranous labyrinth and its neural connections. (After Lindeman, 1969.)

apparatus is used to stabilise eye position and to preserve vision during transient head movements, while signals from detectors of retinal motion project to the vestibular nuclei and contribute both to postural adjustments and to the perception of body attitude and motion.

16.1. THE VESTIBULAR APPARATUS

Structure

It is now necessary to describe in more detail the morphology of the end-organ and to relate structure to function, namely the transduction of angular and linear accelerations.

The gross structure of the inner ear is shown in semi-diagrammatic form in Fig. 16.1. There is a clear anatomical and functional differentiation between the membranous labyrinth (containing the

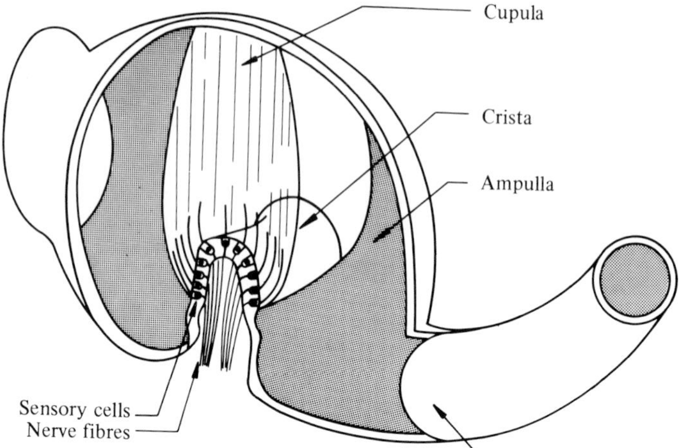

Fig. 16.2. Cut-away view of an ampulla of a semicircular duct. (After Lindeman, 1969.)

vestibular apparatus) and the spiral convolutions of the cochlea (the organ of hearing), though they share the same sculptured cavity of the petrous temporal bone, they have interconnecting fluid systems, and they both relay to the central nervous system by the eighth cranial nerve.

The membranous labyrinth, which is about the size of a pea, is made up of three curved tubes, the semicircular ducts, that are approximately orthogonal the one to the other. These open into the sac-like structure of the utricle which communicates inferiorly with another sac, not inappropriately called the sacculus. These interconnecting structures are filled with a fluid, endolymph, having a low viscosity and an ionic composition in some respects characteristic of an intracellular fluid, i.e. a high potassium and a low sodium content. The membranous labyrinth is supported within, yet separated from, the walls of the bony labyrinth by perilymph. This fluid has a high sodium and a low potassium content, like extracellular fluid, and is in ionic and osmotic equilibrium with the cerebrospinal fluid.

In each semicircular duct there is a dilatation, the ampulla, where the neuro-epithelium is located on a saddle-shaped ridge, the crista ampullaris, which lies across the floor of the ampulla (see Fig. 16.2). Cilia of the specialised sensory cells of the neuro-epithelium (described in detail later) are invested by a gelatinous mass, the cupula, which

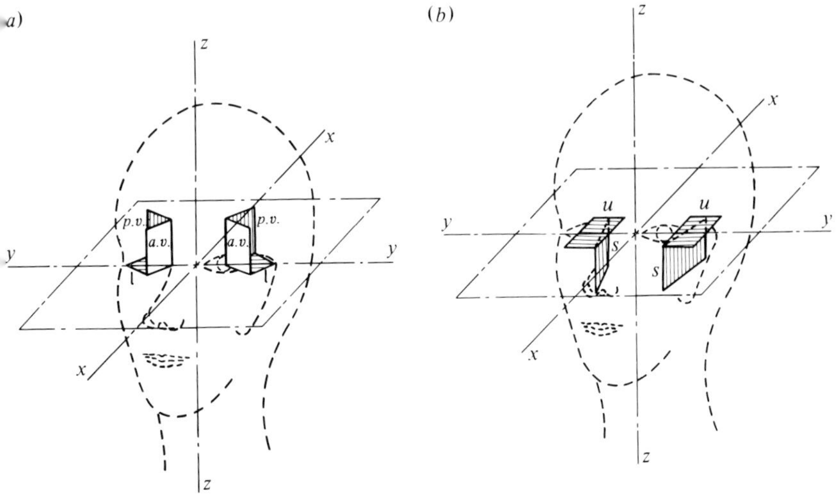

Fig. 16.3. Diagram to show the principal planes (*a*) of the semicircular ducts and (*b*) of the maculae. In *a*, *a.v.* and *p.v.* identify the anterior and posterior vertical ducts, the lateral duct (not labelled) is orthogonal. In *b*, *u* and *s* identify the utricular and saccular maculae respectively. Also shown are the principal reference axes of the head, the anterio-posterior (x), the transverse (y) and the longitudinal (z).

extends transversely across the ampulla from the crista so that its peripheral edge is in contact, all the way round, with the internal wall of the ampulla. The cupula, which is made up of mucopolysaccharides within a keratin framework, forms a watertight diaphragm across the ampulla and prevents the free circulation of endolymph within the semicircular duct. It has sufficient compliance, however, to allow a small displacement of endolymph within the duct and with it a corresponding deflection of the cilia of the sensory cells.

The three semicircular ducts are identified as anterior, posterior and lateral. In man, the plane of the lateral duct lies in the horizontal plane when the head is tilted forward about 30°; with the head in this attitude, the planes of the anterior and posterior canals are approximately vertical and lie at about 45° to the sagittal and coronal planes (Fig. 16.3). The sensory cells of a particular ampulla are optimally stimulated by an angular (i.e. rotational) acceleration acting in the plane of the associated duct. Thus any angular movement of the head will alter the activity of the ampullary receptors of at least one pair of ducts.

The utricle and saccule contain the other vestibular end organs, commonly referred to as the otolith organs, which are functionally adapted to transduce linear accelerations and to provide information about orientation of the head with respect to gravity (Fig. 16.4). Located on the inner wall of both the utricle and the saccule there is a 'spot', the macula, composed of sensory cells. Each macula is of complex shape which may be likened to the double curvature of a cupped hand. The utricular macula, which in outline is somewhat kidney-shaped (reniform), is located on the anterior–inferior wall of the utricle. In man, the principal plane of the macula is roughly parallel to that of the lateral canal. The saccular macula is hook-shaped, and is located on the medial wall of the sacculus; its major plane is more or less parallel to the saggital plane of the head and hence perpendicular to that of the utricular macula (see Fig. 16.3).

The ciliated sensory cells of the maculae are similar to those of the crista, and like them are invested by a gelatinous structure. But in the otolith organs mucopolysaccharides form only a membrane over the neuroepithelium as it carries in its apical surface a packed layer of calcium carbonate (calcite) crystals, the statoconia. This glistening calcareous frosting, the otolithic membrane, on the gelatinous cake is an obvious macroscopic feature when the utricle or saccule is opened on dissection.

Neuroepithelium

Histologically, the sensory cells of the neuroepithelium of the cristae and the maculae are very similar and hence will be described collectively. The neuroepithelium consists of sensory cells partitioned from subepithelial tissue by a basement membrane, each sensory cell being surrounded by several supporting cells. Under the electron microscope two types of sensory cells can be identified (Fig. 16.5): the amphora-shaped Type I cell, and the roughly cylindrical Type II cell. A bundle of 60–100 cilia project from the apical surface of both types of cells; hence they are called hair cells. A transverse section through the cilia reveals a more or less hexagonal pattern, and at just one point of the hexagon a lone cilium of more complex structure, like that of a motile cilium, is to be found; this is the kinocilium. The numerous other cilia, the stereocilia, are of simpler structure and of differing and graded lengths, the longest being close to the kinocilium, the shortest most distal.

Afferent connections from the sensory cells to the peripheral axons of the bipolar cells of the vestibular (Scarpa's) ganglion are made via

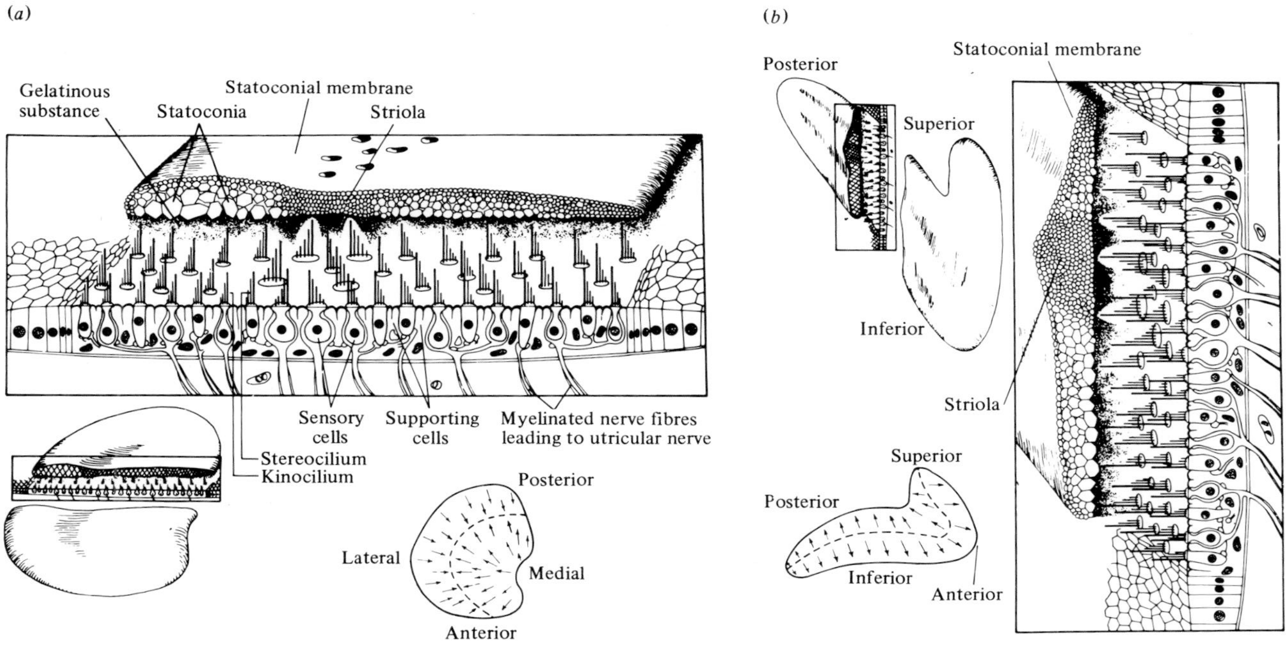

Fig. 16.4. Diagrammatic representations of the morphology (*a*) of the utricular and (*b*) of the saccular maculae. The outline drawings, with arrows, show the directional sensitivity (morphological polarisation) of the hair cells over the surface of the maculae. (After Lindeman, 1969.)

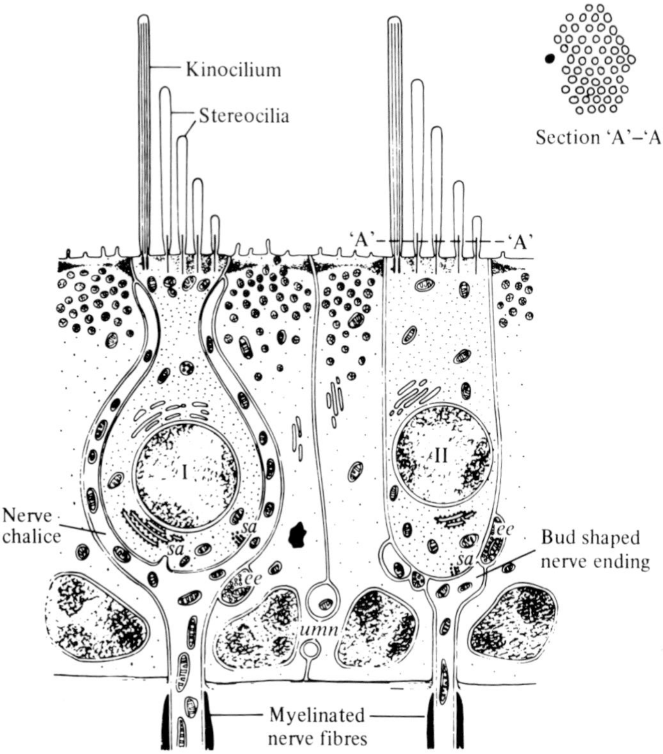

Fig. 16.5. Fine structure of the sensory epithelium of the cristae and the maculae showing: Type I amphora-shaped cells and Type II cylindrical cells, (*ee*) efferent nerve endings, (*sa*) synaptic area, (*umn*) unmyelinated nerve endings. The inset, at the upper right of the figure, shows a cross-section of the cilia; the kinocilium is represented by the filled circle.

a complicated neural plexus. Chalice-shaped nerve-endings enclose the basal surface of the Type I cells while only bouton type endings are found on Type II cells. Some afferent fibres connect with a single cell but the majority receive inputs from several sensory cells of both Types I and II. In addition to these afferent connections, efferent fibres terminate in granulated boutons on the nerve calyx of Type I cells and directly on the basal area of Type II cells.

16.2. MECHANISM OF TRANSDUCTION

Recordings from primary vestibular afferents in a variety of species (ranging from elasmobranch fish to squirrel monkey) have shown that the majority of neurons have a resting discharge which is increased when the cilia of the sensory cells are deflected towards the kinocilium and is decreased when the hairs are deflected away from the kinocilium. Modulation of primary afferent activity is reflected by the membrane potential of the sensory cell: depolarising currents increase the resting discharge whereas hyperpolarisation has the opposite effect. But the precise mechanism by which minute deflections of the gelatinous cupula or otolithic membrane modify the generator potential of the sensory cells is still a matter for speculation.

There is much less doubt, however, about how the hair cells of the semicircular canals are mechanically stimulated by angular movements of the head and those of the otoliths by linear accelerations.

The semicircular canals; transduction of angular movement*

Each semicircular duct is in free communication at either end with the utricle and so may be considered as a closed tube (more precisely a toroid) obstructed by the compliant flap of the cupula (Fig. 16.6). The membranous wall of this fluid-filled toroid is secured within the bony labyrinth and hence to the skull; so it moves with the head. Consider now the mechanical response of this duct–cupula–endolymph system when a rotational movement of the head occurs in the plane of the toroid. At the beginning of the movement, when there is an angular acceleration of the head, a force is developed across the cupula because of the inertia of the ring of endolymph. If there were no cupula and no obstruction to the flow of endolymph within the duct, the endolymph would initially retain its position in space, because of its inertia, as the walls of the duct begin to rotate. With continuation of the rotation, viscous drag between the walls of the duct and the fluid would impart angular motion to the fluid which eventually would rotate with the containing structure.

These properties of the ring of endolymph, inertia and viscous friction, are still operative even though the relative movement of endolymph is restricted by the flap-like cupula. At the beginning of

* Strictly, the semicircular canal is the perilymph-filled cavity within the petrous temporal bone that contains the semicircular duct. The term ‘semicircular canal’ is commonly, albeit inaccurately, used when referring to the duct–cupula–endolymph system and its associated neuroepithelium.

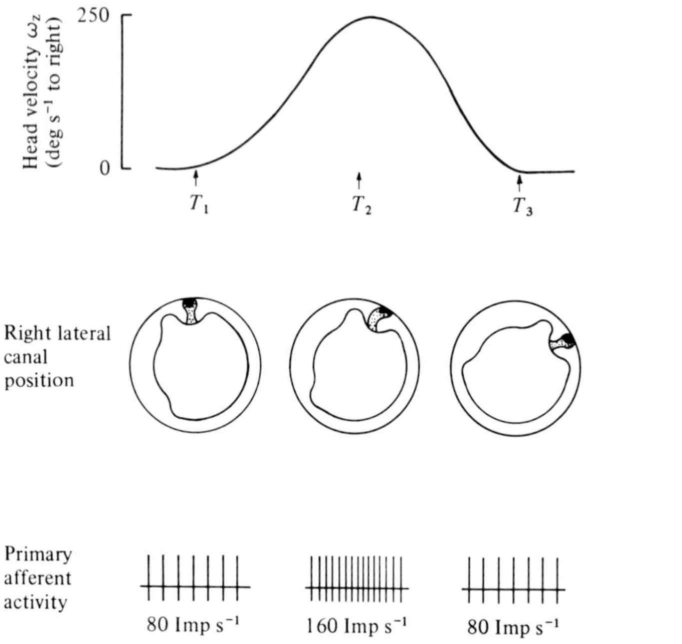

Fig. 16.6. Representation of the angular trajectory of the head during a natural turn to the right in 2 s. The position of the right lateral canal and cupula, and the response of a primary afferent at the beginning (T_1), at peak velocity (T_2) and at the end of the movement (T_3) are shown.

rotation, the cupula is deformed because its motion is resisted by the inertia of the endolymph and there is a small relative movement of endolymph within the duct. Provided the rotational head movement is of short duration, as natural head movements commonly are, the angular deceleration associated with the cessation of the movement drives the cupula back to its resting position. In a particular ampulla all the sensory cells have the same orientation of their kinocilia with respect to the stereocilia (i.e. they have the same directional sensitivity);* so there is an overall increase in afferent activity when the cupula is deflected towards the kinocilium, and the converse when it is deflected away from the kinocilium.

Deflection (or distortion) of the cupula is dependent upon an angular acceleration in the plane of the duct, but the motion of the cupula within the ampulla is heavily damped by viscous forces,

* The directional sensitivity of the sensory cells (determined by the position of the kinocilia) is referred to as 'morphological polarisation' in the literature.

primarily those associated with the flow of endolymph in the narrow duct that must accompany any movement of the cupula within the ampulla. This heavy damping in relation to a weak intrinsic restoring force of the cupula makes the duct–cupula–endolymph system act as a leaky integrator of angular acceleration. Hence, during a normal head movement, in which the duration of acceleration and deceleration is short (< 5 s) in relation to the time constant of integration of the system (*c*. 20 s in man), the cupula deflection (and associated change in activity of ampullary hair cells) is related to the instantaneous *angular velocity* of the head and not to its angular acceleration. In other words, the semicircular canals act as angular speedometers by virtue of being integrating angular accelerometers. It is only when subjected to sustained angular acceleration, an artificial stimulus unlikely to be experienced outside the laboratory, that the afferents signal head acceleration.

It should be noted that the duct–cupula–endolymph system transduces only that angular velocity acting in the plane of the duct. If the plane of the duct lies at an angle θ to the plane of rotation (at an angular velocity, ω) then the angular velocity 'seen' by the semicircular canal is $\theta \cos \omega$. Thus when the plane of the canal is normal to the plane of rotation, and $\theta = 90°$ ($\pi/2$ rad), the receptors are not stimulated by the rotational movement. As each labyrinth comprises three mutually orthogonal semicircular ducts, an angular movement of the head, provided it is of sufficient intensity, will always stimulate the receptors of at least one ampulla of each labyrinth.

The dynamics of the semicircular canals

There is reasonable agreement between neurophysiologial studies of primary afferent activity and mathematical models of the hydrodynamics of the duct–cupula–endolymph system. The methods of investigating the dynamic behaviour owe much to the techniques of electronics and control system engineering where it is common practice to examine the response of a system to sinusoidal forcing functions over a range of frequencies (see Chapters 1, 8 and 13). In order to examine the stimulus–response relationships of the ampullary receptors it is necessary to record from a primary afferent while the experimental animal or preparation (isolated elasmobranch labyrinths have been extensively studied) is exposed to sinusoidal angular oscillation in the plane of the canal whose neural activity is being sampled.

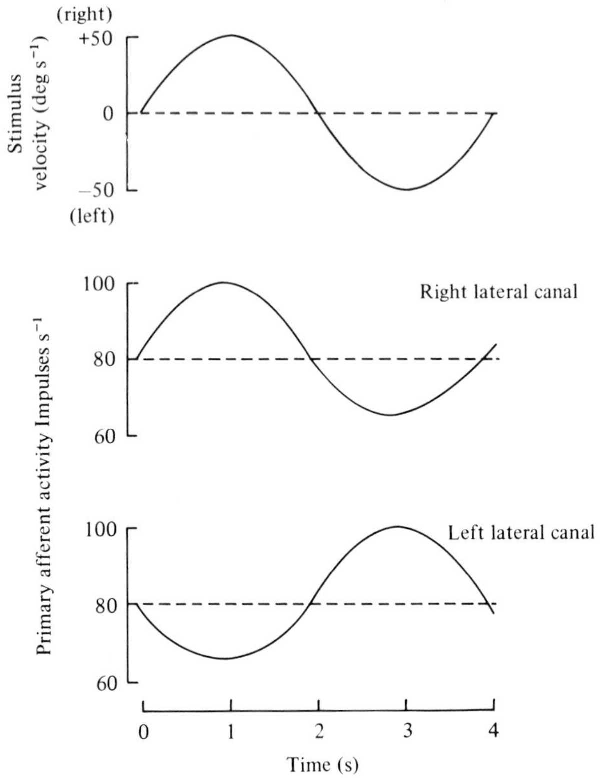

Fig. 16.7. Diagram to show the pattern of primary afferent activity of right and left lateral canal units during an angular oscillation of the head about the *z* axis. Note that there is a larger change in the frequency of firing during the excitatory than during the inhibitory phase of the cycle. (Data from Goldberg & Fernandez, 1971.)

A typical response of a primary afferent from the right lateral canal to oscillation at a frequency and intensity within the 'physiological' range is shown in Fig. 16.7. Note that during the half cycle in which the head was being turned to the right (clockwise when viewed from the vertex) the rate of firing increased above the resting discharge rate; conversely it decreased during the half cycle when rotating to the left. Note also that the maximum and minimum rates of discharge are more or less coincident with the peak velocity to the right and left respectively. In other words, there is little temporal disparity (i.e. no phase difference) between the trajectories of firing frequency and

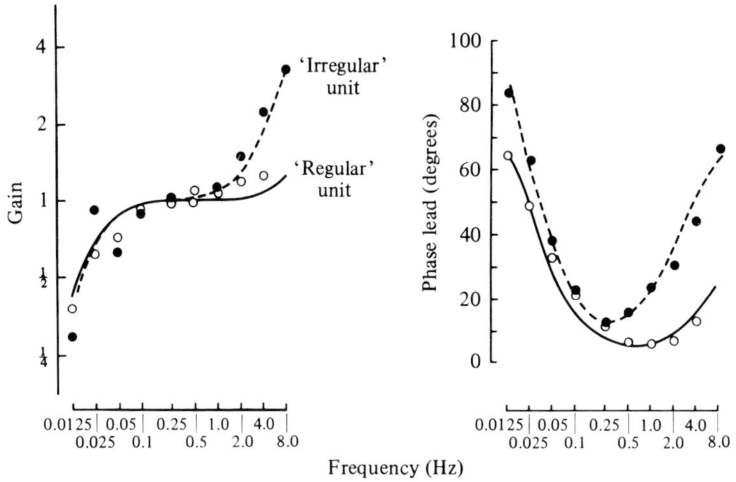

Fig. 16.8. Bode plots of primary afferent activity of two semicircular canal units, one 'regular' (solid line) the other 'irregular' (interrupted line). Both gain and phase are referred to the velocity of the forcing angular oscillation. (Data from Goldberg & Fernandez, 1971.)

stimulus velocity. A recording from the lateral canal on the left side also shows a sinusoidal modulation of activity, but in the opposite sense. It mirrors the response of the right semicircular canal, because rotation to the right causes a deflection of the cupula towards the kinocilium in the right lateral canal and away from the kinocilium in the left lateral canal. This is the general pattern of response of primary afferents from the ampullae; however, individual units have identifiable differences in their resting activity and in their dynamic response. A distinction is made between those units that have a relatively steady resting discharge (tonic or regular units) and those afferents with a more irregular rate of firing (phasic or irregular units).

Measures at different stimulus frequencies of the changes of firing rate per unit change in angular velocity and of the phase relationship between stimulus and response, allow a frequency response or Bode plot to be drawn. For the squirrel monkey this is of the form illustrated in Fig. 16.8. Over the frequency band 0.05–1.0 Hz the sensitivity (or gain) of a 'regular' unit is essentially constant and there is little phase error. The afferent is signalling head velocity. But at frequencies below 0.05 Hz there is a progressive reduction in sensitivity and an increase in phase error (lead) with decreasing frequency. At these low frequencies the afferent signal is more related to head acceleration than head velocity. The corner frequency at which the end organ changes from being a transducer of angular velocity to one of angular acceleration is determined, principally, by the relationship between the intrinsic restoring forces of the cupula and the viscous drag of the endolymph; (in addition, adaptation of the sensory cells to a sustained unidirectional stimulus contributes to the loss of sensitivity at low frequencies). These variables are in part determined by the dimensions of the semicircular duct. Measurements in a wide variety of species have shown that critical duct dimensions (internal radius and radius of curvature) are positively correlated with the mass of the animal. This observation suggests that the frequency response of the semicircular canals is matched to the frequency spectrum of the head movements made by the animal during its normal activity.

At the upper end of the frequency range (*c.* 0.5 Hz) the gain of the system increases, this being more obvious for 'irregular' than for 'regular' units, and there is an associated development of phase advance with increasing frequency. This implies that the transducer becomes increasingly sensitive to the rate of change of velocity (i.e. acceleration) at these higher frequencies. Hydrodynamic theory predicts that the sensitivity of the system should fall and phase lag develop above a certain frequency, governed by the inertia of the endolymph and the viscous drag of the cupula–endolymph system. However, this upper limit to the dynamic response has yet to be determined by experimental studies of primary afferents, though theory, and vestibulo-ocular responses in man, suggest that it is above 10 Hz and probably nearer 30 Hz.

Another widely used method of testing the dynamic response of a system is to examine its response to an impulse. A stepwise change in angular velocity such as is produced by a sudden stop from sustained rotation at constant speed approximates to an impulse of angular acceleration and has long been used, both experimentally and clinically, to assess semicircular-canal function. The response of 'regular' and 'irregular' primary afferents to such a step velocity stimulus in the plane of the innervated canal is shown in Fig. 16.9. With the rapid change in angular velocity there is an almost immediate change in the discharge rate of the ampullary receptor from the resting level as the cupula is deflected. Once the acceleration ceases and the experimental preparation is at rest (or at some other steady angular velocity) the discharge rate begins to return to its

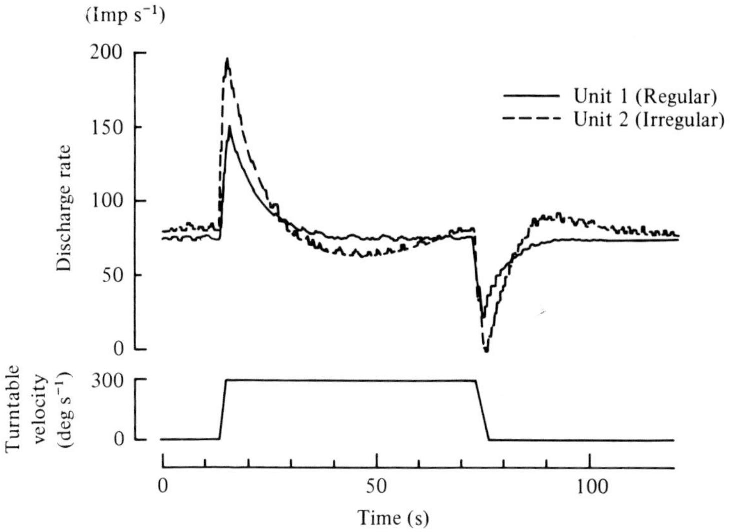

Fig. 16.9. Change of activity of 'regular' and 'irregular' semicircular canal primary afferents in response to a brief angular acceleration to a constant velocity of 300° s^{-1} and subsequent deceleration. (Data from Goldberg & Fernandez, 1971.)

resting level. This approximately exponential decay of activity reflects the slow restoration of the deflected cupula to its rest position and is governed, principally, by the intrinsic elasticity of the cupula and the viscous drag of the endolymph. It may be noted that these same two factors determine the lower 'corner' frequency of the Bode plot. This relationship may be summarised as $\pi/\Delta = T_1 = 1/\omega_1$, where π is the coefficient of viscous drag of the cupula–endolymph system, Δ is the elastic restoring coefficient of the cupula–endolymph system, T_1 is the time constant of decay (s) of impulse response, and ω_1 is the lower break (corner) frequency of sinusoidal response (in rad s^{-1}). In squirrel monkeys T_1 is about 3–5 s, but in man, a more ponderous animal, measures of evoked vestibular responses, rather than of primary afferents, indicate that T_1 is considerably greater, being about 15–20 s. This figure corresponds to a lower 'corner' frequency of about 0.06 rad s^{-1} or 0.01 Hz.

Fig. 16.9 illustrates differences in the behaviour of 'regular' and 'irregular' units: the latter have a greater dynamic sensitivity and are more 'adaptive', as is evidenced by the more rapid decay and subsequent 'overshoot' following the step-velocity stimuli.

The otolith organs; transduction of linear movements

Precise measurements in the pigeon have shown that the densities of the cupula and endolymph do not differ by more than 1 part in 2000. Thus the cupula, coupled as it is to a closed duct of endolymph, should not be deflected by *linear* accelerations of the head. There is some evidence from experiments on amphibians that the discharge of ampullary receptors is not invariant with the orientation of the head to gravity, but the effect is relatively small and its functional value is doubtful. In contrast, the receptors of the saccular and utricular maculae respond rapidly to small changes in the magnitude or direction of the linear acceleration acting on the head. The sensitivity of macula receptors to linear acceleration arises because the compliant cilia support a structure, the otolithic membrane, whose density is more than twice that of the endolymph in which it lies.

Consider the behaviour of an idealised utricular macula when subjected to a linear acceleration in the horizontal plane, such as experienced by a passenger in a car or train that is accelerating from rest (Fig. 16.10). With the onset of the linear motion the head moves, and with it the membranous labyrinth. The dense statoconia of the statoconial (otolithic) membrane also have to be accelerated, and the force required must be transmitted by the cilia of the macular cells and their gelatinous integument. This shearing force, acting parallel to the plane of the macula, causes a minute displacement of the surface of the otolithic membrane relative to the apical surface of the sensory cells and the cilia are deflected from their equilibrium position. Unlike those of a crista, the cells of a given macula vary in directional sensitivity; so some cells will be excited, some inhibited and some not affected at all. The otolithic membrane remains displaced as long as the acceleration continues, but once constant velocity is achieved, the shearing force disappears and the otolithic membrane is restored to its rest position by the intrinsic elasticity of the cilia.

The otolith organs provide information not only about transient linear accelerations but also about the orientation of the head to the earth's gravity, which, dimensionally, is indistinguishable from a linear acceleration (the Einstein equivalence principle). Displacement of the otolithic membrane occurs as a function of head position because the macula cilia must support the effective weight of the statoconial layer. When the plane of a macula is horizontal, with the

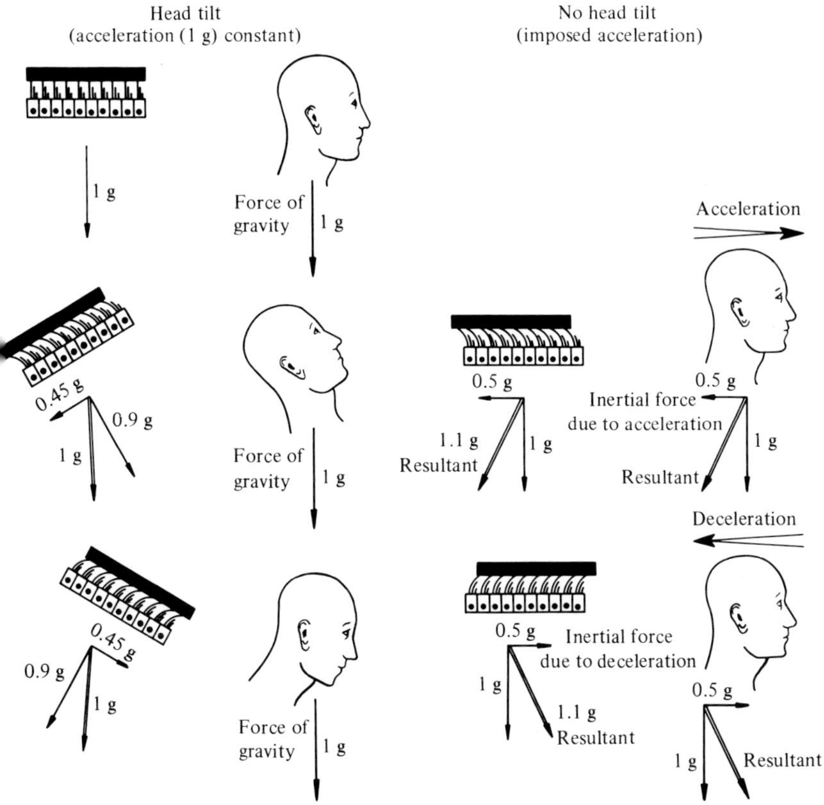

Fig. 16.10. Comparison of the forces acting on the head during tilt about the *y* axis, and during linear acceleration and deceleration in the *x* axis of the head. The displacement of the otolithic membrane of the utricular macula is purely diagrammatic and grossly exaggerated.

cilia pointing either directly upwards or downwards, there will be little distortion of the cilia because of their intrinsic stiffness to axial forces.* But when the plane of a macula is tilted out of the horizontal the cilia will be deflected, because the effective weight of the statoconia appears as a force in shear. The magnitude of this shear

* The sensitivity of the macular receptors to forces acting in shear can be simply illustrated by placing your hand, to represent the statoconial membrane, on the surface of a large soft brush whose bristles represent the cilia of the sensory cells. The bristles are easily deflected by a small force parallel to the surface of the brush (that is a force in shear) but a force normal to the surface must be much greater before the bristles are bent.

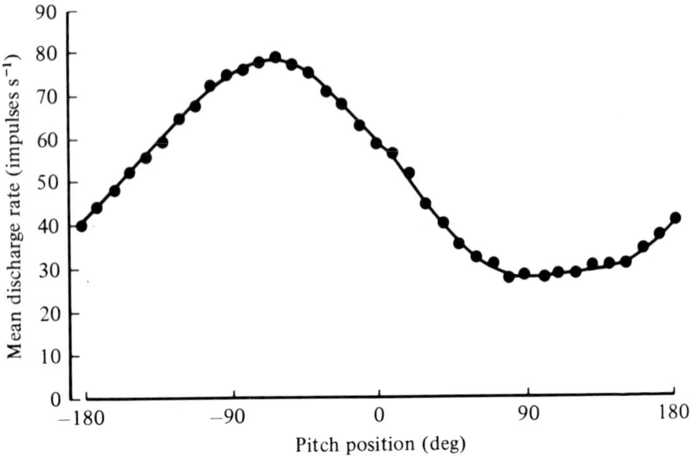

Fig. 16.11. Alteration of primary afferent activity of an otolithic unit during a slow rotation of the head in pitch (i.e. about the y axis). (Data from Loe, Tomko & Werner, 1973.)

force (F) is given by the expression $F = M\mathbf{g} \sin \theta$ (where M is the effective mass of the statoconial membrane, $\mathbf{g}$ is acceleration of gravity and θ the angular deviation of the head from the vertical position). This simple expression is validated by the behaviour of primary afferents, which show an approximately sinusoidal modulation of activity as the head is slowly rotated about a horizontal axis (Fig. 16.11).

Dynamics of the otolith organs

Recordings from the macular primary afferents have shown that many of these receptors, like those of the crista, have a resting discharge though those of the otolith organs show a greater variability both in the regularity of firing and in their response to stimulation. Nevertheless, on exposure of the preparation to a sinusoidally varying linear acceleration there is a well defined modulation of activity from which measures of sensitivity (change in discharge rate per unit change in acceleration) and of the phase relationship of the response to the stimulus can be determined. Bode plots (Fig. 16.12) illustrate the dynamic characteristics of two units, one a regularly firing ‘tonic unit’, the other a ‘phasic’ unit having an irregular resting discharge. Both units exhibit an increase in sensitivity with frequency which means that their activity is determined both by the

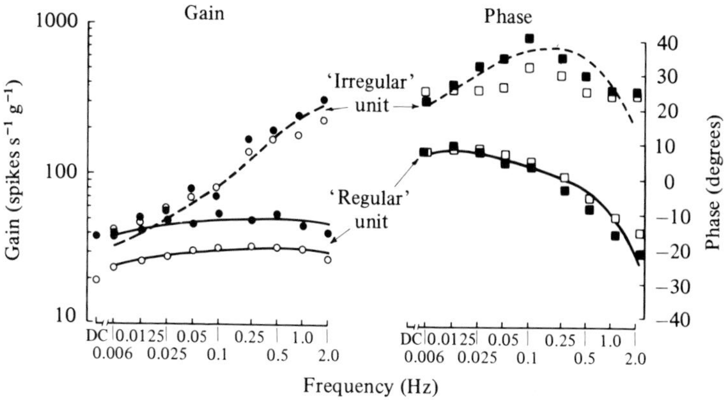

Fig. 16.12. Bode plots of primary afferent activity of two otolithic units, one 'regular' (solid line), the other irregular (interrupted line). Both gain and phase are referred to the acceleration of the forcing function. Solid and open circles identify values obtained during excitatory and inhibitory sine waves respectively. (Data from Goldberg & Fernandez, 1976.)

acceleration and by the rate of change of acceleration (i.e. 'jerk'). It is the 'irregular' units that show the greater change in gain and greater phase advance with increasing frequency, so they are considerably more sensitive to 'jerk' than the 'regular' units.

Differences in the dynamic characteristics of units are also shown by their responses to step acceleration stimuli (Fig. 16.13). 'Regular' units show only a small overshoot in activity before the frequency of discharge settles at a level appropriate to the magnitude of the imposed acceleration in shear. Thus as transducers of acceleration they show relatively little adaptation or jerk sensitivity. On the other hand, there is considerable alteration of the activity of 'irregular' units immediately following the stimulus. Indeed, some units signal only a change in acceleration and have negligible sensitivity to steady-state acceleration.

The preceding characterisation of macular units, as predominantly static or phasic receptors, is an oversimplification which should not obscure the wide variation in dynamic characteristic between otolithic units. The older classification of labyrinthine receptors, in which the semicircular canals and otoliths were identified as having 'dynamic' and 'static' functions respectively, disregards the essential role of the otoliths as detectors of a *change* in head position, whether this be a translational movement or a reorientation of the head to gravity.

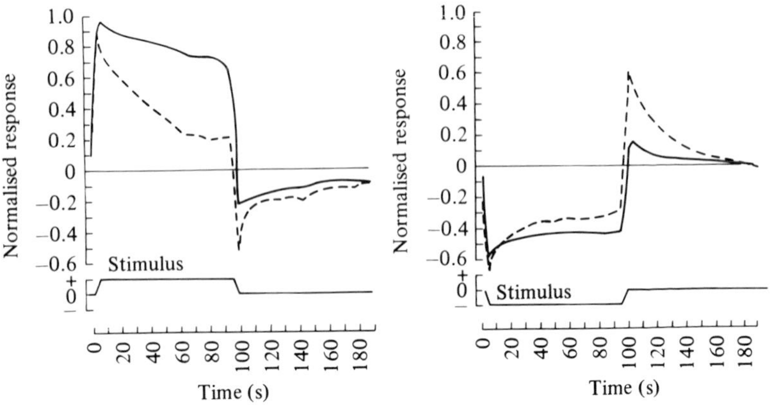

Fig. 16.13. Change in primary afferent activity of otolithic units in response to excitatory (on left) and inhibitory (on right) linear acceleration stimuli having a trapezoidal trajectory with onset and offset durations of 4–5 s. The graphs show the mean normalised responses of 21 'regular' units (solid line) and 14 'irregular' units (interrupted line); normalisation was based on the initial response to the excitatory stimulus. (Data from Goldberg & Fernandez, 1976.)

16.3. CENTRAL VESTIBULAR PATHWAYS

The projections of primary vestibular afferents to the vestibular nuclei and cerebellum have been mapped in some detail, as have the ramifications of the ascending, descending and decussating projections of the vestibular nuclei and cerebello-vestibular pathways. The neuro-anatomical picture is one of considerable complexity, of which only an outline (Fig. 16.14) can be given here.

First-order afferents from the cristae of the semicircular ducts and the saccular and utricular maculae terminate in the ipsilateral vestibular nuclear complex and send collaterals that traverse the superior vestibular nucleus to enter the cerebellum where the majority terminate as mossy fibres in the ipsilateral flocculus, nodulus and ventral part of the uvula. Other primary afferents terminate in the ventral and dorsal paraflocculus and in the dentate nucleus; in addition, a small number project to the contralateral 'vestibular' cerebellar cortex.

The vestibular nuclear complex is composed of four principal nuclei (lateral or Deiter's, medial, superior, and descending) and several minor cell groups. The distribution of primary vestibular

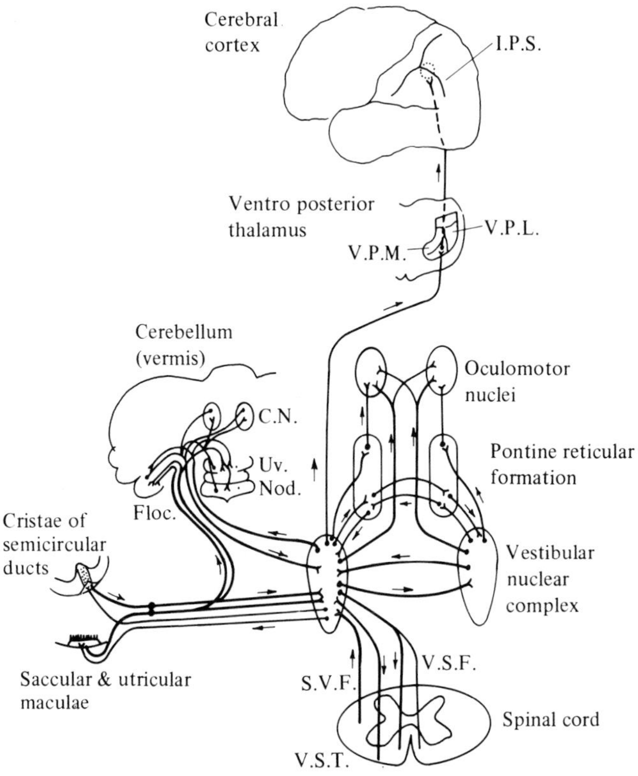

Fig. 16.14. Diagrammatic representation of major pathways of the vestibular sensory system. Abbreviations: I.P.S., Interparietal sulcus; V.P.M., V.P.L., medial and lateral parts of the ventroposterior thalamus; C.N., cerebellar nuclei; Uv., uvulus; Nod., nodulus; Floc., flocculus; S.V.F., spino-vestibular fibres (ascending in dorsal spino-cerebellar tract); V.S.F., vestibulo-spinal fibres (descending in sulcomarginal fasciculus); V.S.T., vestibulo-spinal tract.

afferents within these integrative centres is circumscribed insofar as the dorsal part of the lateral nucleus and the medial part of the medial nucleus are devoid of direct afferent input. The rest of the nuclear complex receives a varied distribution of primary afferents such that certain areas, like the superior vestibular nucleus, have a unique representation of canal and otolithic afferents, whereas in other areas inputs from both groups of receptors converge. In addition, afferent projections are received from the cerebellar cortex and cerebellar nuclei, from the reticular formation and from ascending spinal tracts.

Primary afferents do not decussate, but second-order neurons from the contralateral vestibular nuclei project widely within the vestibular nuclear complex. Some of these commissural fibres make inhibitory connection with corresponding cell groups (e.g. a cell excited by afferents from the ipsilateral lateral canal is inhibited on excitation of the contralateral lateral canal). This reciprocal innervation is the substrate for the differential (i.e. push–pull) operation of complementary pairs of canals and ensures that when there is equality of the afferent discharge from each vestibular apparatus, sensations of turning and reflex responses are not evoked.

The vestibular nuclear complex is thus a complicated integrative centre in which signals from the vestibular apparatus, conveying information about head motion and orientation, are combined with signals from joint receptors related to the position and movement of the trunk and limbs. Information about movement from retinal receptors also reaches the vestibular nuclear complex, probably via cerebellar afferents, though this is but a facet of the powerful control exercised by the cerebellum over the vestibular nuclei.

Efferent fibres arising within the vestibular nuclei form ascending, descending and commissural fibre systems. These pass to the vestibular cerebellar cortex and fastigial nuclei, to the reticular formation, and to mesencephalic nuclei. A large fibre system, arising principally in the lateral vestibular nucleus, projects caudally via the vestibulo-spinal tract on to spinal motoneurons, and mediates the equilibratory reflexes of vestibular origin. Another group of fibres from the superior and medial vestibular nuclei ascend by way of the medial longitudinal fasciculus to the oculomotor nuclei of the III, IV and VI cranial nerves. These are the efferent pathway of the vestibulo-ocular reflex, which stabilises the eyes and preserves vision during rapid head movements.

Neuro-anatomical evidence for ascending fibre systems rostral to the posterior commissure is equivocal. However, the demonstration of a short latency (2.5 ms) evoked potential in the central posterio-inferior nucleus of the thalamus on stimulation of the contralateral vestibular nerve, implies that second-order vestibular neurons relay directly to the thalamus.

In the cerebral cortex of the rhesus monkey, responses evoked by vestibular stimulation are localised to a small area of the post-central gyrus at the lower end of the intraparietal sulcus. This is at the face level of the somatosensory field in somatic area 2 where the vestibular field (area 2v) has a distinct cytoarchitectonic structure. The majority of neurons in the vestibular projection have a short latency (4 ms) to vestibular nerve stimulation and, like those in the thalamus, show convergence of vestibular and kinaesthetic (limb movement) afferents. It thus seems likely that the primary vestibular projection to area 2v is by a thalamo-cortical pathway.

The location of the vestibular field in the human cerebral cortex is probably similar to that in the monkey, for it has been reported that electrical stimulation of the upper lip of the intraparietal sulcus elicits sensations of rotation, bodily displacement and apparent motion of the visual scene. Likewise, vertigo and other typically vestibular sensations feature in the aura of the epileptic fits suffered by patients with focal lesions in this cortical area.

16.4. PERCEPTION OF BODY ATTITUDE AND MOTION

The vestibular apparatus, despite its elegant functional adaptation as a transducer of head motion and attitude, is but one receptor system providing the organism with cues that allow it to determine its orientation and motion within a gravito-inertial reference system. Thus in any experimental situation where the contribution of vestibular mechanisms to the perception of body orientation (or other sensory experiences) is to be assessed, it is desirable to exclude or control these non-vestibular cues. The exclusion of vision presents no problem, but controlling the input from somatosensory receptors, notably cutaneous receptors, pressure receptors in supporting tissues and capsular joint position receptors, is more difficult. Yet all these mechanoreceptors are influenced by forces acting on the body. Apart from a few studies carried out on patients with high spinal transections or on subjects totally immersed in water, most psychophysical experiments involving changes in the force environment examine the combined response of the otoliths and the more generally distributed mechanoreceptors. Selective stimulation of the semicircular canals is more easily achieved, because the subject, seated on a turntable free to rotate about a vertical axis, experiences little change in his force environment, unless he is seated off-centre or the turntable is accelerated at a high rate.

Perception of angular motion

Psychophysical data on the perception of angular motion accord quite closely with end-organ dynamics, insofar as the threshold for the detection of an angular movement of short duration is determined

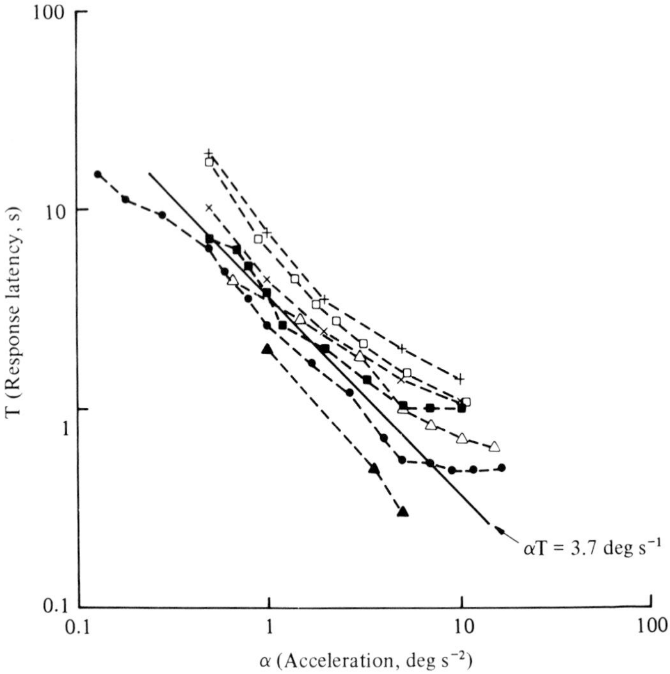

Fig. 16.15. Relationship between the time taken to detect rotation, and the magnitude of the step angular accelerations to which the subjects were exposed. The different symbols give the results of experiments in which angular acceleration was applied about different body axes, by various workers.

by the magnitude of the change in velocity rather than of the acceleration achieved by the stimulus. Measures of the time (t) taken by subjects to detect motion when exposed to different sustained angular accelerations (α) have shown that the product αt is a constant, provided the duration of the stimulus is less than about half the long time constant (T_1) of the semicircular canals (Fig. 16.15). The product αt has the dimension of angular velocity and has a mean value of $3.7° s^{-1}$.

Sustained rotational stimuli with prolonged angular accelerations do not occur during natural locomotor activity, but can be experienced in aircraft or space vehicles. Laboratory experiments have shown that, when such long duration stimuli are presented, the sensory threshold is determined by the magnitude of the angular acceleration, rather than the velocity change, achieved. The median value of the threshold for angular accelerations about the *z* axis of the head is 0.32° s^{-2} with a range of 0.05–2.2° s^{-2}.

Most natural head movements involve phasic changes in angular velocity that are well above detection threshold. The passive nodding of the head that accompanies running or walking commonly has a peak angular velocity in excess of $\pm 10°$ s^{-1} at 1–2 Hz; the volitional head movements, such as occur when the head is moved to allow fixation on an object detected in the peripheral visual field, usually have a peak velocity of at least 100° s^{-1} and may reach angular rates as high as 400° s^{-1}.

Perception of supra-threshold rotational movements is reasonably precise, even though the subject is required to integrate velocity information from the canals and report angular displacement. There is some variation between subjects in the slope of the line that relates perceived to imposed angular displacement, but there is no significant deviation from a linear stimulus–response relationship. Other studies, employing step velocity stimuli, have demonstrated that the subjective scaling of supra-liminal stimuli follows a linear, as opposed to a logarithmic or power, law; in this respect the vestibular sensory system is unlike most other sensory modalities (compare Chapter 13).

Perception of linear motion

The dynamic response of the otolith organs suggests that the detection threshold for translational (linear) movements would be determined by some combination of acceleration and rate of change of acceleration (jerk) of the stimulus. When experimental subjects are exposed to low-frequency sinusoidal linear oscillations the detection threshold, expressed as the peak acceleration of the stimulus, does indeed decrease with increasing frequency. The average threshold value is about 0.2 m s^{-2} at 0.2 Hz and falls to about 0.03 m s^{-2} at 1 Hz. The slope of the line relating threshold to frequency is of similar magnitude, but of opposite sign, to that relating the average sensitivity of otolithic units to frequency. This implies that jerk contributes to, but does not dictate, the sensory threshold for this type of oscillatory stimulation. Threshold data at frequencies above 1 Hz are highly discordant, probably because of differences in the way cutaneous receptors were stimulated by the vibration in the varied test situations.

The frequency dependence of the detection threshold would suggest that the detectability of a translational motion, starting from

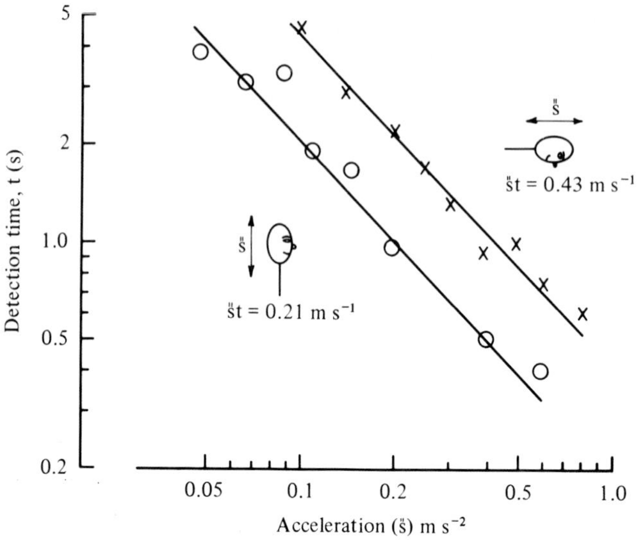

Fig. 16.16. Relationship between the time taken to detect motion and the magnitude of the step linear accelerations to which the subjects were exposed. Crosses and circles indicate data from different workers.

rest with a constant acceleration, would also be determined by the magnitude of the acceleration and its rate of onset; but, surprisingly, this is not so. Experiments, analogous to those on the detection of step angular acceleration stimuli, except that subjects were accelerated along a vertical or horizontal track rather than a turntable, have shown that the detection threshold is determined by the velocity achieved rather than by the rate of change of velocity (Fig. 16.16). As with the detection of angular acceleration steps, the time to detect the motion (t) is inversely proportional to the acceleration ($\ddot{s}$), such that $\ddot{s}t = k$. The linear velocity (k) has a magnitude of 0.2–0.4 m s^{-1} for accelerations acting in the x and z body axes. Somewhat higher values of z-axis threshold were found in subjects who were accelerated on a horizontal track than in those who were accelerated up or down a vertical track.

The ability of the perceptual mechanism to extract the changing component of a linear motion stimulus from the steady background acceleration of gravity is illustrated by the sensations of movement reported by subjects exposed to linear oscillation on a horizontal track. Subjects without functional labyrinths and without prior knowledge of the motion of the device experienced a cyclically varying tilt as they sensed the changing direction of the resultant acceleration vector. But normal subjects accurately perceived the plane and direction of the motion, as well as its velocity and amplitude, though at low frequencies (0.1 Hz) detection of the reversal of motion was phase advanced upon the stimulation by up to 90° ($\pi/2$ rad).

Perception of the vertical

Studies of the ability of man to perceive his orientation to the gravitational vertical, or to a resultant linear acceleration vector (as in a centrifuge), are more numerous than those in which dynamic linear motion stimuli were employed. Experiments carried out using a seat free to tilt in either the x or y axis have shown that the mean positioning error was, typically, less than 1° from the true vertical. The arc of uncertainty, in which 75% of the judgements fell, was $\pm 2.2°$ for both x and y axis tilts, so it may be inferred that the threshold for detection of tilt from the vertical is of the order of 2°, an angular deviation from the vertical that produces a shear acceleration of 0.35 m s^{-2} in the plane of the utricular macular.

It is tempting to take these data as indicants of otolith function, but experiments, on patients with defective vestibular function and on normal subjects immersed in water, indicate that in the absence of vision, somatosensory cues play an important, if not dominant, role in the perception of orientation to the gravitational vertical. When subjects were immersed in water the arc of uncertainty was $\pm 31°$ in pitch and $\pm 27°$ in roll. The variability is thus more than an order of magnitude greater than that found in those tilt-chair studies (see above) in which no attempt was made to reduce somatosensory cues. In contrast, the perception of the vertical by labyrinthine defective subjects, not deprived of pressure cues by water immersion, is only slightly inferior to that of normal subjects.

16.5. VESTIBULO-OCULAR RESPONSES

Earlier it was pointed out that one of the important functions of the vestibular apparatus was to preserve vision when the head moves. The essential compensatory nature of the eye movements evoked by stimulation of vestibular receptors was recognised more than a century ago, though it is only in the last decade that the ability of the vestibulo-ocular 'reflex' accurately to stabilise the eyes during rapid, high-frequency head movements has been appreciated.

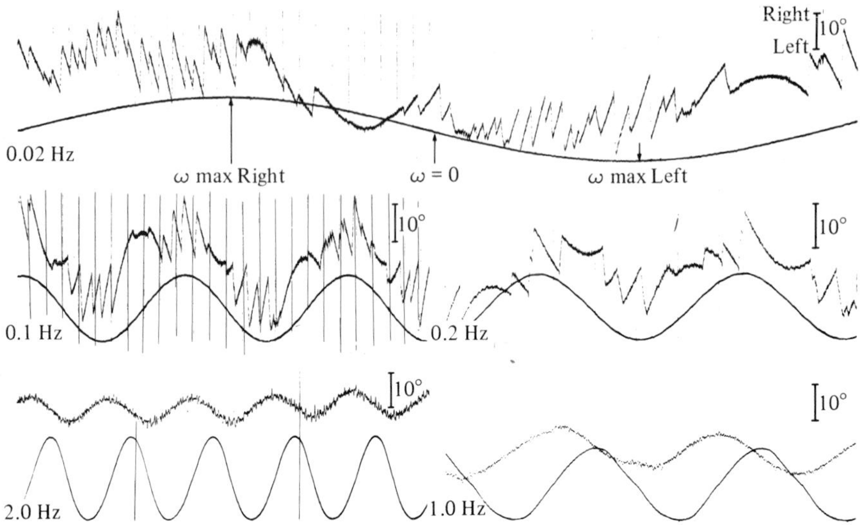

Fig. 16.17. Lateral eye movements of a subject exposed to yaw (z) axis angular oscillation in darkness. The upper trace in each section of the figure shows eye displacement, recorded by electro-oculography; the lower trace is the stimulus velocity with peak values of $\pm 30° \text{ s}^{-1}$. The vertical time bars occur at 1 s intervals.

Adequate stimulation of either the otoliths or the semicircular canals produces eye movements but only rotational movements, stimulating the semicircular canals, engender eye movements that effectively compensate for the head movement. Essential features of the vestibulo-ocular responses to angular motion stimuli are shown in fig. 16.17. Note that at high frequencies of oscillation (1 and 2 Hz) the eye movement is a sinusoidal displacement that mirrors the stimulus; as the head turns to the right the eye moves to the left, so as to stabilise the direction of gaze in space and thus preserve visual acuity. At lower frequencies of oscillation, which entail a larger angular displacement of the head, the oculomotor response is more complex; the slow, compensatory movements are interrupted by fast eye movements, saccades,* which displace the eye in the opposite direction. During each slow phase of the eye movement the eye is space-stabilised, but with each saccade there is a change in gaze position.

An alternating pattern of slow and fast eye movements is called

* Saccade; from the French, a jerk or jolt; see also Chapter 11.

nystagmus and may be readily observed in a subject who is suddenly stopped after he has been spinning for some time (say to the right) at a constant speed. On stopping, the semicircular canals will signal a turn to the left, so the slow phase of the nystagmus will be compensatory (i.e. to the right) with fast phase to the subject's left. Since the vestibular signal produced by the stopping stimulus is inappropriate, the eye movements are also inappropriate: vision is blurred and there is apparent movement of the visual world because of movement of the retinal image. If there is an object upon which the subject can fixate, the nystagmus is suppressed and normal vision restored after a few (5–10) s, though this suppression by fixation may be severely impaired by drugs (ethanol being a common example) or by neurological lesions, in particular those affecting the cerebellum. The study of post-rotational nystagmus can be a valuable aid in the diagnosis of vestibular and neurological disorders.

The functional benefit afforded by the vestibulo-ocular reflex is that it ensures foveal fixation of an object fixed in space during angular movements of the head having a frequency or velocity beyond the dynamic range of the fixation reflex alone. The ability of man, with head still, to discriminate fine detail of a visual target oscillating at 0.5 Hz is already impaired, and at 2 Hz the target is just a blur. In contrast, when the head is oscillated (with a comparable relative angular velocity between head and visual target) the target is seen clearly with little loss of acuity until a frequency of 9–10 Hz is exceeded. As the dominant frequencies of the involuntary movements of the head during normal locomotor activities lie in the 2–4 Hz domain, the functional benefit of the vestibulo-ocular response is apparent.

Stimulation of the otolith organs also evokes eye movements though these, unlike the canal-dependent responses, are not truly compensatory. The eye movement is in a compensatory direction and in the appropriate plane, but it is rarely more than a tenth of the head movement in any plane of tilt.

The automatic nature of the eye movements induced by vestibular stimulation and the di-synaptic pathway between end-organ and oculomotor nuclei rather suggests that the vestibulo-ocular mechanism is a simple reflex. In comparison with other neural control systems, it is; nevertheless, the mechanism exhibits considerable neural processing and plasticity. The reader may care to speculate upon how the velocity-coded signal from the semicircular canals is integrated, with respect to time, to produce a pattern of excitation of the oculomotor nuclei appropriate to the required eye position. A second astonishing aspect of the vestibulo-ocular response is its attenuation and ultimate reversal under experimental conditions such that it no longer stabilises the image on the retina when the head rotates. In these experiments, human subjects were persuaded to wear spectacles containing a prism system that reverses the seen world, so that objects to the left appear to the right and vice versa. After a difficult adaptation period lasting more than a week, the subjects could move around almost normally, making all the appropriate visual responses. To ensure stability of the retinal image when the head was rotated, the eye now had to rotate in the direction opposite to that normally required; and measurements of eye rotation in the laboratory showed that these new compensating 'reflexes' had in fact been achieved.

16.6. MOTION SICKNESS

Motion sickness is not a pathological condition but a normal response of man, and many animals, to certain types of motion stimuli. Typically, exposure to provocative motion engenders first a slight feeling of malaise, then nausea of increasing severity and eventually vomiting. These symptoms are commonly accompanied by feelings of warmth, sweating and pallor, and more variably, by headache, dizziness, drowsiness and depression. This gallimaufry of signs and symptoms constitutes the motion sickness syndrome.

There are many different types of provocative motion (corresponding for example to sea sickness, air sickness, swing sickness, camel sickness and, more recently, space sickness) but what they have in common is the generation of discordant information from the sensory systems transducing body orientation and motion. The mismatch of sensory cues can consist in the concurrent signalling of incompatible information by the semicircular canals and gravireceptors, or more generally in the failure of sensory cues to correlate with those expected from past experience.

The concept that motion sickness is caused by discordant motion cues was first put forward in the nineteenth century. However, the observation that deaf mutes and others without functional labyrinths were immune to motion sickness, as were dogs deprived of their vestibular cerebellum, fostered the idea that vestibular 'overstimulation' was the principal aetiological factor. The deficiency of this theory is that it cannot explain why certain quite weak vestibular stimuli (e.g. cross-coupled stimulation; see below) induce motion sickness, whereas much stronger stimuli (e.g. repeated stops from rotation at high speed) evoking a much larger change in afferent activity, have little effect. Nor does it explain the reduction in the susceptibility – the adaptation – that is a common feature of prolonged exposure to provocative motion, nor the recurrence of sickness – the *mal de débarquement* – seen in individuals who, having adapted to an atypical motion environment, return to a familiar one.

A prime example of a canal–otolith mismatch is afforded by the cross-coupled* stimulation of the canals that occurs when a rotational movement of the head is made about one axis while the subject is rotating at a steady velocity about another axis. Stimuli of this type have been used extensively in laboratory studies of motion sickness and are a cause of sickness in aircraft as well as fairground roundabouts and other so-called amusements. A simplified representation of cross-coupled stimulation produced by a head movement in the coronal plane (i.e. about x-axis of the head) during steady rotation about a vertical (z) axis is shown in Fig. 16.18. The three pairs of semicircular canals are represented as just three canals orientated in the coronal, sagittal and transverse planes of the head to transduce roll, pitch and yaw motions respectively. Now, with steady rotation about the z-axis and the head vertical, there is no stimulation of the canals and no sensation of rotation; but as soon as the head is tilted in roll (say through 90°) to the left-ear-down position, the sagittal (pitch) canal is brought into the plane of rotation, and the transverse (yaw) canal is taken out of the plane of rotation. There is an increase in the angular velocity in the plane of the sagittal canal, which accordingly signals rotation about the vertical axis; the transverse canal, however, experiences a decrease in velocity, just as if the z-axis rotation had suddenly stopped. Thus, an illusory sensation of rotation in the transverse plane of the head, is evoked and as such is reported as a sensation of pitching forward because the head is in the left-ear-down position. It is this erroneous signal of rotation in pitch that conflicts with information from the otoliths and other gravireceptors, for they signal that the head is tilted to the left and is not rotating in the vertical plane. Of course, if the head movement is made as rotation about another axis begins, both angular movements are sensed veridically; there is no conflict with otolithic information and no motion sickness.

The vertigo and nausea following the intemperate consumption of

* Sometimes referred to as Coriolis stimulation.

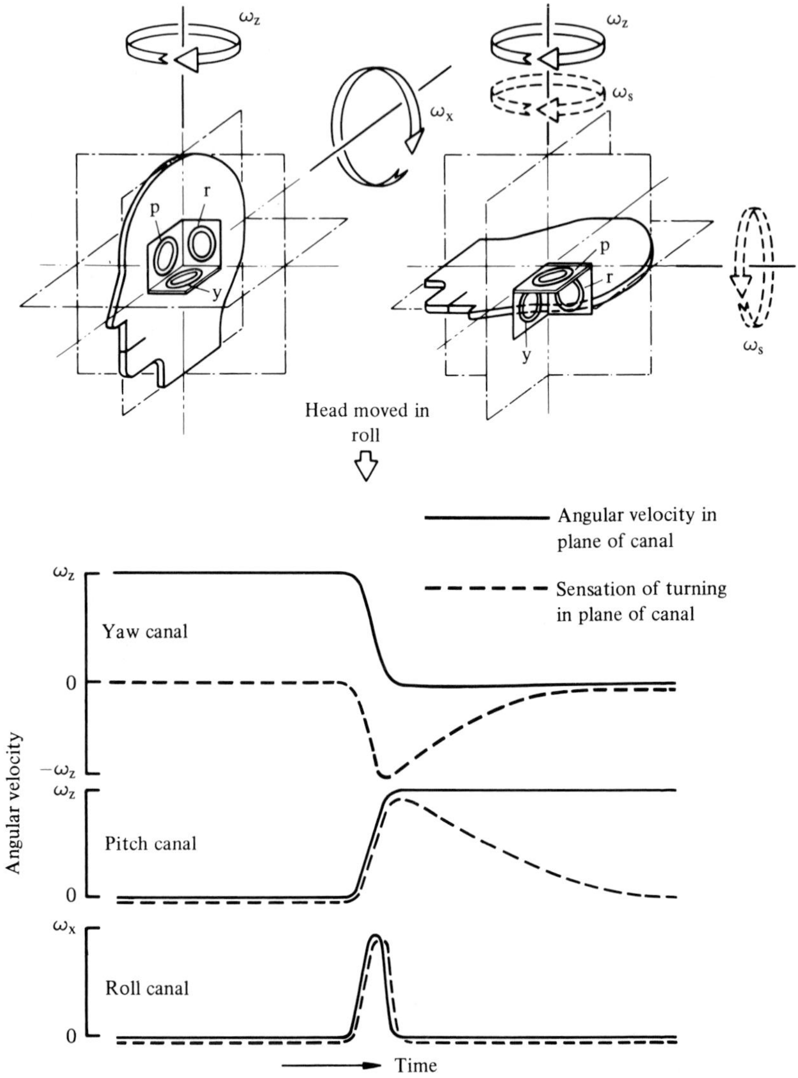

Fig. 16.18. Cross-coupled stimulation. Diagrammatic representation of the stimuli received by three 'semicircular canals' (orientated in the orthogonal planes of the head) during a 90° tilt of the head in roll (about x axis) while rotating at constant speed about a vertical (z) axis. The three idealised canals are identified as pitch (p), yaw (y) and roll (r). Note that on completion of the head movement there is an illusory sensation (ω_s) of rotation about the z axis of the head which is perceived as a pitch-forward movement.

alcohol (ethanol) may be caused, at least in part, by a canal–otolith mismatch. There is evidence that alcohol alters the relative density of cupula and endolymph, such that the semicircular canals lose their normal insensitivity to gravity. The atypical discharge of ampullary receptors, which varies with head position, is responsible for the characteritic, and perhaps familiar, position-dependent vertigo and nystagmus.*

The mismatch consequent to the complex motions of many types of transportation is caused primarily by atypically slow changes in the direction and magnitude of the linear acceleration vector which engender gravireceptor signals that are not accompanied by correlated signals from the canals. For example, the occupant of a car that is travelling fast round a bend experiences a substantial y-axis acceleration with rotation of the resultant acceleration vector through an angle substantially greater than the rotation of the head. The mismatch may be further intensified by active or passive head movements which, in the atypical force environment, cause otolithic signals differing from those produced by the same head movement in a normogravic (1 **g**) environment.

A mismatch between visual and vestibular cues can also induce motion sickness and in many forms of transportation is a contributory, though not an essential, aetiological factor. For example, vestibular mechanisms allow the passenger in a cabin of a boat or aircraft to sense its motion, but visual cues carry no information about the motion of the vessel, only about the relative movement of the occupant within the cabin.

16.7. ADAPTATION

Passing reference was made above to the modification of vestibular responses that occurs when stimuli are prolonged or repeated. These adjustments, governed by the character of the sensory stimulus and the task to be performed, are behaviourally advantageous to the organism. Various types of modification can be distinguished by their differing temporal and causal characteristics and this suggests that several distinct mechanisms are involved. At the highest level of integration there are the rearrangements of sensory–perceptual and

* Deuterium oxide (heavy water) taken in sufficient quantity also causes positional vertigo and nystagmus, though significantly the direction of the nystagmus and the sensation of turning, for a particular orientation of the head to gravity, is opposite to that of the positional responses seen in the initial phases of alcohol intoxication.

sensory–motor relationships that develop progressively following an alteration in the patterning of afferent activity. The time scale of this rearrangement is measured in days, even weeks, and depends on repeated or prolonged exposure to the atypical sensory input. The first 3 to 4 days of adaptation to abnormal motion environments, such as is exhibited by those who attempt to carry out normal activities whilst living in a rotating room, aboard a heaving ship, or in the weightless environment of orbital flight, is characterised by the progressive decline of non-veridical sensations, the replacement of inappropriate motor responses and the elimination of motion sickness. Animal experiments suggest that these plastic changes, sometimes referred to as habituation or central compensation,* are dependent upon the integrity of the floccular lobe of the vestibular cerebellum and its projections via the inferior olive to vestibular and other brain-stem nuclei.

In contrast, there are short-term adaptive changes that are characterised by suppresion of behaviourally inappropriate responses, one example being the accelerated decay of post-rotational nystagmus and vertigo when veridical visual cues are present. These more or less immediate response modifications are probably mediated by inhibitory projections from the vestibular cerebellum to the vestibular nuclei. The vestibular efferent system might also be involved as this, too, appears to be principally inhibitory in its action. However, alteration of end-organ sensitivity by efferents cannot be the sole adaptive mechanism, as the various elements of the response (like nystagmus, postural adjustment and sensations of turning) to a particular stimulus commonly exhibit adaptation at differing rates as well as differential degrees of suppression with repetition of the test stimulus.

These two processes, the one that modifies rapidly the adaptive coupling between end-organ and response, and the other, a slower process (akin to learning) that is responsible for sensory rearrangements, are not mutually exclusive; for in those situations where a stereotyped stimulus is repeated, the adaptation may be quite specific without any generalised alteration of vestibular stimulus–response relationships. One example of such specific adaptation is the ability of the trained ice skater or ballet dancer to maintain balance and to suppress normal post-rotational eye movements, on suddenly stopping from a high-speed spin. But in darkness the post-rotational

* This term is commonly used to describe the adaptation exhibited by patients with unilateral loss or derangement of vestibular function.

nystagmus and sensations are little different from those of an untrained subject: the skater falls to the ground if spin recovery is attempted whilst blindfolded.

There is yet another adaptive mechanism, shared by the vestibular system with other modalities, which involves the selection of sensory information to be processed – to be attended to – by cognitive centres within the nervous system. Limitation of the informational load, by some form of attentional filter, is required because of the limited capacity of the central processor. Thus sensory information generated by unchanging or repeated stimuli becomes classified as familiar or inconsequential and loses priority in the competition for cognitive processing. The sensory information may still make an important contribution to the behaviour of the organism, but now the processing is relegated to more stereotyped routines in non-cognitive centres. This would account for our lack of awareness of vestibular information in everyday life. As noted at the beginning of this chapter, it is only when we are exposed to unusual motion stimuli, or when labyrinthine function is deranged by disease, that vestibular sensations obtrude. Yet all the time, the specialised mechano-receptors of the vestibular apparatus are providing the central nervous system with information about the motion and attitude of the head that is integrated with visual, proprioceptive and somatosensory information to maintain postural equilibrium and ocular stability.

16.8. SUGGESTIONS FOR FURTHER READING

General references

Guedry, F. E. & Correia, M. J. (1978) *The Vestibular System.* Chapter 9. Basic Biophysical and Physiological Mechanisms, and Chapter 10 Vestibular Function in Normal and Exceptional Conditions. In *Handbook of Behavioural Neurobiology.* Vol. 1. Sensory Integration, ed. R. B. Masterton, pp. 311–407. New York & London: Plenum Press. (A useful review of basic and applied topics.)

Howard, I. P. & Templeton, W. B. (1966) *Human Spatial Orientation.* New York & London: J. Wiley & Sons. (A good source for earlier work on relevant psychophysics; a new edition is in preparation.)

Kornhuber, H. H. (ed.) (1974) *Handbook of Sensory Physiology*, vol. 6. Vestibular System. Part 1. Basic Mechanisms. Part 2. Psychophysics, Applied Aspects and General Interpretations. Heidelberg: Berlin: New York. Springer. (A massive, pp. 1356, text in two volumes. Parts C, D and E are relevant to this chapter.)

Wilson, V. J. & Melville Jones, G. (1979) *Mammalian Vestibular Physiology*, pp. 365. New York & London: Plenum Press. (A comprehensive and readable review of the relevant neuroanatomy and neurophysiology.)

Special references

Microscopic anatomy of vestibular apparatus. Lindeman, H. H. (1969) Studies on the morphology of the sensory regions of the vestibular apparatus. *Ergebnisse der Anatomie*, **42**, 1–113.

Neurophysiology of primary afferents. Goldberg, J. M. & Fernandez, C. (1971) Physiology of peripheral neurons innervating semicircular canals of the squirrel monkey, Parts 1, 2, 3. *Journal of Neurophysiology*, **34**, 635–84. Fernandez, C. & Goldberg, J. M. (1976) Physiology of peripheral neurons innervating otolith organs of the squirrel monkey, Parts 1, 2, 3. *Journal of Neurophysiology*, **39**, 970–1008.

Neuro-anatomy of central pathways. Brodal, A., Pompeiano, O. & Walberg, F. (1962) *The vestibular nuclei & their connections, anatomy & functional correlations*, pp. 193. Edinburgh: Oliver & Boyd.

System dynamics. Barnes, G. R. (1980) Vestibular control of oculomotor and postural mechanisms. *Clin. Phys. Physiol. Meas.* **1**, 3–40.

Vestibulo-ocular responses & visual-vestibular interactions. Henn, V., Cohen, B. & Young, L. R. (1980) Visual-vestibular interactions in motion perception and the generation of nystagmus. *Neurosciences Research Progress Bulletin*, **18**, 459–651.

Perception of motion and the gravitational vertical. Clark, B. (1970) The vestibular system. *Annual Reviews of Psychology*, **21**, 273–306. Guedry, F. E. (1974) Psychophysics of vestibular sensation. In *Handbook of Sensory Physiology*, vol. 6, pt 2, pp. 3–154, ed. H. H. Kornhuber. Heidelberg: Berlin: New York. Springer.
Gundry, A. J. (1978) Thresholds of perception for periodic linear motion. *Aviation Space Environmental Medicine*, **49**, 686–97.

Motion sickness. Reason, J. T. & Brand, J. J. (1975) *Motion Sickness*, pp. 310. London: Academic Press.

Effect of alcohol and heavy water. Money, K. E. & Myles, W. S. (1974) Heavy water nystagmus and effects of alcohol. *Nature*, **247**, 404–5.

Adaptation of vestibular responses. Collins, W. E. (1970) Habituation of vestibular responses: an overview. In Fifth Symposium on the Role of the Vestibular Organs in Space Exploration, pp. 157–94. SP–314. Washington D.C.: N.A.S.A.

CHAPTER SEVENTEEN

Cutaneous sensory mechanisms

A. IGGO

17.1. PERIPHERAL MECHANISMS

The skin provides a major source of sensory input to the central nervous system and can provide information both about the environment at close quarters, by contact, and about more remote sources by the effect of radiation on skin temperature. The former input comes from the mechanically sensitive receptors (mechanoreceptors) and the latter principally from temperature-sensitive ones (thermoreceptors). If the environmental hazards are severe or potentially damaging a third set of receptors become active (nociceptors). The existence of these three basic and major groupings of receptors has been challenged from time to time, but recent morphological and physiological studies leave no doubt about their separate existence in the skin of man. The organs filling the inner part of the body (the viscera) also receive an afferent innervation via the visceral nerves. These afferent fibres are usually engaged in reflex regulation of visceral activity, but activity in them can also enter consciousness.* The visceral sensory receptors, in contrast to cutaneous receptors, are less numerous but any personal experience of abdominal pain would be enough to convince the sceptic of their existence.

The expression 'receptor' or 'sensory receptor' is used interchangeably for the morphological structure in which a sensory nerve ends, for the nerve terminal itself, for that part of the morphologically distinct structure that acts as the transducer and also for the whole functional unit including the afferent nerve fibre. This confusion can be reduced if the term 'sensory unit' or 'afferent unit' is used to describe the neuron (cell body in the dorsal root ganglion, central and peripheral branches of the axon of the neuron, and the nerve terminals) together with the associated morphologically distinct

* The terms 'afferent' and 'sensory' are sometimes distinguished, the former being applied to all fibres carrying information towards the CNS, while the latter is restricted to those that result in a sensation; but in comon usage 'sensory' is often used simply as a synonym for 'afferent'.

structure that, combined with the peripheral nerve terminals, forms the characteristic receptor in the skin. For example the SA I mechanoreceptor described on p. 378 is an afferent unit, whose morphological receptor contains Merkel discs (the nerve terminals) and Merkel cells (associated transducer cells). The term 'afferent unit' will be used in this chapter when the whole complex (as defined above) is discussed, but the term 'receptor' will continue to be used as a general expression where no ambiguity arises.

The cell bodies of the afferent nerves that end in somatosensory and visceral receptors are in the dorsal (posterior) root ganglia, and their axons typically have two branches, one directed peripherally to the skin via spinal and peripheral nerves, and the other directed to the spinal cord. The majority of the latter enter the central nervous system via the dorsal (posterior) roots (see p. 391). The major sensory innervation of the head is via the fifth cranial nerve (trigeminal). Major peripheral nerves, such as the median or the sciatic, are a mixture of sensory, motor, and sympathetic postganglionic axons, and from them muscular, articular and cutaneous nerves arise. Even the cutaneous nerves, such as the saphenous and sural, are mixed since they contain both sensory and sympathetic axons. Injury or direct stimulation of any nerve may, therefore, evoke a variety of unrelated actions. Although this chapter deals with sensory mechanisms, it is as well to remember that the peripheral cutaneous nerves also have important trophic actions on the skin. These nerves and the peripheral receptors associated with them are the only source of sensory input from the skin to the central nervous system. If a cutaneous nerve is cut and prevented from regenerating there is complete and immediate loss of sensory input from the skin that was supplied by that nerve. There may be a slow partial recovery of sensation, but this is due to ingrowth of sensory nerves from surrounding innervated skin; and there may also be abnormal sensations associated with activity arising in the central stump of the cut nerve. The central nervous system itself lacks any sensory innervation, except for that associated with the blood vessels and the meningeal linings.

The cutaneous sensations can be divided into several qualities, such as touch, pressure, warmth, cold and pain, and this chapter gives an account of the physiological mechanisms involved. The skin contains the variety of receptors already stated and there are at least 15 functionally and morphologically distinct kinds of afferent units. The afferent input in these is the raw material from which our

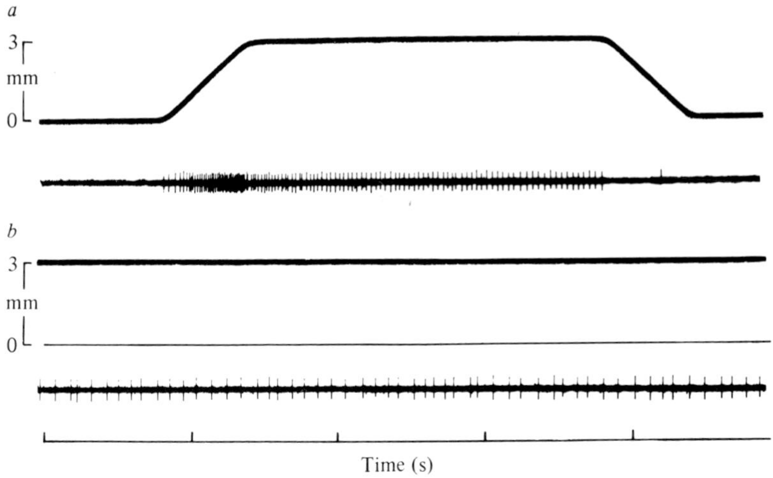

Fig. 17.1. The responses of a slowly adapting mechanoreceptor. Single unit records from the median nerve of a conscious human subject (A.I.) taken with a tungsten microelectrode inserted through the skin and into the nerve. The slowly adapting receptor (SA I) was in glabrous skin of a finger-tip. The record in (*a*) shows the response before, during and after indentation of the skin and the bottom record in (*b*) shows the continuous discharge during steady displacement of the skin, which had been maintained for 60 s.

cutaneous sensations are derived and which enable us to know the location, extent, size and duration of a stimulus to the skin. The sensation of light touch for example requires an input in sensory units different from those needed for the feeling of cold. It is now possible to record the impulse traffic in afferent fibres in conscious human subjects, while applying tactile, thermal or other stimuli to the skin. This is done by inserting fine recording microelectrodes through the skin and into the peripheral nerves. The kind of result obtained is illustrated in Fig. 17.1, which shows the discharge of impulses in a single fibre in the median nerve when the skin of the middle finger was touched. The subject, in the case illustrated the author, felt a steady sensation of light pressure on one finger and it was possible to correlate his sensory report with the activity in the nerve fibre. However, exact correlation of sensation with afferent input still eludes us since it is not yet possible to restrict the sensory input to the nerve fibre whose activity is recorded. It is known that the stimulus used in Fig. 17.1 would excite many afferent units so that further study is needed to establish beyond question the sensory roles of the different afferent units (see p. 381). However, the conclusions drawn from the detailed study of cutaneous sensory mechanisms in animals are in general confirmed by the more limited studies so far made on man.

The sensory acuity of different regions of the body is not equal. Sensory tests such as two-point discrimination (the ability to recognise two stimuli as a pair and not as a single stimulus) show that the back is least sensitive and regions such as the lips, tongue and finger tips have a very high acuity. These differences in tactile acuity are in part attributable to the lower density of sensory innervation on the back, but in addition, the receptive field size, i.e. the area of skin innervated by a mechanoreceptive afferent unit, is much smaller on the finger tips – by analogy with a photographic emulsion, the grain is finer on the finger tips. A further factor is the kind of receptor present in different body regions. Finger-tip skin contains several kinds of very sensitive mechanoreceptors and therefore can provide a range of information about tactile stimuli in contrast, say, to the glans penis and glans clitoris. The skin of the glans contains fewer kinds of receptors with mechanical thresholds that are higher than for cutaneous mechanoreceptors elsewhere. There are no Pacinian corpuscles or Merkel receptors. Thus an explanation is available for the curious fact that the glans is relatively insensitive to light touch, although the sensation aroused may be compelling. Thus, some of the sensory variation in different regions of the body can be accounted for by variations in the properties of afferent units, but the central nervous system has considerable power to modify the messages ascending to the higher levels of the brain. There is evidence that this modification is particularly important for messages concerned with pain. It is stressed here that, although sensation depends on an input through the skin nerves, the central nervous system also plays a vital role in fashioning the sensations that are actually experienced.

Axonal composition of cutaneous nerves

The myelinated axons range in diameter from 14 μm to 1.3 μm, and the unmyelinated from 1.3 μm to 0.24 μm. The conduction velocity range is from 84 m s^{-1} to less than 0.4 m s^{-1}. The myelinated axons, of which a nerve such as the sural contains 8000, are about equally divided into two groups on the basis of their size/diameter, one with a mean diameter of 8 μm and the other with a mean diameter of 2–2.5 μm. The non-myelinated axons are at least four to five times more numerous. The largest myelinated axons, the group > 5 μm in diameter, predominantly innervate mechanoreceptors, whereas the smaller myelinated and non-myelinated afferent axons innervate a diversity of receptors. The non-myelinated fibres also include sympathetic postganglionic axons, which are of course efferent, not afferent.

17.2. CUTANEOUS RECEPTORS

The sensory fibres entering the skin usually end in, or near, the epidermis. On their way they form plexuses in the dermis, often associated with the smaller blood vessels, but although they may branch en route they do not lose their individual identity. Each afferent fibre ends in at least one receptor. Some of these are morphologically complex, although this is not essential to achieve sensory specificity and there are receptors with very simple, even bare, nerve terminals. Receptor structure has been greatly clarified by electron microscopy and several distinct categories can be distinguished. Physiological studies that examine the discharge of afferent impulses from the receptors have provided information about their functional characteristics. The combination of electrophysiological and morphological methods has been particularly successful in correlating the functional and structural properties of the peripheral cutaneous receptor mechanisms.

Since the cutaneous receptors differ in their sensitivity to 'natural' stimuli, the afferent fibres are actually 'labelled-lines'. These special properties are summed up in the term 'specificity' which can be considered from two points of view. The first takes account of the peripheral or biophysical specificity of the afferent units in responding differentially to 'natural' cutaneous stimuli, the second assesses the specificity of the sensation in human subjects in response to the same stimuli. The latter, psychophysical method requires a number of assumptions to be made about the processing of sensory information by the nervous system.

Peripheral specificity. When a single cutaneous afferent unit is tested with a variety of different kinds of 'natural' stimuli (mechanical, thermal and chemical), a discharge of impulses might be evoked by each of them. Most kinds of cutaneous receptors can be excited by several kinds, or qualities, of stimuli and it can be argued on these grounds that they are all non-specific. The main problem is that a direct comparison of the intensity or force of stimulation cannot be made when comparing a mechanical with, say, a chemical stimulus. Instead it is necessary to apply a range of intensities of each kind of stimulus to representative samples of the various kinds of afferent unit to assess the responses to different 'qualities' of stimuli as well as the relative potency of each 'quality' in exciting different kinds of afferent unit. As an example the slowly-adapting SA I mechanoreceptors with thick myelinated axons (see p. 378) can be driven to a maximum discharge of about 1200 impulses s^{-1} by a mechanical stimulus, to not more than about 100 impulses s^{-1} by a strong thermal stimulus and to about 40 impulses s^{-1} by the chemical, histamine. Conversely, 'cold' receptors with thin axons fire maximally at 150 s^{-1} during rapid cooling of the skin, are almost unresponsive to severe mechanical stimuli and are indifferent to chemicals, such as histamine. The maximal rate at which the afferent fibres can discharge impulses must also be taken into account. In the large myelinated fibres (12 μm diameter) it is greater than 1000 s^{-1} while in the non-myelinated axons it is only 300 s^{-1}.

On the basis of these kinds of tests, three main categories of cutaneous receptor can be distinguished: mechanoreceptors, thermoreceptors and nociceptors.

Mechanoreceptors

The mechanoreceptors have high sensitivity to indentation or pressure on the skin, or to movement of the hairs. There are two main kinds, rapidly adapting and slowly adapting.

Rapidly adapting (RA) mechanoreceptors. These are excited by movement and code the velocity of movement as a frequency of discharge. If a ramp and plateau stimulus is given as in Fig. 17.2 they respond only during the movement and not when the skin is held constant at a new position. At a constant velocity of indentation the RA afferent units will discharge a stream of impulses at a frequency that is proportional to the velocity of displacement. The relationship fits a power function of the form $R = aS^b$ where R is the response, S is the stimulus intensity, a is a constant, and b is the exponent. Receptors of this kind are present in hairy and glabrous skin.* They include Pacinian corpuscles, which are present in both; hair follicle afferent units, which, as their name implies, are found only in hairy skin; Meissner's corpuscles in glabrous skin of primates; and Krause's end bulbs in non-primate glabrous skin.

* Glabrous skin is the smooth, hairless, fine-ridged skin of the hands and feet.

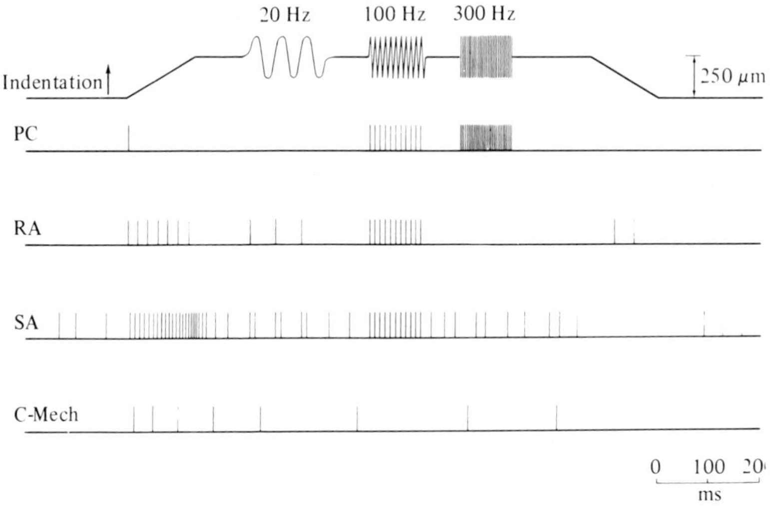

Fig. 17.2. Afferent discharge in different types of cutaneous mechanoreceptor in response to identical displacements of the skin, indicated in the top tracing. The Pacinian corpuscle (PC) is excited by the initial deflection at the start of the ramp displacement and at stimulus frequencies of 100 and 300 Hz, whereas the other rapidly adapting receptors (RA) fire repetitively during the ramp, are silent like the PC at constant displacement and are excited by low frequency (20 Hz) and middle frequency (100 Hz) vibration. The slowly adapting (SA) receptors have a resting discharge, and respond during steady displacement as well as during the ramp, and their responses are synchronised by low and middle frequencies of vibration. The C-mechanoreceptor (C-Mech) is less sensitive than the SA receptors, which have myelinated axons, and fires at a low rate during displacement. It is not synchronised by the vibratory stimuli.

Pacinian corpuscles (Fig. 17.3). These are a special class of RA receptors. They are small grey pearl-shaped structures in the deeper skin layers, 0.5–2 mm long and with an onion-like lamellar structure, formed of non-nervous tissue; they contain an elongated nerve terminal at the core. The afferent fibre is myelinated, with the last node of Ranvier inside the encapsulated receptor. The nerve terminal is ovoid in cross section, contains a circumferential array of mitochondria and an abundant fine tubular reticulum just beneath the receptor membrane. A mechanical function for the lamellation has been proposed: it is suggested the successive layers slip over each other so that a sustained deformation applied to the outer surface of the corpuscle causes only a transient deformation of the sensory

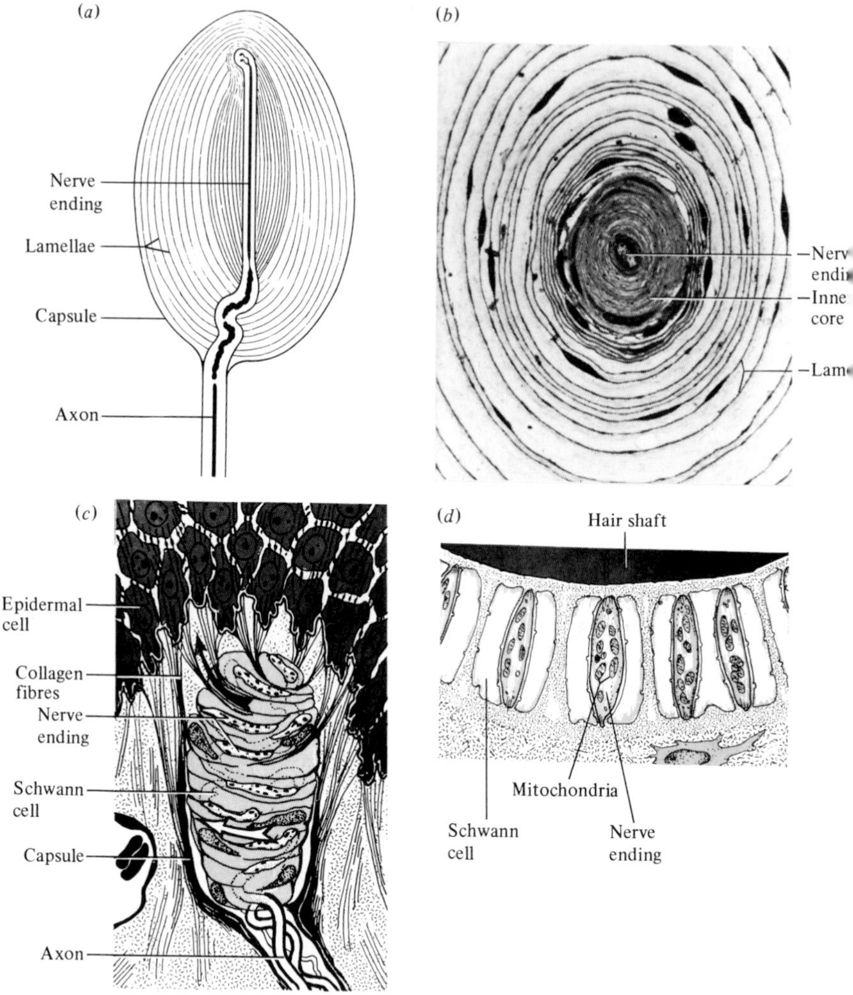

Fig. 17.3. Morphology of cutaneous mechanoreceptors. Three examples of encapsulated, rapidly adapting receptors. (*a*) Pacinian corpuscle (diagrammatic); (*b*) Pacinian corpuscle in cross section, showing the concentric lamellation with terminal part of axon at the centre; (*c*) Meissner corpuscle from glabrous skin. The overlying epidermal cells are connected to the corpuscle by collagen fibres; (*d*) Hair follicle afferent terminals in a cross-sectional view of part of a hair follicle. The nerve terminals are rich in mitochondria and are surrounded by Schwann cells.

terminal itself. Thus the Pacinian corpuscles in the soles of the feet (or skin of the buttocks) remain sensitive to vibrations, even when deformed by the weight of the whole body. The most important physiological feature of Pacinian corpuscles (and Pacini-like receptors such as the Golgi–Mazzoni ending) is this ability to detect mechanical vibrations recurring at high frequencies. The threshold is 1 μm or even less at frequencies between 200 and 400 Hz, but the receptor responds over a frequency range of 70–1000 Hz, although the threshold may be 10–75 times greater at the extreme frequencies (Fig. 17.5). The corpuscle is unable to detect even large stimuli at frequencies below about 50 Hz and cannot, therefore, signal a steadily maintained deformation of the skin. It is highly specific in its sensitivity and does not discharge impulses when the skin temperature is changed.

Meissner corpuscle (*Fig. 17.3*). This is an encapsulated RA receptor with a myelinated afferent fibre. It lies in the dermis of human glabrous skin, tucked into the dermal papillae that fill the grooves formed by epidermal ridges. The capsule is formed of several distinct layers of non-nervous tissue. The nerve endings are formed from helical sheets or laminae oriented at right angles to the long axis of the corpuscle and separated by sheets of Schwann cells. Collagen fibres connect the distal half of the receptor to the overlying epidermis which is modified structurally to provide an effective mechanical link that transmits movements of the epidermis to the nerve endings in the receptor. The Meissner corpuscle is a rapidly adapting or velocity-sensitive mechanoreceptor and discharges impulses only during movement of the skin (Fig. 17.2). The effective frequency range of a vibratory stimulus is lower than for the Pacinian corpuscles, from 10 to 200 Hz (Fig. 17.5).

Krause end bulbs, which exist in two varieties, cylindrical and globular, are also found in glabrous skin, most commonly in non-primate mammals. They have a simple lamellated capsule and the axon ends within the capsule of the receptor as a single rod-like extension in the cylindrical form and as an intertwined spiral in the globular form. The Krause receptors, once mistakenly thought to be cold receptors on the basis of the nineteenth-century psychophysical studies, have now been shown to be tactile receptors by correlative morphological and functional studies. Like the Meissner corpuscles, they are velocity detectors, being most sensitive over a frequency range of 10–100 Hz and giving no response to steadily maintained indentation of the skin.

Hair follicle receptors. Hair follicles are innervated by myelinated afferent fibres which end in a circumferentially arranged complex of endings around the hair-root sheath just below the sebaceous glands. The nerve terminals of hair follicles have the general form of elongated rods running parallel to and encircling the hair and epithelial root sheath (Fig. 17.3). Each ending is sandwiched between, or enclosed in, Schwann cells and is in contact with the basement membrane of the root sheath on one side and the collagen of the corium on the other. The hair follicle afferent units are rapidly-adapting mechanoreceptors and discharge impulses only when the hair is moved. There are several kinds that differ in sensitivity, rate of firing and size of receptive field (Types D, G, T). They serve the same role in hairy skin as the Meissner corpuscles do in glabrous skin.

In many species there are specialised hairs that form the vibrissae and shorter tactile hairs around the mouth. They are complex sensory organs with a rich afferent nerve supply that includes both rapidly and slowly adapting mechanoreceptors. The various afferent units in these hair follicles can between them signal the amplitude and rate of mechanical displacement, at both low and high frequencies, as well as its direction.

Slowly adapting (SA) mechanoreceptors These displacement mechanoreceptors, like the velocity detectors, respond during displacement of the skin, but in addition they sustain a discharge of impulses when the skin is held in a new position (Figs. 17.1, 17.2). In contrast to RA receptors they can provide information about the longer-term changes in mechanical conditions in the skin. During the plateau of indentation (Figs. 17.1, 17.2) the rate of firing declines at a rate characteristic for the kind of SA receptor, and the tissues in which it lies, to reach a rate that is proportional to the amount of indentation. Once again this satisfies a power function, $R = a\,S^b$, where R is the rate of discharge of impulses and S the indentation, so that the SA receptors can code the indentation as a frequency of discharge. There are two kinds of slowly-adapting mechanoreceptors, SA I and SA II, which differ in the time constants of adaptation, SA I adapting more quickly than the SA II. The SA I units are normally silent in the absence of an applied mechanical stimulus, can respond at very high frequencies if the skin surface is stroked rapidly and fire impulses in an irregular stream. The SA II, in contrast, often carry a resting discharge at 2–20 s^{-1}, are easily excited by stretching the skin, and discharge their impulses in a regular stream. The responses of SA receptors to a steadily maintained deformation can be enhanced by cooling the skin, but the dynamic response is brief, lasting only a few seconds, in contrast to that of the sensitive thermoreceptors whose dynamic discharge may continue for a minute. This effect of skin temperature on the responses of SA receptors accounts for Weber's illusion, the observation that cold objects feel heavier than warm ones of equal weight, when they are placed on the skin. Weber (1795–1878) used thalers, since he was a German, but no doubt 50 p or dollar pieces would do.

The *Merkel cells* and associated nerve terminals (Fig. 17.4) form the receptor for SA I afferent units. The Merkel cells are at the base of the epidermis. They have polylobulated nuclei, conspicuous granular vesicles on the dermal side of the cell and small rod-like projections (microvilli) on the epidermal surface of the cell. At the base of each Merkel cell is a disc-shaped expansion of a branch of a myelinated sensory axon, the Merkel disc.

The *Ruffini ending* like the Krause corpuscle was also mistakenly identified in the nineteenth century as a thermoreceptor, the warm receptor. Recent correlative functional and morphological studies have established that it is the receptor of the SA II cutaneous afferent unit (Fig. 17.4); it can provide a continuous indication of the intensity of steady pressure or tension within the skin. The encapsulated receptors (Fig. 17.4) are in the dermis of both hairy and glabrous skin, often lying more deeply than the Meissner and Krause endings. Similar receptors are found in, or near, the capsule of diarthrodial joints.

C-mechanoreceptors. The C-fibre mechanoreceptors have small receptive fields, about 3×2 mm, in hairy skin and can give a slowly-adapting discharge when the skin is indented (Fig. 17.2) or the hairs moved. If they are stimulated repetitively, however, there is a rapid fall in excitability and after 20 or 30 s the receptors fail to respond because the receptor terminals have become inexcitable. This behaviour is in striking contrast to that of the SA I and SA II mechanoreceptors with myelinated axons, which can sustain a response almost indefinitely. The functional role of these C-fibre mechanoreceptors is enigmatic, although they may contribute to the sensation of itching. The skin receives a rich supply of non-myelinated afferent axons, at least five times as numerous as the myelinated axons, but they do not end in microscopically visible end-organs. In electron micrographs they can be followed with considerable difficulty and end in the dermis and epidermis especially at the dermo-epidermal

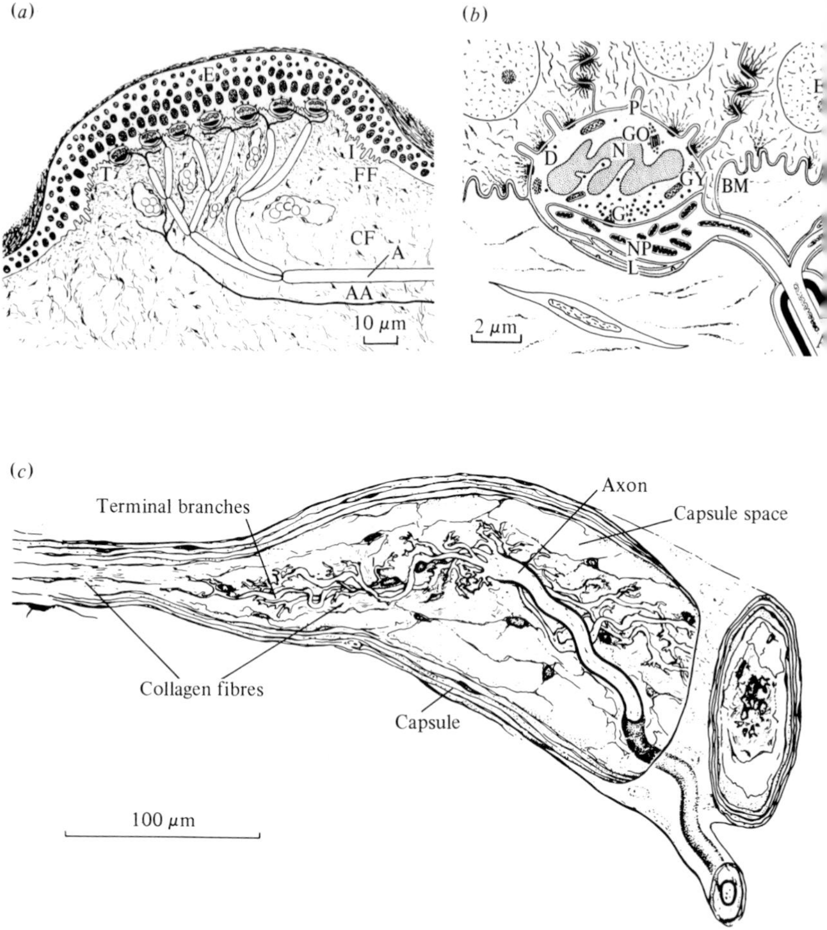

Fig. 17.4. Morphology of slowly-adapting mechanoreceptors. (*a*) SA I receptor showing the Merkel cells at the base of the epidermis (E) innervated by a myelinated axon (A). (*b*) Details of a Merkel cell containing a nucleus (N), Golgi bodies (GO), glycogen (GY), desmosomes (D) and the associated Merkel disc (NP) which is an expanded terminal of the myelinated axon (A). The Merkel cell and disc are on the epidermal side of the basement membrane (BM). (*c*) A diagrammatic view of a Ruffini ending. The myelinated nerve breaks up in the core of the receptor to form a fine brushwork of terminals.

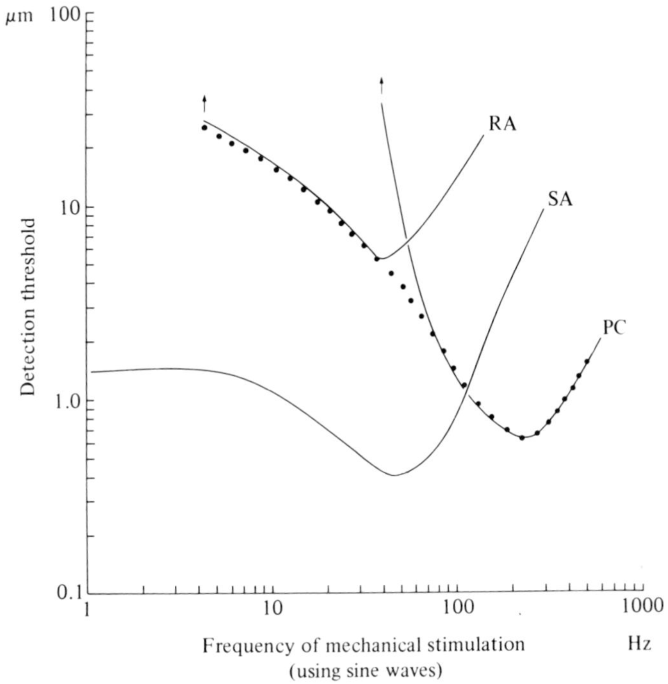

Fig. 17.5. Detection thresholds (solid lines) for three kinds of cutaneous receptor (SA, slowly-adapting; RA, rapidly-adapting; PC, Pacinian corpuscles) at various frequencies of sinusoidal stimulation, and the detection thresholds (dots) for the sense of flutter (5–60 Hz) and of vibration (50–600 Hz) in human subjects. There is a close fit of the RA and flutter thresholds and of the PC and vibration thresholds.

junction. The axons and the terminals are sheathed in Schwann cells, except for small patches of the terminals which are exposed to the fluid-filled interstitial spaces. Although the terminals do not have any microscopically visible specialisation they are able to function as transducers of natural stimuli delivered to the skin.

Sensory role of cutaneous mechanoreceptors. Different mechanoreceptors may contribute in varying degree to a given sensation such as the sense of flutter-vibration evoked by vibrating the skin. This sensation is experienced over a frequency range from about 5 to 400 Hz, and the threshold amplitude of vibration at different frequencies changes in a systematic manner, being highest at the extreme frequencies and lowest at 200–300 Hz (Fig. 17.5). Parallel studies of cutaneous mechanoreceptors show that the upper and lower frequency ranges are matched by different kinds of afferent unit. The RA units (Meissner corpuscles in glabrous skin and hair follicle afferents elsewhere) correspond very closely in their thresholds with the human sensory thresholds at the lower frequencies and the Pacinian corpuscle (PC) units match at the higher frequencies. This strong evidence for particular sensory roles for the RA and PC units is further strengthened by the existence in the somatosensory cortex of cortical neurons that respond either to sinusoidal cutaneous stimulation at 5–80 Hz (RA input), at 80–400 Hz (PC input) or to steady indentation of the skin (SA input).

Thermoreceptors

The characteristic properties of thermoreceptors are a continuous discharge of impulses at a given constant skin temperature, an increase or decrease when the temperature changes and an insensitivity both to mechanical stimuli and to pain-producing chemicals. The receptive fields of individual axons are spot-like, usually covering an area less than 1 mm^2. These spots correspond to the 'cold' and 'warm' spots found during sensory testing in human subjects. There are two classes of thermoreceptors – '*cold*' *receptors* with maximal response in the range 25–30 °C and '*warm*' *receptors* which respond maximally at 40–42 °C. The effective stimulus is the temperature of the receptor.

The firing frequency accelerates in 'cold' receptors when the temperature is falling and vice versa for the 'warm' receptors. This dynamic sensitivity is high and enables the receptors to respond to quite slow (less than 1 °C in 30 s) and small changes in skin temperature. The records in Fig. 17.6(*a*, *b*) show for cold receptors a) a continuous firing, related to temperature when the conditions are static b) an acceleration of discharge during a fall in temperature and c) inactivity during a rise in temperature. It is noteworthy that a cold unit may become inactive as the temperature is rising through a thermal range at which the unit would be active if the temperature was held constant. The general relationship of the responses at constant temperature for a cold receptor is given in Fig. 17.6. The dynamic responsiveness (i.e. the rate of firing during temperature changes) shows a similar relationship, but the firing rate can be ten times higher, depending on the rate of change of temperature. The discharge patterns of primate thermoreceptors are distinctive. Cold fibres show a conspicuous grouping of the impulses into bursts, the

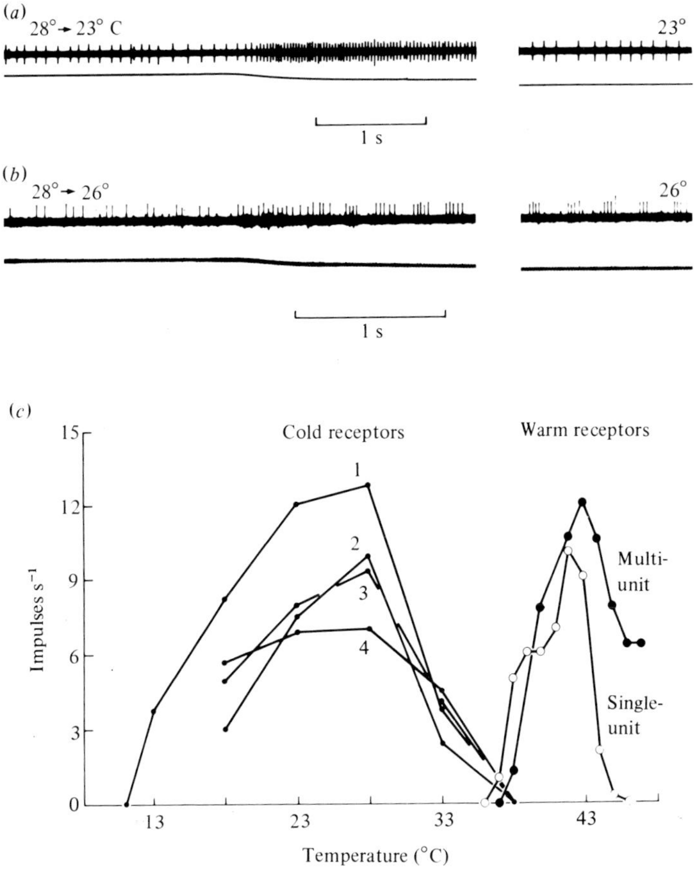

Fig. 17.6. Cutaneous thermoreceptors. (*a*) Afferent discharge from a cold receptor in the cat when the skin, initially at 28°, is cooled to and then held at 23°. (*b*) Cold unit in monkey, showing the typical grouping of impulses in bursts, especially at the steady temperature of 26 °C. (*c*) Cold and warm receptor responses at constant skin temperatures (static sensitivity curves). Maximal sensitivity of the cold receptors is about 28 °C and of the warm receptors about 42 °C.

presence and number of which show a temperature dependence, whereas the warm receptors discharge impulses at more regular intervals.

The afferent fibres of the thermoreceptors are small. In non-human primates the cold units have myelinated afferent fibres, 1–2 μm diameter, and the warm fibres are non-myelinated.

Nociceptors

When damaging or potentially damaging intensities of natural stimulation are applied to the skin the mechano- and thermoreceptor systems can be powerfuly excited, but careful quantitative studies have established that they are actually driven to maximal activity by less intense (i.e. innocuous) stimuli. Such receptor systems therefore cannot mediate pain, although they may contribute to the sensory quality of the pain experienced. The receptor systems that detect and signal high intensities of stimulation form a distinct class of sense organs, termed nociceptors, which are the peripheral 'pain' detectors. They have unencapsulated or 'free' nerve endings. The common analytical methods used to study these systems are the use of severe mechanical and thermal stimuli and algogenic ('pain-producing', from algos = pain and gennon = producing) chemicals, combined with single unit recording from peripheral nerve fibres. There are two main categories of nociceptor: a) *mechanical nociceptors*, which respond to pin-pricks, to squeezing and to crushing of the skin, and b) *thermal or mechanothermal nociceptors*, which respond to severe mechanical stimuli and to high and/or low skin temperature. Pain-producing chemicals can also excite the nociceptors, but a clear systematic classification based on the excitatory actions of such agents is difficult to establish because the chemicals excite, in varying degree, a variety of receptors including both mechanoreceptors and nociceptors. The polymodal nociceptors, which are examples of mechanothermal nociceptors, are so called because they can be excited by severe mechanical and thermal stimuli as well as by topical application of strong acids and pain-producing chemicals such as histamine. Sensitive mechanoreceptors, such as the SA I and SA II, also respond, although to different degrees, to a similar range of stimuli.

Mechanical nociceptors in primates have myelinated or unmyelinated afferent fibres. An effective stimulus is firm pressure by squeezing the skin or penetration of the epidermis by sharp objects such as serrated forceps or sharp pins. The receptors are quite unaffected by lightly stroking or brushing the skin; nor are they excited immediately by high or low skin temperatures unless these are very high (> 50 °C) and repeated several times. Even then there is a delay or more than 30 s before the receptors begin to fire. These receptors can give very rapid warning of physical injury, since the fastest axons conduct at velocities of 50 m s^{-1} (Group II afferent fibres) and form the signalling system for 'first' pain.

The thermal/mechanical nociceptors can also be excited by similar mechanical stimuli, but the distinguishing sign is an immediate vigorous and sustained response to severe thermal stimuli with a theshold of 43–45 °C. These receptors have unmyelinated fibres in the periphery, although a small proportion gain a myelin sheath more proximally in the peripheral nerve. Even so the maximum conduction velocity is less than 10 m s^{-1}. These more slowly conducting pain fibres signal 'second' pain. Nociceptors also respond to a very low skin temperature, as low as 2–3 °C, and can sustain a continuous discharge at these low temperatures. The afferent fibres of these particular nociceptors are non-myelinated. An interesting feature of these afferent units, which parallels subjective experience, is an acceleration of discharge when the skin is being re-warmed from very low temperatures.

Inflammation. A characteristic feature of the inflammatory response of the skin to injury is hyperalgesia. Hyperalgesia means an enhanced painful response to a noxious stimulus (from Greek hyper = over, above, and algos = pain). It is caused by the intra-cutaneous release of chemicals such as 5-hydroxytryptamine (5-HT), histamine, bradykinin and prostaglandins. The first three are known to alter the excitability of cutaneous mechanoreceptors so they cannot be regarded as purely algogenic, i.e. exciting only the nociceptors, although they do enhance their excitability. Injection of 5-HT and histamine into the arterial supply of the skin will excite a low frequency discharge in thermal nociceptors and may lower their threshold for noxious stimulation. Repetition of a noxious thermal stimulus may also lead to a fall in threshold of the thermal nociceptors, by several degrees, from, say, 45 °C to 38 °C. This develops slowly and several minutes may elapse before it becomes apparent. It also subsides slowly and is probably due to the formation and release of chemicals, as part of an inflammatory reaction, that alter the sensitivity of the receptors to both thermal and mechanical stimuli. These kinds of action account for the hyperalgesia that follows thermal injury of the skin. There is a parallel hyperaesthesia which is an increased sensory awareness of tactile stimuli, and is probably due to the action of the inflammatory agents on the mechanoreceptors.

Anti-inflammatory agents such as aspirin (acetylsalicylic acid), indomethacin and other non-steroidal anti-inflammatory agents reduce inflammation and relieve pain. Aspirin and indomethacin block the action of the enzyme, prostaglandin synthetase, and thus reduce the formation of prostaglandins. For this reason it is proposed that the prostaglandins are involved in the genesis of inflammation and the causation of pain. They do not directly excite nociceptors when they are injected close-arterially into the skin, but have a potentiating or enhancing effect on the action of 5-HT and bradykinin so that the latter become more painful. Aspirin probably exerts its analgesic action indirectly by preventing the formation of the potentiating agents, rather than by acting directly on the nociceptors. Some anti-inflammatory agents may, in addition, act directly on the afferent units and reduce their excitability.

17.3. VISCERAL RECEPTORS

The viscera are much more sparsely innervated than is the skin. The afferent fibres travel to the central nervous system in the visceral nerves, together with sympathetic or para-sympathetic nerve fibres, but complete their journey in the dorsal roots, for the majority, or the ventral roots for a minority of the population. Their cell bodies are in dorsal root or cranial nerve ganglia and the axons are either thinly myelinated or non-myelinated and therefore have low conduction velocities.

The visceral sensory nerve endings do not have a conspicuous encapsulation, with the exception of the Pacinian corpuscles, which are the most abundant in the mesentery, and other small encapsulated mechanoreceptors in the mesentery and in the wall of the urethra. All viscera have an afferent innervation, but the afferent activity in them usually enters consciousness only as an ill-defined sensation of fullness or of pain. The visceral afferent units are necessary for the reflex regulation of visceral activity, e.g. chemoreceptors and baroreceptors in the carotid body and carotid sinus detect the oxygen and hydrogen ion content of the blood and arterial pressure levels respectively, and contribute to the regulation of breathing and of blood pressure. Activity in these particular afferents may never enter consciousness.

Mechanoreceptors

(a) *Slowly adapting 'in series' stretch receptors* are excited by distension or stretch of the hollow viscus in whose wall they lie and are even more powerfully excited by contraction of the smooth muscles of the visceral wall. They provide ambiguous sensory information about the physical state of the viscera, since they do not distinguish between passive distension and active contraction of the organ. The afferent fibres from the upper and lower alimentary canal enter the central nervous system via the vagi or the pelvic nerves, whereas from the mid-region of the canal (intestines and to a lesser degree the stomach) they also travel centrally in the same nerves as the sympathetic outflow; these are principally the splanchnic nerves.

The receptors are called 'in series' stretch receptors because they are excited both by passive and by active stretch. The dual excitation is only possible if the receptors are arranged in the tissues so that active shortening of the smooth muscle cells actually stretches or elongates them. They are analogous to the Golgi tendon organs in skeletal muscles and are in contrast to the muscle spindle receptors which are unloaded by active muscle contraction. A characteristic feature of smooth muscle is its ability to elongate or shorten slowly to maintain a more-or-less constant intraluminal pressure, or at least to minimise increases in pressure as a viscus fills. There are parallel changes in the level of activity in the 'in series' stretch receptors, which are much more active if a viscus is filled rapidly, than if it is filled slowly.

In several organs the smooth muscle is rhythmically active under isotonic conditions, in which there are changes in volume of the viscus without large changes in pressure. In such organs the force of the contraction and the intramural tension can increase several-fold if the onward movement of the contents is obstructed and the conditions become isometric. The afferent discharge then becomes much more powerful and prolonged. A familiar example is provided by the urinary bladder. If, during normal emptying of a full bladder, the flow of urine is suddenly arrested then a powerful isometric contraction of the bladder can ensue, causing powerful activation of the receptors (Fig. 17.7), accompanied by sensations of severe discomfort. A similar kind of response accounts for the severe pain caused by intestinal obstruction, which may lead to powerful localised contractions resulting in the sensation of colic.

(b) *Peritoneal and serosal mechanoreceptors*. These receptors are in the peritoneal structures and the afferent fibres travel in the sympathetic nerves. Any individual afferent fibre may have a widespread distribution with up to nine separate receptor terminals in the serosa. The receptors are slowly adapting mechanoreceptors and are excited by local stretching or tension, but, in contrast to the 'in series' stretch receptors there is no consistent relation between either bowel movements or volume and activity in the afferent units. These splanchnic afferent units have properties that do not fit them for reflex regulation of visceral activity and they may have a role in the

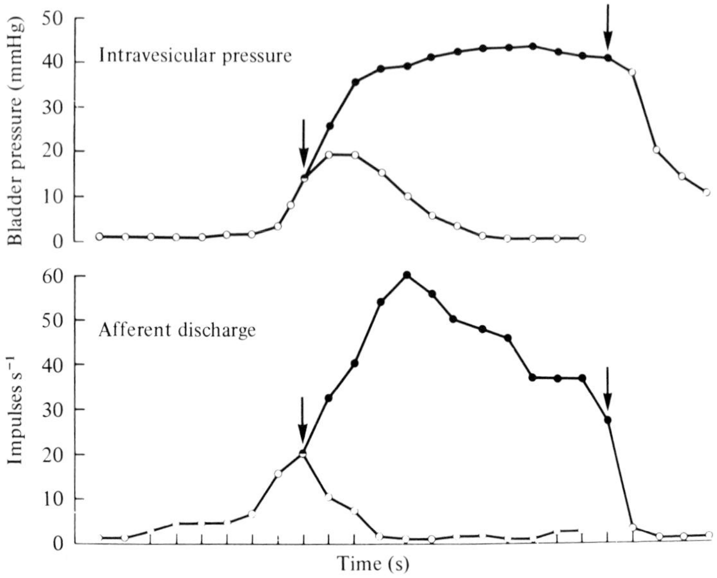

Fig. 17.7. 'In Series' receptor in the wall of the urinary bladder. The open circles show the response of the afferent unit and the intravesicular pressure during an isotonic contraction and the filled circles the effect of occluding the outflow of fluid from the bladder at the first arrow, and release of the occlusion at the second arrow.

mediation of painful stimuli. They are, for example, excited by bradykinin which is an algogenic chemical.

(c) *Pacinian corpuscles.* The properties of these receptors in the viscera are identical with those in the skin. They are present in the mesentery, particularly of the cat. Their functional role is obscure, although a vascular regulatory action has been suggested for them.

(d) *Flow receptors* are present in the urethra and are rapidly-adapting mechanoreceptors. They become active when fluid is flowing through the urethra and the rate of firing is proportional to the flow rate. The discharge disappears promptly when the flow stops. The afferent fibres are probably myelinated and enter the sacral spinal cord via the pelvic nerves.

(e) *Mucosal mechanoreceptors.* Afferent nerve endings penetrate the mucosa and although some of them are pH receptors there is also evidence for mechanically sensitive mucosal receptors. They may be spontaneously active and are excited by the movement of contents along the lumen, particularly by solid contents. The receptors are in the muscularis mucosae and they are not excited by powerful local intestinal contractions, in contrast to the intramural 'in series' stretch receptors.

Visceral chemoreceptors

The pH and osmolarity of the gastro-intestinal contents exert reflex and sensory effects.

pH receptors. The gastric mucosa contains receptors that are excited by high or low concentrations of hydrogen ions. The afferent units excited by low pH have thresholds about pH 1.5 and will sustain a continuous slowly adapting discharge if fluids at this pH are placed in the stomach. Neutralisation of the fluid stops the discharge. The receptors can also be excited by a firm stroke or stretch of the mucosa, but not of the serosal surface of the stomach. This may be a non-specific effect, comparable to the excitation of retinal receptors by pressure on the eyeball, and the mucosal receptors are much less sensitive to mechanical stimulation than the intra-mural 'in series' stretch receptors. Chemoreceptors similar to these have also been reported in the non-glandular stomach. The acid-sensitive pH receptor may mediate the discomfort and pain associated with gastric hyperacidity, particularly if the acid gastric contents enter the oesophagus or are not adequately neutralised when they enter the duodenum.

Glucoreceptors. These are present in the gastric and intestinal mucosa. They are excited by simple sugars (e.g. glucose and lactose) in concentrations ranging from 0.25 to 0.5 M. The discharge is non-adapting in the sense that it is continuous if the sugar solution is left in contact with the mucosa.

17.4. SPINAL CORD MECHANISMS

The majority of the cutaneous and visceral afferent fibres enter the spinal cord through the dorsal roots. In primates they are organised so that the axons of larger diameter are medially placed in the roots, where these enter the spinal cord, and the finer myelinated and non-myelinated axons are lateral (Fig. 17.8). The nerve fibres from large mechanoreceptors therefore enter the spinal cord separately from the nociceptor and thermoreceptor afferents. The largest afferent fibres branch after entering the spinal cord and may send a large main branch cranially in the dorsal columns to end in the dorsal

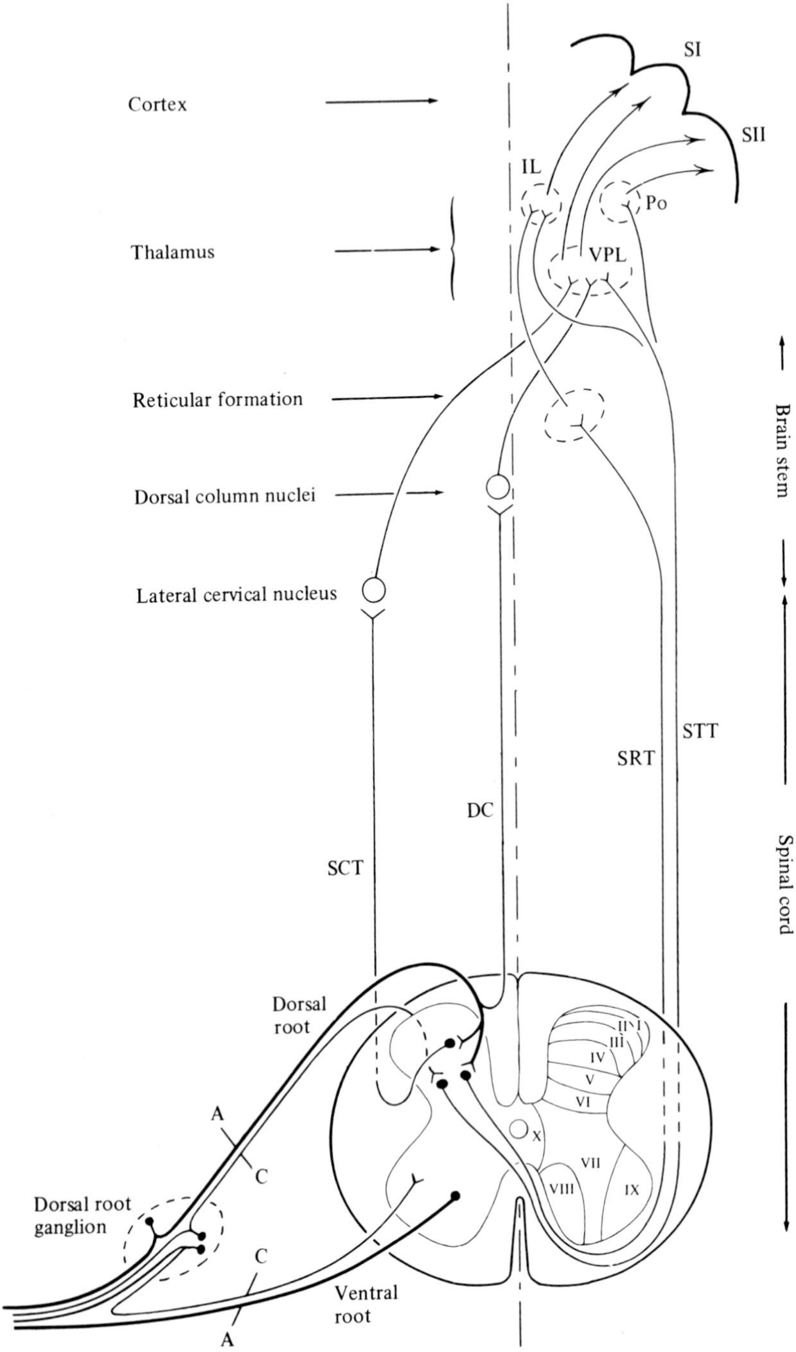
Cortex
Thalamus
Reticular formation
Dorsal column nuclei
Lateral cervical nucleus
SI
SII
IL
Po
VPL
Brain stem
Spinal cord
STT
SRT
DC
SCT
Dorsal root
II
I
III
IV
V
VI
X
VII
VIII
IX
A
C
Dorsal root ganglion
C
A
Ventral root

column nuclei (cuneate and gracile nuclei in the medulla oblongata). Some also send a main branch caudally. Collaterals from these dorsal column branches enter the dorsomedial border of the dorsal horn, penetrate its deeper layers and terminate synaptically on dorsal horn neurons. Current work using horseradish peroxidase (HRP) injected intracellularly into the axons of afferent units has made it possible to describe the anatomical organisation of individual afferent fibres. The collateral branches of the hair follicle afferents, after descending through the grey matter, sweep dorsally to end in flame-like arborisations that synapse with large neurons in laminae III and IV whereas the SA I and PC afferents branch and end in laminae IV and V (SA I afferents) or in III, IV and V (Pacinian corpuscle).

The small afferent fibres enter Lissauer's tract, a longitudinal bundle of afferent and propriospinal fibres at the dorsolateral margin of the spinal cord, and then penetrate and end in the superficial grey matter of the dorsal horn. None of these small fibres travels very far from its segment of entry into the spinal cord.

The classical pathway for sensory fibres entering the nervous system is the dorsal (posterior) roots. Renewed attention is now being given to those afferent units that send their afferent fibres into the ventral roots (Fig. 17.8), although the cell bodies are in the dorsal root ganglion. About 30% of the non-myelinated axons in the ventral roots are afferent fibres of this kind. Most are visceral afferents, but their functional role has not yet been clearly established. A very small proportion (< 1%) of the myelinated afferent fibres may also enter the central nervous system in the same way. The particular interest

Fig. 17.8. Afferent pathways into the spinal cord and to the cerebral cortex. The entry of large myelinated (A) fibres and non-myelinated (C) fibres through the dorsal roots is shown, as also are the aberrant C fibres that have cell bodies in the dorsal root ganglion, but which loop back to enter through the ventral roots. The lamination (I to X) of the grey matter is shown on the right-hand side of the cross section of the spinal cord. The ascending sensory pathways in the spinal cord are: SCT, spinocervical tract; DC, dorsal column afferent collaterals; SRT, spinoreticular tract; STT, spinothalamic tract. The latter two are crossed tracts. The former end in the lateral cervical and dorsal column nuclei and then cross over the midline to enter the ventroposterior (VPL) nucleus of the thalamus. The STT goes to three thalamic nuclei (VPL), (Po) Posterior group and (IL) Intralaminar. The sensory pathways then continue on to the S I and S II somatosensory areas of the cerebral cortex.

of these afferent fibres in the ventral roots is that they may provide an alternative sensory pathway if the dorsal spinal roots are damaged.

Second-order neurons in the sensory pathways

Collaterals of the afferent fibres synapse on neurons in the dorsal horn. These include *propriospinal* neurons with short axons that are distributed locally near their cells of origin or in adjacent segments of the cord, and *tract* neurons that send axons to more remote parts of the spinal cord or the brain. The dorsal horn of the cord can be the scene of complex interactions, with an interplay of both presynaptic and postsynaptic events. From the sensory viewpoint, prime importance can be given to the tract neurons which, together with the afferent collaterals in the dorsal column destined for the dorsal column nuclei, are the major pathways for sensory information travelling on to the brain from the spinal cord.

Dorsal horn neurons. Electrophysiological recording with microelectrodes, combined with quantitative natural stimulation of the skin, has established the properties of many individual dorsal horn neurons. Four broad groupings are recognised among the larger neurons on the basis of their responses to cutaneous stimuli and their location in the dorsal horn. *Class 1 Dorsal horn neurons* are excited by cutaneous mechanoreceptors only. The usual response is either a brief burst or a sustained train of impulses when the skin is stroked or lightly pressed. These neurons are most common in laminae IV and V of the dorsal horn (see Fig. 17.8). *Class 2 Dorsal horn neurons* are also excited by mechanoreceptors, but in addition can be powerfully excited by the nociceptors, especially those with non-myelinated afferent fibres. In suitable conditions the discharge may continue at high frequency for many seconds after withdrawal of the noxious stimulus. These neurons, which are relatively large, are the most numerous and are abundant in laminae IV and V. *Class 3 Dorsal horn neurons* are excited only by the nociceptors and are unresponsive to the mechanoreceptors. These neurons are smaller than the Class 1 and 2 neurons and are most abundant in lamina I. Some of them, Class 3a, are excited only by myelinated nociceptors and others, Class 3b, by both the myelinated and the non-myelinated nociceptors, so that noxious thermal stimuli powerfully affect only the Class 3b (Fig. 17.9). *Class 4 Dorsal horn neurons* are driven specifically by thermoreceptors. So far only 'cold' sensitive units have been reported in the spinal cord. These neurons are in lamina I.

All dorsal horn neurons are probably in monosynaptic relation with their respective afferent fibres, but there is always some degree of convergence by the afferent units via inter-neurons. The receptive fields from which the individual dorsal horn neurons can be excited are topographically organised and tend to be smaller in the extremities than in the proximal parts of limbs and/or the trunk. A well-organised topographical map can be distinguished: for example in the sixth and seventh segments of the lumbar spinal cord the extremity of the foot is represented medially and the leg laterally.

Substantia gelatinosa. This is a layer of very small and densely packed neurons that lies in lamina II between the laminae containing the cells described above (see Fig. 17.8). The S.G. neurons include cells which, in contrast to the larger neurons, have a distinctive continuous background activity that can be modified by an input from cutaneous mechanoreceptors and nociceptors. The receptive fields are similar in their functional organisation to those of the Class 1 to 3 dorsal horn neurons and the S.G. neurons can be put into similar classes, except that they are inhibited rather than excited by the relevant stimuli. For that reason they have been called $\bar{1}$ (Inverse 1), $\bar{2}$ (Inverse 2) and $\bar{3}$ (Inverse 3). It has further been proposed that there is a functional linkage between the S.G. cells and the larger dorsal horn cells which leads to reciprocal inhibitory interactions between mechanoreceptive and nociceptive afferent input. This interaction may enhance sensory contrast by suppressing the central effect of low levels of afferent activity and curtailing the activities of the more weakly excited neurons. The more deeply placed of the dorsal horn neurons (Laminae VI and VII) are polysynaptically activated by cutaneous neurons and show signs of much greater convergence, with more prominent responses also to muscle afferents.

Inhibition of neurons in the dorsal horn

Inhibition in the dorsal horn strongly influences the transmission of information from the cutaneous receptors into the sensory pathway. Neurons can be inhibited *postsynaptically*, by synapses on their cell bodies and dendrites that release an inhibitory transmitter which causes an inhibitory postsynaptic potential (IPSP), or *pre-synaptically* by synapses that are formed on the pre-synaptic terminals and which may depolarise the terminal, causing primary afferent depolarisation

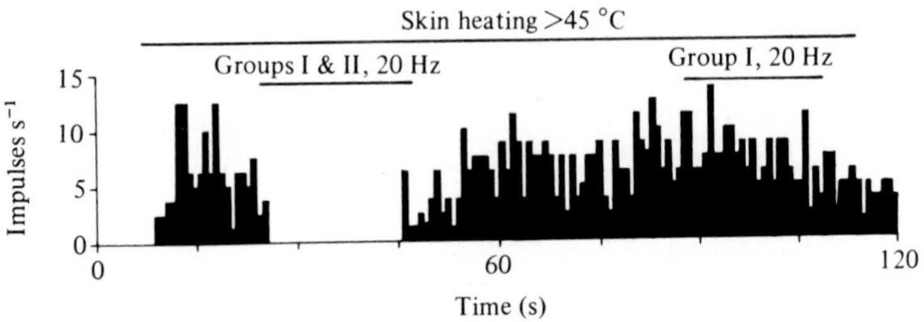

Fig. 17.9. Responses of a nociceptor-driven Class 3 dorsal horn neuron to heating the skin and the effect on this vigorous response of stimulation of cutaneous mechanoreceptor afferent fibres. The muscle afferent fibres alone (Group I) had no inhibitory action, but when the cutaneous afferent fibres were also excited (Groups I and II) there was complete inhibition of the nociceptor-evoked discharge.

(PAD). The quantity of transmitter chemical released at a synapse is proportional to the size of the pre-synaptic action potential and since PAD causes a reduction in its size the amount of excitatory transmitter released by an active excitatory terminal is thereby reduced. This in turn, diminishes the excitatory postsynaptic potential (EPSP) and therefore weakens the postsynaptic responses.

Segmental inhibition. The large mechanoreceptive afferents that excite Class 1 and 2 neurons both monosynaptically and via inter-neurons, also have an inhibitory action on the same neurons that limits the intensity of the response to cutaneous input. This action may restrict spike discharges to the most powerfully excited neurons and thus have the effect of enhancing or increasing contrast by suppressing the activity of neurons on the fringe. Inhibition of this kind is even more powerfully effective in selectively blocking the otherwise vigorous excitation of Class 2 neurons by nociceptor input. There is a similar action (Fig. 17.9) on the nociceptor-driven Class 3 neurons on which, of course, there is no direct excitatory action by the mechanoreceptors. This segmental inhibition provides a simple explanation for the familiar relief of pain or itch which comes from rubbing a sore spot, and from the use of counter-irritants – the low intensity nociceptor input is prevented by segmental inhibition from exciting the tract neurons. These mechanisms are also exploited in the use of transcutaneous or peripheral-nerve electronic stimulators in clinical practice to relieve some pains of peripheral origin.

Somaesthetic pathways

Major sensory tract systems ascend in the spinal cord, but they are not anatomically distinct and none of them can be identified with a simple morphological tract. The dorsal columns (Dorsal funiculi) comprise the ascending branches of afferent fibres, together with second-order fibres originating in the dorsal horn and the dorsal column nuclei. The spinothalamic and spinocervical systems, are second-order tracts, that arise from neurons in the dorsal horn (see Fig. 17.8).

Dorsal columns. Most of the nerve fibres in this tract are collaterals of the larger myelinated axons from the cutaneous mechanoreceptors, and this tract has considerable importance as a pathway for tactile information. These axons end in the dorsal column nuclei in the lower medulla in a precise somatotopically ordered map. In the dorso-medially placed gracile nucleus the trunk and hind limbs are represented with the hind leg in a medial position, the toes in a dorsal position and with the trunk ventral. The more laterally placed cuneate nucleus receives its afferent input from the forelimbs. Two cytoarchitecturally distinct zones occur in the nuclei: the rostral and deeper regions are 'reticular', comprising a mixture of large and small neurons, whereas in the caudal half the cells are more uniform in size and are grouped in clusters. The primary afferent fibres are distributed preferentially to the 'cluster' region. The second-order fibres in the dorsal funiculus come from the Class 1 and 2 dorsal horn neurons and are distributed to the 'reticular' zone of the nuclei, with considerable overlapping. The nuclear cells are excited both by low-threshold cutaneous tactile inputs and by proprioceptive stimuli to joints and muscles.

The ascending fibres from the gracile and cuneate nuclei project via the medial lemniscus to the ventrobasal nuclei of the thalamus (VPL in Fig. 17.10) and a well-ordered representation of the body surface is maintained in this projection. The ventrobasal nuclei in the thalamus are therefore concerned with tactile information originating in the sensitive cutaneous mechanoreceptors, and the neurons are capable of extracting and preserving information with a high spatial discriminative quality. In their turn these nuclei project to the somatosensory cortex (see Fig. 17.10).

Spinothalamic tract (STT). This reaches its greatest development in primates, including man. It lies rather superficially in the anterolateral

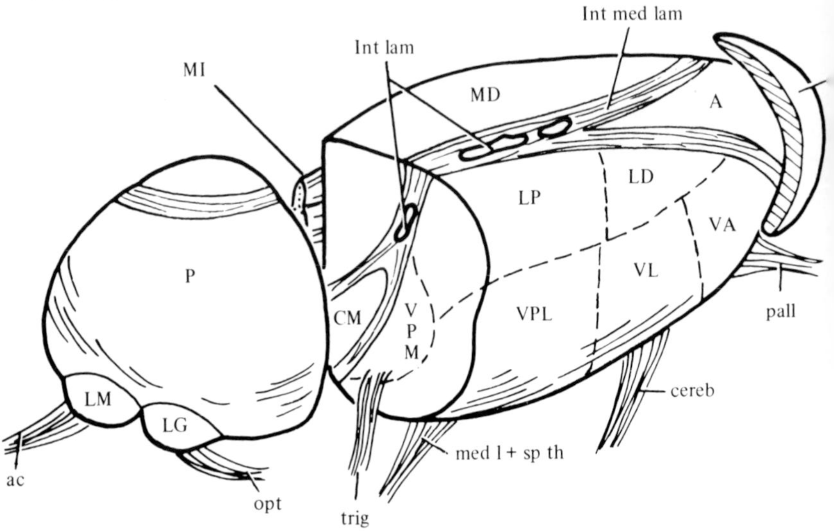

Fig. 17.10. Diagrammatic representation of the dorsolateral aspect of the right human thalamus in a partly exploded view. The principal thalamic nuclei are indicated, together with major input tracts. The major nuclei are: A, anterior; CM, centromedian; Int. lam., intralaminar; LD, LP, lateralis dorsalis and posterior; LG, lateral geniculate body; LM, medial geniculate body; MD, dorsomedial; Ml, midline; P, pulvinar; R, reticular; VA, ventralis anterior; VL, ventralis lateralis; VPL and VPM, ventralis posterior lateralis and medialis. The main afferent contingents are: ac, acoustic; opt, optic; trig, trigeminal; med l+sp.th., medial lemniscal and spinothalamic; cereb, cerebellar; pall., pallidal. (from Brodal A. (1969). *Neurological Anatomy in relation to clinical medicine.* New York: Oxford University Press.)

quadrant of the spinal cord and contains axons ascending from neurons in the contralateral dorsal horn. In primates these neurons, which are in either the very superficial grey matter (in Lamina I) or more deeply placed in Lamina VII and VIII, include Classes 1, 2, 3 and 4 dorsal horn neurons so that information about all kinds of natural cutaneous stimuli can be transmitted via the spinothalamic tract. Of particular importance is the projection of nociceptor-driven and of thermosensory neurons in the STT. Fig. 17.10 shows that the ascending axons cross to the opposite side of the spinal cord. This cross-over occurs within two segments of the level of entry of the afferent fibres into the spinal cord. Electrophysiological studies are now in good agreement with the older clinical findings that accidental or surgical transection of the STT can lead, at least temporarily, to a loss of both pain and temperature sensations from the opposite side of the body. The presence of Class 3 (nociceptor-driven) neuronal projection into the STT would appear to be sufficient explanation for the role of this tract in the mediation of pain. However, the Class 2 neurons also project into it and these may provide additional spatial information.

The STT passes through the bulbar brain stem, although not recognisable as a specific anatomical tract, and ends in the thalamus in three major somaesthetic areas (Figs. 17.8, 17.10): in the ventro-posterior (VPL) nucleus, in an ill-defined posterior (PO) group of nuclei and in the intralaminar (IM) nuclei. In all these nuclei there are interactions between different sub-modalities of cutaneous sensory input and in the VPL and PO the STT projection interacts with activity arriving over the other major sensory tracts.

Spinocervical tract (*SCT*). This lies in the dorsolateral quadrant of the spinal cord and arises from dorsal horn neurons in Laminae IV. It is an ipsilateral pathway in the spinal cord. Class 1 and 2 dorsal horn neurons project into the SCT, with a preponderance of neurons excited by hair follicle afferents. Many of the SCT axons come from Class 2 neurons, which are excited by both mechano- and nociceptor afferent inputs. The sensory role of the tract, which though present in primates is most highly developed in carnivores, is uncertain. It projects via the dorsolateral funiculus on to the lateral cervical nucleus, that lies in the upper cervical spinal cord. Ascending axons from this nucleus decussate and travel in the medial lemniscus, along with ascending axons from the dorsal column nuclei, to end on the contralateral ventroposterior region of the thalamus (VPL), Fig. 17.10).

Spinoreticular tract. Neurons in deeper laminae (VII and VIII) of the spinal cord project to the contralateral reticular formation of the brain stem. The receptive fields are much more complex than for the spinothalamic and spinocervical tract neurons. They may be bilateral, cover large areas of skin, include mechanoreceptors and nociceptors and be both excitatory and inhibitory. These properties are due to convergence of the more superficial dorsal horn neurons on to the spinoreticular tract neurons. The ascending axons travel in the contralateral ventrolateral quadrant of the spinal cord and overlap the spinothalamic tract axons (Fig. 17.8). The sensory function of the SRT is uncertain, but it is involved in arousal and alerting reactions that indirectly influence sensory awareness.

17.5. TRIGEMINAL SENSORY MECHANISMS

The face and oral structures, including the teeth, are innervated by the trigeminal (fifth cranial) nerve and all the nerve fibres of this very large sensory supply pass to the trigeminal sensory nuclei in the brain stem. The cutaneous receptor systems are similar in their functional organisation to those found elsewhere in the body. The teeth are a special case. They have a sensory innervation through the tooth pulp that is distinctive in giving rise to an almost pure sensation of pain. The sensory nucleus of the trigeminal system can be separated into several sub-nuclei. The principal sensory nucleus corresponds to the dorsal column nuclei and receives an afferent input principally from the large mechanoreceptor afferents. It projects to the ventrobasal nucleus of the contralateral thalamus (Fig. 17.10) and thence to the somatosensory cortex. The caudal trigeminal sensory nucleus corresponds to the dorsal horn of the spinal cord. The superficial lamina (equivalent to lamina I) receives an afferent input from nociceptors and from specific thermoreceptors. The deeper laminae (analogous to lamina IV and V of the spinal cord) receive excitatory inputs from mechanoreceptors and nociceptors. The caudal nucleus therefore contains neurons corresponding to Classes 1 to 4 of the dorsal horn and also projects to the contralateral thalamus (VPM in Fig. 17.10) by a pathway corresponding to the spinothalamic tract. The more highly discriminative aspects of tactile sensation are handled by the principal sensory nucleus, whereas the nociceptive and thermal aspects are dealt with by the caudal nucleus.

17.6. CEREBRAL MECHANISMS

Cerebral cortex

The cerebral cortex is the final destination of cutaneous and visceral sensory inputs. In primates and man (Fig. 17.11) the post-central gyrus, comprising areas 3a, 3b, 2 and 1 according to Brodmann's classification, is the primary somatosensory receiving area (SI). Most of its subcortical afferent fibres come from the ventroposterior nucleus of the thalamus, with a slight input from the intralaminar group of nuclei. The cortex of SI is related to the opposite half of

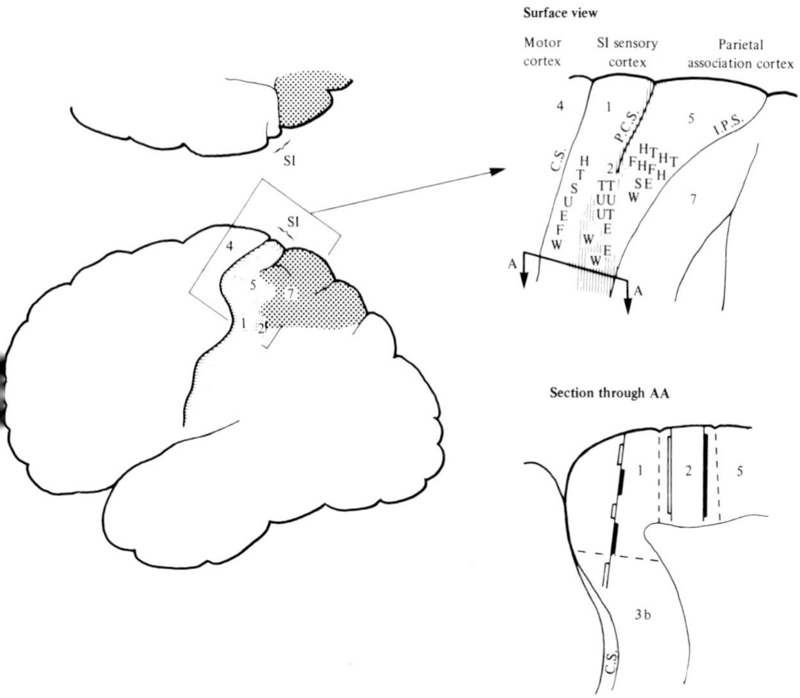

Fig. 17.11. The somatosensory area of the cerebral cortex in man, showing the S I area, comprised of Brodmann's area 3, 1 and 2, and the association areas 5 and 7. There is a lateral view of the left cortex and a partial midline view. The surface view is an expanded representation of part of the post-central cortex in a monkey to show the detailed representation of part of the body surface in area 1 and the progressive loss of the topographical identity in areas 2 and 5. The lower right-hand diagram shows a sectional view through the S I cortex at A–A, to indicate the location of area 3b which is buried in the central sulcus and also the columnar arrangement. Two columns orthogonal to the surface are shown in area 2, and as indicated by the filled and hollow bars along 2 electrode tracks, contain neurons that respond to different modalities of cutaneous stimuli, whereas the electrode track in area 1 and 3b passes through several columns, since the path of the track is not orthogonal to the cortical surfaces. SI, primary somatosensory cortex; C.S., central sulcus; P.C.S., post central sulcus; I.P.S., inferior parietal sulcus; H., hind limb; T., trunk; S., shoulder; U., upper arm; E., elbow; F., forearm; W., wrist.

the body and to both sides of the scalp and the face. There is a well-ordered cortical representation or map of the body surface as well as of the thoracic and abdominal viscera. The most detailed representation is of the digits and lips – regions that are used for active exploration and finely controlled movements. Individual neurons are activated from specific receptive fields and by specific physiological stimuli. There is a columnar organisation (Fig. 17.11). The neurons are organised in columns that are orthogonal with respect to the surface. All cells in a column respond to similar, if not identical, modalities or sub-modalities of stimuli. Taken together with the topographical mapping of the body in the SI area, the result is that individual columns in the primary sensory receiving area are functional modules having a high degree of specificity for both the modality and the peripheral location of skin stimulation. This modular organisation of the columns is a characteristic feature of the neuronal architecture of the cerebral cortex. Each module has an input of specific afferent fibres, of commissural fibres from the corresponding module in the opposite hemisphere of the cortex and from other modules in its own hemisphere.

The primary projection of the specific cutaneous receptors is to area 3b, and the columnar modules here have both a functional and spatial specificity – any column is excited by a particular sub-modality, even perhaps by a particular kind of receptor from a given place on the skin (Fig. 17.11). Area 3a, on the other hand, is the primary projection area for the joint and muscle receptors. The strip of cortex adjacent to area 3 is area 1 and the neurons here are converged on by both cutaneous and deeper receptor systems (joint and muscle). Area 2 lies posterior to area 1 and here the convergence is still more marked and the cells develop additional properties, such as feature interaction. These SI areas project in turn to the parietal association areas, first to area 5 and then to area 7. The latter also receives a projection from other cortical sensory areas, and neurons in area 7 can be excited by cutaneous, visual and auditory stimuli. These associated areas are involved in the cerebral processing of somaesthetic afferent input.

There is another region of the cortex that receives a somaesthetic afferent input, *the second somatic sensory area* (*S II*), which is on the superior wall of the lateral (Sylvian) sulcus in primates. It is smaller than S I and the individual cells show signs of extensive convergence in terms of the kinds of afferent input as well as of receptive fields. There is a bilateral representation of the body. The anterior part of

S II is excited by tactile stimuli, whereas the proprioceptors in the deeper tissues are less effective excitants. The afferent input comes from the same ventroposterior nuclei of the thalamus that supplies S I. The posterior region of S II receives an input from the posterior group (PO) of thalamic nuclei and although, like the anterior region there is a bilateral representation, the neurons are activated by stimuli with a nociceptive or painful quality.

Intercerebral connections. There are well-organised inter-connections of corresponding areas on the two sides of the cerebral cortex, made by association fibres that travel via the corpus callosum. Corresponding parts, except for the representations of the distal limbs, are connected. S I receives a projection from S I of the other cortical hemisphere whereas S II has an input from both S I and S II.

Sensory functions of the cerebral cortex

A wide variety of sensations are experienced when natural stimuli are applied to the skin. Surgical ablation of appropriate parts of the peripheral and central nervous system will abolish or impair these normal sensory responses although there is no simple relationship (see p. 397 for the effects of a spinal cord lesion). Since they are elaborated at a cortical level it is instructive to examine the responses of conscious human subjects to non-injurious electrical stimulation of the exposed cerebral cortex during neurosurgical treatment. When the hand–wrist–forearm region of the primary somatosensory cortex (Fig. 17.11) is stimulated electrically at rates greater than 15 s^{-1} for durations greater than 0.5 s^{-1}, a variety of sensations are elicited. These range from paraesthesias (described as tingling; electric shock; pins and needles; pricking; numbness) to 'natural-like' sensory responses. The latter include responses such as, 'something moving inside'; feeling of movement of the hand, wrist or forearm; deep pressure; surface touch; vibration; warmth or coldness. The sensations are 'natural-like' and the subjects have difficulty sometimes in describing them verbally. Electrical stimulation of the primary somatosensory cortex is thus capable of initiating activity that ensues in sensory responses, although the results above do not establish that only the S I cortex is involved in the response. It is of particular interest that several sub-modalities or qualities of sensation can be elicited – thermal as well as tactile. One striking result of investigations of responses to electrical stimulation of the cerebral cortex is that although most cutaneous modalities are reported, there is an almost complete absence of reports of pain when the S I cortex is stimulated.

There are also no convincing electrophysiological reports from studies on experimental animals of S I cortical neurons responding only to noxious stimuli although there is now abundant evidence to show that there are spinal and thalamic neurons with such properties. Pain can be evoked in conscious human subjects by electrical stimulation of the secondary somatosensory cortex (S II).

Descending control of the sensory pathways

The somatosensory pathways provide a well-ordered mechanism for the central transmission of sensory information originating in the cutaneous receptors, subject, however, to continuous control from supraspinal centres. At every synapse in the pathway in the dorsal horn, the dorsal column nuclei and the thalamus the onward transmission of information can be modified by both presynaptic and postsynaptic action, with inhibition more prominent than excitation. The somatosensory cortex sends many efferent fibres to lower levels. Those in the pyramidal tracts can have direct actions on the sensory nuclei in the medulla and in the spinal cord, whereas more indirect actions are exerted via the brain stem. There are also direct cortico-thalamic efferents. By these means the cortex is able to control which afferent input reaches it, that is, the cortex can attend to some inputs and ignore others and this switching of attention can be recognised at several relay stations in the sensory pathways. These mechanisms are also important in improving sensory contrast. Tactile spatial acuity and recognition depend, of course, on the afferent input from both cutaneous and deeper receptors and the passive movement of an object over the skin will improve perception. The active manipulation of the object – 'active touch' – is, however, even more effective.

Descending control of 'pain' pathways. The brain can control the inflow of impulses along the pain pathways. The brain stem contains neural mechanisms capable of making conscious animals and human subjects completely indifferent to painful stimulation of the skin.

Brain stimulation analgesia is evoked by electrical stimulation through electrodes placed in the periaqueductal grey matter, or in the median raphé nuclei. The analgesia develops with a short latency and may long outlast the period of repetitive stimulation that induces it. The mechanism is selective and during the analgesia there are normal sensory and reflex responses to tactile stimulation. The analgesia is, in part, due to excitation of neurons in the median raphé that send their axons down to the dorsal horn. The neurons contain the chemical transmitter 5-HT, which is released in the dorsal horn, and leads to the selective inhibition of nociceptor-driven neurons. The peptide, Substance P, is a putative excitatory transmitter in the dorsal horn that may interact with 5-HT.

Morphine, acting at both brain-stem and spinal-cord levels, can produce analgesia that can be prevented by naloxone, a specific morphine antagonist. Since opiates such as morphine are addictive, there is considerable interest in the possibility of using other drugs that share the analgesic, but not the addictive properties of the opiates. A pentapeptide, enkephalin, has recently been isolated from the brain. It binds to the same opiate-binding sites in the brain as morphine and therefore has the same analgesic action as morphine. Enkephalin may be the naturally occurring opiate agonist. The action can also be blocked by naloxone. The enkephalin pentapeptide is the terminal region of a much larger polypeptide, beta endorphin, that is present in the pituitary gland and which may have other functions unrelated to analgesia. This new knowledge is pointing the way to a fuller understanding of the mechanisms of brain-induced analgesia, with future clinical implications.

17.7. SENSATION

The final product of the operation of the cutaneous sensory system is the perception of distinctive sensations, such as touch, vibration, warmth, cold and so on. It is now possible to correlate some sensations with activity arising from particular kinds of cutaneous receptors. Experimental analysis has established that the sensory consequences of vibratory stimulation of the skin are due to the excitation of Pacinian or Meissner receptors. Fig. 17.5 plots the mechanical sensitivity of three cutaneous mechanoreceptors (Pacinian corpuscles, Meissner corpuscles and SA I receptors) in glabrous skin measured at different frequencies of vibration, from 0 to 800 Hz. There are clearly marked differences. The sensory thresholds in conscious subjects for the sensation of flutter (5–50 Hz) and of vibration (50–600 Hz) are also plotted. There is a remarkably close agreement between the Pacinian corpuscle thresholds and the vibratory sensory thresholds at high frequencies; and between the Meissner receptor and 'flutter' sensory thresholds at intermediate frequencies. There is no correlation between either vibratory or 'flutter' sensory thresholds and the thresholds of the SA I slowly adapting receptors. The results provide compelling evidence for the existence of highly specific and discriminative sensory pathways from the functionally distinctive cutaneous receptors to the cerebral cortex and can with confidence be extended to other qualities of cutaneous sensation, although these have not yet been analysed to the same extent.

Sensory localisation

Tactile stimulation of the skin is accurately localised by a blindfold subject and this ability is readily accounted for by the abundance of mechanoreceptors in the skin, and the well-ordered projection of their afferent fibres on to neurons in the tactile pathways. The general features of this system have already been described. Warmth and cold are less well localised, even though there are warm and cold spots in the skin, probably because of the more widespread convergence of the afferent fibres on to the tract neurons in the thermosensory pathway. Nevertheless, the localisation is reasonably accurate, and what is deficient is an ability to recognise two separate thermal stimuli if they are close together on the skin, so that the spatial acuity of the thermosensory system is low. Painful stimuli are even less well localised, partly because pain, even that evoked by a pin-prick, has a radiating or spreading quality. Superficial pin-pricks are localised with the greatest accuracy, whereas painful stimuli to deeper tissues, or noxious thermal stimuli delivered to the skin, are poorly localised. This is unexpected when considered from the viewpoint of the receptors since all kinds of nociceptor have small receptive fields. The paradox is resolved when the receptive fields of neurons in the sensory pathways are considered. There is extensive convergence from cutaneous afferent units on to spinothalamic tract neurons, some of which are excited by noxious stimuli delivered anywhere on a limb. The segmental and descending inhibitory control systems may have an important role in aiding the process of localisation in the spinothalamic system by suppressing the more weakly excited neurons on the fringe of the most powerfully excited. The effect of removal of the descending control systems is seen when the spinal cord is blocked: large increases in excitability of dorsal-horn neurons develop in the isolated part of the cord, especially the Class 2 dorsal horn neurons, (see p. 392) which receive excitatory convergence from both mechanoreceptors and nociceptors.

The localisation of painful stimuli may also be aided by the concurrent excitation of mechanoreceptor and thermosensory afferent pathways. Noxious stimuli always exceed the thresholds of the mechanoreceptors or the thermoreceptors; which will therefore be excited concurrently with the nociceptors. It is probably that the tactile systems are of greater use in this respect, since they possess a good inherent capacity for spatial localisation.

Referred pain

Skin pain, or cutaneous hyperexcitability, can exist in the absence of any evident stimulus to the skin. This pain is often quite well localised to a particular place. A well-known example is angina pectoris, in which the sensation of pain or discomfort is felt on the anterior aspect of the thorax and may extend down the inner aspect of the forearm. The afferent input actually has its origin in ischaemic cardiac muscles and yet the pain is not experienced only in the heart, it is also *referred* to the chest wall. Similarly the sensations aroused by the afferent input from receptors in a full bladder may be referred to the external genitalia.

In all such instances the origin of the sensation is the visceral receptors and the mis-reference arises because of convergence of visceral and cutaneous afferent input on to common neurons in the sensory pathway, at spinal, thalamic and cortical levels. Since, during development of sensory and perceptual images of the body, each individual will most commonly associate a particular sensation with stimuli delivered to the skin, the simplest explanation for referred pain is that the activity of particular elements in a sensory pathway becomes labelled with a reference to the skin. Subsequently, the afferent input from an injured or diseased viscus converging on one of these shared interneurons will be perceived as arising in the skin, rather than, or as well as, in the viscera. The stability of the reference is due to the segmental origin of the afferent fibres. Thus the cardiac afferents, which are the seat of angina pectoris, enter the lower cervical and upper thoracic segments of the spinal cord and hence the sensation is referred to the cutaneous regions that send their afferents to this part of the cord.

Parallel paths and processing

Information from the cutaneous and visceral and proprioceptive receptors may reach the cerebral cortex by two or more major pathways with a minimum of two neurons en route. In addition to these direct paths there are other polyneuronal pathways via reticular nuclei of the brain stem. The cutaneous somaesthetic system is thereby protected against complete disability due to trauma, since no single pathway has an exclusive sensory function. In a surgically verified case of a spinal-cord injury that, except for part of one antero-lateral quadrant, completely and cleanly severed the spinal cord at the third thoracic segment, the patient suffered a complete motor paralysis below the level of the lesion. The sensory loss was restricted to an absence of temperature sensations and of the pricking component of pain sensations on the opposite side, thus confirming the essential role of the STT for thermal sensations. Vibration sensitivity was also absent, as would be expected since the dorsal columns are the ascending pathways. The unexpected results were that the patient could recognise and localise tactile and pressure stimuli on both sides of the body although at a higher threshold than normal, that passive movement could be detected on the same side as the lesion, and that pain (but not pricking sensation) could be evoked on both sides. It is clear from cases such as this that parallel pathways can be important in preserving some degree of sensory capacities when some of the pathways are damaged.

Surgical ablation of the S I cortex itself causes sensory deficits in, for example, two-point discrimination and point localisation. The sensory impairment may be short-lasting, and animal experiments indicate that surgical ablation of both S I and S II may be followed by recovery of sensory discriminative capacity, although there is an initial loss as an immediate consequence of the operation.

17.8. SUGGESTIONS FOR FURTHER READING

General reviews

Angel, A. (1977) Processing of sensory information. In *Progress in Neurobiology*, **9**, 1–122.

Anderson, D. J. & Matthews, B. eds. (1977) *Pain in the trigeminal region.* Amsterdam: Elsevier Scientific Publications.

Bonica, J. J., Lindblom, U. & Iggo, A. (1983) *Advances in Pain Research and Therapy*, vol. 5. New York: Raven Press.

Brown, A. G. & Réthelyi, M. (Eds) (1981) *Spinal Cord Sensation.* Edinburgh: Scottish Academic Press.

Cervero, F. & Iggo, A. (1980) Substantia gelatinosa of the spinal cord: a critical review in *Brain*, **103**, 717.

Gordon, G. Ed. (1977) Somatic and visceral sensory mechanisms, *British Medical Bulletin*, **33**, part 2.

Gordon, G. Ed. (1978) *Active Touch.* Oxford: Pergamon Press.

Hubbard, J. I. Ed. (1974) *The peripheral nervous system.* Chapters 12–16. New York and London: Plenum Press.

Iggo, A. Ed. (1973) *Handbook of Sensory Physiology*, vol. II. *Somatosensory system.* Heidelberg: Springer Verlag.

Iggo, A. & Andres, K. H. (1981) Morphology of cutaneous receptors in *Annual Review of Neuroscience*, **4**.
Hamman, W. & Iggo, A. (1984) *Sensory Receptor Mechanisms.* Singapore: World Scientific Publishing.
Laverack, M. S. & Cousins, D. J. Eds (1981) *Sense Organs.* Glasgow: Blackie.
Mountcastle. V. B. (1974) Neural mechanisms in somesthesia. In *Medical Physiology* (13th edn), vol. 1, pp. 307–47. St Louis: Mosby.
Neil, E. Ed. (1972) *Handbook of sensory physiology. III/I Enteroceptors.* Heidelberg: Springer-Verlag.
Willis, W. D. & Coggeshall, R. E. (1978) *Sensory mechanisms of the spinal cord.* New York: John Wiley & Sons.
Zotterman, Y. Ed. (1976) *Sensory functions of the skin in primates.* Oxford: Pergamon Press.

Special references

Pacinian corpuscle. Loewenstein, W. R. (1971) *Handbook of Sensory Physiology*, Vol. I, pp. 269–90. Heidelberg: Springer-Verlag.
Morpho-functional correlation. Iggo, A. & Muir, A. R. (1969) *Journal of Physiology*, **200**, 763–96; Chambers, Margaret, R., Andres, K. H., von Düering, Monika & Iggo, A. (1972) *Quarterly Journal of Experimental Physiology*, **57**, 417–45; Iggo, A. & Ogawa, H. (1977) *Journal of Physiology*, **266**, 275–96.
Weber's thaler illusion. Stevens, J. C. & Green, B. G. (1978) *Sensory Processes*, **2**, 206–19.
Glans penis. Cottrell, D. F., Iggo, A. & Kitchell, R. L. (1978) *Journal of Physiology*, **283**, 347–67.
Human sensory recording. Vallbo, Å. B. & Hagbarth, K.-E. (1968) *Experimental Neurology*, **21**, 270–89. Vallbo, Å. B., Hagbarth, K.-E., Torebjörk, H. E. & Wallin, B. G. (1979) *Physiological Reviews*, **59**, 919–57.
Thermoreceptors. Hensel, H., Andres, K. H. & von Düering, Monika (1974) *Pflügers Archiv*, **352**, 1–10; Hensel, H. & Zotterman, (1951) *Acta physiologica scandinavica*, **23**, 291–319; Iggo, A. (1969) *Journal of Physiology*, **200**, 403–30.
Non-myelinated afferents. Iggo, A. & Kornhuber, H. H. (1977) *Journal of Physiology*, **271**, 549–65.
Nociceptors. Bessou, P. & Perl, E. R. (1969) *Journal of Neurophysiology*, **32**, 1025–43; Iggo, A. (1959) *Quarterly Journal of Experimental Physiology*, **44**, 362–70; Perl, E. R. (1968) *Journal of Physiology*, **197**, 593–615.
Prostaglandin and algogenic chemicals. Beck, P. W. & Handwerker, H. O. (1974) *Pflügers Archiv*, **347**, 209–22; Collier, H. J. & Schneider, C. (1972) *Nature New Biology*, **236**, 141–3; Moncada, S., Ferreira, S. H. & Vane, J. R. (1975) *European Journal of Pharmacy*, **31**, 250–69; Vane, J. R. (1971) *Nature New Biology*, **231**, 232–5.
Visceral receptors. Iggo, A. (1957*a*) *Quarterly Journal of Experimental Physiology*, **42**, 130–43; Iggo, A. (1975*b*) *Quarterly Journal of*

Experimental Physiology, **42**, 398–409. Morrison, J. F. B. *Journal of Physiology*, **233**, 349–61.

Afferents in spinal cord. Brown, A. G., Rose, P. K. & Snow, P. J. (1977) *Journal of Physiology*, **272**, 779–97.

Substantia gelatinosa. Cervero, F., Iggo, A. & Molony, V. (1979*a*) *Quarterly Journal of Experimental Physiology*, **64**, 297–314; Cervero, F., Iggo, A. & Molony, V. (1979*b*) *Quarterly Journal of Experimental Physiology*, **64**, 315–25.

Segmental inhibition and dorsal horn neurons. Cervero, F., Iggo, A. & Ogawa, H. (1976) *Pain*, **2**, 5–24; Handwerker, H. O., Iggo, A. & Zimmermann, M. (1975) *Pain*, **1**, 147–65.

Spinothalamic tract. Trevino, D. L., Coulter, J. D. & Willis, W. D. (1973) *Journal of Neurophysiology*, **36**, 750–61.

Pre-synaptic inhibition. Schmidt, R. F. (1973) In *Handbook of Sensory Physiology*, vol. II, *Somatosensory System*, ed. A. Iggo, 151–206. Heidelberg: Springer-Verlag.

Cerebral cortical modules. Szentágothai, J. (1975) *Brain Research*, **95**, 475–96.

Sensory correlation with receptors. Mountcastle, V. B., La Motte, R. H. & Carli, G. (1972) *Journal of Neurophysiology*, **35**, 122–36. Ochoa, J. & Torebjörk, E. (1983) *Journal of Physiology*, **342**, 633–54.

Acupuncture and electroanalgesia. Andersson, S. A. (1979) In *Advances in Pain Research and Therapy*, vol. 3, eds. J. J. Bonica, J. C. Liebeskind & D. G. Albe-Fessard, pp. 569–85. New York: Raven Press. Callaghan, M., Sternbach, R. A., Nyquist, J. K. & Timmermans, G., (1978) *Pain*, **5**, 115–27.

Brain stimulation analgesia and opioid peptides. Fields, H. L. & Anderson, S. D. (1978) *Pain*, **5**, 333–49; Fields, H. L. & Basbaum, A. I. (1979). In *Advances in Pain Research and Therapy*, *3*, eds. J. J. Bonica, J. C. Liebeskind & D. G. Albe-Fessard, pp. 427–40. New York: Raven Press. Mayer, D. J. (1979) In *Advances in Pain Research and Therapy*, vol. 3, ed. J. J. Bonica, J. C. Liebeskind & D. G. Albe-Fessard, pp. 385–410. New York: Raven Press.

CHAPTER EIGHTEEN

Chemical senses: Smell

E. B. KEVERNE

The chemical senses, and the chemical communication they make possible, have played a crucial part in mammalian evolution. Sexual behaviour, defensive and territorial behaviour, mother–infant interactions, and feeding behaviour have all been fundamental to the survival of a species, and in all these behaviours the chemical senses have played an important role. It is not, therefore, surprising to find that in most mammalian species a relatively large area of the brain is given over to the analysis of chemical stimuli. The chemical senses comprise olfaction and taste, the former providing exteroceptive information on odour cues in the environment, whilst taste requires ingestion for chemical analysis to proceed before the material is swallowed.

In the higher primates and man, the olfactory areas of the brain are relatively small, and this fact, taken together with the smaller area of *olfactory epithelium* (the primary receptor surface for smell) has led a number of neuroanatomists to suggest that the olfactory chemical sense is relatively unimportant for man and infra-human primates. This view was reinforced by the long-held belief that the olfactory sense is an exception in that it does not have access, via the thalamus, to neocortical areas.

It is true that man depends little on the chemical senses for survival, an emancipation related not so much to the dimensions of the smell brain, but to man's exceptional capacities to learn, to reason, and to generate and understand speech. Yet this is not to say that the chemical senses are unimportant. Learnt chemosensory discriminations allow modern man to identify many of the artificial compounds (some toxic or explosive) that he has introduced into his environment. Even our aesthetic preferences among chemical stimuli, our preferences for wines or perfumes, are in part determined by our cultural background. Thus, although it may have been the development of the neocortex that freed man from dependence on the chemical senses, it is also the development of the neocortex that has permitted such finer appreciations of the trained nose or palate. The discovery by Powell and his collaborators in Oxford that olfactory pathways do in fact project to the neocortex via the thalamus now provides an anatomical basis by which the old 'smell brain' has access to these integrative capacities of the neocortex.

18.1. DEVELOPMENTAL ORIGINS OF THE CHEMICAL SENSES

Although functionally similar, the chemical senses are developmentally quite distinct. The primordia of the olfactory system appear in late somite embryos as two ectodermal thickenings called the olfactory placodes. Neuroblasts in the epithelium of each nasal placode give origin to olfactory nerve fibres which grow towards the apical region of the corresponding cerebral hemisphere, which in turn elongates to form the olfactory bulb. Cells in the olfactory bulb around which the olfactory nerves terminate, and with which they synapse, give origin to the secondary olfactory fibres. Thus, the fibres of the olfactory nerves differ from those of other nerves in that they are entirely of ectodermal origin, and their cell bodies remain in the epithelium.

Taste receptors are also epithelial in origin but unlike the olfactory receptors they do not give rise to centrally projecting neurons. On the contrary, it is thought that the taste receptors are themselves induced to develop by the already established nerve supply (VII and IX cranial nerves). The taste buds develop in papillae, of which those responding to sour and bitter are the first to appear, under the influence of the glosso-pharyngeal nerve on the upper surface of the posterior part of the tongue. The papillae responding to sweet and salty tastes appear rather later on the anterior portion of the tongue and develop under the influence of the chorda tympani branch of the facial nerve. The taste receptors, therefore, are not true neurons and have no axon, but establish synaptic contact with the peripheral end of the sensory nerve whose cell body is located in the sensory ganglia of either the seventh or the ninth cranial nerve.

18.2. PERIPHERAL OLFACTORY SYSTEM

The structure of the nose varies from a simple sac found in salamanders, frogs and tortoises, to a more complex structure in mammals, almost completely occupied by the turbinate bones. The ethmoturbinals (superior nasal concha) carry the olfactory epithelium containing the olfactory receptor neurons. The olfactory epithelium differs from the surrounding respiratory mucosa in the presence of Bowman's glands and a characteristic yellow pigmentation. A third difference is that, although cilia are present, they do not show rhythmic beating like the cilia of the respiratory regions. Estimates of this area within the nasal cavity are of the order of 2–4 cm^2 in man, while the dog has 18 cm^2 and the cat 21 cm^2.

Olfactory epithelium

The olfactory mucosa is a relatively simple structure when compared with the retina or the inner ear, and is composed mainly of three cell types: receptor cells, supporting cells and basal cells arranged as in Fig. 18.1*b*. The epithelial surface is itself covered by a layer of mucous in which receptor cilia and the microvillae of supporting cells are embedded. The mucous layer is formed by the secretions of Bowman's glands, and may vary in thickness. Since odorant molecules must pass through it in order to reach the receptors, the absorptive properties of this mucous layer may influence the time course, the pattern and the degree of odorant access to the receptors.

The olfactory receptor

The receptor cells are arranged like a mosaic between supporting cells and overlie a single layer of basal cells. The receptor cells are flask-shaped, with apical cilia entering the mucous layer, and their unmyelinated axons cross the basal lamina en route to the olfactory bulb. Le Gros Clark described many morphological differences among the olfactory receptors and proposed that this could indicate functionally different units. It now appears more likely that these differences are related to different stages of maturation and subsequent degeneration, since evidence is available that olfactory neurons are continually replaced.

Cilia have long been established as a feature of the olfactory neuron (Fig. 18.1*b*, *c*), and emphasis has been placed on their role in the perception of odours. They clearly increase the area of the receptor-cell surface, and are believed to contain the receptor sites for different odour molecules. There is growing evidence that the receptor sites are proteins, and electron microscopy of freeze-fracture preparations reveals membrane particles on the sensory cilia of olfactory receptor cells (Fig. 18.1*c*). The membranes of the olfactory cilia are appreciably richer in protein particles than those of the adjacent motile respiratory cilia. So it seems reasonable to suppose that these intra-membraneous particles might be the receptor sites responsible for binding odour molecules.

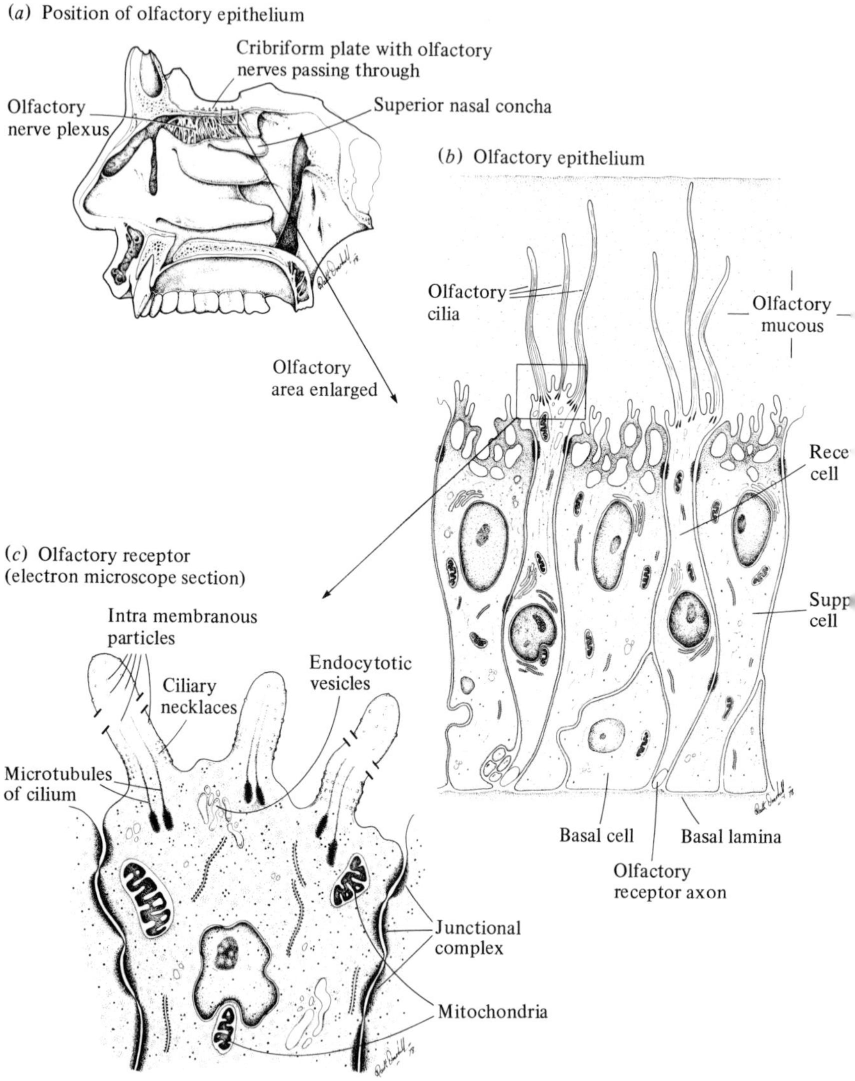
(a) Position of olfactory epithelium
Cribriform plate with olfactory nerves passing through
Olfactory nerve plexus
Superior nasal concha
Olfactory area enlarged
(b) Olfactory epithelium
Olfactory cilia
Olfactory mucous
Rece cell
Supp cell
Basal cell
Basal lamina
Olfactory receptor axon
(c) Olfactory receptor (electron microscope section)
Intra membranous particles
Ciliary necklaces
Endocytotic vesicles
Microtubules of cilium
Junctional complex
Mitochondria

Fig. 18.1. The peripheral olfactory system.

Peripheral analysis and coding of olfactory information

It is thought that the initial stage of signal generation occurs in an olfactory receptor when an odorant molecule is bound to a protein site on the cilium. However, electrophysiological records from individual receptor units usually fail to show specificity to single odorants: often a given unit is responsive to a large number of different odorants and even the grouping of specificities is not consistent from receptor to receptor. This suggests that many different receptor proteins are present on the cilium of a given receptor cell. How then are particular odours identified by the olfactory system?

One speculation has been that variations in the spatial distributions of different molecules might be produced by the mucous sheet (see above) through which stimulus molecules must pass to reach the receptors. In particular, there is evidence that different odours generate different anteroposterior *gradients* of activity across the receptor sheet and that the gradients for different compounds are inversely related to their retention times as measured by gas chromatography. It is possible to imagine neural mechanisms that measured such gradients – in the sort of way that comparisons are made between points on the retina or the basilar membrane. According to this hypothesis, the mucous and the receptors embedded in it would be the analogue of the fixed phase in chromatography and the moving air stream during a sniff would be the analogue of the moving phase. However, a process of this kind cannot be the primary basis for olfactory analysis, since there are compounds (such as R and S carvone) that have very similar retention times but distinct odours. Moreover, subjects can still distinguish and correctly recognise odours when they are introduced by the retronasal route.

Since, then, receptors are not intrinsically specific to single odorants and do not have absolute specificity imposed on them by the mucous, later stages of the olfactory system must recognise the pattern of activity (the *across-fibre firing pattern*) produced by that set of receptor cells that bear receptor proteins specific to molecules of a given odour. As an illustration, we represent in Fig. 18.2 three partially overlapping groups of receptors, which respond to three different odours, A, B and C. If group A receptors fire, we recognise odour A; if group B fire, odour B is perceived; and so on. Other odours may lead to activity in two of the groups of receptors or in all three.

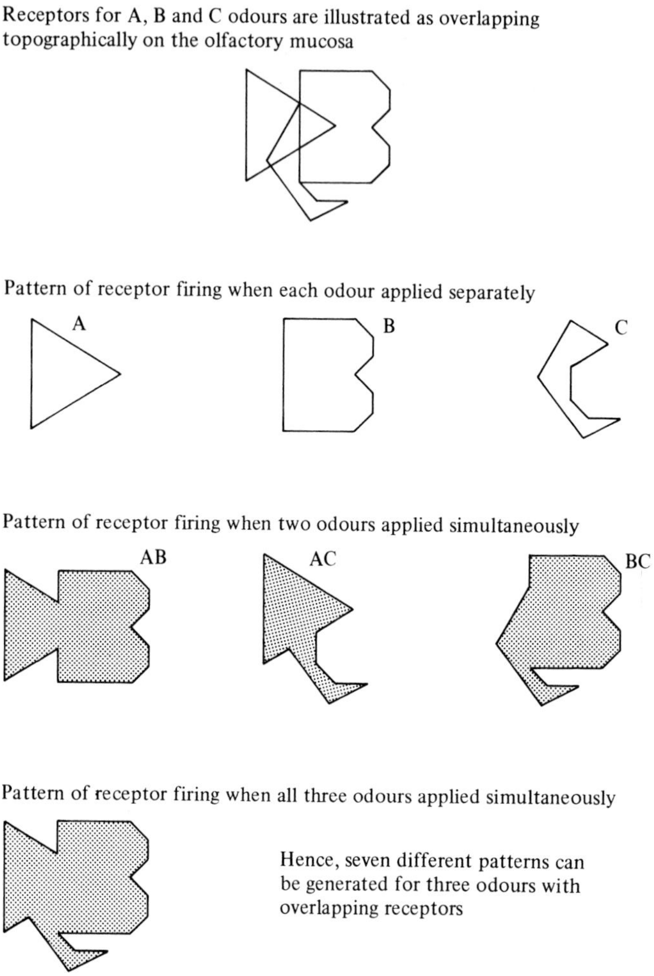

Receptors for A, B and C odours are illustrated as overlapping topographically on the olfactory mucosa
Pattern of receptor firing when each odour applied separately
A
B
C
Pattern of receptor firing when two odours applied simultaneously
AB
AC
BC
Pattern of receptor firing when all three odours applied simultaneously
Hence, seven different patterns can be generated for three odours with overlapping receptors

Fig. 18.2. Olfactory coding: a pattern theory.

This model of olfactory function is of course simplified. As represented, we have seven different types of receptor firing pattern for a simple A, B, C odour model. In reality, receptors are found experimentally to respond to as few as two odours and to as many as 24, both these figures being underestimates because of the experimental limits on the number of odours that can be tested during electrophysiological recording from a single unit. Since the grouping of specificities is not consistent from receptor to receptor, the number of possible patterns is very great.

The multiple specificity of individual receptors may provide an explanation for the psychophysical phenomenon of *cross-adaptation*, the alteration in the perceived quality of one odour according to the identity of a preceding odour. Suppose that there is a subset of receptors that contributes to the total patterns of response to both odours. Cross-adaptation may arise if the odour A left this subset of receptors with altered sensitivity, either because some of the sites of the receptors were still occupied by molecules of A or because these cells were left less responsive by their recent activity. The total pattern of response to B would thus be distorted.

When more than one odour is presented simultaneously, we are often unable to identify the components of the mixture, rather as we are unable to distinguish the component wavelengths of white light. This is perhaps because the pattern generated by each individual odour is lost in the total pattern produced by the odour complex. It is true that skilled perfumers can analyse odour complexes but it is likely that they have learnt cognitively to recognise the sensations that given odours produce when mixed.

Depending on the odour molecule, there may be more than one active site on the molecule which interacts with different binding proteins. If these binding proteins vary in their affinity for the molecule and are distributed along the cilia of different receptor cells, then even a pure odour is capable of generating different patterns of receptor activity according to the concentration. This would account for the psychophysical phenomenon of a given molecule (e.g. indoles) smelling differently at increasing concentrations. By introducing the idea of multiple recognition sites for a given odour we can imagine a coding system at the receptor level of immense specificity. Likewise, if a receptor requires two, three or more sites to be bound (allosteric receptors) before it fires, this permits a totally different firing pattern for odour complexes over and above that produced by each of the component odours. Taken together, odours with multiple recognition sites and receptors with multiple binding sites, some allosteric, provide almost unlimited potential for olfactory coding.

18.3. THE OLFACTORY BULB

The olfactory bulbs are those parts of the preserved human brain that are usually missing, but whose absence is rarely noticed by medical students. Though relatively unobtrusive parts of the human brain, in lower mammals the olfactory bulbs protrude conspicuously from the forebrain, occupying a significant proportion of the cranial cavity. The olfactory bulb is the first relay station for olfactory inputs, some fifty million unmyelinated olfactory afferents converging into each bulb in a species such as the rabbit. When the olfactory bulb is sectioned, histologically distinct layers can be seen, and the several types of neurons therein were originally described by Ramon y Cajal in 1911 (Fig. 18.3); his original description still holds true today. The principal neuron of the olfactory bulb is the *mitral cell,** the axons from which run caudally and merge together to form the major output of the bulb, the *lateral olfactory tract*. Most of the dendrites of the mitral cell terminate in the external plexiform layer, but the apical dendrite terminates in a *glomerulus*, a discrete, roughly spherical region of neuropil. It is within this glomerulus that the mitral cell makes synaptic contact with the incoming olfactory nerves. There are about 2000 glomeruli in the rabbit olfactory bulb and a convergence in the order of 1000 to 1 for olfactory input to mitral cells, each glomerulus containing some 25000 olfactory axons and the apical dendrites of some 25 mitral cells. Surrounding the glomeruli are the cell bodies of the short-axon *periglomerular cells*, which also send their branching dendritic tree into the glomerulus. The periglomerular cell distributes its axon to neighbouring glomeruli.

In addition to the periglomerular cells, there are two further types of intrinsic neuron within the olfactory bulb. They are the granule cells, which form a thick layer (see Fig. 18.3), and the less common short-axon cells, which are usually found in the same layer as granule cells and which branch into the external plexiform and granule cell layers. The granule cells themselves are found deep to the mitral cells and have a superficial process that terminates in the external plexiform layer, forming numerous connections with the lateral dendrites of the mitral cells. The granule cells lack a true axon and their processes resemble dendrites in their fine structural features, being covered with numerous small spines (gemmules).

Smaller versions of the mitral cells, the tufted cells, are found in the external plexiform layer and appear to send axons with the mitral cell axons in the lateral olfactory tract.

The olfactory bulb has been likened to the retina, in that the histology suggests a direct pathway (primary receptor fibre → mitral cell) that is subject to modulation by two tiers of horizontal connections (corresponding respectively to the periglomerular and

* So named because of a resemblance to a bishop's head-dress.

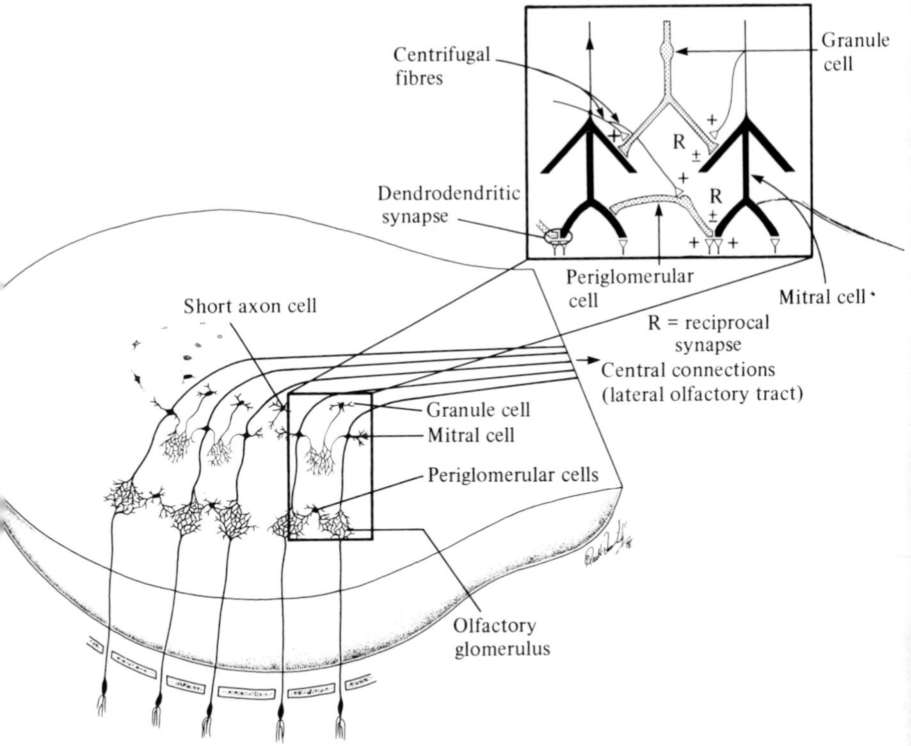

Fig. 18.3. The olfactory bulb.

the granular cells). Unlike the case of the retina, nothing is really understood of the informational coinage in which the system deals; but at a strictly physiological level something can be said of the organisation of the bulb and a tentative circuit diagram is incorporated in Fig. 18.3. The incoming axons of the olfactory nerve enter the glomeruli and synapse on the dendrites of mitral, tufted and periglomerular cells. Electron microscopy reveals a complex synaptic triad in which the olfactory nerve fibres form what are thought to be excitatory synapses with both the apical dendrite of the mitral cell and the dendrites of periglomerular cells; the dendrites of the periglomerular cells are thought to be inhibitory to the apical dendrite of the mitral cell. (The latter type of synapse is described as *dendro-dendritic*; notice that the same neuronal process is both pre- and postsynaptic, an arrangement not permitted in the typical neuron of the textbooks but one in fact not uncommon in the CNS, cf. Fig. 2.4). Thus the periglomerular cell offers a pathway by which information from other glomeruli could influence the signals transmitted onwards by the mitral cell. The second tier of horizontal interaction in the olfactory bulb is provided by the granule cells, which make dendro-dendritic synapses with the mitral cells in the external plexiform layer. There is evidence that the granule cell is inhibitory to the mitral cell. The granule cell itself is postsynaptic to collaterals of mitral cell axons and to centrifugal fibres from other parts of the brain. Thus there is a pathway by which the activity of a given mitral cell can be influenced by that of other mitral cells and by descending influences.

18.4. THE CENTRAL OLFACTORY CONNECTIONS

The central connections of the olfactory system are more readily understood by distinguishing the cortical pathways from those that project directly into the limbic brain. The primary cortical projections of the main olfactory bulb pass ipsilaterally in the lateral olfactory tract to the pyriform cortex (solid lines in Fig. 18.4*a*). In man this occupies a relatively small area on the anterior end of the hippocampal gyrus known as the uncus. This part of the cortex lacks the typical granular appearance characteristic of the other primary sensory areas and is called allocortex. Recent techniques of autoradiography and electron microscopy suggest that the primary projection of the pyriform cortex is to the thalamus (medialis dorsalis) which then projects to the neocortex (orbito-frontal cortex). Thus the olfactory system is similar to other sensory systems in that cortical access is achieved by way of a specific nucleus relaying in the thalamus (Fig. 18.4*a*, solid line).

The olfactory system also has a direct monosynaptic link with the limbic brain. In most animals, but not in primates, this neural connection is made with the amygdala (cortico-medial nuclei), a part of the brain concerned with sexual and aggressive behaviour, and itself only one synapse removed from that part of the hypothalamus responsible for neuroendocrine regulation (Fig. 18.4.*b*). This limbic projection is specialised to the extent of having its own system of receptors (the vomeronasal organ) and a spatially and histologically distinct region of the olfactory bulb. The vomeronasal organ is a blind-ending sac situated at the base of the nasal septum and ensheathed in a scroll of cartilage, communicating with both oral and

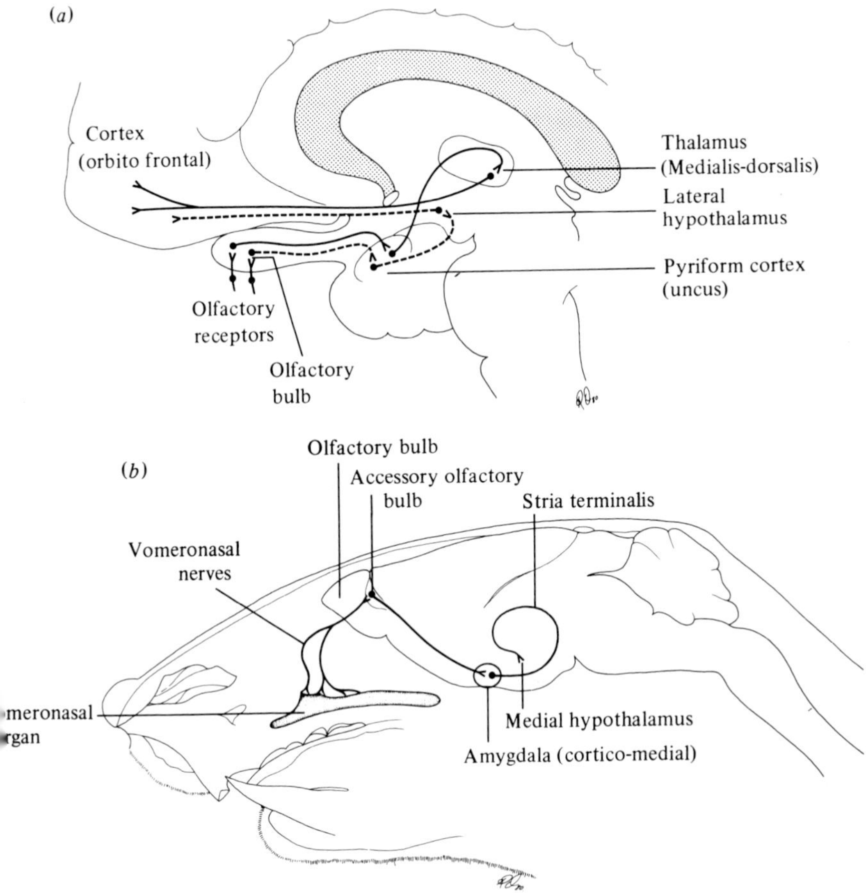

Fig. 18.4. (*a*) The central olfactory projections (characterisic of higher primates). (*b*) The accessory olfactory projections (characteristic of rodents).

nasal cavities. The vomeronasal organ and its neural connections are often referred to as the accessory olfactory system and unlike the main olfactory system this does not gain access to the thalamus and in turn to neocortical regions. Primates have olfactory pathways that project indirectly to the limbic brain but these are not associated with a specialist group of receptors like those of the vomeronasal organ, and have their cell bodies in the olfactory bulb (Fig. 18.4*a*, broken line).

18.5. CENTRIFUGAL OLFACTORY PATHWAYS

Centrifugal control of afferent input is common in sensory systems and the olfactory system is no exception. Anatomical and electro-physiological investigations reveal the primary neuron of the olfactory bulb, the mitral cell, as a focal point for both incoming olfactory and centrifugally directed information. Centrifugal fibres entering the mammalian olfactory bulb originate from the anterior olfactory nucleus, and from the midbrain aminergic (5-HT, NA) projections. The anterior olfactory nucleus is a focus for basal forebrain projections and is the origin of commissural projections to the contralateral olfactory bulb. The monoamine projections originate in the midbrain raphé nucleus and the locus coeruleus, and are known to be serotoninergic and noradrenergic respectively. A cholinergic fibre projection to the glomerular and mitral-cell layers of the olfactory bulb originates in the lateral preoptic area. Histofluorescence studies have demonstrated that serotoninergic centrifugal fibres appear to be directed solely to the glomerular layer whereas the noradrenergic fibres are directed largely to the granule cell layer. Centrifugal influences in the olfactory bulb are mainly inhibitory. Other parts of the brain (for example, the reticular formation) can therefore influence the overall level of activity in the olfactory bulb. It was originally suggested that such centrifugal projections may serve to 'wipe clear the system' so that new olfactory messages may be written out. More recent evidence suggests that centrifugal pathways, by a process of disinhibition, may selectively influence incoming olfactory information. For example, electro-physiological experiments have demonstrated selective modulation of olfactory input in correlation with the internal alimentary state. Briefly, it has been shown in rats that, when they are hungry, the multi-unit discharge of the mitral cell layer is enhanced in response to food odours. The activity of the mitral cells is not similarly enhanced by food odours in satiated rats, or by non-food odours in hungry rats. The increased mitral cell activity in response to food odours is abolished by section of centrifugal fibres. Moreover, if a non-food odour (eucalyptol) accompanies the food together with aversive conditioning (sickness induced by apomorphine) there is a significant decrease in the palatability of this food, together with modulation of the mitral cell activity to the eucalyptol odour.

An interesting *psychophysical* variation in sensitivity occurs during the menstrual cycle in women. Sensitivity to the musk-like odour of pentadecalactone exaltolide an ester used as a fixative in perfumes, is highest around ovulation and lowest around menses. (Many men cannot smell this pentadecalactone at all). The results seem to represent a true change in sensitivity, and not a shift in criterion (see Chapter 12, p. 234 and Fig. 1.1) and the variation does not occur for some other odorants tested, such as the ester, amyl acetate. The mechanism is not yet known. The variation *could* be explained by centrifugal control at the level of the olfactory bulb, but equally it could reflect a neural change elsewhere in the olfactory system or, for example, an effect of oestrogen on the glands producing olfactory mucus. The last hypothesis is not unreasonable, since the esters, such as pentadecalactone for which sensitivity is known to vary, are relatively involatile ones that would be predicted (from gas chromatographic measurements) to diffuse less readily through the olfactory mucus. In the monkey, histological changes are known to ocur in the supporting cells of the olfactory epithelium during the menstrual cycle and may also provide a structural basis for altered olfactory sensitivity.

18.6. PSYCHOPHYSICS

Specific anosmia

A matter for everyday remark is the striking variation in olfactory sensitivity between individuals. Not well known, but of especial interest, is the existence of *specific anosmia*, a disproportionately low sensitivity to a single odorant, or group of related odorants, in a subject who enjoys normal sensitivity to a wide range of other odorants. For example, one person in ten is reported to be insensitive to the poisonous gas hydrogen cyanide, and one in a thousand insensitive to the butyl mercaptan odour of the skunk.

The understanding of colour vision has been much advanced (and, quite often, much confused) by the analysis of the genetic deficiencies that apparently arise from the absence or mutation of individual classes of receptor (Chapter 9). Specific anosmia has been seen as an analogous tool for the analysis of olfaction. J. E. Amoore has studied a group of individuals selected for being insensitive to the sweaty smell of isobutyric acid and his results, shown in Fig. 18.5, offer the nearest olfactory analogue to the psychophysically estimated sensitivities of single channels in other modalities (see Chapters 1 and 9). As usual, the psychophysical results do not directly give the true

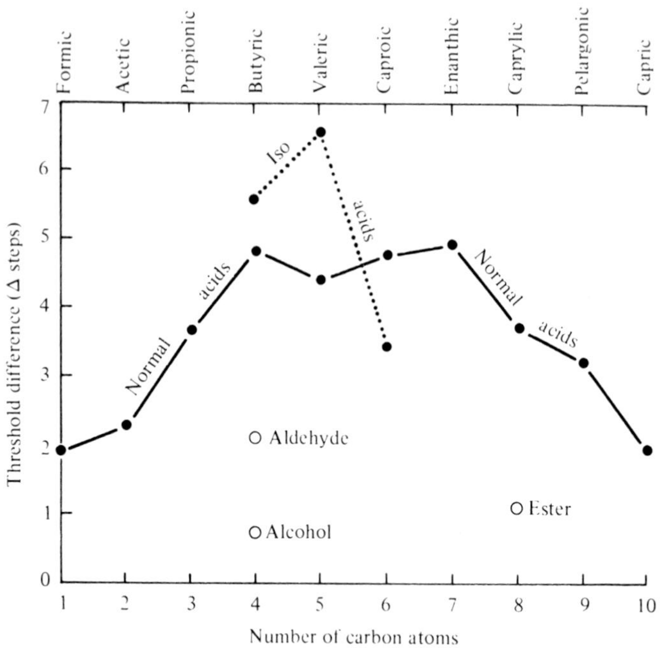

Fig. 18.5. Specific anosmia to carboxylic acids. The ordinate represents the logarithm of the increase in concentration required for detection by the anosmic observers relative to that required by normal observers. The abscissa is the number of carbon atoms in the molecule. The alcohol, aldehyde and ester are the isobutyl derivatives. The greatest loss of sensitivity is seen for the structural isomer of valeric acid, the branched-chain isovaleric acid. The same two groups of observers were used throughout; the 'anosmics' were selected for their insensitivity to isobutyric acid. (After J. E. Amoore.)

tuning curve of the putative channel. The ordinate of Fig. 18.5 represents the logarithm of the increase in concentration required for detection by the 'anosmic' group; each division of the ordinate corresponds to a halving of the dilution. The abscissa represents the series of carboxylic acids tested, arranged according to the number of carbon atoms in the molecule; these acids have a common rod-like shape but differ in their lengths. The sensory deficiency, the degree of anosmia, is low for one and two carbons (formic and acetic acids), increases to a plateau for molecules with 4 to 7 carbon atoms, and then falls again for larger molecules. Molecular shape also seems significant: isobutyric and isovaleric acid show a greater deficiency than the corresponding normal acids of the same formula.

The anatomical site of the channel suggested by these results is not known; the anosmia *could* arise from the absence of a particular receptor protein but it might equally correspond to the absence of some central channel that followed transformation of the primary data. (Only recently has the analogous question in colour vision been settled.) Of course the shape of the psychophysical curve (Fig. 18.5) must depend not only on the channel for which isovaleric acid is the optimal stimulus, but equally on channels with maximal sensitivity at adjacent positions on the sensory dimension. Nevertheless, the specific anosmias are intriguing and may hold clues as to the dimensions along which smell is analysed.

Directional smelling

Animals not only need to be able to identify chemical stimuli but also need to know which way they should turn in order to move up or down the gradient of concentration. Judgement of the direction of an olfactory source could depend on successive sampling (as when, say, the head was turned from side to side), but the sensory physiologist G. von Békésy has reported that human observers can discriminate direction even when their heads are fixed. This ability is analogous to our capacity to localise sounds (Chapter 15) and appears to depend on analogous differences in intensity and in time-of-arrival at the two nostrils. When an odorant was released for 1 s from a source 8 cm from the nose, the observer could locate the source with a precision of 7–10° when it was near the medial plane (the vertical plane passing though the ridge of the nose). In analytic experiments, stimuli of carefully controlled concentration and time-of-onset were concurrently delivered to each nostril by means of two separate cannulae. A 10% difference in concentration between the nostrils or a difference of the order of 1 ms in presentation time was sufficient, von Békésy claimed, to displace the localisation of the smell from the medial plane.

18.7. OLFACTION AND BEHAVIOUR

Karlson and Butenandt introduced the term 'pheromone', a word derived from the Greek *pherin* (to transfer) and *hormon* (to excite), to denote substances secreted (externally) by an animal, which then can exert effects on the behaviour or physiology of another individual of the same species. Although originally defined in the context of insect communication, the word has since been extended to include chemical communication in all species. The ability to detect and exploit odours appears to reach its highest degree of development among the mammals, whose responses to pheromones are not as stereotyped as those of the insects.

Three kinds of response to pheromones are now recognised among mammals. (1) Primer pheromones alter the physiology and thereby the behaviour of the receptor animal. This type of response is slow to develop and demands a prolonged stimulation which is mediated through the nervous and endocrine systems. (2) If the pheromone produces a more or less immediate change in the behaviour of the recipient, it is said to have a releaser effect. Sex attractants constitute a large and important category of the releaser pheromones. (3) A third response, olfactory imprinting, has also been described in some mammalian species: exposure to odour cues at a critical stage in the development of the neonate may result in permanent modification of sexual preferences in the adult.

One species in which olfactory influences on sexual behaviour are clearly established is the golden hamster. When the female golden hamster is sexually receptive, an abundant, highly odorous substance exudes from her genitalia. The male hamster sniffs and licks this vaginal discharge before and intermittently throughout mating. If this secretion is applied to the fur of a male, other males respond to him as though he were a female and attempt to mate with him. Moreover, male hamsters no longer mate if their olfactory bulbs are removed; merely anaesthetising the olfactory mucosa interferes with the male's mounting behaviour.

The vaginal discharge itself seems not be critical for the maintenance of the mating sequence but rather for its initiation. A male hamster without a sense of smell placed with a receptive female does not begin mating. If copulation has already begun, however, and then anosmia is induced by rapid occlusion of the nostril, the mating sequence is not interrupted. Such sexually experienced anosmic males can therefore mate, but this depends on their having perceived some olfactory information concerning the female.

In the main, most of the ways in which olfactory cues influence sexual behaviour have been demonstrated in male mammals. Olfaction is important in the female too, but perhaps not in the same way. Attractant pheromones appear not to be so important in female sexual behaviour, but primer pheromones, on the other hand, seem in most instances to act only on the female to bring about changes in reproductive physiology. The best example of primer pheromone activity in mammals is shown in the control of the oestrous cycle of the laboratory mouse, where cycles of four to twelve or more days can be induced with some predictability, depending upon the olfactory environment. Lengthening of the oestrous cycle of the laboratory mouse is apparently produced by urine odour from other females, while the induction of oestrous in females with suppressed cycles occurs in the presence of male urine. The effects of male urine appear to be induced by a small molecule that is androgen-dependent, is species-specific, and may act at remarkably low concentrations. This odour can override the suppressive effects of female odour, and therefore initiate the oestrous cycles and hence synchronise ovulation among a group of females.

The odour of male urine also has an additional effect of producing an olfactory block to pregnancy. This occurs if urine from a male other than the stud, and preferably of a different strain, is introduced to the female within the first three days after mating. These females fail to show implantation and return to oestrus within five days. Hence, odour cues in certain mammalian species can induce predictable physiological changes by acting on neuroendocrine mechanisms. This action is analogous to that of light cues, in the form of daylength changes, which affect circadian and seasonal endocrine rhythms; and that of tactile cues that induce reflex ovulation or pseudopregnancy. Those mammals that respond to primer pheromones possess specialised olfactory receptors in the vomeronasal organ and since these, via the accessory olfactory system, are in close anatomical relationship with the neuroendocrine hypothalamus, this is almost certainly the primary pathway by which environmental odour cues affect reproductive status. Lesions to the vomeronasal system have the predictable effect of blocking the suppression of oestrus and of preventing the olfactory block to pregnancy. The mechanism by which olfactory signals are translated into neuroendocrine events appears to involve the hypothalamic dopamine system and in turn its effects on prolactin and luteinising hormone.

Man and the higher infra-human primates do not possess a vomeronasal system, and there is no evidence that olfactory cues influence neuroendocrine state in higher primates. Olfactory cues can certainly influence female attractiveness but since the behavioural response can be readily suppressed in the complex social experiential environment, neural circuitry must be available for integration with experiential and other sensory information. It is not unreasonable to assume such complex integration is a function of the neocortical olfactory projection. Although no direct olfacto-limbic projection similar to that of the accessory olfactory system of lower mammals has been found in primates, it would be wrong to assume that the limbic brain is deprived of olfactory information. Electrophysiological studies have demonstrated an olfactory projection via the pyriform cortex and lateral hypothalamus of the monkey before reaching the orbitofrontal cortex. This indirect but subthalamic projection via the limbic brain to the neocortex may prove to be a neural pathway by which olfactory cues could influence mood states and behaviour. If a sub-thalmic olfactory projection is also present in man, then a phenomenon akin to 'blind sight' might also exist for the olfactory sense. If this is ever shown to be true in man then the olfactory system might be considered to have privileged access to that part of the brain subserving emotional behaviour.

18.8. SUGGESTIONS FOR FURTHER READING

General reviews

Beidler, L. M. (1971) *Handbook of Sensory Physiology*, vol. IV, 1, *Olfaction*. Berlin: Springer.

Carterette, E. C. & Friedman, M. P. (1978) *Handbook of Perception*, vol. VIA, *Tasting and Smelling*. New York & London: Academic Press.

Moulton, D. G. (1978) In *Handbook of Behavioural Neurobiology*, vol. I, ed. R. B. Masterton. New York & London: Plenum.

Monnier, M. & Holroyd, E. (1975) *Functions of the Nervous System*, vol. 3, *The Sense of Smell*. Amsterdam, Oxford, New York: Elsevier Scientific Publications.

Denton, D. A. & Coghlan, J. P. (1975) *Olfaction and Taste*, vol. V. New York & London: Academic Press.

Doty, R. L. (1976) *Mammalian olfaction, reproductive processes, and behaviour*. New York & London: Academic Press.

Specific references

Structure of olfactory mucosa and receptors. Graziadei, P. P. C. & Metcalf, J. F. (1971) *Zeitschrift für Zellforschung Mikroskopische Anatomie*, **116**, 305. Kerjaschki, D. & Hörandner, H. (1976) *Journal of Ultrastructural Research*, **54**, 420. Menco, B. P. M., Dodd, G. H., Davey, M. & Bannister, L. H. (1976) *Nature*, **263**, 597.

Olfactory coding. Mozell, M. M. (1970) *Journal of general Physiology*, **56**, 46. O'Connell, R. J. & Mozell, M. M. (1969) *Journal of Neurophysiology*, **32**, 51. Adrian, E. D. (1950) *British medical Bulletin*, **6**, 330–4. Gesteland, R. C., Lettvin, J. Y. & Pitts, W. H. (1965) *Journal of Physiology*, **181**, 525. Moulton, D. G. (1976) *Physiological Reviews*, **56**, 578.

Physiology of olfactory bulb. Mori, K. & Takagi, S. (1978) *Journal of Physiology*, **279**, 569; 589.

Olfactory bulb structure. Shepherd, G. M. (1972) *Physiological Reviews*, **52**, 864. Willey, T. J. (1973) *Journal of Comparative Neurology*, **152**, 211.

Central olfactory connections. Powell, T. P. S., Cowan, W. M. & Raisman, G. (1965) *Journal of Anatomy*, **99**, 791. Scalia, F. & Winans, S. S. (1975) *Journal of Comparative Neurology*, **161**, 31. Heimer, L. (1972) *Brain, Behaviour and Evolution*, **6**, 484. Tanake, T., Tino, M. & Takagi, S. F. (1975) *Journal of Neurophysiology*, **38**, 1284. Giachetti, I. & MacLeod, P. (1977) *Brain Research*, **125**, 166.

Centrifugal olfactory connections. Price, J. L. & Powell, T. P. S. (1970) *Journal of Anatomy*, **107**, 215. Dahlström, A., Fuxe, K., Olson, L. & Ungerstedt, U. (1965) *Life Sciences*, **4**, 2071. Broadwell, R. D. & Jacobowitz, D. M. (1976) *Journal of Comparative Neurology*, **170**, 321. Pager, J. (1974) *Physiology and Behaviour*, **12**, 189.

Psychophysics and perception. Engen, T. (1982) *The Perception of Odors*. New York: Academic Press.

Effects of menstrual cycle. Vierling, J. & Rock, J. (1967) *Journal of applied Physiology*, **22**, 311. Mair, R. G. *et al.* (1978) *Sensory Processes*, **2**, 90.

Specific anosmia. Amoore, J. E. (1967) *Nature*, **214**, 1095; and in Beidler, L. M. (1971) *Handbook of Sensory Physiology*, vol. IV, 1. Berlin: Springer.

Directional discrimination. von Békésy, G. (1964) *Journal of applied Physiology*, **19**, 369.

Olfaction and behaviour. Winans, S. S. & Powers, J. B. (1977) *Brain Research*, **126**, 325. Keverne, E. B. (1978) In *Biological determinants of Sexual Behaviour*, ed. J. B. Hutchinson, p. 727. Chichester: John Wiley & Son. Reynolds, J. & Keverne, E. B. (1979) *Journal of Reproduction and Fertility*, **57**, 31. Tanabe, T., Yarita, H., Tino, M., Ooshima, Y. & Takagi, S. F. (1975) *Journal of Neurophysiology*, **38**, 1269.

CHAPTER NINETEEN

Chemical senses: Taste

E. B. KEVERNE

Physiologically the specialised receptors of the tongue have been regarded as the taste organs, but it should be remembered that our total taste perception is considerably more complex, involving the smell of food, its texture and its temperature.

Although the tongue may vary in shape in different animal species and serve not only for ingestion, but also for mastication and even grooming, there are certain morphological characteristics that all tongues have in common. The tongue is a muscular organ covered in mucous membrane and situated in the floor of the mouth. It may be divided into *oral* and *pharyngeal* parts, a line of *circumvallate papillae* marking off the junction between oral and pharyngeal areas. Papillae are found on the dorsal surface of the tongue, and fall into four main classes: the filiform, fungiform, circumvallate and foliate papillae. The *filiform* ('thread-like') papillae are mechanical, non-gustatory papillae. *Fungiform* papillae are flattened, raised structures that resemble button mushrooms (hence the name); they contain one or more taste buds in the surface epithelium and respond mainly to sweet and salty tastes. They are usually restricted to the oral part of the tongue, and can be identified with the naked eye as red spots, owing to their rich blood supply. The *circumvallate* papillae are sunken papillae with a trough separating them from the surrounding walls. In the walls of the papilla are tiers of taste buds which respond mainly to sour and bitter tastes. Secretory glands associated with the circumvallate papillae help to rinse the taste buds located on their walls. This is aided by a tongue movement, which forces liquid in and out of the trough. *Foliate* papillae also contain taste buds and lie on the lateral border of the tongue just anterior to the line of circumvallate papillae. This area of the tongue is particularly sensitive to sour taste, whilst sweet sensitivity is greatest at the tip of the tongue. Taste buds are not restricted to the tongue, and in man taste buds are present on the epiglottis, on the upper third of the oesophagus, and on the palate. Taste buds in the tongue are innervated by the cranial nerves seven and nine. The VIIth cranial nerve supplies the anterior two-thirds of the tongue by way of the chorda tympani nerve, which conveys mainly the sensation of sweet and salty from the fungiform papillae. The glossopharyngeal nerve (IXth cranial) serves the posterior two-thirds of the tongue and carries mainly sour and bitter sensations from the circumvallate and foliate papillae. The Xth cranial nerve innervates those additional taste buds found on the epiglottis and palate and the upper part of the oesophagus.

19.1. STRUCTURE OF THE TASTE RECEPTOR

The peripheral gustatory organs illustrated in Fig. 19.1 are the taste buds, which are composed of epithelial cells modified into several different cell types and numbering from 30 to 80 in each taste bud. The term 'bud' appropriately describes the appearance of the taste receptors, but the developmental changes that bring about renewal of taste cells within the bud occur in an inward direction. Hence the outermost cells, which surround the taste bud and are crescentic in profile, are thought to differentiate into basal cells. The basal cells in turn give rise to the more centrally situated principal taste cell, of which there are three main types. These have been characterised according to their cytoplasmic inclusions, microvilli projections, and synaptic connections with unmyelinated nerve fibres. The apical parts of these cells extend as narrow projections toward the gustatory pore. This pore, together with the pit just beneath, is filled with secretions.

The sensory nerve fibres, which have their cell bodies in ganglia of either VIIth, IXth or Xth cranial nerves, establish contact with the taste cells. The taste receptors do not have axons, but form a complex functional relationship with their nerve supply involving both efferent and afferent synapses. A single fibre may innervate several different taste cells, and a single taste cell may be innervated by more than one nerve fibre.

Of the three principal cell types found in the taste bud, two have been long recognised and characterised as 'light' or 'dark' cells according to the density of their cytoplasmic inclusions. The light cells (also known as Type II) do not have dense cytoplasmic inclusions and the tubular bundles which extend deep into their apical region together with the extensive contact these cells make with nerves would suggest they are true gustatory receptors. The predominant (Type 1) principal cell or dark cell (Fig. 19.1) is thought

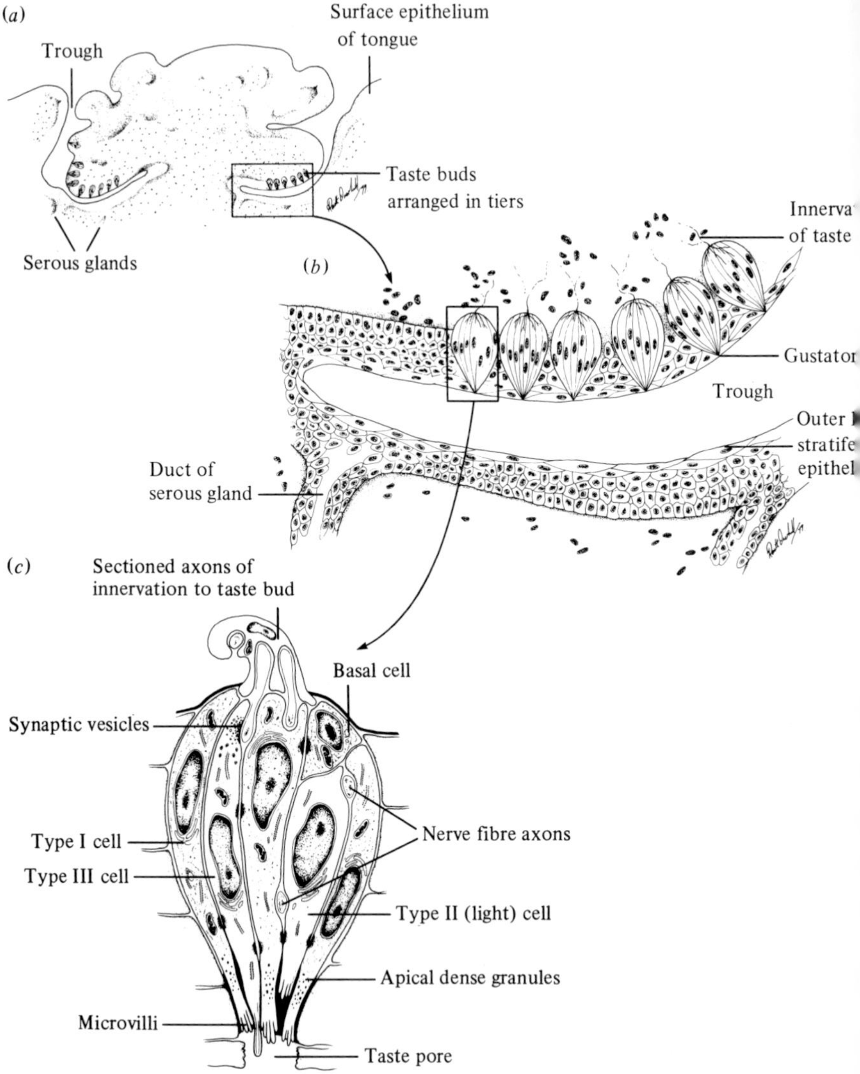

Fig. 19.1. The peripheral taste receptors. (*a*) Circumvallate papilla of tongue. (*b*) Enlargement of (*a*); (*c*) reconstruction of taste bud, based on electron microscope studies (after Murray, 1973.)

to be supportive rather than a true gustatory receptor. These cells have nerve terminals in intimate contact with their nuclei but these endings may be trophic in nature, stimulating differentiation into definitive taste cells. Another distinctive feature of this cell type is the presence of dense granules in the apical region, which are thought to be the precursors of the mucous seen in the pit.

A third, but less numerous type of principal cell has recently been characterised using electron microscopic techniques. Because of the nature of its synaptic contacts with the complex interweaving unmyelinated nerve fibres, it is also thought to be gustatory. These synapses of Type III cells have the appearance of typical chemical synapses, with the arrangement of vesicles indicating transmission from cell to nerve.

19.2. TRANSDUCTION

Before the taste stimuli can elicit a response from the receptors, the chemical cues have first to penetrate the layer of mucous overlying them. The apical pit of the taste bud is filled with a dense muco-substance into which penetrate the tips of the taste cells where the receptor sites are thought to be located. It has been suggested that the mucous protects the receptors against drying or osmotic damage, but its role in the transduction process is uncertain. A logical assumption might be that the mucous secretion serves to rinse out the taste pit in order for reactivation of the adapted receptors. Admittedly, in neural terms this would be a slow process, but then taste sensations tend to linger; one only has to remember the unpleasant aftertaste of sipping sour milk or accidentally biting into a penicillin capsule. Secretion of the mucous substance is thought to be from the supportive cells (Type 1) which, because of their large numbers and distribution, completely surround the true receptors. These supportive cells also receive a nerve supply, the so-called trophic feedback, that may well serve to active the secretory process following chemical stimulation of the receptors.

Taste receptors are thought to be excited by adsorption of the tastant to some elements of the receptor cell surface. It is thought that taste stimuli interact with molecules in the membranes of the receptor cells and do not enter the cell, since many toxic substances can be tasted without damaging the taste system. The receptor site may be a protein, and the formation of a protein–substrate complex could be part of the adsorption process. Adsorption of taste stimuli on the

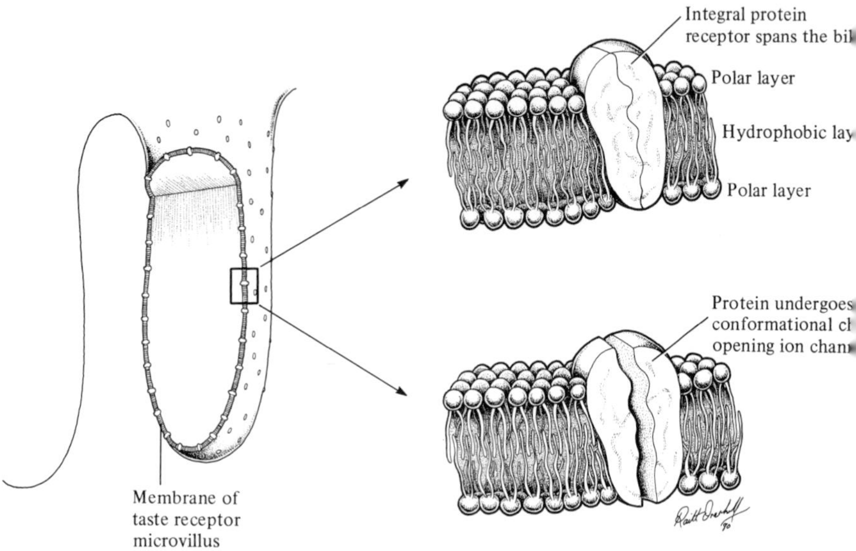

Fig. 19.2. Hypothetical arrangement of receptors on taste cell microvillus.

receptor membrane is then thought to change the conformation of receptor macro-molecules, which in turn may be responsible for a change of permeability to ions (Fig. 19.2). This transduction mechanism is, however, still hypothetical. Since there are four primaries described for taste (sweet, bitter, salty and sour), models for stimulus–receptor adsorption have been based on the physical or chemical similarities of molecules with similar taste qualities.

There has been some success in identifying structural similarities among molecules of sweet-tasting compounds. Sweetness varies inversely with the degree to which sugar hydroxyl groups form hydrogen bonds with receptor sites. The fact that a molecular structure common to all sweet-tasting substances exists suggests that the sweet receptor site has a complementary configuration. Recently, a protein fraction has been extracted from bovine tongue epithelium which forms complexes with sweet-tasting stimuli and probably represents the receptor site for sweet taste. However, the existence of a receptor protein that interacts with sweet-specific molecular configurations does not explain why there are differences in the firing rates of individual taste fibres to a given sweet substance. These differences could be accounted for by assuming more than one receptor protein for sweet, but with differing binding affinities.

Substances that taste salty are mainly ones that dissociate in an

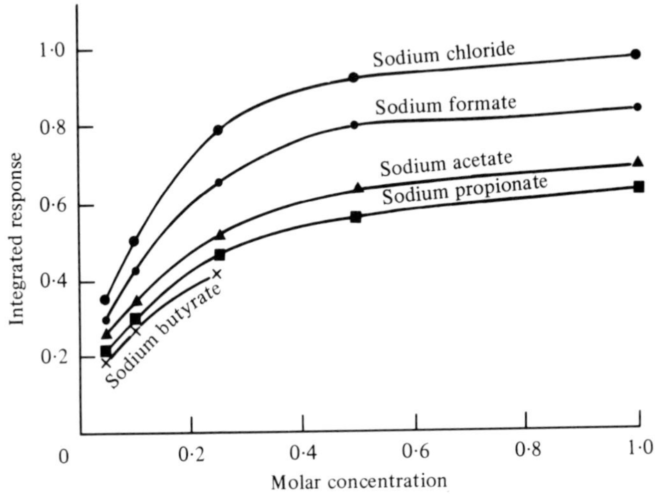

Fig. 19.3. Integrated response of chorda tympani to various concentrations of salts applied to the surface of the tongue, in rats. (After Beidler.)

aqueous solution into positive and negative ions. Beidler suggested that cations are excitatory and anions are inhibitory in taste. He based this conclusion on the finding that the stimulatory efficiency of salts depends primarily on the cation although the anion may modify somewhat the salty taste of the attendant cation. Thus, similar molar concentrations of sodium chloride, sodium bromide and sodium nitrate have the sodium cation in common and produce neural responses of the same magnitude, while similar molar concentrations of sodium chloride, magnesium chloride, ammonium chloride and potassium chloride have the chloride anion in common and produce neural responses of very different magnitudes. The latter two salts (ammonium chloride and potassium chloride) are not efficient in stimulating the same single taste fibre that responds well to sodium chloride, and none of the last three salts taste very salty. Beidler's suggestion that anions are inhibitory in taste stems from his finding that salts with a common cation produce neural responses of diminishing magnitude as the size of the anion increases, for example sodium chloride to sodium butyrate as in Fig. 19.3. Proteins have long been thought to be the receptor molecules and since the ability of anions to bind with protein increases with the size of the anion Beidler suggested that the anion might inhibit the action of the cation.

An alternative to the protein receptor model for transduction of salty tastes has recently gained favour. It has been suggested that the crucial step for the detection of salty stimuli involves the binding of cations to phospholipids in the membrane. Ion pairs compete with cations in this binding, and thus many of the interactions between anions and sodium ions in the transduction of salty tastes might be accounted for quantitatively. While such a mechanism may provide an alternative model for taste transduction of saltiness, the idea has not yet been applied successfully to the transduction of other tastes.

The sour tastes are thought to result from the dissociated hydrogen ion in acids. Not all acids are equally sour even when pH and normality are controlled, and other factors may therefore play a role in determining the intensity of sourness. Beidler's model of acid receptors (sour taste) is very similar to that for salt; the binding of hydrogen ions produces the neural response and the anions merely inhibit the action of the hydrogen ions. The receptor membrane becomes more positive as an increasing number of hydrogen ions are adsorbed, which in turn partially inhibits the addition of more hydrogen ions. This can be overome by simultaneous adsorption of the negative anions of the acid. Thus the ability of a particular anion of an acid to bind to the membrane helps to determine the number of hydrogen ions bound at a given pH and the resultant degree of sourness.

Bitterness, like sweetness, is produced by a number of organic compounds and by some inorganic ones. A number of naturally occurring bitter substances are found in plants (alkaloids, glycosides, diterpenes) and their toxicity may have provided the selective pressure to evolution of aversion to bitter tastes. No single structural component can be related to bitter substances in the way that it can for sweetness. PTC (phenylthiourea) and other compounds containing N—C=S groups are bitter, but different individuals show marked differences in their thresholds to such bitter molecules. The sensitivity to their bitter taste is mediated by a single Mendelian dominant gene, and 'non-tasters' inherit the recessive complement.

The initial event of bitter stimulation is thought to be quite different from other tastes and may actually involve penetration of bitter compounds into hydrophilic regions of the lipids of the receptor membrane. Bitter-tasting compounds reduce intracellular cyclic AMP levels of taste receptors by activating phosphodiesterase. This in turn decreases the receptor cells' permeability. The reduced permeability may account for why some compounds such as quinine still taste bitter at high concentrations which in vitro act as phosphodiesterase inhibitors. As the stimulant concentration rises, the permeability of the cell decreases by activation of phosphodiesterase, and hence the physiological activity of phosphodiesterase is not inhibited in vivo because of the reduced membrane permeability. The ease with which a compound can enter the cell is suggested to be related to its lipid solubility so that those compounds of the greatest lipid solubility would be the most bitter.

19.3. NEURAL CODING IN TASTE

It is commonly assumed that animals share with man the same taste qualities and preferences, but in fact a wide range of differences are found. Thus man and the laboratory rat like both sucrose and saccharin solutions, while the dog accepts sugar but rejects saccharin, and the cat is indifferent to both. Different species appear, therefore, to live in distinct taste worlds which may or may not overlap. Beidler (1955) has reviewed the work on electrophysiological responses to various taste stimuli in a number of animals and no consistent pattern emerges between different species. Since our understanding of the neural coding of taste stems from electrophysiological recordings, these species differences have to be borne in mind. Four primary gustatory qualities can be distinguished by the human tongue: salt, sour, sweet and bitter. But the taste specificity of a nerve fibre is not absolute: a single taste fibre can respond to a variety of chemical stimuli. However, if the recordings are made from a large population of fibres, then the response pattern across the population to a given stimulus is unique (see Fig. 19.4). On the basis of these findings, Pfaffmann suggested that the sensory qualities of taste must be separated by a logical type of analysis performed on the *pattern* of responses in a population of fibres. This is basically different from the arrangement in, for example, the skin, where each class of fibre responds only to a restricted range of stimuli and qualities can be separated without further analysis (see Chapter 17). We have met an analogous hypothesis in the case of olfaction (see above). Experimental support of this theory came from work by Erickson who measured the 'across-fibre-response' patterns of a set of chorda tympani neurons to different taste solutions. He recorded responses to ammonium chloride, potassium chloride and sodium chloride in 13 single fibres of the rat chorda tympani. Many of these fibres differed from each other in the number of impulses generated to each of these stimuli. However, when the overall pattern which each

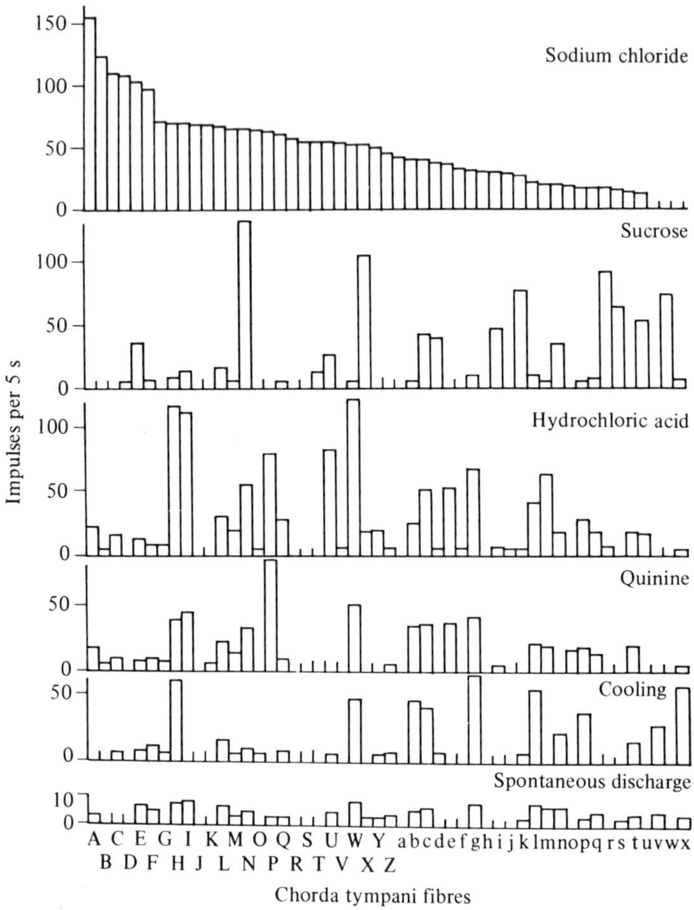

Fig. 19.4. Response profiles of 50 chorda tympani fibres in the rat. Fibres are arranged in order of responsiveness to 0.1 M NaCl (top) and responses to sucrose, acid, bitter, and temperature are shown below, together with the spontaneous discharge. (After Ogawa, H., Sato, M. & Yamashita, S. (1968). *Journal of Physiology*, **199**, 223–40.

A single chorda tympani nerve fibre is not specifically sensitive to a particular gustatory stimulus, but there are quantitative differences in activity among the fibres.

population of fibres generated was plotted, those for potassium chloride and ammonium chloride were similar to each other but very different from that of sodium chloride. In the case of an experimental rat, the similarities between tastes could be determined only by use of discrimination tasks, and Erickson tested for this using shock avoidance as the incentive for learning. Erickson found that rats indeed generalised more rapidly for those taste cues that generated similar across-fibre-patterns. Rats shocked for drinking potassium chloride avoided ammonium chloride and vice versa but neither group avoided sodium chloride.

If a characteristic pattern of activity *is* to be generated across a number of neurons by a given taste stimulus, not all taste stimuli can affect a given neuron to the same extent. Some neurons show a much larger response to a given stimulus than others, and are referred to as 'salt-best', or 'sucrose-best' etc. In the rat, the chorda tympani nerve, supplying the front of the tongue, carries sucrose-best and acid-best fibres, while quinine- or bitter-best fibres are mainly restricted to the glossopharyngeal nerve. Hence the peripheral taste neurons can be segregated into four classes on the basis of their relative sensitivity to four basic stimuli, which to man taste sweet, salty, sour or bitter. This is often referred to as a 'labelled line' theory as opposed to the 'pattern' theory. Indirect support for this theory comes from work by von Békésy in which single fungiform papillae were stimulated on the human tongue and most were found to be sensitive to only one taste quality. It should be added, however, that other scientists have chemically stimulated single papillae producing qualitatively ambiguous sensations with a minority of papillae being reported to respond to only one of the taste qualities.

The question thus arises whether it is the activity of individual fibres or the pattern of activity in a set of fibres that is used by the brain in gustatory analysis. In psychophysical terms, the across-fibre pattern theory suggests that stimulus combinations may be handled in a synthetic manner, fusing into a single message in which the original components may be lost. Thus complex taste qualities are not perceived in terms of the components sweet, sour, bitter or salty. An analogy can be made with colour vision: we perceive white light when spectral blue and yellow simultaneously strike the same spot on the retina. Likewise in taste we perceive a complex flavour and not a composite of sweet, sour, bitter or salty. Neural recordings from the taste cortex support this, since no discrete or even 'best' topographical locations for the sweet or bitter sensitive neurons have been found at this level. In the peripheral gustatory nerves a few fibres respond almost exclusively to one of the primary stimuli (see Fig. 19.4) and these rapidly conducting fibres may be used by the animal for reflexive approach or avoidance behaviour. Studies on drinking behaviour have shown that the rat can reject an aversive stimulus by licking less than 5 μl of solution for as little as 50 ms. The rapidity of the rat's behavioural response is very important theoretically, since it establishes that the very earliest component of the neural signal is used by the rat for such avoidance behaviour. The neural responses recorded 'across-fibres' by Erickson omit the first 300 ms from the analyses and are presumably not used by the rat in stereotyped taste-avoidance behaviour. This later portion of the neural response may contain the taste information that the rat uses for quality coding. Hence, the taste stimuli of foods may be analysed at two levels. First its components (sweet, bitter, salt, sour), may be analysed rapidly and result in reflexive sucking, licking and swallowing, or rejection if aversive. Second the firing pattern of a population of taste fibres may be what is responsible for the conscious experience of taste.

19.4 NEURAL INFLUENCES ON THE FORMATION OF TASTE BUDS

It has been demonstrated by the use of isotope labelling that cells of the taste buds are constantly renewed, and on average live only ten days. It has further been suggested that perigemmal cells surrounding the taste buds undergo mitotic division, and some of the daughter cells enter the taste bud, are innervated and themselves differentiate into specialised taste cells. The interaction between nerve and tongue tissue that results in the formation of taste buds is a trophic nerve effect, as shown by the dependence of taste buds on their nerve supply. Taste buds disappear after denervation and reappear with normal structure and function following nerve regeneration. Moreover, nerves that normally innervate taste buds (VII, IX and X) may substitute for each other and cause taste-bud regeneration in any of the tongue's papillae, but no such trophic effects can be demonstrated with other sensory, motor or autonomic nerves. These findings are in accordance with a view expressed by Torrey some 45 years ago that the neural influences on taste-bud formation originate as a neurosubstance in the nerve-cell body, which passes down the nerve fibre to the tongue where it transforms epithelial cells into taste cells.

While the importance of the trophic influence of the nerve has been established, it is pertinent to ask whether specific properties of the epithelium may also be important. For example, would the innervation of any epithelial tissue by a gustatory nerve result in the appearance of taste buds? To answer this question the sensory vagal ganglion together with grafts from the ventral surface of the tongue or ear have been transplanted into the eye chamber. Normally, taste buds are absent from these epithelial areas, but in this experimental situation taste buds appeared in the tongue graft but not the skin graft. Although taste buds were induced to develop on non-gustatory regions of the tongue, their failure to develop in skin suggests that an epithelial specificity is involved in taste-bud formation. Moreover, the tongue tissue itself, rather than the type of nerve that supplies it, seems to determine what type of taste buds will form. Thus, when taste buds are cross-innervated (e.g. when IX nerve is directed to innervate fungiform papillae on the anterior part of the tongue), the nerve, when it regenerates, is changed in its relative responsiveness to different taste stimuli. Electrical recordings from the IX nerve transplanted to the anterior region of the tongue now show a relatively good response to sodium chloride (salt) and a poor response to quinine (bitter). In other words, the regenerated taste buds maintain the characteristics of those taste buds normally present in that particular region of the tongue. Thus the difference in responsiveness to taste stimuli at the front and back of the tongue is an intrinsic property of the taste receptor cells themselves, and not a result of the trophic influences from their nerve supply.

19.5. AFFERENT TASTE PATHWAYS

Taste is similar to other sensory systems in having access to both neocortical and sub-cortical brain regions. The nerve fibres conveying afferent responses from the taste buds are the peripheral processes of unipolar nerve cells situated in the ganglia of cranial nerves VII, IX and X. The chorda tympani branch of the VIIth cranial nerve, which carries taste fibres from the anterior two-thirds of the tongue, has primary cell bodies in the geniculate ganglion. The circumvallate taste fibres, which originate in the posterior one-third of the tongue and are carried in the glossopharyngeal nerve, have their primary cell bodies in the inferior glossopharyngeal ganglion. The vagus nerve carries some special visceral afferent fibres from the extreme dorsal part of the tongue and the superior surface of the epiglottis.

All taste fibres are first distributed to the ipsilateral nucleus of the solitary tract in the medulla. All three cranial nerves enter the solitary tract in a similar manner and terminate in a successive rostrocaudal order with a moderate degree of overlap. Most gustatory input is transmitted via the chorda tympani and glossopharyngeal nerves; hence the rostral, or prevagal, part of the solitary tract forms the main gustatory relay.

Cortical taste afferents

Like all sensory pathways, taste afferents (Fig. 19.5*a*) are relayed via the thalamus to the cortex. Lesions, electrophysiological stimulation and, more recently, autoradiography, have identified the specific thalamic nucleus as part of the ventrobasal complex, the main relay for somatosensory information, in the division known as ventralis medialis (VM). The projection from the solitary nucleus to thalamus is ipsilateral in the cat and monkey and there is some overlap in VM with thermal and tactile afferents from the tongue. The posterior part of the ventromedial thalamus lying subjacent to the centrum medianum has been conclusively defined as the thalamic nucleus responsible for relay of gustatory input.

The neural pathway from the thalamic taste nucleus (VM) passes into and through the internal capsule to two separate areas of the cortex in the monkey. There is a taste projection to the *parietal cortex* (SMI). A second taste projection is to the *insular cortex*, which although not a part of somatosensory area II, is situated just rostral to it. No cortical responses are evoked in this area by tactile stimulation, and it probably represents a pure cortical taste area. Taste nerve relays to the parietal cortex (SMI) may be responsible for spatial localisation of the gustatory stimulus.

Subcortical gustatory projections

For some years, because of the influence of taste on autonomic function, it has been assumed that taste fibres from the solitary nucleus have a direct subcortical projection (see Fig. 19.5*b*). Further evidence for this supposition stems from tests with chronic *decerebrate* rats. These show the same stereotyped reflexive responses to taste stimuli as do normal rats, with ingestion of sweet and salt, but rejection of sour and bitter solutions. Moreover, *anencephalic* neonate children also register the characteristic facial expression to sour and bitter solutions. It appears therefore that certain discriminative responses to primary tastes result from integrative mechanisms that

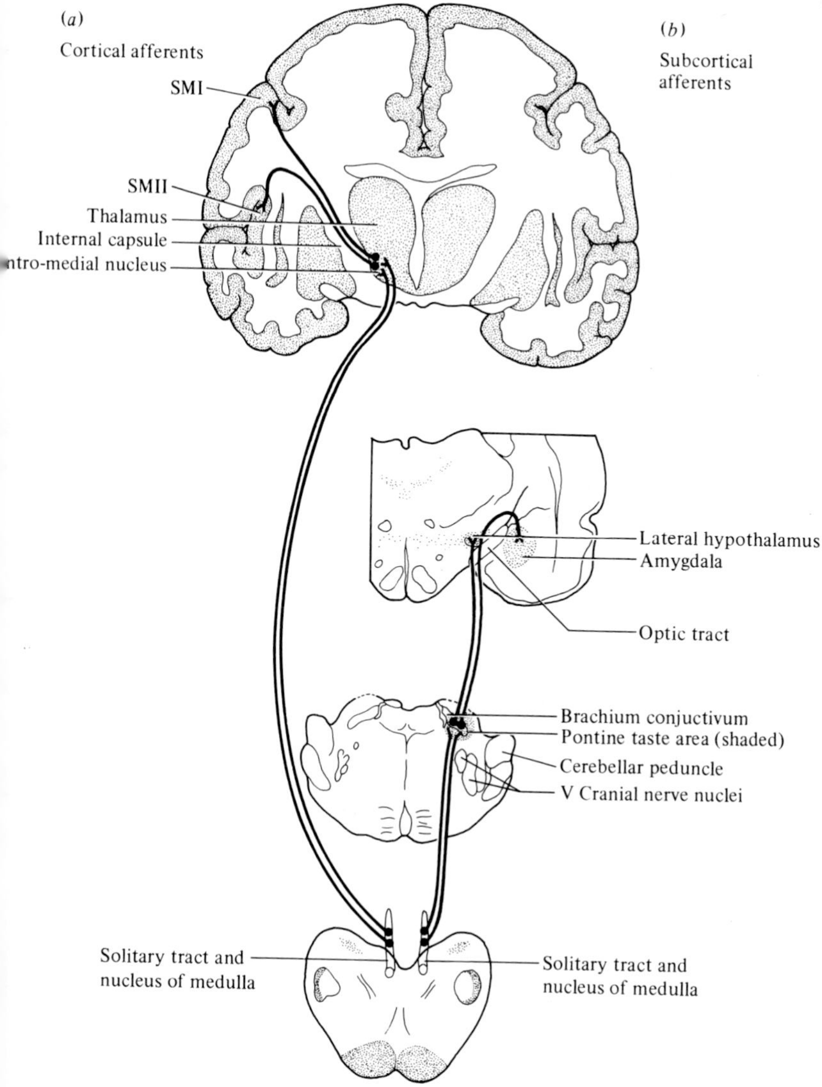
(a)
Cortical afferents
SMI
SMII
Thalamus
Internal capsule
ntro-medial nucleus
(b)
Subcortical afferents
Lateral hypothalamus
Amygdala
Optic tract
Brachium conjuctivum
Pontine taste area (shaded)
Cerebellar peduncle
V Cranial nerve nuclei
Solitary tract and nucleus of medulla
Solitary tract and nucleus of medulla

Fig. 19.5. Central taste projections.

do not primarily involve cortical structures. The sub-cortical neural pathways for taste have only recently been explored.

Microelectrode recordings have been used to locate taste-responsive neurons in the *pons* of the cat. The neurons were located by electrical stimulation of the chorda tympani and were shown to respond when the tongue was stimulated with standard taste solutions. The taste neurons were found among a group of cells lying outside and dorsomedial to the principal sensory nucleus of V. A more precise location of the pontine taste area by unit recordings has picked up gustatory responses in the brachium conjunctivum as well as dorsal and ventral to it. In the region of the pontine taste area there is a great deal of overlap among populations of neurons which connect with cranial nerves VII and IX, although individual neurons seem to maintain their identity (either VII or IX).

The projections of the third order gustatory relay have not been separately traced from these discrete pontine taste areas. Autoradiography injection of isotope into the whole of the pontine taste area has revealed projections not only to the thalamic taste area but to the central nucleus of the amygdala and lateral hypothalamus. The precise functions of these areas in relationship to feeding behaviour have received some attention: lesions of the central amygdaloid nucleus in the rat reduce prey-killing behaviour, while gastric secretion and visceral mobility are elicited from stimulation of a rather widespread area of the amygdala. The lateral hypothalamus has long been implicated in feeding behaviour since electrical stimulation of this area will cause an animal to increase its daily food intake, while lesions have the opposite effect, causing aphagia which may result in the death of an animal.

19.6. TASTE AND BEHAVIOUR

Some years ago it was shown that infants and young children grow normally and remain in good health when left alone to select their own diet from a wide variety of natural unprocessed foods. Rats likewise, given access to a large choice of purified diet, spontaneously select foods roughly in accordance with their dietary requirements for growth and development. The factors governing which foods are selected include smell and texture, but it is the taste that allows an animal to select the nutrients demanded by internal homeostatic mechanisms or to avoid the sickness caused by ingested toxins. With taste, as opposed to smell or texture, there appears to exist a relationship between the palatability and nutritive consequences of eating certain foods. In general, sweet substances usually have beneficial nutritional consequences, while bitter substances are aversive and often toxic if eaten. Salty tastes are usually beneficial if they taste palatable, but at unpalatable concentrations are often metabolically toxic. Food preferences vary according to physiological needs, and the term 'specific hunger' refers to an animal's preference for a particular food-stuff in response to a nutritional deficiency.

Specific hunger for salt

Before the days of salt tablets, industrious labourers from the steel mills and coal mines were renowned for lavishly flavouring their beer and food with salt after finishing their work shift. This habit made their food unpalatable to anyone else, and even to themselves at any time other than immediately after work. Specific hunger for salt has also been reported clinically. For example, a young boy with severe adrenal deficiency survived by virtue of consuming table salt by the handful, but died when his salt intake was restricted. Among animals, several species are known to seek out and preferentially ingest food and water with a high salt content when their diet has been low in sodium. Experimentally, specific hunger for salt has been produced by bilateral adrenalectomy, by inserting a parotid fistula that results in rapid sodium depletion through salivary loss, or simply by placing the animals on low sodium diets. In all cases, salt intake is regulated according to the size of the deficit and appears to be an innate rather than a learned response. Thus, animals deprived of salt for the first time show an immediate preference for sodium salt solutions within 10 s, when they are given a choice between solutions one of which is low in sodium salts. In both sheep and rats, the initiation as well as satiation of sodium appetite does not depend directly on plasma concentrations of sodium ions, but a taste-dependent mechanism appears to provide directly for intermediary regulation. Thus, in sheep, the salt-deficient animal will drink some two to three litres of hypertonic sodium bicarbonate during as little as two to three minutes, and this is followed by a marked loss of interest in the solution. An increase in the concentration of plasma sodium does not however follow until 15–30 min later. This rapid satiation of salt appetite following oral ingestion contrasts with the very slow decrease in salt appetite when sodium solutions are administered systemically. Thus, intracarotid infusions of sodium chloride for 15 min, which produce a rapid rise of sodium ions in the blood, do not consistently affect sodium appetite in the short-term. Taste stimulation, by itself, would seem, therefore, to be important in satisfying a salt appetite.

Specific hunger for sweet-tasting substances

A number of studies have shown that rats deprived of food, or rendered hypoglycaemic by insulin, markedly increase their intake of nutritive and non-nutritive sweet-tasting solutions. The increased intake of non-nutritive sweet-tasting substances, like saccharine solutions, by hungry rats suggests that the specific appetite for sweet is directly related to the taste of the solution, and not to its beneficial consequences. This has been confirmed by an experiment in which insulin-treated rats were presented with a choice between water, a glucose solution, and a much sweeter fructose solution which is, however, less capable of relieving the deficit produced by insulin. On the first day of access to these solutions, most of the insulin-treated rats prefer the fructose solution. On the second day, however, the experimental animals switch their preference. It appears that the initial appetite for the sweeter solution is an innate response to the deficit, but that by the second day the animals have learned to take what they need.

Taste preferences to sweet or sugar are not restricted to the laboratory rat, and such preferences are displayed by a wide range of organisms starting at the bacteria, including many insects and invertebrate species, and moving through most mammalian species to the primates and man. This evolutionary development of a specific hunger, or preference for sweet-tasting substances, especially in the face of deficiency, is highly adaptive since sweet substances in general tend to be major source of calories, and usually, sweet taste is associated with a greater caloric density.

Aversion to bitter substances

Natural aversions to bitter substances have also been acquired by a wide variety of species including protozoa, coelenterates, birds, mammals and man. The rejection of bitter may therefore represent a phylogenetically ancient natural response, a view supported by the finding that many animals and plants have themselves become bitter-tasting, fending off predators by taking advantage of this widespread natural rejection of bitter. The skin of toads is covered with glandular cells that secrete bitter (and toxic) materials, while many plant species contain toxic alkaloids and toxic glycosides and other plants have evolved harmless bitter flavouring, mimicking a toxin.

The natural aversion that many species show towards bitter substances, which are often toxic, has clearly favoured their survival. This natural aversion can, however, be overcome by associative learning, giving what we might call 'an acquired taste'. I suppose beer and tonic water fit this category of 'acquired taste' for ourselves, while Indian labourers who have eaten only poor-quality food have been reported to describe bitterness as increasing in pleasantness with increasing concentration. In animal experiments, modification of a natural aversion to bitter has also been accomplished. Guinea-pigs receiving, as the only source of fluid, a bitter solution of sucrose octoacetate, eventually accept this normally aversive solution even when offered a choice with uncontaminated water.

Conditioned taste aversion

Just as the natural aversion to bitter tastes can be overcome, taste aversions can also be acquired if the food or drink consumed induces illness. For example, if an animal is injected with apomorphine, it experiences a sudden, but brief, bout of nausea and illness. If the animal is given sweet (saccharin)-flavoured food before the onset of the illness, then that animal will shift its taste preference away from sweet. In other words, it will develop a sweet-taste aversion, just as we show an aversion to a variety of foods consumed coincidentally before a bout of illness. Similarly a meal of hamburger laced with 6 g of lithium chloride in capsule form causes nausea in coyotes, and is sufficient to make these animals reject hamburger the next time it is presented. Similar taste aversions have been produced in rats, mice, cats, monkeys, ferrets, birds, fish and reptiles, and such associative learning is so powerful that taste aversions may be acquired in a single learning trial. Conditioned taste aversion probably does not depend on cognitive learning, but on a more automatic changing of palatability, since its involuntary nature has been demonstrated by acquisition of taste aversion in an animal whose experimental illness was imposed under general anaesthesia. Research into the neural mechanisms of conditioned taste aversions has examined the effects of neural lesions. The effects of gustatory cortex ablation, functional decortication, hypothalamic and limbic lesions on conditioned taste aversion indicate that both cortical and subcortical structures participate in this type of learning. Unfortunately such approaches do not permit a distinction to be made between regions of the brain responsible for processing the changed significance of taste signals and regions that execute the effects of such gustatory processing. Other approaches have employed electrophysiological recording from different areas of the brain following taste stimulation in aversion conditioned and normal animals. Differences between such animals are found in the responses of units in the ventromedial and lateral hypothalamus, areas of the brain strongly implicated in feeding behaviour.

The fact that taste aversions can be acquired without any apparent cognitive awareness suggests that this complex taste reflex is phylogenetically ancient. Its biological significance resides in endowing the animal with the ability to select nutrients that are not prejudicial to the maintenance of bodily integrity. However, many plants and animals have succeeded in evolving unpalatable flavours which only mimic toxins. The development of neocortical taste mechanisms may be viewed as an adaptation by higher mammals to overcoming the selective advantage acquired by plants and animals possessing a harmless flavour mimicking a toxin. Higher mammals are thus able cautiously to explore a large variety of novel, potentially dangerous foods and even acquire a taste for unpalatable foods that are not poisonous.

19.7. SUGGESTIONS FOR FURTHER READING

General references

Beidler, L. M. (1971) *Handbook of Sensory Physiology*, vol. IV, 2, *Taste*. Berlin: Springer-Verlag.

Bartoshuk, Linda M. (1978) Gustatory System (Chapter 13) in *Handbook of Behavioral Neurobiology*, vol. I, ed. R. B. Masterton. New York & London: Plenum.

Wolf, R. & Monnier, M. (1975) *Functions of the nervous system*, vol. 3, *Sensory Functions and Perception – The sense of taste*. Amsterdam & Oxford: Elsevier Scientific Publications.

Wolstenholme, G. E. W. & Knight, J. (1970) *Taste and Smell in Vertebrates*, Ciba Foundation Symposium, London: Churchill.

Special references

Structure of taste receptor. Farbman, A. I. (1965) *Journal of Ultrastructural Research*, **12**, 328. Murray, R. G. & Murray, A. (1967) *Journal of Ultrastructural Research*, **19**, 327.

Transduction. Price, S. (1973) *Nature*, **241**, 54. Dastoli, F. R. & Price, S. (1966) *Science*, **154**, 905. Shallenberger, R. S. & Acree, T. E. (1967) *Nature*, **216**, 480. Jackinovitch, W. (1976) *Brain Research*, **110**, 481.

Beidler, L. M. (1954) *Journal of General Physiology*, **38**, 133. Price, S. & De Simone, J. A. (1977) *Chemical Senses Flavour* **2**, 427.

Coding. Beidler, L. M., Fishman, I. Y. & Hardiman, C. W. (1955) *American Journal of Physiology*, **181**, 235. Pfaffmann, C. (1941) *Journal of Cellular and Comparative Physiology*, **17**, 243. Erickson, R. P. (1963) in *Olfaction & Taste*, vol. I, ed. Y. Zotterman, 205–14. London: Pergamon Press. Von Békésy, G. (1966) *Journal of Applied Physiology*, **21**, 1. Frank, M. (1975) in *Olfaction & Taste*, vol. V, ed. D. A. Denton & J. P. Coghlan, pp. 59–64. New York, San Francisco, London: Academic Press.

Neural influences on taste bud formation. Zalewski, A. A. (1974) *Annals of the New York Academy of Sciences*, **228**, 344–9. Oakley, B. (1967) *Journal of Physiology*, **188**, 353.

Taste pathways. Rose, J. E. & Mountcastle, V. B. (1952) *Journal of Comparative Neurology*, **97**, 441. Benjamin, R. M., Emmers, R. & Blomquist, A. J. (1968) *Brain Research*, **7**, 208. Norgren, R. & Pfaffmann, C. (1975) *Brain Research*, **91**, 99. Norgren, R. (1976) *Journal of Comparative Neurology*, **166**, 17.

Taste and behaviour, specific hungers. Garcia, J., Hankins, W. G. & Rusiniak, K. W. (1974) *Science*, **185**, 824. Le Magnen, J. (1967) in *Handbook of Physiology*, ed. C. F. Code, section 6, vol. 1. Washington: American Physiological Society. Rozin, P. (1968), *Journal of Comparative Physiology and Psychology*, **66**, 82. Denton, D. A. (1967), in *Handbook of Physiology*, ed. C. F. Code, section 6, vol. 1. Washington: American Physiological Society.

Conditioned taste aversion. Best, P. & Zuckerman, K. (1971) *Physiology and Behaviour*, **7**, 317. Burešová, O. & Bureš, J. (1975) *Journal of Comparative Physiology and Psychology*, **88**, 47. Rozin, P. & Kalat, J. W. (1971), *Psychological Reviews*, **78**, 459.

CHAPTER TWENTY

The development of sensory systems and their modification by experience

J. ATKINSON, H. B. BARLOW AND O. J. BRADDICK

The sensory systems of the adult are finely adapted for obtaining the information we need from the environment we live in. Philosophers, psychologists and biologists have argued for a long time how far this functional fitness is the consequence of a design that is already specified (even if not fully expressed) at birth, and how far it comes about because the development of the senses can be directed by the pattern of input that they receive – a controversy sometimes referred to as Nature vs Nurture, or Nativism vs Empiricism. How sensory systems develop is not only a matter of scientific or philosophical interest: many clinically important disorders of vision, particularly of binocular vision, seem to be the result of development taking the wrong course.

Much of our knowledge of adult sensory processes rests on the human ability to report sensory experiences. We cannot obtain such reports from young infants, and so we must use less direct kinds of information to understand what sensory mechanisms are like soon after birth. These include physiological findings from animals whose developmental course may well be different from the human, behavioural responses which are necessarily of a rather limited nature, and anatomical observations whose functional significance may be uncertain. Only by looking at these different sources of information together are we likely to achieve a reliable picture of how the senses develop in early life.

20.1. NEUROPHYSIOLOGICAL EVIDENCE

The central connections of sensory neurons have been described in several places in this book (Vision, Chapters 1, 2 and 10; Hearing, Chapter 14; Touch, Chapter 17); in many cases a sensory surface, such as the skin, the cochlea, or the retina, is mapped in a topographical representation, point-by-point, on to the cerebral cortex through intermediate stations such as the LGN and cochlear nuclei, and also on to parallel systems such as the optic tectum (the superior colliculus in mammals). How does a dorsal root neuron that picks up from a receptive area in a particular cutaneous region find the appropriate place to establish its central connections in, say, the cuneate nucleus? How does an VIIIth nerve fibre innervating a point in the cochlea find its appropriate, properly ordered, position to terminate in the cochlear nucleus? Of course this is no more mysterious than the establishment of any other system of ordered neural connections, such as those that underlie a reflex or the generation of the controlled rhythmical movements of respiration, but topographical sensory maps present a nice opportunity for experimental investigation. An obvious way to attempt to find out how they are established is to find out what happens when the mapping is disrupted or interfered with. Much of this experimental work has been performed in amphibians and fish on the mapping of the retina on to the optic tectum, the mid-brain visual centre. The cortico-geniculate system of mammals has also been used to reveal factors that interfere with the establishment of a normal pattern of connections to individual cells.

Mapping of the visual field on to the optic tectum in amphibia and fish

The optic tectum is the main visual centre of lower vertebrates. It is responsible for guiding the feeding behaviour of frogs and other amphibians, and in man and higher vertebrates its homologue, the superior colliculus, is thought to be concerned mainly with the fixation of the eyes; stimulation of a locus corresponding to a particular part of the visual field causes a coordinated movement of the eyes that directs the fovea to that particular part of the visual field (see Chapter 11). The colliculus probably has little capacity for pattern recognition, but it may be important in determining the 'where' of visual stimuli, even though it has little to say about the 'what'.

In amphibians and fish, unlike mammals, regeneration of the optic nerve fibres occurs after their section, or the transplantation of an eye, thus extending the range of experiments possible. Some of the early experiments indicated that normal function followed such regeneration, even though the paths of the regenerating fibres suggested that the mapping was likely to have been deranged, and this fact supported the notion that some form of 'functional regulation' occurred through experience and training. This possibility was decisively rejected by experiments of Roger Sperry in the early 1940s.

The function of the tectal map can be tested by observing how a newt directs itself towards a lure dangled in its visual field, and the first important result was to show that no adjustments to these responses occur when an eye is surgically inverted, leaving the optic nerve intact, even though the responses then *misdirect* the newt. If functional regulation occurred, it should have abolished these responses and substituted appropriate ones, in the same way that human subjects adapt within a few weeks to seeing the world through prisms that invert the images on their retinae.

Sperry also showed that the responses following regeneration of the optic nerve were appropriate and essentially normal, provided that the eye from which the optic nerve fibres regenerated was in its normal orientation in the animal's head. If, on the other hand, the eye was rotated when the optic nerve was cut, the inverted, inappropriate responses were obtained after regeneration. This indicated that the connections established by a fibre depended upon the position of its ganglion cell in the retina, and not on the part of the visual field which the ganglion cell corresponded to, as would have been the case if functional regulation had occurred. On this basis Sperry proposed that the specificity of connections was established by some form of *chemo-affinity* between the ganglion cells at a particular retinal locus and the cells in the tectum with which they had to establish functional contact.

Later the map of the visual field on the tectum was determined electrophysiologically by placing an electrode at many positions in turn on the tectum, and finding the positions in the visual field that caused maximum activity for each of them. Sperry's conclusions were mainly confirmed, but with this technique many new experiments became possible, and these showed that the original chemo-affinity idea required modification. By various tricks, such as removing part of an eye or part of the tectum, it is possible to find what happens when part of the retina regenerates into the whole of the tectum, or the whole retina into part of the tectum. These experiments showed that ordered point-to-point mapping occurs even when fibres regenerate to positions with which they do not normally connect; the order and neighbourhood relations in the map are preserved even though the retinal ganglion cells make connections to different tectal cells. Thus position relative to other terminating fibres is apparently more important in establishing the map than the absolute position of the terminations on the optic tectum, and this cannot depend on cell-to-cell specificity. Nevertheless, up to this point there had been no suggestion that use, or visual experience, had any effect. The findings on the ipsilateral projection, however, do provide such evidence.

In amphibians the direct connections from the retina all cross over and go to the contralateral optic tectum. Now a stimulus in the visual field in front of the animal, where the fields of view of the two eyes overlap, can be shown to excite the same point on the tectum through both the ipsi- and the contralateral eye. The connection from the ipsilateral eye is indirect, through nuclei in the mid-brain caudal to the tectum, and when Gaze and Keating studied it they found clear evidence of functional regulation. If one eye is rotated, its contralateral map rotates exactly as Sperry had described, but its ipsilateral map stays aligned with the map of the other, unrotated, contralateral eye. This, however, is only true if the animal is kept under conditions such that both eyes are used. If it is kept in the dark, the ipsilateral projection remains anatomically unchanged, and activity from the unrotated contralateral eye is needed to cause the ipsilateral map to stay aligned with it. Here then is a case where a map is changed as a result of the pattern of visual experience that the animal receives. The effects of experience can be demonstrated in mammalian systems also, but here it is a matter of changes in the properties of individual cells, and there has been no evidence for the establishment of a new map.

Development and experience in mammalian visual pathways

It was pointed out in Chapter 10 that many people with otherwise normal vision have poor stereopsis. There is in addition a large group of people who suffer from *amblyopia*: one eye has poor visual performance even though no pathology is visible, except (in many cases) strabismus (see also Chapter 3). The functional abnormalities in these cases have long been attributed to abnormal visual experience in early life, and for this reason Hubel and Wiesel started investigating the effects of ocular occlusion and strabismus on the lateral geniculate nucleus and visual cortex very soon after they had established the normal properties of single neurons recorded in these places (see Chapter 1, p. 27; Fig. 2.6; and Chapter 10). Their first experiment was to occlude, by suturing the eyelids, one or both eyes of kittens as soon as they opened on about the tenth day. Several weeks or months later they opened the occluded eye and recorded from single neurons in the lateral geniculate nucleus or cortex. Their most important result is shown in Fig. 20.1. It will be recalled that the

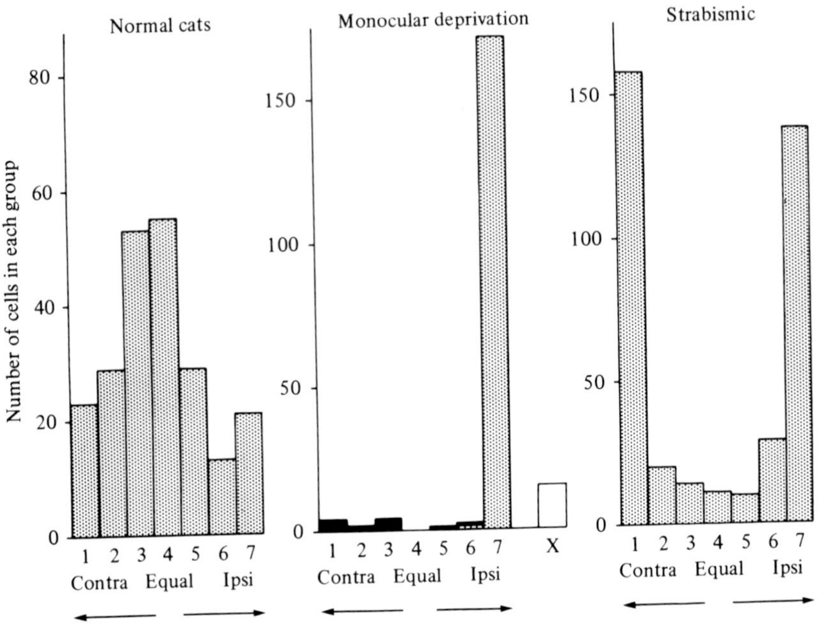

Fig. 20.1. The effects of monocular occlusion and strabismus on the ocular dominance distributions of cells recorded in primary visual cortex of kittens. The cells were grouped according to the relative influence of input from each eye. Group 1 has input from the contralateral eye exclusively; group 7 has input from the ipsilateral eye exclusively; the others have intermediate degrees of dominance, group 4 representing equal input from the two eyes. Cells with normal properties are shown in stippled columns, and it will be seen that these occur in all groups of the normal cat. The centre histogram shows the results recorded from the left cortex of 5 kittens aged 8–14 weeks in which the right eyes were closed by lid suture at 10–14 days; the cortical cells became strongly dominated by the non-occluded, ipsilateral eye, and most of the small number of cells connected to the right eye, shown in black, had abnormal properties because they lacked the usual orientational selectivity. In these animals there were also a few cells (group X) that could not be excited visually at all. The right histogram shows results from four kittens that had been made strabismic; in these animals the two eyes do not receive properly aligned images, so the inputs from the two eyes to a given cortical region do not agree with each other. It will be seen that each eye dominates many cells, but compared with normal cats there are few cells that receive input from both eyes. (Adapted from Hubel & Wiesel, 1965.)

partial decussation of fibres at the optic chiasma (see Fig. 2.5) ensures that the lateral geniculate neucleus of one side receives information about the contralateral hemifield through the retina of both eyes, and the fibres of the optic radiation then relay this information to the primary visual cortex.

When cells in the normal cortex are recorded from it is found that some can be excited only from the contralateral eye. The number of these cells is indicated by the height of column 1 in the histograms of Fig. 20.1. Other cells can be excited only through the ipsilateral eye, and these are classified as group 7 and plotted in the appropriate column. The middle column, group 4, consists of those cells equally well driven from either eye, and the other groups correspond to cells judged to have intermediate degrees of dominance of each eye. It is important to realise that, when the two eyes are in their normal positions, the receptive fields of all cells at a given cortical locus are in nearly the same positions in the visual field, irrespective of the eye hrough which one is driving them.

The middle histogram of Fig. 20.1 shows results from cats in which one eye had been closed by suturing at the time of eye opening, so that it never received the normal excitation from a patterned image falling on its retina. The result was to diminish dramatically the proportion of cortical neurons that could be driven through that eye at the time when its eyelids were opened for the purpose of the recording experiment several months later. Apparently disuse of an eye weakens the connections it makes, through the lateral geniculate nucleus, to cortical neurons, and usage of it is required to keep them strong.

The right-hand part of Fig. 20.1 shows the result obtained in cats which had the external rectus muscle of one eye surgically removed in order to give them an artificial strabismus. The eyes were open, but misaligned, so each obtained normal excitation by patterned retinal images, but at any one moment the pattern at a point in one eye would not agree with the pattern at the equivalent point in the other. The histogram shows that each dominated a higher proportion of cortical neurons than in the normal, and the number of cells that received connections from *both* eyes was greatly reduced. The experiment can also be done by alternately occluding each eye: they each receive visual experience, but never concurrently, and the result is similar to that for strabismus.

One paradoxical feature of these results should be mentioned. Hubel and Wiesel examined their material anatomically, as well as by determining the functional performance of the cells neurophysiologically. What they found was a reduction of size in the cells of the laminae of the LGN fed by the deprived eye, whereas functionally these cells responded more or less normally. In the cortex on the other hand, they could not find any anatomical abnormality in the cells they recorded from, even though this was where they consistently found functional abnormalities. It has since been shown that there are changes in the width of the ocular dominance stripes (see Fig. 2.6) for each eye following monocular deprivation, but the nature of the intimate changes in synaptic connections that underlie the dominance shift, and the reason for the shrinkage of geniculate neurons, remain undiscovered.

Since usage of the pathway appears to be necessary to preserve normal function it might be thought that suture of both eyelids, or total visual deprivation by rearing in the dark, would produce even greater changes than those shown in Fig. 20.1. Ocular dominance histograms of such cats are, however, scarcely different from those of normal animals. It can be concluded that the pathway from one eye to a cortical neuron is not disconnected unless there is another pathway from the other eye competing for control of the same cortical neuron. However one cannot conclude from the normality of the ocular dominance histogram that the totally deprived cortex develops perfectly normally in all respects, and indeed this is not the case; after a few weeks of total deprivation many visually unresponsive cells are found, and few respond in the way expected in the cortex of a normal animal of the same age. Ocular dominance histograms display only one particular characteristic of the visual cortex.

Some progress has been made in determining the extent to which other properties of cortical neurons require appropriate experience to become established or to be preserved. Cortical neurons respond selectively to bars or edges of the appropriate orientation, different cells responding to different orientations (see Fig. 1.7), and attempts have been made to determine the effects of rearing kittens in an environment where they see contours of only one orientation. These have not been uniformly successful, perhaps because very complete deprivation is needed to produce clear-cut effects, and it is easier to deprive an eye of all pattern vision than it is to deprive it completely of vision for one orientation while allowing it to see others. If kittens are kept mainly in the dark, but for a few hours a day wear goggles permitting them to see only a grating, then their cortex shows reliable distortion of the normally uniform distribution of orientational

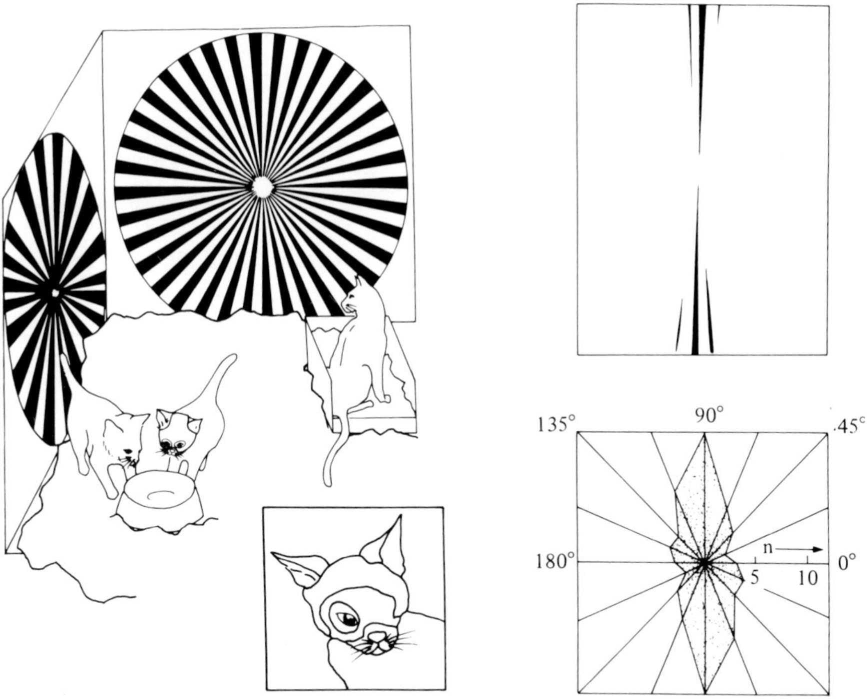

Fig. 20.2. Kittens can be prevented from seeing lines and edges of a range of orientations by fitting them with masks holding strong cylindrical lenses that blur contours parallel to the axis of the cylinder. The lower central inset shows such a kitten with one eye completely occluded and the other with a lens over it. Kittens were exposed in cages for a few hours a day, wearing a variety of such masks and lenses (top left). The star-shaped patterns in the cage, when seen through a cylindrical lens with its axis horizontal, would appear as at top right. Directly below is shown the distribution of the preferred orientations of 46 neurons from the visual cortex of a kitten that had been reared with such lenses over both eyes. The majority of the preferred orientations are similar to those that the kitten had seen. By comparing the effects of different patterns of exposure it was concluded that presynaptic activity by itself was insufficient to bias the distribution; the patterns seen also had to cause postsynaptic activation, as is required for modification of Hebb-type synapses. (From Rauschecker & Singer, 1981.)

tuning. However exposure in cylindrical drums painted with horizontal or vertical stripes has produced conflicting results. Fig. 20.2 shows the results for a third method: a kitten wore goggles with strongly astigmatic, cylindrical, lenses which blur all lines and edges parallel to the cylinder, leaving a sharp view of those at right angles. It will be seen that the cortex contained an excess of cells tuned to the seen orientation, and a deficiency tuned to other orientations.

Experiments have been performed with many combinations of total deprivation, normal vision, and exposure through astigmatic lenses (as shown in Fig. 20.2). It was concluded from these that all known examples of functional regulation in mammals can be accounted for by a simple postulate. This is that the synapses from geniculate neurons on to cortical cells have a property proposed in order to account for learning and memory by the psychologist D. O. Hebb many years ago. Such a 'Hebb synapse' increases in potency if presynaptic activity is frequently successful in causing postsynaptic firing, and decreases in potency if presynaptic activity often occurs without postsynaptic activity. Perhaps the growth and shrinkage of LGN neurons reflects the influence on the cell bodies of a 'synaptic rewarding factor' that is released by activated cortical neurons and picked up by the synaptic terminals which caused the activity, thus promoting growth of the terminals and their parent cells. Ophthalmologists have long been aware of the benefits of 'exercising' the visual pathways in preventing the development of amblyopia or reversing its progress. We now see that it is not sufficient to exercise the presynaptic pathway: the presynaptic activity must also successfully excite the postsynaptic neurons in the visual cortex.

The sensitive or critical period

If the experiments that have just been described are repeated on adult animals one finds that there are no effects. The connections can be modified only within a certain age range, and this is known as the *sensitive*, or *critical* period. For kittens it extends from about three weeks, just after eye-opening and the clearing of the optic media, to beyond three months, the peak of sensitivity occurring at four to six weeks and thereafter declining. Similar effects to those described above in the cat can be demonstrated in rhesus monkeys. Their eyes are open and their optic media clear at birth, and the sensitive period probably starts a little earlier and extends a bit later than in cats. Indirect evidence suggests that for humans the time scale is slowed down, perhaps by a factor of about 4, as indicated by the measurements of acuity shown in Fig. 20.3. Visual resolution is poor when the system is immature and increases towards the adult values during the sensitive period. The measurements on human infants require a variety of special methods (see below and Fig. 20.4), but they would be compatible with a sensitive period extending to about two years.

Experiments have also been done in which kittens were exposed to patterns moving in one direction but not the other, and their cortex was found to contain a preponderance of cells selectively responsive to the direction of motion that had been seen. In these experiments some evidence was obtained that the sensitive period for determining directional selectivity finished earlier than that for determining ocular dominance, for if a kitten was exposed to moving patterns through one eye which was then sutured shut while the other was opened, and the kitten was then exposed to stripes moving in the other direction, it was found that the ocular dominance switched to the newly opened eye, but the most frequent directional preference failed to switch and corresponded to that seen earlier by the other eye. Perhaps there is not one single sensitive period, but a succession occurring at different ages.

It is clear from these experiments that functional regulation within the sensitive period plays some role in establishing the connections in the primary visual cortex of adult cats, monkeys, and presumably man. But how important is this role? How well would the cortex function without any previous experience? A key piece of evidence here is the performance of cortical neurons in very young, visually inexperienced, animals. When recorded from neurophysiologically, some cells in the cortex of 10–14 day old kittens show properties like those of the adult, including selectivity for orientation, size, and direction of movement. One can therefore say with confidence that cortical neurons do not require experience in order to possess these pattern selective properties. The same is true for the columnar structure (Fig. 2.6), the clustering of cells with similar ocular dominance or orientational properties: signs of this also appear early and do not require experience. From this evidence alone one might jump to the conclusion that the whole intricate structure and function of the cortex is laid down by innate developmental processes, and that experience is only necessary to preserve it, but this would be wrong. Only rudiments of the normal selectivity and columnar structure have been described in immature or visually deprived cortex; furthermore there are many cells with abnormal properties in such cases.

These new results on the interplay of visual experience and innate developmental factors in creating the visual cortex form a fascinating continuation of the ancient nativist/empiricist controversy. The balance between innate factors and experience can perhaps be summarised tentatively as follows. The crude mapping of the retina on the cortex through the geniculate is innate, and so are the simple forms of pattern selectivity of cortical neurons. But the fine adjustments which are necessary for high acuity (Fig. 20.3) and for stereoscopic vision can only be achieved by the functional regulation of the synaptic potencies of the incoming geniculate fibres. No one doubts that changes in the connections of cortical cells in animals are closely connected with the abnormalities of stereopsis in humans and with the causation of amblyopia but the exact relations are unknown. The ocular dominance histograms of the developmental neurophysiologists tell one little about the very precise binocular connections that underlie stereopsis (see Chapter 10), while the human conditions cannot of course be analysed neurophysiologically. Above all it must be recognised that we do not fully understand the role of the primary visual cortex in vision and therefore cannot interpret the developmental changes with security. There is more to seeing than cortical neurophysiology has yet told us.

Critical periods in other systems

The mammalian visual cortex is not the only system that is sensitive at one particular time in development and not at others. Another well-known example is imprinting in birds: those species that leave the nest soon after hatching treat the first object that has remotely appropriate characteristics as 'mother', and proceed to follow it to the exclusion of all other objects, including, sometimes, their real mother. Another avian example is the formation of birdsong in species that form small communities. Some crude form of song is generated by adults that mature in isolation but the details of the typical full-blown song are modelled on those of the particular conspecifics they hear between 14 and 50 days after hatching. Some time later, when they reach sexual maturity, they proceed to reproduce this type of song. This results in considerable variation between the songs of different flocks of the same species: each has its own self-perpetuating dialect.

An example in mammals that is well-known to dog-breeders is the 'socialisation' of pups. If they are not handled by humans before about 6–8 weeks of age, and particularly if human contact is delayed while they stay with their mothers, they will tend to be shy of humans all their lives. Apparently fear of the unfamiliar tends to develop in pups after a phase of curiosity and exploration, and if familiarisation does not occur early, fear tends to dominate all encounters with humans and is likely to prevent the formation of the typical bond with a human master or mistress.

Obviously one cannot expect all mental attributes to develop simultaneously, and this example shows how important the sequence may be. One should not, however, think of critical periods as rigidly defined intervals that come to an abrupt end, but rather as stages in development when particular experiences have their greatest effects. It is clear that they may be immensely important in emotional development, and in education in general, which is one reason why great interest attaches to the physiological processes underlying the simple examples in sensory systems. Possible examples of critical periods in human perception of phoneme borders and human linguistic development will be mentioned below.

20.2. DEVELOPMENT IN HUMAN INFANTS

Studies in the first months of life show the senses to be capable of discriminating stimuli on most of the basic dimensions that will be used in mature perception. The ability to distinguish qualities and intensities of taste and smell is important for the infant's regulation of food and fluid intake. The behaviour that in the adult is controlled by vision and hearing is largely undeveloped in the young infant, but presumably the information provided by these senses is needed in order for that behaviour to be acquired.

Structural development of the visual pathway

In the monkey, and probably in man, all the neurons of the visual pathway have been generated before birth. This is not to say that they have achieved their ultimate interconnections, functions, or even positions. In the retina the limited evidence available suggests that the fovea is undifferentiated for the first few months of life. Incomplete myelination of the optic nerve in the first months slows transmission of visual information to the cortex but the functional consequences of this are uncertain. The layers of the LGN serving the two eyes are clearly segregated at birth. As we have seen, the columnar segregation of neurons according to ocular dominance and orientation is present at birth in monkeys, and soon after in cats, although it is subject to modification by experience (Fig. 20.1). Histology confirms that the richness of interconnections in the human visual cortex increases greatly over the first few months of life.

Optics and accommodation

The optics of the human eye, unlike the cat's, are clear at birth. A further requirement for the formation of a sharp retinal image is that the eyes should be appropriately focussed. Refraction of infants with accommodation relaxed by a cycloplegic drug generally shows some hypermetropia (Chapter 4). However, the state of active accommodation is more relevant to the quality of vision. Newborn infants often show appropriate accommodative responses to stimuli at distances of 75 cm and closer. Errors of accommodation seem to result from fluctuations, indicating that the muscular capacity to change accommodation is present but that it is not well controlled. Between birth and six months of age normal infants develop accurate and consistent accommodation over the adult range. It should be emphasised that this development may depend not only on the accommodative mechanism itself, but also on the development of the mechanisms necessary to detect whether the retinal image is in or out of focus, and on the ability to concentrate attention on stimuli at different distances.

About 50% of infants under one year of age show significant astigmatism. In the great majority of cases this astigmatism disappears in the first two years of life, leaving 5–10% of adults astigmatic. When we understand the origin and developmental significance of infant astigmatism, we may have a better understanding of how the growth of the eye is regulated to achieve a good optical result.

Another optical change accompanying the growth of the eye during infancy is a decrease in the angle alpha between optical and visual axes (see Chapter 3).

Eye movements

Saccadic eye movements (see Chapter 2) occur from birth and are directed to centralise targets in the visual field, although infants under two months of age commonly require several saccades to achieve fixation. In contrast, smooth pursuit of a target does not occur until six to eight weeks of age. However, optokinetic nystagmus, including the slow phase which resembles pursuit movements, can be elicited by full field movements from birth, and vestibular reflex eye movements (see Chapters 11 and 16) are also present. In infants under about two months old, with monocular viewing, optokinetic nystagmus is asymmetric; only temporal-to-nasal field movements elicit it. Only temporal-to-nasal field movements elicit reflex following in amphibia, and the nasal-to-temporal direction in mammals is believed to depend on a pathway to pretectum via binocular visual cortex. Presumably some part of this pathway is immature in very young infants.

The eye movements of young infants are of practical importance because almost all behavioural evidence of infants' visual capacities comes from their ability to fixate preferentially, or to track, a stimulus that they can detect.

Developmental of visual acuity

Fig. 20.4 shows one method of determining whether an infant can resolve a grating. An observer watches the infant and has the task of deciding which side a target, such as a grating is on, even though the target is visible only to the infant. Infants seem to prefer looking at gratings rather than blank screens, so their preferred direction of gaze will show the observer which side a grating is on, provided that the infant can resolve the grating. The observer can decide correctly which side the grating lies only if the infant can resolve it, and thus prefers to look at it. This method (called preferential looking) works well for infants six months old, or younger. Older infants can only be tested in this way for a very short period of time, after which they become bored and want to look at something different. Other methods, such as recording evoked potentials or looking at tracking eye movements may prove more successful with these older infants.

The general picture that emerges from these varied studies is shown in Fig. 20.3. At ages under one month acuity is between 0.8 and 1.5 cycles deg^{-1}, and it increases rapidly to 2 to 4 cycles deg^{-1} at two months and 4 to 5 cycles deg^{-1} at three months. At six months, estimates diverge widely; results from evoked potentials suggest acuities of at least 15–20 cycles deg^{-1} at this age. It is extremely difficult to make reliable measurements in the age range between six months and two years, so it is not yet known when adult acuity is generally achieved. Contrast sensitivity (see Chapter 8) shows a developmental trend similar to that for acuity; an interesting feature is that the relative loss of sensitivity at low spatial frequencies, generally taken to indicate the working of lateral inhibition (Chapter 8), is not present in infants before two months of age. Neurophysiological results in monkeys and cats suggest that development of the neural pathway, probably at cortical levels, underlies these changes

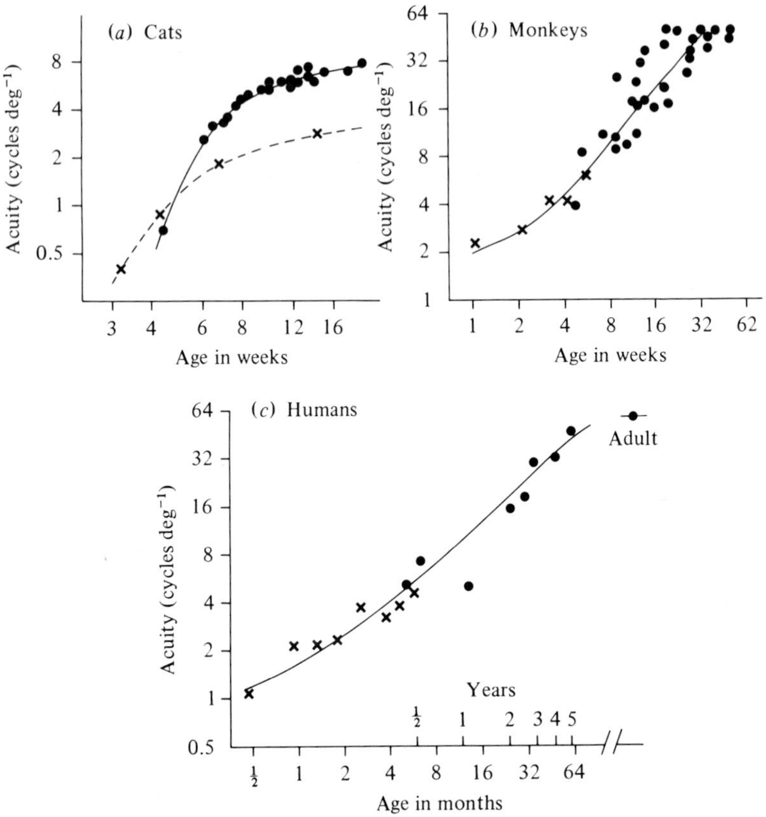

Fig. 20.3. Visual acuity increases with age. (*a*) In cats the increase starts at about 3 weeks and is almost complete by 12 weeks; the crosses show values determined by using the electrical response evoked from the cortex, which can be done in young animals but does not yield such high values as a behavioural test (filled circles). (*b*) Monkeys have higher acuity and the improvement continues to at least 6 months; here again two methods were used, crosses showing values obtained by a preferential looking method (see Fig. 20.4 and text), dots an operant technique. (*c*) For humans the time scale of the figure is slowed fourfold compared with monkeys, eightfold compared with cats; the crosses and dots show results from preferential looking and operant techniques, as before. The values of acuity are comparable to those of monkeys, but the rate of improvement is much slower and in these observations continues even beyond 2 years. (*a* adapted from Giffin & Mitchell, 1978 and Freeman & Marg, 1975; and *b*, *c*, from Teller, 1981.)

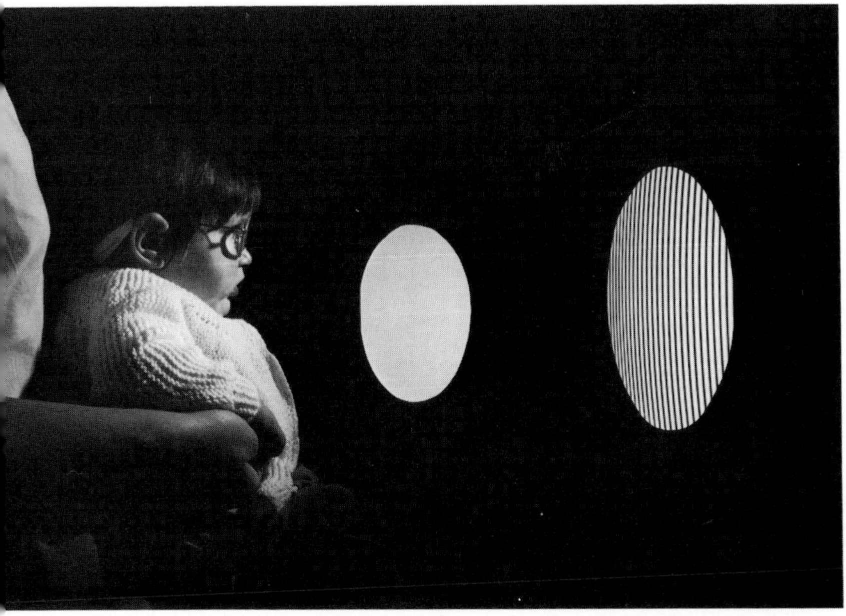

Fig. 20.4. Testing an infant's visual acuity by the method of preferential looking. The observer is concealed from the infant's view and is unaware, except by means of observing the infant's behaviour, on which of the two screens the grating is presented. The average luminance of the two screens is equal. This photograph shows the method in a clinical application: the child has a refractive error and her acuity while wearing a spectacle correction is being tested.

in spatial vision. The optical improvement in accommodation is not great enough to account for them.

It should be emphasised that the acuities of young infants, although very poor by adult standards, allow the detection of most of the information about nearby objects that is likely to be important to them. Their acuity would be inadequate for reading fine print or driving a car, but there are few visual demands of this kind in the infant's life.

Colour vision in infants

Newborn infants show characteristic photopic components in the electroretinogram, and photopic spectral sensitivity curves have been demonstrated at two months. These results prove that cones start functioning early. Experiments that test two-month-old infants' detection of one wavelength against a background of another show that they are using at least two cone mechanisms; it seems likely, then, that some form of colour vision is present from birth, although the existence of separately functioning cones does not prove that they can be used for distinguishing hues.

Binocular vision in infants

Newborn infants make movements of the two eyes that are mostly conjugate, and they show some systematic changes of convergence for targets at different distances (see Chapters 10 and 11). The degree of binocular coordination and vergence control continues to improve up to six months. This oculomotor coordination implies the possibility of stereoscopic vision. A number of tests have indeed demonstrated that at least some two to three month-old infants can detect stereoscopic disparities, and thus can be supposed to have a functioning system of binocular interaction in the visual cortex.

Perceptual development

To state the visual acuity or other elementary visual capacities of a developing child does not tell us what use the child can make of these capacities. For example, the ability to detect disparities does not necessarily imply that the infant can use them to derive information about the third dimension.

Six-month-old infants do show an appreciation of visual depth, though the contributions of different sources of depth information have not been separated. Below six months it is difficult to obtain responses that would be valid indicators of depth perception, but there is evidence that infants as young as two months old recognise the constant three-dimensional shape of an object viewed from different angles, and this must require the use of depth information. There are also reports that even younger infants will show startle or avoidance responses to the visual pattern of a looming approaching object, although some of these results have not been confirmed.

Besides the problem of interpreting visual information, we must also consider what visual information the infant *selects*. For example, infants of six weeks and under can discriminate faces and geometrical forms that differ in outline, but apparently they make little use of internal details of these patterns, even when their acuity is quite adequate for the task. This limitation of selective attention is overcome by two to three months of age, but there are likely to be other ways in which infants differ from adults in the ways in which they select from the available visual information.

Auditory development

The structures of the human middle and inner ear appear to be adult-like in form and dimensions at birth, except for the outermost row of hair cells which continue to develop postnatally. Measurements of acoustic impedance (see Chapter 14) in the neonate imply that the ear is clear of fluid or tissue and has a conductive loss of no greater than 5–10 dB relative to the adult. Evoked potentials from the brainstem indicate thresholds for click stimuli that are 15–20 dB above adults. Behavioural responses and cortical evoked responses are only found at higher intensities, possibly because the central auditory pathways are less mature than the periphery. There are no very satisfactory measures of infants' auditory frequency response, but there is evidence that they are relatively less sensitive than adults to high frequencies (e.g. 4000 Hz and above).

For infants as for adults, human speech is the most important acoustic stimulus, and infants show abilities to discriminate speech sounds remarkably early. Three-week-old infants can discriminate one individual's voice from another, and a wide variety of discriminations between phonetic categories (e.g. voiced and unvoiced stop consonants) have been demonstrated at one to three months. It has been argued that these experiments show the existence in the infant of mechanisms that are specialised for processing linguistic categories. However, it can be shown that infants make similar categorical discriminations for non-speech sounds, and that non-human species make them for human speech sounds, so it is more likely that the discriminations depend on general-purpose acoustic mechanisms which are useful for speech sounds among others.

There is some evidence for plasticity in the human auditory system limited to a critical period, of the type that has been shown for the acquisition of birdsong and in the visual system of mammals. The best-known example is the great difficulty that the Japanese have in distinguishing the sounds of 'l' and 'r'. The distinction is not made in the Japanese language, and those who have not learned to make the distinction by the time they are about six years old can only learn to do so with extraordinary difficulty when they reach maturity. It has been suggested that there is a more general type of critical period for language development, because deaf children who miss the early experience of speech often fail to develop clear articulation, although their syntax, vocabulary, and comprehension can reach the same level as normal children.

20.3. INTER-SENSORY AND SENSORY-MOTOR CO-ORDINATION

It would be a mistake to consider the development of the different senses in isolation from motor development. To use sensory information effectively the child must link visual information with that of the other senses and relate it to the bodily movements which he has initiated. Some of these links appear to be present at birth. For example, a newborn infant will turn head and eyes towards a sound source, and localise it quite accurately if enough time is allowed for the response. Curiously, this localising behaviour declines after birth, to return later at around five months of age. It has been suggested that the initial behaviour reflects a primitive sense of space which does not differentiate between sensory modalities, and that the temporary decline is due to the infant developing separate representations of visual and auditory space which can subsequently be coordinated.

Sensory-motor links are plastic not only in infancy but also in adulthood. Many experiments have artificially altered the relationship between visual and motor space, for example by the use of goggles that displace or invert the visual image (see also Chapter 16). Adult subjects can adapt to the new relationship; the extent and locus of this adaptation depends on the modification and on the tasks undertaken. The learning of new sensory–motor relationships both in the infant and the adult requires actively controlled movement by the individual, rather than just the passive experience of visual changes coupled to motions of the body. The physiological mechanisms underlying this learning have not been determined.

20.4. CONCLUSIONS

We started this chapter by pointing out that the adult gains his knowledge of the environment through complex and finely adapted sensory systems. These are formed by sensory experiences interacting with innate, genetically formed, neural mechanisms and it is difficult to generalise about which influence is predominant. Thus the nativist/empiricist controversy sprang from asking the over-simple question 'Nature *or* Nurture' when in fact both must always be involved. However the over-simple question is still a fruitful one to ask, for on specific issues it can sometimes be answered, as for instance with colour blindness (almost entirely Nature) and stereo-blindness (probably mostly Nurture). But perhaps the greatest interest attaches to the cases where no such simple answer is possible, because these cases demonstrate the intimate relations between the two. To use a modern idiom, Nature writes the program, but early experience can enter some key parameters that have a profound influence on how it handles input in later life.

20.5. SUGGESTIONS FOR FURTHER READING

General references

Jacobson, M. (1978) *Developmental Neurobiology* (2nd Edn) London: Plenum. (A comprehensive reference book.)

Barlow, H. B. (1975) Visual experience and cortical development. *Nature*, **258**, 199–204. (Review of Nature *versus* Nurture controversy about mammalian visual cortex.)

Salapatek, P. & Cohen, L. (1987) *Handbook of Infant Perception*. New York: Academic Press (Covers a wide range, both in vision and audition.)

Atkinson, J. & Braddick, O. J. (1989) Development of basic visual functions. In A. Slater & G. Brenner (eds.) *Infant Development*. London: Erlbaum.

Movshon, J. A. & Van Sluyters, R. (1981) *Annual Review of Psychology*, **32**, 477. (Neurophysiology, with emphasis on deprivation experiments. Critical review that is less dry than some treatments in *Annual Reviews*.)

Special references

History of Nature vs Nurture controversy. Morgan, M. J. (1977) *Molyneux's Question*. Cambridge: Cambridge University Press.

Chemoaffinity theory for amphibian optic tectum. Sperry, R. W. (1951) Mechanisms of neural maturation. Ch. VII in *Handbook of Experimental Psychology*, ed. S. S. Stevens. New York: Wiley.

Functional regulation of ipsilateral projection in amphibian optic tectum. Keating, M. J. (1977) *Philosophical Transactions of the Royal Society*, *B*, **278**, 277.

Effects of monocular occlusion and strabismus on ocular dominance histograms. Wiesel, T. N. & Hubel, D. H. (1965) *Journal of Neurophysiology*, **28**, 1029. Hubel, D. H. and Wiesel, T. N. (1965) *Journal of Neurophysiology*, **28**, 1041.

Effects of visual deprivation on LGN cells. Wiesel, T. N. and Hubel, D. H. (1963) *Journal of Neurophysiology*, **26**, 978.

Visual deprivation, stripe rearing, and Hebb synapses. Rauschecker, J. P. & Singer, W. (1981) *Journal of Physiology*, **310**, 215.

Sensitive or critical period in kitten visual cortex. Hubel, D. H. & Wiesel, T. N. (1970) *Journal of Physiology*, **206**, 419.
Critical period for motion selectivity in kittens. Daw, N. W. & Wyatt, H. J. (1976) *Journal of Physiology*, **257**, 155.
Improvement of acuity with age in kittens, monkeys, and humans. Giffin, E. & Mitchell, D. E. (1978) *Journal of Physiology*, **274**, 511. Freeman, E. N. & Marg, E. (1975) *Nature*, **254**, 614. Teller, D. Y. (1981) *Trends in Neuroscience*, **4**, 21. Jacobs, D. B. & Blakemore, C. (1988) *Vision Research* **28**, 947–58.
Imprinting in birds. Bateson, P. P. G. (1966) *Biological Reviews*, **41**, 177.
Sensitive period for primary socialization in pups. Scott, J. P. & Fuller, J. L. (1965) *Genetics and social behaviour of the dog.* University of Chicago Press.
Critical period for song-learning in birds. Konishi, M. (1978) Ethological aspects of auditory pattern recognition. Chapter 9 in vol. VIII (*Perception*) of *Handbook of Sensory Physiology*, ed. R. Held, H. W. Leibowitz & H–L. Teuber, pp. 289–309. Berlin, Heidelberg, New York: Springer-Verlag.
Formation and migration of cells in the visual pathway. Rakic, P. (1977) *Philosophical Transactions of the Royal Society*, **B**, **278**, 245.
Accommodative responses in infants. Braddick, O., Atkinson, J., French, J. & Howland, H. C. (1979) *Vision Research*, **19**, 1319. Banks, M. S. (1980) *Child Development*, **51**, 646.
Pursuit eye movements in infants. Aslin, R. (1981) In *Eye movements: Cognition and Visual Perception*, ed. D. F. Fisher, R. A. Monty & J. W. Senders, Hillsdale, N.J.: Lawrence Erlbaum.
Asymmetry of OKN. Atkinson, J. (1981) In *Eye movements: Cognition and Visual Perception*, ed. D. F. Fisher, R. A. Monty & J. W. Senders. Hillsdale, N.J.: Lawrence Erlbaum.
Infant acuity and contrast sensitivity. Dobson, V. & Teller, D. Y. (1978) *Vision Research*, **18**, 1469. Atkinson, J. & Braddick, O. (1981) In *The development of perception: Psychobiological perspectives*, ed. R. N. Aslin, J. R. Alberts & M. R. Petersen. New York: Academic Press.
Infant colour vision. Werner, J. S. & Wooten, B. R. (1979) *Infant Behaviour and Development*, **2**, 241.
Infants' vergence movements. Aslin, R. N. (1977) *Journal of experimental Child Psychology*, **23**, 133.
Binocular function in infants. Aslin, R. N. & Dumais, S. T. (1980) In *Advances in Child Development and Behaviour*, vol. 15, ed. L. Lipsitt & H. Reese. New York: Academic Press. Braddick, O. J. & Atkinson, J. (1983) *Behavioural and Brain Research*, **10**, 141.
Depth perception in infants. Yonas, A. (1979) In *Perception and its Development*, ed. A. D. Pick. Hillsdale, N.J.: Erlbaum.
Dominance of external contours in neonates' perception. Milewski, A. E. (1976) *Journal of experimental Child Psychology*, **22**, 229.
Infants' auditory thresholds. Hecox, K. (1975) In *Infant Perception: from Sensation to Cognition*, ed. L. B. Cohen & P. Salapatek. New York: Academic Press.

Infants' recognition of speech sounds. Kuhl, P. K. (1978) In *Communicative and Cognitive Abilities: Early Behavioural Assessment*, ed. F. D. Minifie & L. L. Lloyd. Baltimore: University Park Press. Mehler, J. & Bertoncini, J. *Psycholinguistics Series* 2: *Structures and Processes*, ed. J. Morton & J. C. Marshall. London: Elek.

Infants' auditory localisation. Muir, D., Abraham, W., Forbes, B. & Harris, L. (1979) *Canadian Journal of Psychology*, **33**, 320.

Co-ordination of vision and touch. Rock, I. & Harris, C. S. (1967) *Scientific American*, **216** (5), 96.

The l–r distinction in Japanese. Miyuwaki, K., Strange, W., Verbrugge, R., Liberman, A. M., Jenkins, J. J. & Fujimura, O. (1975) *Perception and Psychophysics*, **18**, 331.

Active and passive movement in sensory-motor development. Held, R. (1965) *Scientific American*, **213** (11), 84.

Index

Page numbers in italic type indicate reference to tables or figures.